Breeding Major Food Staples

Breeding Major Food Staples

Edited by

Manjit S. Kang and P.M. Priyadarshan

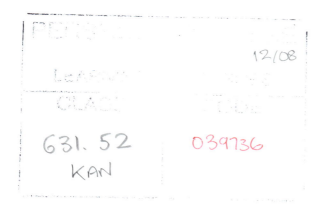

Blackwell
Publishing

Manjit S. Kang, Ph.D., is Professor (Retired) of Quantitative Genetics, Louisiana State University and is now Vice Chancallor of Punjab Agricultural University at Ludhiana.
P. M. Priyadarshan, Ph.D., is a plant breeder at the Rubber Research Institute of India.

Blackwell Publishing Professional
2121 State Avenue, Ames, Iowa 50014, USA

Orders: 1-800-862-6657
Office: 1-515-292-0140
Fax: 1-515-292-3348
Web site: www.blackwellprofessional.com

Blackwell Publishing Ltd
9600 Garsington Road, Oxford OX4 2DQ, UK
Tel.: +44 (0)1865 776868

Blackwell Publishing Asia
550 Swanston Street, Carlton, Victoria 3053, Australia
Tel.: +61 (0)3 8359 1011

Authorization to photocopy items for internal or personal use, or the internal or personal use of specific clients, is granted by Blackwell Publishing, provided that the base fee is paid directly to the Copyright Clearance Center, 222 Rosewood Drive, Danvers, MA 01923. For those organizations that have been granted a photocopy license by CCC, a separate system of payments has been arranged. The fee codes for users of the Transactional Reporting Service is ISBN-13: 978-0-8138-1835-1/2007.

First edition, 2007

Library of Congress Cataloging-in-Publication Data
Breeding major food staples / edited by Manjit S. Kang and P.M. Priyadarshan.—1. ed.
     p.   cm.
  Includes bibliographical references and index.
  ISBN-13: 978-0-8138-1835-1 (alk. paper)
  ISBN-10: 0-8138-1835-4 (alk. paper)
1. Food crops—Breeding.   2. Crop improvement.   I. Kang, Manjit S.   II. Priyadarshan, P. M.

  SB175.B74 2007
  631.5′2—dc22

2007008143

The last digit is the print number: 9 8 7 6 5 4 3 2 1

Table of Contents

Foreword

Humans are guests of green plants on this planet. Green plants are a source of food, clothing, fuel, construction materials, and medicines. As ornamentals, they are aesthetically pleasing. Considering their importance for human survival and advancement of civilization, plants have been constantly improved since their domestication starting about 10,000 years ago. One has to compare modern varieties with their wild relatives to get an idea of the improvements that have been made. This conscious and unconscious selection has resulted in numerous primitive varieties, landraces, pure lines, and improved varieties of our crop plants. Plant breeders have utilized these genetic resources to develop highly productive modern varieties adapted to diverse growing environments.

The advancements in plant-breeding technology during the last century, based primarily on Mendelian genetics, have been employed for developing the productive crop varieties for feeding ever-increasing world population. The latest example of power of plant breeding is the green-revolution varieties of rice and wheat, which led to major increases in food grain production. During a 40-year period (1960–2000), world population doubled from 3.0 to 6.0 billion people, but food grain production increased threefold. The abundant food availability is the basis of unparalleled prosperity and political stability in Asia.

In spite of these advances in food grain production, 840 million people in the world still go to bed hungry everyday. World population is likely to increase from 6.5 billion now to 8 billion in 2025 and 9 to 10 billion in 2050. Moreover, as the standard of living improves, people start eating high-value foods, such as meat, milk, and eggs. This leads to increased demand for cereals as livestock feed. Considering this scenario, food production must double during the next 30–40 years. To meet this challenge, we need crop varieties with higher yield potential and greater yield stability.

The time-tested methods of classical plant breeding will be continuously employed, but modern advances in cellular and molecular biology and genomics are being increasingly utilized.

This book provides excellent reviews of present status of breeding food grain and tuber crops as well as bananas. These crops provide 60–65 of the calories consumed by the world population. Thus, their importance can hardly be overemphasized. The first two chapters are an authoritative review of classical and modern techniques of crop improvement. Worldwide, more than 2.5 billion people suffer from micronutrient deficiencies. During the past 10 years, importance of breeding food crops with dense micronutrients, such as iron, zinc, and vitamin A, has been emphasized and a new term "biofortification" has been coined. Therefore, the third chapter on biofortification is a useful addition. The fourth chapter deals with bioinformatics for managing voluminous data. None of the books on plant breeding has chapters on bioinformatics. Thus, this chapter is another welcome addition.

I hope this volume will prove useful for students, teachers, and researchers working on crop improvement.

Gurdev S. Khush
Former Principal Plant Breeder
International Rice Research Institute
Philippines
World Food Prize Laureate—1996

Preface

Population growth and food production are inextricably linked, food being one of the very basic needs of humans. In his 1798 publication "*An Essay on the Principle of Population*,"[1] Thomas Robert Malthus wrote, ". . . the power of population is indefinitely greater than the power in the earth to produce subsistence for man." The Green Revolution of the 1960s/70s and the fact that Malthus' dire predictions of population outrunning food supply by the middle of the 19th century failed reflects human ingenuity to develop and continually enhance scientific tools to increase food production. In addition, population growth has been slowed down through education and governmental policies.

Current world population is about 6.6 billion and increasing by the second, as indicated by the population ticker at http://www.irri.org/.[2] The current population growth rate of about 1.2% or 77 million annually is substantially less than the 2% rate of the early 1970s.[3] According to the U.S. Census Bureau[3], world population is expected to reach 9–10 billion by 2050. That food production would need to be doubled in the next 30 years and tripled in the next 50 years to feed people is an enormous challenge facing all involved in agriculture.

Poverty, hunger, food insecurity, and malnutrition remain serious problems around the world. An estimated 840 million people, which is 13% of the world population, remain chronically malnourished, with most of them living in less developed countries. In a speech on April 11, 2006 delivered to Biotechnology Industry Organization (www.bio.org), Chicago, Illinois, former U.S. President Bill Clinton eloquently highlighted this problem, "Keep in mind, half the world's people live on less than $2.00 a day; a billion people live on less than $1.00 a day; a billion people go to bed hungry every night . . ."[4]

Remarkable progress in production and productivity has been achieved during the past 50 years. Advances in agricultural technology, brought about by the re-discovery and application of Mendel's laws of heredity, discovery of the structure of DNA, and science-based agricultural research during the 20th century, were the underpinnings of the progress.

In addition to various national agricultural research institutes and universities, international institutes have been at the forefront of solving the world food problem. The Consultative Group on International Agricultural Research (CGIAR), which has 15 centers around the world, has been charged with the mission of achieving sustainable food security and reducing poverty in developing countries. The CGIAR's motto is "Nourishing the future through scientific excellence."[5]

Cereals have been and are expected to remain the most important calorie-providing staple food in the world. Rice, wheat, and maize are the leading cereals. Many CGIAR research centers are devoted to reducing poverty and hunger in the world, and to ensuring secure access to food in Third World countries. For example, the mission of the International Rice Research Institute (IRRI) in the Philippines—a CGIAR center—is "To reduce poverty and hunger, improve the health of rice farmers and consumers and ensure that rice production is environmentally sustainable."[2] Similarly, the mission of the International Maize and Wheat Improvement Center (CIMMYT) in Mexico is "To use knowledge and technology to increase food security, improve the productivity and profitability of farming systems and sustain natural resources."[6] Other CGIAR centers also have similar missions.

The rapid pace at which the world population continues to increase dictates the continuing need to produce greater quantities of staple crops, such as rice, wheat, maize, potato, sweetpotato, cassava, soybean, barley, and banana to feed the people of the world. This publication, *Breeding Major Food Staples,* covers improving yields and quality of these crops through breeding, biofortification, and the use of molecular genetic tools, such as transformation (gene transfer), genome mapping, and bioinformatics. *Breeding Major Food Staples* brings together the state-of-the-art technical information through prominent experts in plant genetics and breeding, as well as bioinformatics. The book is divided into three sections: (a) general topics, which apply to all crops; (b) cereal crops; and (c) calorie- or carbohydrate-supplying root and tuber crops and bananas. There are two unique chapters in the book (section 1) that are generally not found in plant breeding textbooks: Biofortification and bioinformatics. Biofortification refers to the process of breeding food crops that are rich in bioavailable micronutrients, such as Fe, Zn, and vitamin A.[7] Bioinformatics encompasses the fields of biology, computer science, and information technology.[8]

Chapter 1, "Crop Breeding Methodologies: Classic and Modern," discusses the various traditional and modern (molecular and genetic engineering) plant breeding methodologies. This chapter is not specific to any particular crop.

Chapter 2, "Genetic Enhancement of Polyploid Crops Using Tools of Classic Cytogenetics and Modern Biotechnology," is an authoritative review of polyploidy in crops and of biotechnological/molecular genetic tools used in crop improvement.

Chapter 3, "Biofortification: Breeding Micronutrient-Dense Crops," is authored by two HarvestPlus scientists. HarvestPlus is an international, interdisciplinary research program, whose mission is to reduce micronutrient malnutrition by developing nutrient-dense staple foods through breeding.

Chapter 4 is focused on bioinformatics and plant genomics for staple crops improvement. Bioinformatics—a field yet to mature—plays an essential role in the sequencing and characterization of genomes as well as in trait analysis and optimization of breeding strategies.

Chapter 5, "Breeding Spring Bread Wheat for Irrigated and Rainfed Production Systems of the Developing World," provides detailed breeding strategies to develop disease-resistant wheat and applications of molecular markers in wheat breeding.

Chapter 6, "Rice Breeding for Sustainable Production," highlights, among other issues, the impact of rice breeding in the world, wide hybridization, use of molecular markers in rice breeding, and genetic transformation and transgenics.

Chapter 7, "Barley Breeding for Sustainable Production," provides background information on the barley gene pool, types of barley, and diseases and pests of barley. It highlights important issues, such as doubled-haploid production, breeding for stresses, participatory breeding, and molecular breeding.

Chapter 8, "Corn Breeding in the Twenty-first Century," traces the evolution of corn breeding in the USA into a vast hybrid-corn industry, while noting important technological developments, people who made it happen, and the social and political environment in which the corn industry flourished.

Chapter 9, "Soybean Breeding Achievements and Challenges," provides a comprehensive picture of soybean breeding. It details first nutritional aspects of soybean, and then provides a discussion of soybean diseases and the modern breeding methodologies used in soybean improvement, e.g., genomics and genetic transformation.

Chapter 10, "Breeding Potato as a Major Staple Crop," discusses domestication and evolution of the modern potato crop, genetic resources and transformation in potato, somaclonal variation, and resistances to biotic and abiotic stresses. This chapter also highlights molecular markers, quantitative trait loci (QTL) studies, gene cloning, and molecular marker-assisted introgression.

Chapter 11, "Breeding of Sweetpotato," provides a thorough discussion of floral biology, breeding goals, nutritional quality, and breeding strategies used in sweetpotato improvement. Various biotechnological tools used in sweetpotato breeding, i.e., tissue culture and molecular breeding approaches, such as molecular-marker-assisted selection, are elaborated in this chapter.

Chapter 12, "Cassava Genetic Improvement," describes the breeding objectives and schemes used in cassava improvement. In addition to a discussion of the hereditary aspects of important traits, molecular marker-use is described.

Chapter 13, "Banana Breeding," thoroughly discusses *Musa* breeding strategies, resistance to diseases, insects, and nematodes, and biological control of pests. The accomplishments and strategies used in genetic transformation of *Musa* and the use of biotechnology as an adjunct to classical breeding in the genetic improvement of *Musa* are highlighted.

The monumental task of getting the book ready for printing would not have been possible without the cooperation and dedication of the authors. They deserve much appreciation and gratefulness. A word of thanks also goes to Mr. Justin Jeffryes and Ms. Erica Judisch of Blackwell Publishing, Ames, Iowa, without whose encouragement and support, the project could not have been completed.

We trust *Breeding Major Food Staples* will serve as not only a repository of research on various crops, but will also be useful for teachers and students of modern plant breeding. We hope this publication will stimulate new ideas and directions in research and help plant breeders meet the challenge of feeding almost eight billion people in the near term and possibly 10 billion people in the long term.

Manjit S. Kang
P.M. Priyadarshan

References and Notes

1. *"An Essay on the Principle of Population, as it affects the Future Improvement of Society with remarks on the Speculations of Mr. Godwin, M. Condorcet, and Other Writers."* Printed for J. Johnson in St. Paul's church-yard, London. (The first edition was published anonymously in 1798).

2. International Rice Research Institute, Los Banos, The Philippines (http://www.irri.org/; February 2007).

3. U.S. Census Bureau (www.census.gov; February 2007).

4. AgBioView (see the April 14, 2006 issue at http://www.agbioworld.org).

5. Consultative Group on International Agricultural Research (www.cgiar.org; February 2007).

6. International Maize and Wheat Improvement Center (http://www.cimmyt.org/; February 2007).

7. HarvestPlus (http://www.harvestplus.org/; February 2007).

8. U.S. Environmental Protection Agency (http://www.epa.gov/comptox/glossary.html; February 2007).

Contributor List

Preface

Gurdev S. Khush
Department of Vegetable Crops
University of California
Davis, CA 95616

Chapter 1

Manjit S. Kang
Vice Chancellor
Punjab Agricultural University
Thapar Hall
Ludhiana 141 004, India

Prasanta K. Subudhi
School of Plant, Environmental, and Soil
    Sciences
Louisiana State University Agricultural
    Center
Baton Rouge, LA 70803-2110

Niranjan Baisakh
School of Plant, Environmental, and Soil
    Sciences
Louisiana State University Agricultural
    Center
Baton Rouge, LA 70803-2110

P.M. Priyadarshan
Rubber Research Institute of India
Regional Station
Agartala-799006
India

Chapter 2

Prem P. Jauhar
USDA–Agricultural Research Service
Northern Crop Science Laboratory
Fargo, North Dakota 58105

Chapter 3

Wolfgang H. Pfeiffer
HarvestPlus
c/o International Center for Tropical
    Agriculture (CIAT)
A.A. 6713
Cali, Colombia

Bonnie McClafferty
HarvestPlus
c/o The International Food Policy Research
    Institute
2033 K St. NW
Washington, D.C. 20006

Chapter 4

David Edwards
Australian Centre for Plant Functional
    Genomics
Institute for Molecular Biosciences and
    School of Land, Crop and Food Sciences
Institute for Molecular Biosciences
University of Queensland
Brisbane, QLD 4072
Australia

Chapter 5

Ravi P. Singh
International Maize and Wheat
    Improvement Center (CIMMYT) Apdo.
    Postal 6-641
06600, Mexico, DF

Richard Trethowan
International Maize and Wheat
    Improvement Center (CIMMYT) Apdo.
    Postal 6-641
06600, Mexico, DF
(Present address: University of Sydney,
    Plant Breeding Institute, PMB 11,
    Camden, NSW, 2570, Australia)

Chapter 6

Sant S. Virmani (formerly with IRRI)
Virmani Consulting International LLC
4425 Partney Court
Plano, Texas 75024

M.Ilyas-Ahmed
Directorate of Rice Research
Rajendranagar
Hyderabad-500030, India

Chapter 7

Salvatore Ceccarelli
International Center for Agricultural
    Research in the Dry Areas (ICARDA)
P.O. Box 5466
Aleppo, Syria

Stefania Grando
International Center for Agricultural
    Research in the Dry Areas (ICARDA)
P.O. Box 5466
Aleppo, Syria

Flavio Capettini
International Center for Agricultural
    Research in the Dry Areas (ICARDA)
P.O. Box 5466
Aleppo, Syria

Michael Baum
International Center for Agricultural
    Research in the Dry Areas (ICARDA)
P.O. Box 5466
Aleppo, Syria

Chapter 8

G. Richard Johnson
Department of Crop Science
University of Illinois
Urbana, IL 61801

Chapter 9

Silvia R. Cianzio
Dept. of Agronomy
Iowa State University
Ames, IA 50011-1010

Chapter 10

John E. Bradshaw
Scottish Crop Research Institute
Invergowrie
Dundee DD2 5DA, U.K.

Chapter 11

S.L. Tan
Rice & Industrial Crops Research Centre
Malaysian Agricultural Research &
    Development Institute (MARDI)
P.O. Box 12301
50774 Kuala Lumpur, Malaysia

M. Nakatani
Research Information Office
Ministry of Agriculture
Forestry and Fisheries
Kasumigaseki 1-2-1
Chiyoda-ku
Tokyo 100–8950
Japan

K. Komaki
Department of Field Crop Research
2-1-18 Kannondai
Tsukuba
Ibaraki 305-8518
Japan

Chapter 12

Hernán Ceballos
International Center for Tropical
    Agriculture and National University of
    Colombia—Palmira Campus
Apartado Aéreo 6713
Cali, Colombia

Martin Fregene
International Center for Tropical
    Agriculture
Apartado Aéreo 6713
Cali, Colombia

Juan Carlos Pérez
International Center for Tropical
    Agriculture
Apartado Aéreo 6713
Cali, Colombia

Nelson Morante
International Center for Tropical
    Agriculture
Apartado Aéreo 6713
Cali, Colombia

Fernando Calle
International Center for Tropical
    Agriculture
Apartado Aéreo 6713
Cali, Colombia

Chapter 13

Michael Pillay
International Institute of Tropical
    Agriculture
P. O. Box 7878
Kampala, Uganda

Leena Tripathi
International Institute of Tropical
    Agriculture
P. O. Box 7878
Kampala, Uganda

Breeding Major Food Staples

Part 1

General Topics

Chapter 1

Crop Breeding Methodologies: Classic and Modern

Manjit S. Kang, Prasanta K. Subudhi, Niranjan Baisakh, and P.M. Priyadarshan

Introduction

Plant breeding has been vitally important in feeding and clothing the world's population. The human population has been increasing exponentially worldwide, whereas food production has been increasing linearly, thus creating a "world food problem" (Kang, 2006). Development of new breeding and production methodologies has been the basis of increased crop productivity. In this chapter, we aim to provide a review of crop breeding methodologies that have evolved across pre-Mendelian and post-Mendelian eras.

Plant breeding began long before the laws of genetics became known (Jauhar, 2006). Farmers practiced selection before science-based crop breeding and improvement began; the latter came into being after the rediscovery of Mendel's laws of inheritance in 1900. A classic vs. modern dichotomy is relative. Crop breeding methodology practiced soon after the rediscovery of Mendel's laws of inheritance would be dubbed "modern" in relation to pre-Mendelian breeding practices. By the same token, the breeding practices of the 1900 to 1970 era might be called "classic" today in relation to current genetic technologies, for example, genetic transformation and molecular-marker-based selection.

Many definitions of plant breeding exist. Plant breeding is said to be both an art and a science (Poehlman, 1987; Jensen, 1988).

The art refers to the ability of the breeder to observe differences that may have economic value in plants of the same species (Poehlman, 1987), whereas science refers to the application of principles and methods of genetics and statistics as well as other related disciplines, such as botany, biochemistry, plant pathology, soils, and crop physiology and ecology (Kang, 1998). Crop breeding has also been referred to as the enterprise of providing genetic solutions to impaired plant productivity that arises from changes in climatic and edaphic factors, the altered spectrum of pests, changes in economic and consumer demands, and government policies (Scowcroft, 1988). Gepts and Hancock (2006) regard plant breeding as a vibrant, multidisciplinary science characterized by its ability to reinvent itself by absorbing and using novel scientific findings and technical approaches. They propose that a contemporary plant-breeding curriculum should include studies of inheritance and selection of complex traits; basic biology of plants (e.g., reproductive biology, Mendelian genetics); quantitative genetics and selection theory; plant breeding; and related sciences such as genomics, applied statistics, experimental design, and pest sciences. Gepts and Hancock (2006) quote the definition of plant breeding given by Schlegel (2003): ". . . an applied, multidisciplinary science. It is the application of genetic principles and practices associated with the development of

5

cultivars more suited to the needs of humans than the ability to survive in the wild; it uses knowledge from agronomy, botany, genetics, cytogenetics, molecular genetics, physiology, pathology, entomology, biochemistry, and statistics."

Pre-mendelian breeding: domestication, selection, and hybridization

Pre-Mendelian agriculture evolved from nomadic activity into a more systematic, settled crop culture. The transition from hunting and gathering to agriculture was very slow and gradual. Events leading to crop breeding included domestication of plants, selection, and hybridization. Domestication is a selection process used to adapt plants and animals to better suit the needs of humans (Gepts, 2002); domestication of plants dates back to 9000 B.C., in the hills above the Tigris (Wehner, 2006). Domestication of most of the important food crops was complete by 3000 B.C. and 1000 B.C. in the old world and new world, respectively (Wehner, 2006). Six domestication regions have been identified: Mesoamerica, the southern Andes, the Near East, Africa, Southeast Asia, and China (Gepts, 2002). The ecological origins of some of the major food and feed crops are categorized as: Mediterranean (wheat, barley, oats, and beet); savanna (maize, rice, sorghum, millets, soybean, cassava, sweet-potato, and bananas/plantains); tropical lowlands (coconut); tropical highlands (potato); and tropical forests (sugarcane) (Allard, 1999).

The process of domestication has been going on for 10,000 years (Gepts, 2002). During domestication, selection was practiced for traits such as increased seedling vigor, loss of seed dormancy, loss of seed dispersal, increase in seed number, modification of photoperiod, color, texture, and reduction in toxic chemicals (Harlan, 1992). These traits are called "domestication syndrome traits" (Harlan, 1992; Gepts, 2002). The domestication process operated on spontaneous mutations (generally loss-of-function mutations) that occurred in wild populations (Gepts, 2002).

The concept of center of origin, which was proposed by N.I. Vavilov (1926), relates to the region in which a crop species originated. These areas may also be centers of diversity, i.e., areas with a great deal of genetic variability. According to Harlan (1975), domestication of crop plants could have occurred in centers and non-centers of origin. A center represents a limited geographical area where a crop was domesticated and from where it spread to other areas, whereas a non-center is a broad area where a crop may have been domesticated simultaneously in several different locations (Fehr, 1987).

After domestication, the next step in Pre-Mendelian crop breeding entailed selection of seeds by farmers for planting the next season's crop. The farmer practices did not involve hybridization of plants—an important activity in plant improvement. Hybridization was preceded by the discovery of sex in plants by Camerarius in 1694. Joseph Kölreuter first reported artificial crossing or hybridization of plants in 1761. Gregor Mendel, Charles Darwin, and William James Beal produced plant hybrids in the mid-to-late 1800s (Hallauer, 2004). Hybrids in oat were developed in the United States in 1875 (Wehner, 2006). Detasseling of maize was suggested by Bidwell in 1867 (Wehner, 2006), which constituted an important practice in hybrid maize production in the 20th century. For a detailed account of historical developments relative to plant breeding, see: http://cuke.hort.ncsu.edu/cucurbit/wehner/741/hs741hist.html.

Hybridization is the key to plant and animal improvement. Cook (1937) gave an up-to-date account of hybridization in plants and animals up to the first part of the 20th century.

Post-Mendelian classic breeding methods

Rediscovery of Mendel's laws of inheritance in 1900 laid the foundation for rapid progress in crop improvement. A majority of economically important traits in plants, e.g., grain yield, are quantitative or multigenic in nature (Stuber, 2002). Selection of appropriate breeding methods depends on (1) the reproductive system of crop to be improved (self-pollinated vs. cross-pollinated; sexual vs. asexual reproduction) and (2) complexity of the trait to be improved (qualitative vs. quantitative).

Improvement of maize—a cross-pollinated crop—began with the experiments on inbreeding by George H. Shull and Edward M. East in 1905. Maize breeding and principles of breeding cross-pollinated crops in general are intimately connected with quantitative genetics theory (Hallauer and Miranda, 1988). The application of statistical methods in the improvement of quantitative traits began early in the 20th century (Stuber, 2002). Hallauer and Miranda (1988) have summarized much of the maize breeding progress that has been achieved through the application of quantitative genetic methodologies to maize breeding. Quantitative genetics plays a significant role in self-pollinated crops as well (Meredith, 1984).

Self-pollinated crops

In self-pollinated crops, commercial cultivars are generally pure lines. Hybrids are also produced in some crops, e.g., rice, wheat, and sorghum.

In the beginning of a breeding program, new varieties or cultivars may be developed mainly by means of three methods: introduction, selection, and hybridization (Poehlman, 1987). Introduction is simply bringing in new varieties from various sources and identifying those that are locally adapted. Selection must be practiced to identify desirable types from among a genetic mixture of varieties. Mass selection is one of the most commonly used and oldest procedures in improving plants. This procedure achieves success when applied to highly heritable traits. Several cycles of selection may be needed to achieve a desired level of improvement for a trait. The magnitude of genetic gain from selection depends on heritability and selection intensity. Mass selection was successfully used to improve seed size and seed composition in soybean (Fehr and Weber, 1968). However, mass selection cannot provide desired genetic gain for traits that are not highly heritable.

Because not all varieties possess all of the desired traits, hybridization is practiced to combine the desirable attributes of different varieties. Much information exists on hybridization methods of breeding in classic textbooks on plant breeding (Fehr, 1987; Poehlman, 1987; Allard, 1999). Thus, only the most widely used methods will be briefly discussed here. The methods are (1) bulk breeding, (2) pedigree selection, (3) single-seed descent, and (4) doubled-haploid lines.

Bulk breeding

The invention of the bulk population method is credited to H. Nilsson-Ehle, in 1908 (Jensen, 1988). Florell (1929) published a landmark paper on bulk-population method. The method is simple, inexpensive, and is used to accomplish inbreeding in the segregating generations after hybridization. In the early generations (e.g., F_1 through F_4), a breeder does not practice conscious selection; natural selection operates. Population size increases progressively through the generations (F_1 = 50 to 100 plants; F_4 = 3,000+ plants). In each generation, seed is bulked. Deliberate selection may be practiced in the F_5 generation when a relatively high level of homozygosity has been achieved. Suppose there are five segregating genes in a population, 72% homozygosity will be achieved by the F_5 generation and nearly 100% by the F_{10} generation. Allard (1999) provides details of the underlying genetic theory. If this method were to be used for improving disease resistance, selection must be done in the presence of the pathogen causing the disease. This method may be used in conjunction with mass selection.

Pedigree selection

Following hybridization of two selected parents, selection begins in the F_2 generation. Pedigree records are kept in each generation. In the F_3 to F_5 generations, progeny rows are grown, superior rows are identified, and desirable plants are selected within rows. Preliminary yield trials may be conducted in the F_7 generation. For five segregating genes, 92% homozygosity will be achieved by the F_7 generation. When visual selection is practiced, traits under consideration must have relatively high heritability. This method requires detailed record keeping of pedigrees.

Single-seed descent

Single-seed descent (SSD) is a modified pedigree method, in which many lines are inbred to homozygosity. SSD allows inbreeding in any environment, including off-season nurseries or greenhouses, which cannot be used under the bulk method. The SSD method does not allow natural selection until inbred lines are produced. For clearly identifiable traits, selection can be practiced among single plants.

Procedurally, F_1 seed is bulked and 50+ plants are grown. In F_2, a large population is grown (2,000+ plants). A single seed is harvested from each plant. In F_3 and F_4 generations, harvested seeds from the previous generation are grown and a single seed per plant is harvested. In F_6, progeny rows are

grown from plants selected in F_5. Superior rows are harvested. Each row represents a different F_2 plant. In F_7, preliminary yield trials may be conducted. Because off-season or greenhouse nurseries can be used, the number of years to reach yield trials may be reduced. Record keeping, as practiced in pedigree selection, is not necessary. Soybean is a good example of a crop to which SSD has been successfully applied. Modifications of the SSD method exist (e.g., single-hill method and multiple-seed method).

Doubled-haploid method

Doubled-haploid line development is the most efficient method of creating truly homozygous lines (homozygous at all loci). Doubled-haploid lines may be developed naturally (by means of parthenogenesis) or artificially. In some species, for example, maize, naturally occurring doubled haploids result from spontaneous doubling of chromosomes of haploids. In barley, haploid plants may be produced by chromosome elimination after hybridization (Poehlman, 1987). Pollinating wheat ovules with maize pollen may induce haploid production in wheat. Haploids may be produced in the F_2 generation. Colchicine may be used to double the chromosome number of haploid plants to produce doubled haploids. Preliminary yield trials may be conducted in the F_5 generation.

Backcross breeding

Backcrossing is practiced to correct a deficiency/defect in an otherwise productive cultivar/line (recurrent parent) by introducing a gene from another cultivar/line (nonrecurrent or donor parent). Following the initial cross between the recurrent parent and the donor parent, a series of backcrosses is made to the recurrent parent to recover/reconstitute the genetic complement of the recurrent parent. This method is often used to introduce disease resistance genes into otherwise desirable cultivars/lines. Selection for disease resistance is practiced during backcrossing to ensure the successful transfer of a resistance gene from the donor to the recurrent parent. It takes four to five backcrosses to recover most of the genetic complement of the recurrent parent. The following formula may be used to calculate the percentage of recurrent parent genome recovered during backcrossing:

$$\text{Recurrent parent (\%) genetic complement} = [1 - (\tfrac{1}{2})^{n+1}] \times 100$$

where n is the number of backcrosses. Thus, after the initial cross between a recurrent parent and a donor parent, the percent genetic complement of the recurrent parent will be equal to $[1 - (\tfrac{1}{2})^{0+1}] \times 100 = 50\%$.

After four backcrosses, the percentage of the recurrent parent's genetic complement recovered will be equal to $[1 - (\tfrac{1}{2})^{4+1}] \times 100 = (31/32) \times 100 = 96.875\%$.

Cross-pollinated crops

Cross-pollinated plants are, as a rule, heterozygous in nature. When they are subjected to inbreeding, vigor is generally reduced (inbreeding depression). Open-pollinated maize varieties are heterozygous, heterogeneous, and vigorous. Upon inbreeding, homozygosity increases and vigor is reduced. Inbreeding helps remove undesirable recessive gene combinations. Inbred lines are produced and crossed with suitable testers or among themselves in a diallel design to evaluate their performance in hybrid combinations. In cross-pollinated plants, heterozygosity and concomitant hybrid vigor or heterosis are exploited. Commercial cultivars in cross-pollinated crops are hybrids rather than pure lines. Heterozygosity is associated with heterosis and homozygosity is associated with inbreeding depression. Heterosis may be expressed as: $F_1 - MP$,

where F_1 and MP are F_1 hybrid and mid-parent performances, respectively. There are differences in opinion about the genetic basis of heterosis. One school of thought is that heterosis results from overdominance, whereas another school of thought suggests that heterosis is simply caused by dominance. The following may help shed some light on the nature of heterosis:

Consider population "A" in the Hardy-Weinberg equilibrium with the following zygotic frequencies and genotypic values relative to locus M (see Table 1.1):

Table 1.1.

Genotype	Frequency	Genotypic value
M_1M_1	p^2	$+u$
M_1M_2	$2pq$	au
M_2M_2	q^2	$-u$

$$\text{Population mean} = \mu_A = (p - q)u + 2pqau$$
(Kang, 1994).

Consider a second population "B" relative to locus M (see Table 1.2):

Table 1.2.

Genotype	Frequency	Genotypic value
M_1M_1	r^2	$+u$
M_1M_2	$2rs$	au
M_2M_2	s^2	$-u$

$$\text{Population mean} = \mu_B = (r - s)u + 2rsau$$

$$\text{Mid-parent value for the two populations} = \tfrac{1}{2}[(p - q + r - s)u + 2(pq + rs)au]$$

A cross between Population A and Population B gives (see Table 1.3):

Table 1.3.

Genotype	Frequency	Genotypic value
M_1M_1	pr	$+u$
M_1M_2	$ps + qr$	au
M_2M_2	qs	$-u$

$$F_1 \text{ mean} = (pr - qs)u + (ps + qr)au$$

$$\text{Heterosis} = F_1 - MP$$
$$= [(pr - qs)u + (ps + qr)au] - \tfrac{1}{2}[(p - q + r - s)u + 2(pq + rs)au]$$
$$= [(pr - qs) - \tfrac{1}{2}[(p - q + r - s)]u + [(ps + qr) - (pq + rs)]au$$

Upon further algebraic manipulation, we get:

$$\text{Heterosis} = (p - r)^2 \, au \text{ (for inbred lines,}$$
$$au = 0; \text{ hence no heterosis)}$$

The above relationship indicates that a heterotic response is expected when there is a difference in gene frequencies in parental lines of a cross and some degree of directional dominance at one or more loci.

If heterosis results from overdominance, then heterozygosity at some or all loci is a must:

Inbred parent 1 (Genotype: AABBccDD) ×
Inbred parent 2 (Genotype aabbCCdd),
then F_1 genotype: AaBbCcDd.

If heterosis simply results from dominance, then heterozygosity is not a must:

genotype "AaBbCcDd" = genotype
"AABBCCDD"

On the basis of the dominance theory of heterosis, some suggest that it is possible to have inbred lines that are as good as hybrids. This remains to be seen, particularly in a crop such as maize. Commercial seed companies still sell solely heterozygous F_1 hybrid seed for planting by growers.

Breeding methodology

This section on cross-pollinated crop breeding methodology is divided into two parts: Population improvement, and Inbred and hybrid development.

Population improvement

In cross-pollinated crops such as maize, source populations for inbred line development are created. The source populations may be narrow-based or broad-based. The narrow-based populations may be segregating populations (advanced F_2) derived from random-mating or sib-mating of several single-cross plants. The broad-based populations may be derived from synthetics or double-crosses by means of random-mating or sib-mating. Once a random-mating population has been created—narrow-based or broad-based—it may be subjected to cyclic selection to increase the frequency of desirable alleles and to maintain genetic variability for continued selection by intermating superior progenies for each cycle of selection (Hallauer and Miranda, 1988).

The population size depends on whether the population is of a narrow or a broad genetic base. While approximately 1,000 plants may be sufficient for a narrow genetic-base population, 3,000 to 5,000 plants may be needed for a population of a broad genetic base. Cyclic selection methods may be used to improve such populations.

Mass Selection—Mass selection is phenotypic selection. No controlled pollinations are made. A segregating population is planted and undesirable plants are eliminated preferably before pollination, for example, to improve ear height or develop early flowering in maize. The plants surviving selection are allowed to open-pollinate. If undesirable plants cannot be eliminated before pollination, for example, when selecting for yield (bigger ears) in maize, then pollen from both desirable and undesirable plants can pollinate the selected plants, making selection progress slower than in the former case. Because mass selection is based on phenotype, it is likely to be more successful for traits that have relatively high heritability. Thus, mass selection for reducing ear height is likely to be more successful than for a low heritability trait, such as yield. One or more cycles of mass selection may be needed to achieve a desired level of improvement.

When selection can be made before pollinating (pollen parent control), selection gain or response to selection (R_s) can be predicted by means of the following formula (Bernardo, 2002):

$$R_s = (k_p V_A)/\sqrt{V_P} \quad \text{(pollen control)}$$

where k_p is the standardized selection differential, V_A is the additive genetic variance, and V_P is the phenotypic variance.

When selections cannot be made before pollinating, selection response is halved in comparison to response when selections can be made before pollinating, as shown by the following formula:

$$R_s = \tfrac{1}{2}(k_p V_A)/\sqrt{V_P} \quad \text{(no pollen control)}$$

Family Selection Methods—Mass selection is generally not effective for low heritability traits such as yield. For such quantitative traits, recurrent selection based on family structure and progeny testing is more desirable. Discussed below are some of commonly used recurrent selection procedures for improving quantitative traits.

Half-sib Family Selection—Half-sibs have one parent in common. For example, A × B, A × C, and A × D form a set of half sibs where A is the common parent. Used by maize breeders, the half-sib procedure is also called "ear-to-row breeding" (Poehlman, 1987). In the following season, remnant seed (not the open-pollinated seed) of the selected half-sib families is used for recombination or intermating. Genetic variation among half-sib families = $^1/_4 V_A$. Therefore, selection response formula is as follows:

$$R_s = \tfrac{1}{4}(k_p V_A)/\sqrt{V_P}$$

Modified Ear-to-Row Selection—John Lonnquist advocated this procedure in 1964.

The difference between ear-to-row breeding and modified ear-to-row breeding is that in the latter open-pollinated seeds of selected half-sib families are used for recombination rather than the remnant seed of those families. Thus, selection response will be one-half of that for ear-to-row method:

$$R_s = \tfrac{1}{8}(k_p V_A)/\sqrt{V_P}$$

Because ear-to-row method requires two seasons per cycle and modified ear-to-row selection is accomplished in one season, gains per season are not different. One can improve the gain by practicing mass selection within half-sib families, as suggested by Webel and Lonnquist (1967):

$$R_s = \tfrac{3}{8}(k_p V_A)/\sqrt{V_P}$$

The major advantage of modified ear-to-row method over the original ear-to-row method is that it uses more than one environment (a single replication of families is grown in two or more environments). This helps estimate genotype-by-environment interaction.

Full-sib Family Selection—Here selection is practiced among full-sib families. Assuming inbreeding coefficient is zero, full-sib families capture ($^1/_2 V_A + {}^1/_4 V_D$), where V_D is dominance variance. The selection response will be:

$$R_{s\_fullsib} = \tfrac{1}{2}(k_p V_A) \Big/ \sqrt{\tfrac{1}{2}V_A + \tfrac{1}{4}V_D + V_{PE}/e + V_e/re}$$

where V_{PE} and V_e are, respectively, progeny-by-environment and error variances; and r and e represent number of replications and environments, respectively.

S_1 and S_2 Family Selection—One can increase the selection response by using selfed generations. For example, additive genetic variance increases with increased homozygosity. In S_1 and S_2 generations, additive variance will increase to 1.0 V_A and 1.5 V_A, respectively. In S_1 and S_2, V_D will be $^1/_4$ V_D and 3/16 V_D, respectively (Bernardo, 2002). Thus, selection response equation (numerator and denominator) will be modified accordingly.

The S_1 and S_2 selection schemes operate on homozygous genotypes, which result in greater additive genetic variance than that from half-sib and full-sib families. Because of progressively greater magnitude of additive genetic variance being captured in the S_1 and S_2 generations, selection response will be expected to be greater relative to half-sib and full-sib methods of selection.

Interpopulation Improvement—Comstock et al. (1949) proposed procedures called "reciprocal recurrent selection" (RSS) to improve cross performance between two populations under consideration. The RSS procedure was designed to make maximum use of general and specific combining abilities. The RSS improves not only the two populations but also increases heterosis in their cross. Briefly, the procedure is as follows: Self-pollinate several plants in Population A and testcross to several randomly selected plants in Population B. The process is repeated similarly for Population B, i.e., several plants in Population B are self-pollinated and testcrossed to several randomly selected plants in Population A. This process results in two sets of half-sib families (A testcrossed to B and B testcrossed to A). The performance of the half-sib families is evaluated across environments. Based on testcross performance, selected S_1 progenies are intermated to form the next cycle of selection (Hallauer and Miranda, 1988). Iowa Stiff Stalk Synthetic (BSSS) and Iowa Corn Borer Synthetic #1 (BSCB1) have been used in RSS (Hallauer and Miranda, 1988).

Reciprocal Full-Sib Selection—This scheme is similar to the half-sib reciprocal recurrent selection, but differs from it in that reciprocal crosses are made between plants from the two populations, giving full sibs.

This method is used in populations with prolificacy, i.e., at least two-eared plants. One ear is selfed and the other is reciprocally crossed to a plant in the second population. Thus, both selfed and full-sib seeds are available on the selected plants. This method is more efficient than the reciprocal recurrent selection (half sib) method because two populations are improved simultaneously and information on crosses between them is directly applicable to the development of single crosses (Hallauer and Miranda, 1988).

Inbred and hybrid development

Once a breeder has developed and improved a source population, the inbreeding operation begins. In large commercial maize breeding programs, one would include several narrow-genetic base and broad-genetic base populations. Population size may be 1,000 plants for a narrow-genetic base population and 5,000 plants for a broad-genetic base population. Within each segregating population (known as S_0 population), inbreeding (self-pollination) of agronomically desirable plants is begun. Even though the populations have already undergone population improvement/selection, further selection occurs while selecting plants for inbreeding. In a typical season, in a maize breeding program, one would make 40,000 to 50,000 self-pollinations. After selection at harvest, one would have 20,000 to 25,000 S_1 lines to evaluate in the following year. An ear-to-row system is generally used for each S_1 line. S_1 lines may be visually evaluated at one or two locations in comparison with a check. A general procedure is that two different breeders independently assign a rating to each open-pollinated S_1 line. The ones receiving the highest ratings from the two breeders may be considered for advancing to the S_2 stage. In a breeding nursery, one would make at least three to five self-pollinations in each S_1 row with a goal of harvesting three good ears. The S_2 families (three rows) are planted in the following year. Between and within S_2 family selection is practiced. In the selected S_2 families, the best of the three rows may be used for advancing to S_3 stage. Lines may be used in hybrid development after the S_5 stage (96.875% homozygosity). Once inbred lines have been developed, as in self-pollinated crops, backcrossing can be used to correct defects controlled by one or two genes.

Jenkins (1935) advocated early generation testing to eliminate undesirable families as early as possible. Topcrossing S_1 and/or S_2 lines to a common tester and evaluating their performance in one or more environments helps allocate resources to the families that are most promising. A tester may be an inbred line (to test for specific combining ability) or a broad-genetic base variety (to test for general combing ability). Following theoretical and practical considerations, Bernardo (2002) concluded that early testing could be effective but low heritability would be a limiting factor.

Multi-environment cultivar evaluation

In both self-pollinated and cross-pollinated crops, only limited field evaluation of cultivars or hybrids is done during the developmental phase. Extensive, multi-environment trials are generally conducted to identify superior genotypes—either generally adapted across several diverse environments or specifically adapted to a particular environment. Breeders strive to broaden the genetic base of crop species to prevent problems associated with genetic vulnerability. Because of the emphasis on broadening species' genetic base and the unpredictable climatic factors encountered at different test sites and/or years, differential responses of improved cultivars in different environments are to be expected. The differential genotypic responses to different environments are

collectively referred to as "genotype-by-environment interaction" (Kang, 1998).

Voluminous literature exists on genotype-by-environment interaction, concepts of stability, and methods of identifying genotypes with stable performance across environments (Kang, 1990, 2002; Kang and Gauch, 1996; Yan and Kang, 2003). Genetic improvements reportedly accounted for approximately 50% of the total gains in yield per unit area for major crops during the last half of the 20th century (Simmonds, 1981; Silvey, 1981; Duvick, 1992, 1996). The remainder of yield gain was attributed to improved technology, management, and cultural practices. Data from barley and wheat performance trials from the United Kingdom for the period 1946–1977 revealed that genetic contributions to yield were 30 to 60% (barley) and 25 to 40% (wheat) and genotype-by-environment contributions were 25 to 45% (barley) and 15 to 25% (wheat) (Simmonds, 1981). Genotype-by-environment interaction has many implications, including reduced gains from selection, increased cost of establishing multiple cultivar breeding and testing programs, and potential loss of useful germplasm with limited testing in early selection stages (Kang, 1998). For a more detailed study of the subject, the reader is referred to other publications (Kang, 1990, 2002; Kang and Gauch, 1996; Cooper and Hammer, 1996; Annicchiarico 2002; Yan and Kang, 2003).

Molecular breeding

Molecular marker technology and plant breeding efficiency

Simple cross hybridization is the most prevalent procedure used by plant breeders to create new genetic variants in crop plants. Visual selection and measurements of crop attributes in field trials are used to select the desirable segregants in plant breeding programs. This traditional breeding methodology has made significant progress in the past and is continuing to make steady progress in improving yield, quality, and stability of major food crops. Due to increased demand for limited natural resources, like land and fresh water, concern for the environment (especially the application of fertilizers and pesticides), and skepticism about genetically modified food crops, plant breeders are under continuous pressure to meet the food, fiber, and fuel requirements of the growing world population. There have been continuing efforts to explore and add new tools to the breeder's toolkit. Novel tools, however, should be amenable to easy integration into breeding programs while increasing the precision of selection and reducing the breeding cycle for developing varieties or hybrids for the dynamic market place. Traditional breeding technology has remained static, but the DNA technology to genetically improve crop plants has been evolving rapidly with the addition of new tools and procedures resulting from the advances in molecular biology and genomics. During the course of varietal development, plant breeders are faced with several challenges. Not only do plant breeders have to deal with traits such as adaptation to specific environments, high yield, biotic and abiotic stress tolerance, crop quality, but to suit consumer demand, they must also look for the sources of genes that are often not available in primary crop gene pools. Breeders have continued to explore new methods to characterize and use valuable genes from secondary and tertiary gene pools. Simultaneously dealing with multiple traits requires breeders to design field experiments for evaluation of each trait. If some of these traits can only be evaluated after flowering, breeders lose a growing season in the breeding process. Molecular markers offer several advantages over morphological and isozyme markers in linkage mapping of important agronomic traits in most crop species. Molecular markers are unlimited in number, show higher level of polymorphism, and allow

marker assay at any developmental stage without any environmental interference. Molecular-marker technology promises to help the plant breeders on several fronts accelerate the breeding process. It can be useful in fingerprinting, genetic diversity assessment, mapping of simple and complex traits, marker-assisted selection, and ultimately in cloning genes controlling agronomic traits. Advantages of genotypic or DNA-based selection are: (1) simultaneous selection of multiple traits, (2) marker-based selection for traits that are difficult and expensive to evaluate, (3) selection at the juvenile stage for traits that are expressed after flowering, (4) marker-assisted introgression of useful genes in wide crosses, (5) accelerated backcrossing, and (6) increased precision and speed of selection.

Overview of molecular markers

Soon after the discovery of restriction fragment length polymorphism (RFLP) in humans by Botstein et al. (1980), numerous marker types have been developed and found to be useful in crop improvement (Tanksley et al., 1989; Paterson et al., 1991; Burrow and Blake, 1998; Joshi et al., 1999). The RFLP technique is robust, reproducible, and portable. But its technical complexity, lengthy process, use of radioisotopes, and requirement for a large quantity of DNA necessitated development of several polymerase chain reaction (PCR)-based marker systems, such as Random Amplified Polymorphic DNA (RAPD) (Williams et al., 1990), Amplified Fragment Length Polymorphism (AFLP) (Zabeau and Vos, 1993; Vos et al., 1995), microsatellites or Simple Sequence Repeat (SSR) (Litt and Lutty, 1989), and Sequence Tagged Sites (STS) (Olson et al., 1989). While the codominant nature of RFLPs makes them very reliable and informative for linkage analysis and marker-assisted breeding, the dominant nature of PCR-based AFLP and RAPD markers complicate distinguishing homozygotes from heterozygotes. New software, however, is being developed to identify homozygotes and heterozygotes in AFLP fingerprints (Pot and Pouwels, 2002). The AFLP markers combine the robustness of RFLPs with the simplicity of the PCR technique and are increasingly used for a variety of studies involving genome mapping, diversity, and phylogenetic and fingerprinting studies. The AFLPs show less polymorphism compared with RFLPs and SSRs. On the other hand, growing popularity of SSR markers among plant breeders is due to their abundance, technical simplicity, and higher level of polymorphism despite high developmental expenses (Gupta et al., 1996).

The single nucleotide polymorphism (SNP) (Brookes, 1999) is a more recently developed marker system. It has great potential for deployment for genome mapping and marker-assisted selection (Ye et al., 2001) due to its simplicity, ubiquity, abundance, and potential for automation (Rafalski, 2002). Large-scale genome and expressed sequence tags (EST)-sequencing projects in several field crops have facilitated the discovery and rapid growth of SNP technology (http://www.ncbi.nlm.nih.gov/SNP/) and its utilization in crop improvement.

Application of molecular markers in plant breeding

Fingerprinting and genetic diversity assessment

The molecular tools discussed above can be used in conjunction with the traditional breeding methods to make product development faster, more productive, and more efficient. Molecular markers provide a convenient tool to genetically profile inbreds, varieties, and hybrids and, thus, strengthen the protection of intellectual property rights over plant varieties (Brown and Kresovich, 1996). The latter provides indirect impetus for more private investment in the use of modern plant-breeding techniques to increase agricultural productivity.

Germplasm resources in field crops provide a continued source of genetic variability for plant breeders. With changes in global climatic patterns and emergence of new pests and diseases, both conservation and characterization of plant germplasm resources have become more important than ever before for plant breeders (Schoen and Brown, 1993). Molecular markers can help in identification of duplicate accessions. They can also help select individual accessions of core collections. Genetic similarities estimated through molecular markers can be helpful in identifying the diverse accessions that might possess useful genes for crop improvement. The existence of molecular markers has increased interest in studying the genetic relationship among species by sampling the variability in the whole genome (Aggarwal et al., 1999); this is in contrast to the traditional variables used in the past, such as morphological variation, cross compatibility, cytogenetic relationships, breeding systems, and geographical distribution.

Conservation genetics is an important and emerging discipline. The practical aspects of the strategies and the technologies for conserving plant biodiversity are authoritatively discussed in Henry (2006).

Marker-assisted selection (MAS)

From a breeder's perspective, marker-assisted selection for easily scorable traits should be avoided because identification of linked markers and the marker-assay involving large segregating populations are expensive. Rather, the skilled manpower and resources would be better used to investigate agronomic traits that are difficult or expensive to evaluate. The steps involved in mapping a major gene or quantitative trait loci (QTL) are described in Figure 1.1.

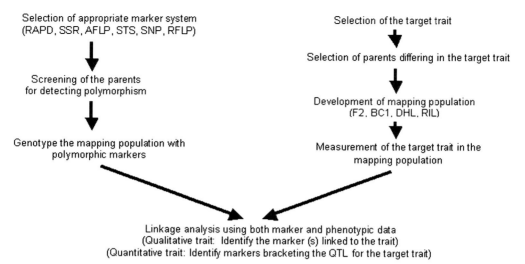

Fig. 1.1.   Steps involved in identifying molecular markers linked to a trait controlled by a major gene or quantitative trait loci (QTL).

Most commonly used mapping populations are doubled-haploid lines (DHL), recombinant inbred lines (RIL), standard F_2 population, and backcross (BC1) population. When breeders have access to near-isogenic lines (NILs) for certain traits, they can be used to rapidly identify markers linked to the gene of interest (Young et al., 1988) with less effort and expense than for other populations. Another strategy advocated to target specific genomic regions is bulked segregant analysis in which DNA from the individuals that are homozygous for the target trait in a segregating population is pooled for quick identification of linked markers (Michelmore et al., 1991). In addition to the most widely used genetic mapping software "Mapmaker" (Lander et al., 1987; Lander and Botstein, 1989), JoinMap (Stam, 1993), QTL Cartographer (Wang et al., 2006), Map Manager (Manly and Cudmore, 1998), and qGENE (Nelson, 1997) are available for map construction, QTL mapping, and marker-assisted breeding. Marker-assisted selection has been successfully performed for many oligogenic traits (Garland et al., 2000; Murai et al., 2001; Jia et al., 2002; Komori et al., 2003). But most agronomic traits are quantitative in nature and are the result of the joint action of several loci on chromosomes (QTL). Tolerance to abiotic stresses, such as salinity, submergence, and drought is difficult to evaluate because the breeder has to either simulate similar stress situation or has to make field evaluation in locations where stress is commonly experienced. Molecular markers can be helpful in identifying individuals with desirable QTL controlling abiotic stress tolerance. Many disease resistance and insect resistance traits are quantitative in nature (for example, blast resistance [Wang et al., 1994] or yellow stem borer resistance [Selvi et al., 2002] in rice). Accumulation of a few major QTL controlling such pest resistances could be rewarding in improving field resistance. If resistance is controlled by recessive genes, transfer process lengthens the variety development process due to the requirement for an additional selfing generation. In such situations, marker-based selection can accelerate the gene transfer process from a donor plant to a desired genotype.

Marker-assisted selection is also appropriate for those traits that can only be evaluated at or after the reproductive stage. For example, grain quality and environment-sensitive genetic male sterility traits are ideal candidates. An SSR marker has been developed to select for amylose content in rice (Ayres et al., 1997). Lang et al. (1999) developed PCR-based markers that can identify thermosensitive genetic male sterility gene *tms3* at seedling stage. Traits like wide compatibility and fertility restoration can only be evaluated in testcross progeny. Molecular markers linked to a wide compatibility gene (Liu et al., 1997) and fertility restorer genes in rice (Akagi et al., 1996; Ichikawa et al., 1997; Yao et al., 1997) have been identified to facilitate gene transfer to appropriate genetic backgrounds.

Further research is needed to devise a new marker-assisted selection (MAS) strategy to suit breeding programs. Ribaut and Bertan (1999 proposed a modified MAS procedure called "single large-scale marker-assisted selection" (SLS-MAS) to improve polygenic traits in maize, where selection pressure for the target gene was combined with maintenance of genetic diversity in the rest of the genome. Zhu et al. (1999) suggested that MAS would be rewarding if optimum combinations of QTL are determined and favorable alleles of those QTL are accumulated instead of pyramiding alleles detected in a reference mapping population.

Analysis of polygenic traits

Most agronomic traits are controlled by many genes, along with significant environmental influences. Contrary to the application of a quantitative genetics strategy, molecular-based dissection of these complex traits, commonly known as QTL analysis, provides much more genetic information with which to make genetic advances. Valuable information obtained through the application of a QTL mapping strategy includes: (1) the number, effect, and chromosomal location of genes affecting traits; (2) the effect of multiple copies of individual genes (gene dosage); (3) the nonallelic interaction between/among genes controlling a trait (epistasis); (4) the pleiotropic effects, i.e., whether individual genes affect more than one trait; and (5) the stability of gene function in different environments (G × E interaction) (Paterson et al., 1991; Tanksley, 1993; Kearsey, 2002). The genes/factors responsible for several quantitative traits in field crops, mapped with greater resolution, could be finally cloned (Frary et al., 2000; Fridman et al., 2000; Yano et al., 2000; Takahashi et al., 2001).

Another strategy to genetically dissect quantitative traits is to use candidate genes (Faris et al., 1999). In this strategy, one should look for certain biochemical pathways with possible involvement in the expression of the target trait. Based on the available sequence information in the database, sequence information for a few genes should be retrieved and tested through mapping if the genes are linked to the QTL underlying target traits. Further linkage between genes and phenotype should be established by means of biochemical and genetic studies. One successful application of this strategy has been demonstrated for corn earworm resistance in maize (Byrne et al., 1996). Availability of high-density genetic and physical maps with linkage to ESTs and gene sequences may accelerate the identification of these candidate genes for success in MAS.

Gene pyramiding

Marker-assisted selection is also appropriate for stacking several genes for single or multiple agronomic traits in a single genotype because it eliminates the need for field-based evaluation that must be independently conducted for selection of an individual trait or gene. Stacking of multiple genes (both major and minor genes) that have resistance to certain biotypes of insect pests or certain isolates of a disease organism to breed field crops with durable resistance is of considerable interest to plant breeders. Because identification of plants with multiple genes in segregating generations simultaneously is practically impossible, the pyramiding process can be facilitated using molecular markers.

Successful pyramiding of multiple genes has been demonstrated in rice to improve resistance to blast (Hittalmani et al., 2000), bacterial blight (Sanchez et al., 2000; Huang et al., 1997a) and gall midge (Katiyar et al., 2001). Three blast resistance genes, *Pi-1, Piz-5, Pi-ta,* were efficiently pyramided into a single genotype using molecular markers, and lines harboring all three genes exhibited an improved resistance response over the lines with single genes (Hittalmani et al., 2000). A similar strategy was followed to successfully develop breeding lines with a broad spectrum of resistances by combining two, three, or four bacterial blight resistance genes, *Xa-4, xa-5, xa-13,* and *Xa-21* (Huang et al., 1997a). Sanchez et al. (2000) developed sequence-tagged site (STS) markers from the RFLP markers tightly linked to these genes and successfully developed three-gene (*xa-5, xa-13,* and *Xa-21*) pyramids in three elite, but bacterial blight-susceptible varieties. This study clearly demonstrated that molecular markers increased both effectiveness and efficiency of selection for recessive genes *xa-5* and *xa-13* in the presence of dominant gene *Xa-21.*

Accelerated backcrossing

Backcross breeding method is commonly used by plant breeders to rectify one or few defects in an otherwise adaptable elite cultivar. This method is often employed when adding genes for resistance to diseases and insect pests. Molecular markers can play an important role in bringing the improved

cultivars to the farmer's field two to three generations earlier compared with the conventional backcross method. While positive selection for the target gene can be used in every backcross generation to identify the desirable individual, negative selection can be performed for the donor genome to accelerate the recovery of a recurrent-parent genome. If wild or exotic species are used for transfer of any target gene(s), linkage drag can be minimized by accurately defining and incorporating the introgressed segments (Brar et al., 1996).

Another strategy is to use marker-assisted backcrossing in conjunction with transgenic breeding. Because success of crop transformation with useful foreign genes in several field crops is genotype-dependent, transformation can be achieved using the variety that is more amenable to crop transformation. The desirable trait in the transgenic line can then be transferred to the elite lines by the accelerated backcrossing procedure. Because the site of integration and number of transgenes influence the expression of the target traits, molecular markers can help in a large-scale screening to identify transgenic lines with the desired level of trait expression.

Marker-based utilization of exotic germplasm

Advanced backcross analysis

Wild relatives and land races of various field crops are invaluable reservoirs of many useful genes that will need to be tapped in the future in efforts to achieve global food security. The wealth of genetic variation residing in those germplasm in gene banks has so far remained un- or under-used. Appearance of an exotic or wild germplasm is not always a good predictor of its genetic potential (Tanksley and McCouch, 1997). New genomics approaches with a focus on genotypic composition rather than visual attributes can now be applied to unlock the genetic potential in those germplasm. Novel strategies, like the advanced backcross QTL method, allow identification of useful genes from wild germplasm (Bernacchi et al., 1998a,b). The advanced backcross QTL method combines QTL discovery with variety development, thus hastening the utilization of unadapted germplasm resources (Tanksley and Nelson, 1996). It can be applied to most self- and cross-pollinated crops with short life cycles, with the exception of highly heterozygous crops or clonally propagated crops in which inbred lines are not developed for crop improvement. In this procedure, deleterious alleles are removed by selection in early backcross generations, and in BC2 or BC3 generation, QTL analysis is performed. This has been useful in identifying beneficial exotic QTL for genetic improvement in many field crop species, such as barley (von Korff et al., 2006), rice (Yoon et al., 2006), bean (Blair et al., 2006), and wheat (Narasimhamoorthy et al., 2006). In an advanced backcross study involving wild rice species *Oryza rufipogon*, Xiao et al. (1996) reported that the *O. rufipogon* alleles at the RM5 locus on chromosome 1 and the RG256 locus on chromosome 2 increased rice yield 18% and 17%, respectively.

Introgression libraries

Another approach to exploit the naturally occurring variation in wild and unadapted species is to construct introgression libraries. An introgression library is a set of lines, each of which harbors a single molecular marker-delineated chromosome fragment from wild species in an elite cultivated genetic background (Zamir, 2001). These individual lines are developed by repeated backcrossing, in conjunction with molecular markers, to define and track the introgressed chromosome fragments in each successive backcross or selfed generation (Eshed and Zamir, 1994). There are several advantages

that can accrue from this modified mapping population. First, it can be used as an immortal mapping population with which to conduct replicated experiments to map any gene with more precision and higher efficiency. Second, the introgressed lines will allow determination of the effect of introgressed fragments in various genetic backgrounds and different environments, and distinguish between additive and dominance effects. Third, it will allow investigations of interactions between QTL, which are difficult to evaluate in traditional mapping populations. However, the most immediate application of introgression lines is to use the derived lines as varieties if any desirable trait can be precisely introgressed to an existing cultivar. Fridman et al. (2004) used an introgression library to identify a QTL for increased sugar yield from wild tomato species *Lycopersicon pennelli*.

In-situ hybridization technique

Plant breeders need new genetic variability almost on a continuous basis for their ongoing breeding programs to address ongoing or new challenges. To use the useful genes from wild relatives, cytogenetic manipulations are usually performed to develop alien addition and alien substitution lines. Molecular cytogenetic tools, such as genomic in situ hybridization (GISH) and fluorescence in situ hybridization (FISH), have been developed to facilitate the development and characterization of alien addition and alien substitution lines. Molecular markers, such as microsatellites and RFLPs, are also used for this purpose. Several traits introgressed from wild relatives have been tagged with molecular markers facilitating their introgression into cultivated genotypic background (Brar et al., 1996). If breakpoints of the alien segments are defined with molecular tags, marker-assisted introgression also can help in removing the unwanted genomic segments of the donor.

Factors affecting successful DNA marker utilization in crop breeding programs

Molecular markers have demonstrated their usefulness in crop breeding programs. The technology is being increasingly adopted by both public and private sector plant breeders for several purposes, as described in earlier sections. However, successful integration of molecular-marker technology into crop breeding programs depends on the following factors:

- **Choosing the right DNA marker system.** Most plant breeding programs operate on modest budgets. Although using MAS offers significant time savings when making genetic advances for certain traits that are otherwise difficult to evaluate (Burr et al., 1983; Tanksley et al., 1989), plant breeders wanting to integrate molecular-marker technology in a crop breeding program lack a simple, low cost, high throughput assay. Choosing the right DNA marker system is critical because it can potentially reduce the expense, time, and effort needed for MAS. Following are the factors that should be considered in determining the suitability of a DNA marker so that large-scale genotyping can be performed in a typical breeding program.
  - DNA quantity needed per assay: RFLP assay requires a large quantity of pure DNA compared with most PCR-based markers, such as RAPD, AFLP, SSR, and SNP.
  - Level of polymorphism: The higher level of polymorphism detected by a particular DNA marker system is a positive attribute because it is useful in QTL mapping and MAS involving closely related parents. Among the molecular markers, SSR markers show a higher level of polymorphism than other markers and should be useful for application in breeding programs.
  - Technological simplicity: RFLP assay involves multiple steps and is, thus, time-consuming and laborious. On the contrary, most PCR-based marker assays, such as SSR, STS, CAPS (Cleaved amplified polymorphic sequence, also known as PCR-RFLP markers), and RAPD, can be easily performed with only minimal technical skills (with the exception of AFLP). They involve a PCR amplification step, using a small amount of genomic DNA, followed by running a gel to detect the size differences in the PCR products. This has made possible quick screening of a large number of segregating individuals in breeding programs. The SSR markers can be screened in special agarose gels and, thus, can be useful for MAS.
  - Reproducibility: RFLP and SSR markers are easy to reproduce and, thus, are easily transferable to other laboratories, whereas RAPD markers have low reproducibility. The RAPDs are a most unreliable marker system and, as such, cannot be used for large-scale genotyping projects.
  - Markers with known map location: Availability of markers with direct links to existing genetic maps can make fine mapping of genes much easier because it allows access to the nearby markers for high resolution mapping. Currently, SSR is the most preferred marker in plant breeding programs because of its simpler assay, lower expense, higher level of polymorphism, and greater abundance in the genome. New marker types, like SNP, may be preferred by breeders in the future as genome sequences for different crop staples become available.
- **Linkage of the marker to the target gene.** This is an important factor for increasing selection efficiency. The tighter the linkage between the molecular markers

and the target gene, the higher the efficiency of selection. Loose linkage between the target gene and the molecular markers give rise to false positives. Incorrect linkage information obtained from the gene mapping experiment may also lead to false positives, thereby, decreasing the efficiency of selection. Before attempting MAS, target traits should be mapped with high resolution using large mapping populations. Field evaluations in replicated trials in multiple environments should be followed to obtain reliable phenotypic data. Sometimes, if the target gene is loosely linked, it is advisable to use flanking markers to increase the selection efficiency.

- **Time and expenses in MAS**. High cost associated with molecular-marker technology is the most important factor limiting its adoption in crop breeding programs. Expenses in performing MAS involve both materials and supplies for the DNA marker assay in addition to the requirement of skilled technical personnel. Another factor is the long delay between processing of tissue samples and data acquisition. Reducing time and expense will require automation of the genotyping process, starting from DNA extraction to marker screening and data acquisition. Some progress has been made in DNA isolation that can advance marker-assisted breeding, SNP discovery and scoring, and analysis of transgenic plants (Shenoi et al., 2002; Gauch, 2002). PCR-based marker systems, like microsatellites and AFLPs, provide simpler DNA assays that can be automated. Another promising marker system, SNP is becoming popular among plant breeders because of its abundance and amenability to high throughput screening (Rafalski, 2002). New technologies should be developed to eliminate the gel electrophoresis step. Tyagi and Kramer (1996) developed a technique, called "molecular beacons," for detecting SNPs.

- **Number of genes that can be handled in MAS.** Plant breeders should decide on the traits for which marker-assisted selection offers considerable savings in both time and effort. As the number of genes being handled or considered in MAS increases, population size correspondingly increases to recover recombinants with all target genes. Mackill et al. (1999) suggested that marker-assisted selection scheme be limited to four or five genes, because further increase in number of genes will increase the population size exponentially.

Challenges and future trends in the application of molecular markers

Molecular-marker technology provides an attractive tool with a wide range of applications to transform the cultivar development process (Subudhi and Nguyen, 2004). Its assimilation into breeding programs, however, is not as widespread as expected. While marker-assisted selection has been successfully used to improve traits controlled by major genes in many field crops, manipulation of quantitative traits, like yield, yield attributes, and tolerance to abiotic stresses through MAS has been a challenge because of epistasis, pleiotropy, and G × E interaction. Success of MAS for quantitative traits depends on reliable mapping results, which can be achieved using a large mapping population and well-distributed linkage map with reasonable genome coverage. The goal should be to capture as much phenotypic variation as possible in the discovered QTL and to obtain accurate estimates of QTL effects for the target trait. Further confirmation of mapping results in different genetic backgrounds and in multiple years and multi-environment trials, along with determination of the effect of individual QTL on the target trait in near-isogenic backgrounds, are essential for successfully implementing MAS. QTL are usually defined as chromosomal segment(s) harboring a single gene or a cluster of linked genes. However, it may not be necessary to fine-map to the single-gene level for MAS unless the objective is cloning the specific QTL (Paterson, 1998). All these activities before MAS are very demanding of time, effort, and expenses. Therefore, plant breeders should identify the potential candidate traits for which MAS would be rewarding for realizing the overall project goal. Some other factors, such as the urgency to develop and release a variety or hybrid with a specific trait, inheritance pattern, and genetic gain compared with conventional selection, should also be considered.

Ideally, MAS will be most effective and efficient when all gene sequences for plant growth and development are known. The sequence information of known genes can be used to analyze allelic diversity among the diverse germplasm and accurate incorporation of favorable alleles for the target trait without any linkage drag. Rapid advances in genomics and related fields are accumulating huge amount of genome information, such as high-density genetic and physical maps, comparative genome maps, ESTs, cloned genes, and gene expression data. The availability of whole genome sequences in two model plants, rice and *Arabidopsis*, now provides opportunities to discover homologous genes in related crop species with larger genomes, e.g., maize, wheat, and barley, through comparative mapping. To use this huge amount of information for crop improvement, efforts should be directed to link the sequence information with phenotype by using new functional genomics, proteomics, and bioinformatics tools (Young, 1999; Wilson et al., 2003).

Due to technological advances, it is now possible to analyze the expression profile of thousands of genes simultaneously using DNA chips or microarrays (Schena et al.,

1995; Lemieux et al., 1998). There have been efforts to link a specific pattern of gene-expression profile to a certain phenotype (Schadt et al., 2003), so that an ideal genotype with a unique trait combination can be developed. Proteomics technology is adding new dimensions to genomics research due to innovations in 2-D electrophoresis and mass spectrometry (Roberts, 2002). Considering the pace of advances in genomics, it is reasonable to hope that new high throughput tools will be available for application in plant breeding programs to select desirable plants with a combination of desirable alleles based on specific pattern of protein or transcript expression (MacBeath and Schreiber, 2000).

There have been tremendous advances in large-scale genotyping technology during the past few years, which has opened up new opportunities for genetic improvement of field crops through genome analysis and marker-assisted selection. A promising high throughput genotyping method called "molecular inversion probe" combines molecular barcoding with the microarray-based detection system (Hardenbol et al., 2003) and is capable of genotyping 10,000 SNPs in a single tube assay (Hardenbol et al., 2005). Therefore, the potential of this new, powerful technology and genomics information and resources to transform the methods of crop development can only be fully realized by close collaboration between the genomics researchers, bioinformaticians, and plant breeders (Stuber et al., 1999; Koncz, 2003).

Tissue culture-induced variation (somaclonal variation) in plant breeding

Genetic variation is the fundamental prerequisite for any breeding method aimed at improvement of crop plants. Conventionally, plant breeders create this variation through sexual crossing (intra- or inter-genic

Table 1.4. Somaclones released as commercial cultivars in selected agronomically important crops (modified from Jain, 2001).

Crop Species	Cultivar	Trait
Banana	Formosana	Fusarium wilt resistance
Wheat	He Zhu No.8	High yield
Tomato	DNAP-17	*Fusarium* resistance
	DNAP9	High solid content
Lathyrus sativus	P-24	Reduced BOAA content in feed, grain, high yield, early maturing
Potato	White Baron	Non-browning
Maize	Yidan No. 6	Variant for grain and forage use
Brassica juncea	Pusa jai Kisan	High yield, shattering resistance
Rice	LSBR-33, LSBR-5	Rhizoctonia resistance
	FR13A	Submergence tolerance
	DAMA	Improved cooking quality, Pyricularia resistance

and/or wide hybridization, mutation, etc.). With the advent of tissue and cell culture technology, variation has been generated among in vitro-regenerated plants of several crop species; this is commonly known as somaclonal variation (Larkin and Scowcroft, 1981). Plant breeders fear that such spontaneous and uncontrolled tissue culture-induced variation could prove detrimental to valuable genetic stocks maintained in vitro. However, such variation has been exploited to realize its potential in plant breeding methods, resulting in a limited number of somaclonal variants being released as cultivars in a few agronomically important crops (see Table 1.4). Point mutations, chromosomal rearrangements, recombination, DNA methylation, altered sequence copy number, and transposable elements are believed to be the basis of somaclonal variation. Several factors, such as genotype, age of donor plants, type of explants, and age of culture (time during subculture), are reported to contribute toward this effect (for a review see Veilleaux et al., 1998; Jain, 2001). Besides, tissue culture itself imposes a stress on the cell, which then tries to adapt through a mechanism called "chromosome reshuffling," resulting in transposition burst (McClintock, 1978).

Somaclonal variation has its own merits and demerits. Because stable mutations occurring in tissue culture are mostly negative, as are mutations induced by classical mutagens, such as irradiation and chemicals, this technique, contrary to expectations, is not regarded as an extra source of novel and unique genetic variation. Nevertheless, somaclonal variation mimics induced mutations and can be induced in both asexually propagated and seed-propagated plants. This still holds promise as a component of plant breeding, along with in vitro mutagenesis. Plant breeders may be convinced of the potential of this variation, if sufficient information is available about molecular understanding and there are reliable molecular markers for the identification—at an early stage of plant development—of hypervariable DNA ("hot spots" that cause some species or genotypes to be more prone to somatic recombination).

Fig. 1.2.   Induction of somatic embryo (embryoids) in callus culture: (A) conversion of the embryoid into atypical plantlet, and (B) high frequency regeneration from the SE in rice. The picture clearly shows the typical globular, heart, and torpedo shaped stages of SE before undergoing conversion to full plantlet.

Somatic embryogenesis

When somaclonal variation is not a goal, especially in transgenic breeding (discussed later in this chapter), somatic embryogenesis and micropropagation offer plant breeders the opportunity to propagate and maintain true-to-type elite germplasm stock. Moreover, hormonal manipulation and duration of callus in culture determine regeneration through either somatic embryogenesis or organogenesis, which could possibly result in variation among the regenerants.

Somatic embryogenesis has been induced and successfully exploited in many crops, including rice. Somatic embryos (SE) could also be useful in the development of synthetic seeds by encapsulating the embryos in sodium or calcium alginate and storing them for future use (Datta and Schmid, 1996). Figure 1.2 is an illustration of induction and high frequency conversion of somatic embryos to complete plantlets in rice. These SE-derived plants were genetically uniform and did not show any observable variation in the morpho-agronomic traits under field conditions (Baisakh and Rao, 1997).

Anther culture and doubled-haploid breeding

Androgenesis is the process of the development of plants from in vitro culture of male gametes such as anthers or microspores instead of using a normal male gameto-phytic pathway that leads to formation of pollen. Anther culture results in the production of haploid plants, which could be diplodized by the use of colchicine to yield di-haploids (DH). But in most crop species, such as rice, wheat, barley, and maize, spontaneous doubling of the chromosome complement occurs during the culture with variable frequency.

Di-haploids offer manifold advantages to the plant breeder because of rapid fixation of homozygosity in just one generation, as against five to six generations in conventional breeding; increased selection efficiency, especially for the recessive traits; and parental purification in hybrid seed research. As well, variation among anther-derived clones (termed as androclonal variation) can be used for selection of improved lines. As mentioned earlier, microspore-derived embryos could also be used for the production of synthetic seed. Encapsulated barley microspore embryos can be stored in cold room and successfully regenerated into normal seed (Datta and Potrykus, 1989). The discovery of anther culture in *Datura* (Guha and Maheswari, 1964) paved the way for the development of di-haploids that have been used to facilitate plant breeding in several crops of economic importance.

Methods of DH development—Androgenesis can be achieved by using the following techniques: (1) culture of the inflorescence containing the uni- to early bi-nucleate stage microspore on liquid or solid media; (2)

Fig. 1.3. Anther culture in indica rice. Anthers producing microspore-derived callus (A), Shoot regeneration of callus (B), development of both green and albino plants (C), and microspore-derived haploid (n), dihaploid (2n), and tetraploid (4n) plants in the greenhouse condition (D).

culture of anthers on liquid media and subsequent embryo regeneration of the shed microspores directly on liquid media containing Ficol (known as shed pollen culture); (3) culture of the mechanically isolated haploid microspore (as protoplast)—although this method is limited by poor regeneration response; and (4) culture of anthers in a solid media for callus/embryoid development and subsequent regeneration into plants. Although all four techniques have been used in crops, especially in cereals, with varying efficiency, the fourth one is the most commonly used in rice (Fig. 1.3).

Factors affecting androgenesis—A recent review discussed the cellular and molecular aspects involved in the induction of androgenesis and embryo formation (Maraschin et al., 2005). There are several reports relative to studies of different parameters that influence anther culture response of various crops. The genotype and the physiological environment of the donor plants, temperature pretreatment, microspore stage, culture medium, and carbon source in the media are some of the important factors contributing to successful DH production. A major bottleneck in the development of dihaploids is albinism, especially in cereals. The albino plants are usually devoid of the 23S and 16S rRNA and have a deleted plastid genome with no *rbcL* gene (see review in Datta, 2005). Again, an optimal combination of the above-mentioned factors has been shown to reduce the albinism, resulting in high rate of green plant regeneration.

Use of DH lines in genome breeding— Because DH lines are homozygous with no segregation in subsequent generations, they have been used in genetic studies involving identification and mapping of major genes/ QTL. This reduces the environmental component of the total variation, thereby allowing precise measurement of quantitative traits by repeated trials (Lu et al., 1996). For example, in rice, DH populations were used to identify QTL for tiller angle (Qian et al., 1999a), top internode length (Tan et al., 1997a), plant height and days to heading, ratooning ability (Tan et al., 1997b), grain yield traits (Benmoussa et al., 2005), seedling cold tolerance (Qian et al., 1999b), root characteristics, such as root vitality (Teng et al., 2000), thickness and root length (Kamoshita et al., 2002), lodging resistance (Mu et al., 2004), seedling salt tolerance (Prasad et al., 2000), grain shape and brown planthopper resistance (Huang et al., 1997b), grain quality (Bao et al., 2002) and several other agronomic traits (Lu et al., 1996). However, segregation distortion (SD) among the DH populations with respect to RFLP markers could limit the success of marker-assisted selection of genes linked to chromosomal regions showing SD (Subudhi and Huang, 1995). The importance of DH lines vis-à-vis an F_2 population for genetic studies of quantitative traits was validated, however (Chen et al., 1996), which demonstrated that the variation among the DH lines and extent of distorted segregation was not significant enough to affect their use.

Androclonal variation and its usefulness—There have been reports of ploidy variations in anther culture of several crops, leading to unexpected variability among progeny plants due to gametoclonal (androclonal in this case) variation. Such variations have been exploited for genetic studies and/or direct release of improved variants as commercial cultivars. For example, aneuploids (e.g., trisomic plants with haploid chromosome and an additional chromosome) have been useful for studies on genic imbalance, chromosome behavior in meiosis, construction of a rice genetic map, and assignment of DNA markers to a specific chromosome (Wang and Iwata, 1991). The production of aneuploids (trisomic, tetrasomic, and pentasomic) is difficult through conventional breeding, whereas these genetic stocks have been produced through anther culture and investigated for genetic and cytological factors (Watanabe et al., 1974; Wang and Iwata, 1991). Similarly a large number of DH lines have been released as varieties in rice and other cereals in China, Korea, and other part of the world (Datta and Torrizo, 2006).

Androgenesis in transgenic research—Androgenesis has a great impact on transgenic breeding in developing homozygous transgenics in a single year. Using haploid microspore culture and protoplasts derived from developing embryos, the first homozygous transgenic *indica* rice was reported in 1990 (Datta et al., 1990). A few more reports are now available for rice (Guiderdoni and Chair, 1992), barley (Jahne et al., 1994), and

Brassica (through microinjection) (Neuhaus et al., 1987). Subsequently, homozygous transgenics have been developed in less than a year through anther culture of primary transgenics of rice (Baisakh et al., 2000; 2001).

However, the fact that many agronomically important crop species are recalcitrant to androgenesis and the lack of proper understanding of the mechanism rendering the microspore cells embryogenic limit the use of such a potential technique (Wang et al., 2000).

Transgenic breeding

It is now well known that conventional breeding methods alone will not be able to keep pace with the growing demands for food of the ever-increasing human population. Moreover, conventional plant-breeding methodologies are often limited by reproductive barriers. Biotechnology (commonly known as genetic engineering) can bolster plant-breeding efforts through the development of transgenics to meet these new challenges in a sustainable way. These developments have opened new vistas in crop improvement, thereby allowing transfer of desirable gene(s) across species and genera. The precise and stable integration of alien gene(s) and the achievement of homozygosity at the transgene locus in two to three generations (or even in one year with the use of anther culture) give a stronger impetus to the development of transgenics than for gene transfer in conventional cross breeding. Transgenics provide the plant breeder with new and broader genetic pools.

Transgenic plants are the product of a transformation technique that delivers alien gene(s) within, among, or across taxa. There are several prerequisites for plant transformation, such as (1) an efficient (preferably genotype-independent) DNA delivery system; (2) available totipotent cells/tissues capable of high regeneration following DNA

Fig. 1.4. Transformation protocol in rice. Immature embryos (IE = explant) arranged for bombardment (A), hygro-mycin resistant callus selected from the bombarded IE (B), shoot and root regeneration from the putative transgenic calli (C & D), hardening (E) and transfer of the putative transgenics in containment greenhouse (F). (Baisakh, 2000).

transfer; (3) a suitable selection system; and (4) skilled personnel.

The method involves the delivery of a gene of interest, along with a selectable marker gene, into a competent tissue, followed by the selection of putatively transformed tissue, and regeneration of the tissue to whole plantlet. A generalized scheme of transformation of *indica* rice using immature embryos as the most competent tissue is shown in Figure 1.4.

DNA transfer method

The first-ever transgenic plant was developed with the *Agrobacterium tumefaciens* method in 1983 (Murai et al., 1983). Subsequently, many different methods have been tested and become available for the transfer of the foreign gene(s).

Agrobacterium tumefaciens-*mediated transformation*

Agrobacterium tumefaciens is a soil-borne, gram-negative bacterium capable of inducing galls in susceptible host plants. The gall-inducing ability of this bacterium is due to a tumor-inducing plasmid (Ti-plasmid), approximately 200 kb in size, which harbors *Vir* loci for virulence and T-DNA having *onc* genes for gall/tumor induction. T-DNA is flanked by *cis*-acting 25 bp imperfect direct repeats, which are involved in direct T-DNA processing and its precise transfer. Unlike other transposons, T-DNA does not encode any products that mediate its transfer. The right border appears to be more crucial for the polar transfer of T-DNA from right through left. The *onc* genes have no role in T-DNA transfer, which offers a unique opportunity to replace them with genes of interest for the construction of a disarmed plasmid vector for plant transformation. *Vir* loci provide *trans* acting products for T-DNA processing and transfer. Several *Vir* genes have been identified in *Octopine* and *nopaline* -type *Agrobacterium* strains. In the presence of phenolic compounds, such as acetosyringone, near the wound sites on host plant, *Vir A* gene encodes Vir A protein, which, upon autophosphorylation, acts as receptor/sensor and activates *Vir G*. The protein dimers encoded by *Vir G* serve as transcriptional activators of all other Vir loci. *Vir C* and *Vir D* are involved in the generation and processing of T-DNA, and especially *Vir D* products function as an endonuclease to nick the T-DNA at the right border and ultimately generate a single-strand copy of T-DNA (ssT-DNA). This ssT-DNA is being protected by Vir E protein that binds to T-DNA and Vir B protein has

a role in the formation of the membrane channel required for the transit of T-DNA to the nucleus of a host cell. Bacterial–plant cell interactions, bacterial chemotaxis, and attachment to host plant cell are mediated by constitutively expressed genes present in the bacterial chromosome.

The large size of Ti-plasmids and the presence of *onc* genes undermine the use of these plasmids for gene delivery. To circumvent these hurdles, two approaches have been developed. One approach uses *co-integrate* vectors wherein the *onc* genes within the borders of T-DNA are replaced by gene(s) of interest by homologous recombination, thus disarming the plasmid. The other approach is by using *binary* vectors. A pair of vectors, viz., mini-Ti and helper Ti, are used in the method. The products of helper Ti cause the transfer of gene(s) of interest in T-DNA of mini-Ti plasmid.

Although monocots were long believed to be recalcitrant to the *Agrobacterium*-mediated gene transfer method, using correct and compatible *A. tumefaciens* strains can effectively transform monocots, too. Several monocot crop species have been transformed with several genes of agronomic importance (see review in Gelvin, 2003). Seed imbibition, flower dipping and infiltration, or plant DNA inoculation are also some possible methods of *Agrobacterium* transformation that alleviate the need for the tissue culture component. However, such methods have had limited success, especially in model plants like *Arabidopsis* and tobacco.

Biolistic-mediated gene transfer

The biolistic method, also called "particle bombardment," "microprojectile bombardment," "particle acceleration," etc., involves using high-velocity metal particles to deliver biologically active DNA into living cells. The high velocity is generated by a device called a "particle gun." Microprojectiles of gold, tungsten, palladium, iridium, and possibly other second and third row transition metal particles are used to carry DNA into the host plant system. Any plant tissue can be used as a target in the biolistic approach, as the microprojectiles coated with DNA penetrate cell walls and membranes to deliver DNA into the cells. Acceleration can be achieved by using a pneumatic instrument to release either compressed air or helium gas. Currently, helium shock wave, in combination with partial vacuum, is commonly used to accelerate the carrier of DNA. This method has high transformation efficiency.

The important variables that influence gene transfer by particle bombardment are the acceleration system, particle size, shape, preparation and binding of DNA into particles, agglomeration and dispersion of particles, particle momentum, and number of particles that impact the target in a given area. Furthermore, the characteristics of the target tissue, such as growth of the explant and tissue pre-treatment, nature, form, and concentration of DNA, are other critical factors affecting transformation efficiency. However, with variable frequency, as mentioned earlier, this method is genotype-independent. A large number of crops have been produced and field-tested using the biolistic method (Altpeter et al., 2005).

Protoplast-mediated gene transfer

Protoplasts (single cells without a cell wall), from which whole plants can be regenerated, have the ability to easily take up foreign DNA (competence). Polyethylene glycol (PEG), electroporation, heat shock, low-dose irradiation, and media ultrafilteration are some of the methods that have been reported to make protoplasts competent for DNA uptake. However, the most commonly used methods are PEG-mediated and electroporation-mediated gene transfer.

Polyethylene glycol increases cell wall permeability and is an efficient agent for chemical fusion of protoplasts in somatic hybridization of various plants and animals. Transformation efficiency and cell viability are affected by PEG concentration. Very high concentrations decrease cell viability, while low concentrations can reduce gene transfer efficiency. In addition, a good embryogenic cell suspension and culture media are important factors directly affecting plant regeneration.

The electroporation method is based on the principle of a transient creation of a niche in the protoplast membrane with a short electric pulse of high-field strength. Optimum electroporation conditions may vary with the size and competence of the protoplast. Amplitude, duration of electric pulse, and media composition also affect transformation efficiency and protoplast viability. This method has been successfully applied to transform maize, rice, and other crops. In rice, the following genes have been successfully transferred: bacterial *hph* marker gene, *hygromycin phosphotransferase* (for hygromycin-B antibiotic resistance) gene, *bar* gene encoding *phosphinothricin* (for herbicide resistance), and *Xa21* gene (for bacterial blight resistance) (see Datta and Datta, 2002).

There are reports on whisker method, silicon carbide method, microinjection, electrophoresis, pollen-tube pathway, and liposome-mediated transformation as alternatives to the afore-described three methods (Rakoczy-Trojanowska, 2002). These methods have been of limited use.

Selectable markers

A gene that helps identify/select cells carrying the transgene is called a "selectable marker" gene. Regardless of the transformation method used, selection of putative transgenic cells or tissues is the key step for final recovery of transgenic plants. The selectable marker can be delivered into a host plant by either cointegrating it in the plasmid harboring the gene of interest or

Table 1.5. Commonly used selectable marker genes in plant transformation (modified after Twyman et al., 2002).

Gene (product)	Source	Selective agent	Remark
aad (aminoglycoside adenyltransferase)	*Shigella flexneri*	Streptomycin, Spectinomycin, Trimethroprim, Sulfonamides	Used mostly for chloroplast transformation
hph (hygromycin phosphotransferase)	*Klebsiella*	Hygromycin B.	Widely used for monocot as well as dicot transformation
nptII/neo/aphII (neomycin phosphotransferase)	*Escherichia coli*	Kanamycin, Neomycin, Geneticin (G148)	Used commonly in dicots
bar (phosphinothricin transferase)	*Streptomyces hygroscopicus*	Phosphinothricin (PPT), a constituent of herbicides Basta™, glufosinate, bialophos	Used in both monocot and dicot transformation
manA/pmi (mannose-6-phosphate isomerase)	*Escherichia coli*	Mannose	Positive selection system
badh (betaine aldehyde dehydrogenase)	*Daucus carota*	Glycine betaine	Used in chloroplast transformation of carrot

by cotransforming with another plasmid containing the marker gene. Recently, some transformation studies have used non-antibiotic selectable marker genes instead of antibiotic resistant genes for the sake of environmental and public health concerns (see Table 1.5).

Reporter genes

Reporter genes are used as assessable markers that give information regarding when, where, and at what level a transgene is being expressed. In addition, if they are fused with signal peptide coding sequences, they can give precise information on targeted proteins. *Chloramphenicol acetyltransferase* (*cat*), *B-glucurodinase* (*gus*), and green fluorescent protein (*gfp*) genes are commonly used as reporter genes. Due to its precision and clarity, the CAT assay is widely used in eukaryotic transformation systems. The *GUS* gene can be localized in tissue sections *in vivo* using histochemical technology, giving it the advantage of being used as both reporter and selectable marker.

This assay is being used extensively for tissue-specific studies and for optimization and/or testing efficiencies of transformation systems. However, because the method is destructive and involves the use of toxic chemicals, it cannot be used for *in vivo* gene expression studies.

The luciferase gene product catalyzes the ATP-dependent oxidation of luciferin in a reaction that produces light and hence the gene can be easily detected and monitored in vivo. However, the LUC gene has not been used much because the light emitted can only be detected using specialized devices, such as luminometers, image enhancing video systems, and X-rays. The gene for green fluorescent protein (GFP), cloned from the jellyfish, *Aequorea victoria*, has the characteristics of a sensitive reporter gene that can be used in in vivo gene expression. GFP is stable and can only be denatured under extreme conditions. The ability to maintain its green fluorescent properties in plant cells has augmented the use of GFP as an attractive alternative in genetic engineering and molecular biology.

Promoters used in plant genetic engineering

Promoters are the DNA sequences that drive the expression of gene(s). The levels and patterns of transgene expression, determined by promoters, depend on the position of promoters in the genome. The widely used promoters are either constitutive or tissue-specific. The transgene is expressed in all tissues at all stages of plant development in the presence of constitutive promoters, whereas its expression is localized to certain tissues in plants under certain conditions at a specific time when the promoters are tissue-specific. Some of the constitutive promoters include cauliflower mosaic virus (CaMV) 35S promoter, rice actin 1 (Act-1) promoter, Ubiquitin (Ubi-1) promoter, and maize ubiquitin alcohol dehydrogenase I (Adh 1) promoter. Phosphoenol pyruvate carboxylase (PEPC) promoter for green tissue-specific expression, Wun2 promoter for wound-specific expression, Hordein, rice glutelin Gt 1, globulin (glb), and prolamin (pro) for endosperm-specific expression, Rcg2 for root-specific expression, and ABA and rd29A promoter for abiotic stress-inducible gene expression are some of the tissue-specific promoters widely used in plant genetic engineering.

Transgenic crops with gene(s) of agronomic importance

Ever since the first transgenic plant was produced in 1983, transgenics have become a reality with the rapid development of genetically modified products for direct or indirect human use. A number of genes of economic importance have been transferred into major crops using different transformation methods (Datta et al., 2004).

Biotic stress tolerance

A vast majority of the transgenics developed and subsequently field-evaluated were for resistance/tolerance against biotic stresses like insect pests, and viral, fungal, and bacterial diseases. During the past decade, remarkable developments have taken place in transgenics with the *cry* (crystal protein) gene from *Bacillus thuringiensis* (*Bt*) showing resistance against lepidopteran, dipteran, and coleopteran insects. In addition, several other insects belonging to hymenoptera, homoptera, orthoptera and also nematodes, mites, and protozoans are being controlled by *Bt* toxin proteins. Several other genes like *gna* (for lectin protein) and *pin2* (protease inhibitor) genes have been successfully used for control of sap-sucking and other insect pests (Baisakh and Datta, 2006).

Similarly, development of transgenic crops against viral diseases involve, in most cases, introduction of a gene encoding a complete or partial viral protein. Various approaches have been used to develop transgenic viral-resistant crops. Structural and non-structural protein genes encoding coat protein and replicase, respectively, have been introduced into crops to confer viral resistance. In addition, anti-sense RNA strategy against coat protein and partial interfering transcripts were also used to control viruses. Transgenic papaya is a classical example of genetically engineered crops in which resistance to ring spot disease has been incorporated. Several other transgenic crops have been successfully deployed to control viral diseases (Dasgupta et al., 2003).

In general, a combinatorial expression of different pathogenesis-related (*PR*) genes, along with *R* genes, would confer an effective, durable, and broad-spectrum resistance to different fungal pathogens or races. The most important PR proteins are chitinases and β-1, 3-glucanases, which act on major structural cell wall components, namely, chitin and glucan, respectively. For example, in rice, transgenics have been developed with many *PR* genes that show considerable levels of resistance against sheath blight fungus (Datta et al., 2002).

Similarly, different classes of *R* genes conferring resistance to bacteria have been isolated, cloned, and tested. Most important in these classes of *R* genes is the *Xa21* gene, which has been widely used to confer resistance to bacterial blight (BB) disease caused by *Xanthomonas oryzae* pv. *Oryzae (Xoo)*. Under natural field conditions as well as in artificial inoculations, the transgenic rice with *Xa21* showed marked resistance against multiple strains and races of the BB pathogen (Tu et al., 2000). Efforts are underway to get an effective and durable horizontal resistance by expressing other novel *R* genes, like *NPR1, RPS1* or members from the same *Xa* family, e.g., *Xa5, Xa13* etc.

Herbicide tolerance

Weeds not only harbor major pests and fungal pathogens of crop plants, but also result in considerable direct yield losses in crops by using available nutrients from soil. Transgenic crops offer a better environment-friendly alternative to herbicides and confer enhanced safety of crops against non-selective herbicide action. Herbicide resistance genes were isolated and transferred to many crops for several herbicides of different modes of action. Diverse crop plants, including rice, wheat, maize, barley, tomato, tobacco, potato, cotton, soybean, rapeseed, peanut, sugarcane etc., have been engineered with herbicide resistance gene(s) and have been successfully evaluated under field conditions. The common herbicides for which herbicide tolerance has been engineered are: glyphosate (Roundup^{TM}), phosphinothricin (Basta^{TM}, Herbiace^{TM}), Bromoxynil (Buctril^{TM}), sulfonylurea (Glean^{TM}, Oust^{TM}), and 2, 4-D.

Resistance to abiotic stress

Abiotic stresses due to temperature (high or low), salinity, and water deficits severely affect crop productivity. A transgenic approach for developing abiotic stress-tolerant crops involves the use of gene(s) that are involved in one or all of the pathways underlying the tolerance mechanism. Although most of these studies were conducted in model dicot systems like *Arabidopsis* and tobacco, in recent years, other monocot crop plants, like rice, maize, tomato etc., have also been studied. The genes that can potentially confer abiotic stress tolerance fall into three categories: osmolyte production (osmoprotectant), ion homeostasis through ion exclusion, and ion compartmentalization. Recently, much progress has been made in the development of transgenics for tolerance against these abiotic stresses in many food crops (for a detailed list, visit www.plantstress.com). However, further progress remains a challenge to the biotechnologists because the mechanism is a complex network in which several individual pathways contribute to stress tolerance. Transgenic crops are being developed for other abiotic stresses as well, including aluminum toxicity, iron deficiency, and heavy metal toxicity.

Targeting productivity

Recently, focus has been slowly shifting toward improving yield levels still further and tapping the maximum yield potential of crops. By improving the source strength, higher yields can be achieved essentially by increasing the net photosynthesis level and minimizing losses incurred during photosynthesis. Through plant genetic engineering, the physiological processes, like stomatal responses, carbohydrate metabolism, leaf senescence, and light accumulation can be altered. However, it is most unlikely that one transgene would effectively achieve all of the above.

In crop plants, like rice and wheat, atmospheric carbon dioxide is fixed by the C_3 photosynthetic pathway and the overall photosynthetic efficiency in C_3 plants is reduced by the oxygenase reaction of Rubisco. Unlike C_3 plants, C_4 plants, like maize and sugarcane, have evolved mechanisms in such a way that local CO_2 concentration around Rubisco is increased. As a result, the oxygenase activity is inhibited, which also is reflected in reduced photorespiration, thereby increasing the net photosynthate. In transgenic rice, the expression of genes, such as *pepc* (phosphoenol pyruvate decarboxylase) and *ppdk* (phosphoenol phosphate dikinase) as well as ME (NADP-dependent malic enzyme) from maize, increased yield. However, detailed studies of such transgenics would provide better understanding of stress under field conditions.

Nutrition enrichment

Approximately 1.2 billion people worldwide suffer from malnutrition, especially from a deficiency of vitamins, minerals, and micronutrients. Conventional interventions, such as diet diversification, supplementation, and fortification, have been used with limited success. For the last few years, production of value-added food products (functional foods) is one important focus of transgenic research aimed at combating "hidden hunger." For example, transgenic rice with provitamin A and high iron accumulation in the milled grains could be an alternative, yet attractive, approach for fighting against the scourge of vitamin A deficiency (VAD) and iron deficiency anemia (IDA). Since the inception of golden rice in *japonica* cultivar (Ye et al., 2000), improved golden rices have been developed in *indica* rices that accumulate higher levels of beta-carotene in the seed endosperm and are potential candidates for field testing (Paine et al., 2005; Swapan Datta, personal communication). Through a transgenic approach, genetically modified rice has also been developed for increased iron retention and its bioavailability in the endosperm (Lucca et al., 2001).

Transgenic crops have also been developed for accumulation of vitamin E, C, and increased protein quality (e.g., lysine in rice), and oil quality (Baisakh and Datta, 2004).

Future prospects

In the future, transgenics, with expression and manipulation of multiple genes, will be better suited to achieving stability and sustainability of the transgene(s) against a rapidly evolving spectrum of diseases and insect pests. As well, multi-gene engineering is needed for installing a complete biosynthetic pathway in transgenic plants. Gene pyramiding could be achieved by stacking transgenes through multiple cotransformations or by conventional cross-breeding of transgenics with one or two genes using transgenes as markers (see Halpin, 2005). In rice, resistance has been achieved against sheath blight, bacterial blight, stem borer, and blast using this strategy (Datta et al., 2002; Narayanan et al., 2004); golden rice was developed using multigene engineering (Datta et al., 2003).

Another topic that needs immediate attention is the development of clean and marker-free transgenic crops. Several strategies, like co-transformation, sexual crossing, cre-lox, FLP-FRT systems, etc., have been reviewed for their potential in generating marker-free transgenics (MFT). In rice, MFTs have been developed with the *Bt* gene (Tu et al., 2003), and with genes for provitamin A biosynthesis (Parkhi et al., 2005; Baisakh et al., 2006). Similarly, a minimal expression cassette bombardment has been produced for developing transgenic rice with clean transgene integration (Fu et al., 2000). Moreover, plastid engineering is an alternative strategy for marker-free transgenics, especially where a large amount of transprotein is required (Daniell, 2002). Although tremendous progress has been made in this area, success in monocots is yet to be realized.

It is now understood that virtually any plant can be transformed with genes of interest. The outputs of the current genomics, proteomics, and metabolomics should be fully exploited with the help of transgenesis. Biotechnology can improve farmer and consumer welfare in many ways. Nevertheless, controversy surrounding GM crops continues in our heterogeneous society. Concerns about issues like horizontal gene flow, food safety, beneficiaries of the technology, and the conflict between plant variety protection (PVP) and intellectual property rights (IPR) need to be addressed. Finally, successful product development requires extensive field trials and public education. Gene technology combined with precise plant breeding and efficient crop management can help meet people's food needs. Once the fruits of genetic engineering in agriculture reach small farms and industrial operations, everyone can benefit from it.

References

Aggarwal RK, Brar DS, Nandi S, Huang N, Khush GS. 1999. Phylogenetic relationships among *Oryza* species revealed by AFLP markers. Theor Appl Genet 98:1320–1328.

Akagi H, Yokozeki Y, Inagaki A, Nakamura A, Fujimura T. 1996. A codominant DNA marker closely linked to the rice nuclear restorer gene, *Rf-1*, identified with inter-SSR fingerprinting. Genome 39:1205–1209.

Allard RW. 1999. Principles of plant breeding. 2nd ed. New York: John Wiley & Sons, Inc.

Altpeter F, Baisakh N, Beachy R, Bock R, Capell T, Christou P, Daniell H, Datta K, Datta S, Dix PJ, Fauquet C, Huang N, Kohli A, Mooibroek H, Nicholson L, Nguyen TT, Nugent G, Raemakers K, Romano A, Somers DA, Stoger E, Taylor N, Visser R. 2005. Particle bombardment and the genetic enhancement of crops: Myths and realities. Mol Breed 15:305–327.

Annicchiarico P. 2002. Genotype × environment interactions: challenges and opportunities for plant breeding and cultivar recommendations. FAO Plant Production and Protection Papers—174. Food and Agriculture Organization of The United Nations, Rome.

Ayres NM, McClung AM, Larkin PD, Bligh HF, Jones CA, Park WD. 1997. Microsatellites and a single-nucleotide polymorphism differentiate apparent amylose classes in an extended pedigree of US rice germplasm. Theor Appl Genet 94:773–781.

Baisakh N, Datta K, Oliva N, Ona I, Mew T, Datta SK. 2001. Rapid development of homozygous transgenic rice using anther culture harboring rice chitinase gene for sheath blight resistance. Plant Biotechnol J 18:101–108.

Baisakh N, Datta K, Rai M, Oliva N, Tan J, Mackill DJ, Khush GS, Datta SK. 2006. Marker-free transgenic (MFT) golden near isogenic introgression lines (NIILs) of indica rice cv. IR64 with accumulation of provitamin A in the endosperm tissue. Plant Biotechnol J 4:467–475.

Baisakh N, Datta SK. 2004. Metabolic pathway engineering for nutrition enrichment in rice. In: Chase CD, Daniell H, editors. Molecular biology and biotechnology of plant organelles. New York: Springer. p. 527–542.

Baisakh N, Datta SK. 2006. Stem borer resistance: *Bt* rice. In: Datta SK, editor. Rice improvement in the genomics era. Binghamton, NY: Haworth Press. (in press).

Baisakh N, Rao GJN. 1997. High frequency somatic embryogenesis in indica rice cultivars. In: Proceedings of the Second International Crop Science Congress, New Delhi, India. p. 1017.

Bao JS, Wu YR, Hu B, Wu P, Cui HR, Shu QY. 2002. QTL for rice grain quality based on a DH population derived from parents with similar apparent amylose content. Euphytica 128:317–324.

Benmoussa M, Achouch A, Zhu J. 2005. QTL analysis for yield components in rice (Oryza sativa L.) under different environments. J Central Eur Agri 6:317–322.

Bernacchi D, Beck-Bunn T, Emmatty D, Eshed Y, Inai S, Lopez J, Petiard V, Sayama H, Uhlig J, Zamir D, Tanksley SD. 1998a. Advanced backcross QTL analysis of tomato. II. Evaluation of near-isogenic lines carrying single-donor introgressions for desirable wild QTL-alleles derived from *Lycopersicon hirsutum* and *L. pimpinellifolium*. Theor Appl Genet 97:170–180.

Bernacchi D, Beck-Bunn T, Eshed Y, Lopez J, Petiard V, Uhlig J, Zamir D, Tanksley SD. 1998b. Advanced backcross QTL analysis in tomato. I. Identification of QTLs for traits of agronomic importance from *Lycopersicon hirsutum*. Theor Appl Genet 97:381–397.

Bernardo R. 2002. Breeding for quantitative traits in plants. Woodbury, MN: Stemma Press.

Blair MW, Iriarte G, Beebe S. 2006. QTL analysis of yield traits in an advanced backcross population derived from a cultivated Andean × wild common bean (*Phaseolus vulgaris* L.) cross. Theor Appl Genet 112:1149–1163.

Botstein D, White RL, Skolnick M, Davis RW. 1980. Construction of a genetic linkage map in man using restriction fragment length polymorphisms. Am J Hum Genet 32:314–331.

Brar DS, Dalmacio R, Rlloran R, Aggarwal R, Angeles R, Khush GS. 1996. Gene transfer and molecular characterization of introgression from wild *Oryza* species into rice. In: Rice genetics III. IRRI, Manila, Philippines. p. 477–486.

Brookes AJ. 1999. The Essence of SNPs. Gene 234:177–186.

Brown SM, Kresovich S. 1996. Molecular characterization for plant genetic resources conservation. In: Paterson AH, editor. Genome mapping in plants. Austin, TX: RG Landes. p. 85–93.

Burr B, Evola SV, Burr FA. 1983. The application of restriction fragment length polymorphism to plant breeding. In: Setlow JK, Hollaender A, editors. Genetic engineering, principles and methods. Vol. 5. New York: Plenum Press. p. 45–59.

Burrow MD, Blake TK. 1998. Molecular tools for the study of complex traits. In: Paterson AH, editor. Molecular dissection of complex traits. Boca Raton: CRC Press. p. 13–29.

Byrne PF, McMullen MD, Snook ME, Musket TA, Theuri JM, Widstrom NW, Wiseman BR, Coe EH. 1996. Quantitative trait loci and metabolic pathways: Genetic control of the concentration of maysin, a corn earworm resistance factor, in maize silks. Proc Natl Acad Sci USA 93:8820–8825.

Chen Y, Lu CF, Xu YB, He P, Zhu LH. 1996. Gametoclonal variation of microspore derived doubled haploids in indica rice: Agronomic performance, isozyme and RFLP analysis. Chin J Genet 23: 97–105.

Comstock RE, Robinson HF, Harvey PH. 1949. A breeding procedure designed to make maximum use of both general and specific combining ability. Agron J 41:360–367.

Cook R. 1937. A chronology of genetics. Year book of agriculture. p. 1457–1477. Electronic Scholarly Publishing Project. 2000. Available at: http://www.esp.org.

Cooper M, Hammer GL. 1996. Plant adaptation and crop improvement. Wallingford, Oxon, UK: CABI Publishing.

Daniell H. 2002. Molecular strategies for gene containment in transgenic plants. Nat Biotechnol 20:581–586.

Dasgupta I, Malathi VG, Mukherjee SK. 2003. Genetic engineering for virus resistance. Curr Sci 84: 341–354.

Datta K, Baisakh N, Oliva N, Torrizo L, Abrigo E, Tan J, Rai M, Rehana S, Al-Babili S, Beyer P, Potrykus I, Datta SK. 2003. Bioengineered *golden* indica rice cultivars with β-carotene metabolism in the endosperm with hygromycin and mannose selection systems. Plant Biotechnol J 1:81–90.

Datta K, Datta S. 2002. Plant transformation. In: Gilmartin PM, Bowler C, editors. Molecular plant biology. New York: Oxford University Press. p. 13–32.

Datta K, Baisakh N, Thet KM, Tu J, Datta SK. 2002. Pyramiding transgenes for multiple resistance in rice against bacterial blight, yellow stem borer and sheath blight. Theor Appl Genet 106:1–8.

Datta SK, Datta K, Potrykus I. 1990. Embryogenesis and plant regeneration from microspores of both Indica and Japonica rice (*Oryza sativa*). Plant Sci 67:83–88.

Datta SK. 2005. Androgenic haploids: factors controlling development and its application in crop improvement. Curr Sci 89:1870–1878.

Datta SK, Baisakh N, Ramanathan V, Narayanan KK. 2004. Transgenics in crop improvement. In: Jain HK, Kharkawal MC, editors. Plant breeding: Mendelian to molecular approaches. New Delhi: Narosa Publishers. p. 333–371.

Datta SK, Potrykus I. 1989. Artificial seed in barley: Encapsulation of microspore derived embryos. Theor Appl Genet 77:820–824.

Datta SK, Schmid J. 1996. Prospects of artificial seeds from microspore derived embryos of cereals. In: Jain SM, Sopory SK, Veilleax RE, editors. Vitro haploid production in higher plants. Vol. 2. The Netherlands: Kluwer Academic Publishers. p. 351–363.

Datta SK, Torrizo L. 20006. Haploid breeding in rice. In: Datta SK, editor. Rice improvement in the genomics era. Binghamton, NY: Haworth Press. (in press).

Duvick DN. 1992. Genetic contributions to advances in yield of U.S. maize. Maydica 37:69–79.

Duvick DN. 1996. Plant breeding, an evolutionary concept. Crop Sci 36:539–548.

Eshed Y, Zamir D. 1994. A genomic library of *Lycopersicon pennelli* in *L. esculentum:* A tool for fine mapping of genes. Euphytica 79:175–179.

Faris JD, Li WL, Liu DJ, Chen PD, Gill BS. 1999. Candidate gene analysis of quantitative disease resistance in wheat. Theor Appl Genet 98:219–225.

Fehr WR. 1987. Principles of cultivar development: theory and technique. New York: Macmillan Publishing Co.

Fehr WR, Weber CR. 1968. Mass selection by seed size and to a soybean cultivar. Crop Sci 35:1036–1041.

Florell VH. 1929. Bulked-population method of handling cereal hybrids. J Am Soc Agron 21:718–724.

Frary A, Nesbitt TC, Grandillo S, Knaap E, Cong B, Liu J, Meller J, Elber R, Alpert KB, Tanksley SD. 2000. fw2.2: A quantitative trait locus key to the evolution of tomato fruit size. Science 289:85–88.

Fridman E, Carrari F, Liu YS, Fernie AR, Zamir D. 2004. Zooming in on a quantitative trait for tomato yield using interspecific introgressions. Science 305:1786–1789.

Fridman E, Pleban T, Zamir D. 2000. A recombination hotspot delimits a wild-species quantitative trait locus for tomato sugar content to 484 bp within an invertase gene Proc Natl Acad Sci USA 97:4718–4723.

Fu X, Duc LT, Fontana S, Bong BB, Tinjuangjun P, Sudhakar D, Twyman RM, Christou P, Kohli A. 2000. Linear transgene constructs lacking vector backbone sequences generate low-copy-number transgenic plants with simple integration patterns. Transgenic Res 9:11–19.

Garland S, Lewin L, Blakeney A, Reinke R, Henry R. 2000. PCR-based molecular markers for the fragrance gene in rice (Oryza sativa L.). Theor Appl Genet 101:364–371.

Gauch S. 2002. Rapid purification of nucleic acids from plant and animal tissues streamlines molecular analysis. Plant, Animal & Microbe Genomes X Conference, January 12–16, 2002, Town & Country Convention Center, San Diego, CA. Available at: http://www.intl-pag.org/pag/10/abstracts/PAGX_W324.html.

Gelvin SB. 2003. Improving plant genetic engineering by manipulating the host. Trends Biotechnol 21:95–98.

Gepts P. 2002. A comparison between crop domestication, classical plant breeding, and genetic engineering. Crop Sci 42:1780–1790.

Gepts P, Hancock J. 2006. The future of plant breeding. Crop Sci 46:1630–1634.

Guha S, Maheswari SC. 1964. In vitro production of embryos from anthers of Datura. Nature 204:497.

Guiderdoni E, Chair H. 1992. Plant regeneration from haploid cell suspension-derived protoplasts of Mediterranean rice (Oryza sativa L. cv. Miara). Plant Cell Rep 11:618–622.

Gupta PK, Balyan HS, Sharma PC, Ramesh B. 1996. Microsatellites in plants: a new class of molecular markers. Curr Sci 70:45–54.

Hallauer AR. 2004. Breeding hybrids. In: Goodman RM, editor. Encyclopedia of plant and crop science. New York: Marcel Dekker. p. 186–188.

Hallauer AR, Miranda JB. 1988. Quantitative genetics in maize breeding. Ames, IA: Iowa State University Press.

Halpin C. 2005. Gene stacking in transgenic plants—the challenge for 21st century plant biotechnology. Plant Biotechnol J 3:141–155.

Hardenbol P, Baner J, Jain M, Nilsson M, Namsaraev EA, Karlin-Neumann GA, Fakhrai-Rad H, Ronaghi M, Willis TD, Landegren U, Davis RW. 2003. Multiplexed genotyping with sequence-tagged molecular inversion probes. Nat Biotecnol 21:673–678.

Hardenbol P, Yu FL, Belmont J, MacKenzie J, Bruckner C, Brundage T, Boudreau A, Chow S, Eberle J, Erbilgin A, Falkowski M, Fitzgerald R, Ghose S, Iartchouk O, Jain M, Karlin-Neumann G, Lu XH, Miao X, Moore B, Moorhead M, Namsaraev E,

Pasternak S, Prakash E, Tran K, Wang ZY, Jones HB, Davis RW, Willis TD, Gibbs RA. 2005. Highly multiplexed molecular inversion probe genotyping: over 10,000 targeted SNPs genotyped in a single tube assay. Genome Res 15:269–275.

Harlan JR. 1975. Geographic patterns of variation in some cultivated plants. J Hered 66:182–191.

Harlan JR. 1992. Crops and man. 2nd ed. Madison, WI: ASA.

Henry RJ. 2006. Genomics and plant biodiversity management. In: Plant Conservation Genetics. New York: The Haworth Press. p. 167–175.

Hittalmani S, Parco A, Mew TV, Zeigler RS, Huang N. 2000. Fine mapping and DNA marker-assisted pyramiding of the three major genes for blast resistance in rice. Theor. Appl. Genet. 100:1121–1128.

Huang N, Angeles ER, Domingo J, Magpantay G, Singh S, Zhang G, Kumaravadivel N, Bennett J, Khush GS. 1997a. Pyramiding of bacterial blight resistance genes in rice: Marker-assisted selection using RFLP and PCR. Theor Appl Genet 95:313–320.

Huang N, Paco A, Mew T, Magpantay G, McCouch S, Guiderdoni E, Xu J, Subudhi P, Angeles E, Khush GS. 1997b. RFLP mapping of isozymes, RAPD and QTLs for grain shape, brown planthopper resistance in a double haploid rice population. Mol Breed 3:105–113.

Ichikawa N, Kishimoto N, Inagaki A, Nakamura A, Koshino Y, Yokozeki Y, Oka M, Samoto S, Akagi H, Higo K, Shinjyo C, Fujimura T, Shimada H. 1997. A rapid PCR-aided selection of a rice line containing the Rf-1 gene which is involved in restoration of the cytoplasmic male sterility. Mol Breed 3:195–202.

Jahne A, Becker D, Brettschneider R, Lörz H. 1994. Regeneration of transgenic, microspore-derived, fertile barley. Theor Appl Genet 89:525–533.

Jain SM. 2001. Tissue-culture induced variation in crop improvement. Euphytica 118:153–166.

Jauhar PP. 2006. Modern biotechnology as an integral supplement to conventional plant breeding: the prospects and challenges. Crop Sci 46:1841–1859.

Jenkins MT. 1935. The effect of inbreeding and selection within inbred lines of maize upon hybrids after successive generations of selfing. Iowa State Coll J Sci 9:429–450.

Jensen NF. 1988. Plant breeding methodology. New York: Wiley Interscience.

Jia YL, Wang ZH, Singh P. 2002. Development of dominant rice blast Pi-ta resistance gene markers. Crop Sci 42:2145–2149.

Joshi SP, Ranjekar PK, Gupta VS. 1999. Molecular markers in plant genome analysis. Curr Science 77:230–240.

Kamoshita A, Zhang JX, Siopongco J, Sarkarung S, Nguyen HT, Wade LJ. 2002. Effects of phenotyping environment on identification of quantitative trait loci for rice root morphology under anaerobic conditions. Crop Sci 42:255–265.

Kang MS. 1990. Genotype-by-environment interaction and plant breeding. Baton Rouge, LA: LSU Agricultural Center.

Kang MS. 1994. Applied quantitative genetics. Baton Rouge, LA: MS Kang.

Kang MS. 1998. Using genotype-by-environment interaction for crop cultivar development. Adv Agron 62:199–252.

Kang MS. 2002. Quantitative genetics, genomics, and plant breeding. London: CABI Publishing.

Kang MS. 2006. Global plant breeding. In: Goodman RM, editor. Encyclopedia of plant and crop science. New York: Marcel Dekker.

Kang MS, Gauch HG Jr. 1996. Genotype-by-environment interaction. Boca Raton, FL: CRC Press.

Katiyar SK, Tan Y, Huang B, Chandel G, Xu Y, Zhang Y, Xie Z, Bennett J. 2001. Molecular mapping of gene Gm-6(t) which confers resistance against four biotypes of Asian rice gall midge in China. Theor Appl Genet 103:953–961.

Kearsey MJ. 2002. QTL analysis: Problems and (possible) solutions. In: Kang MS, editor. Quantitative genetics, genomics, and plant breeding. New York: CABI publishing. p. 45–58.

Komori T, Yamamoto T, Takemori N, Kashihara M, Matsushima H, Nitta N. 2003. Fine genetic mapping of the nuclear gene, Rf-1, that restores the BT-type cytoplasmic male sterility in rice (Oryza sativa L.) by PCR-based markers. Euphytica 129:241–247.

Koncz C. 2003. From genome projects to molecular breeding (editorial overview). Curr Opin Biotechnol 14:133–135.

Lander ES, Botstein D. 1989. Mapping Mendelian factors underlying quantitative traits using RFLP linkage maps. Genetics 121:185–199.

Lander ES, Green P, Abrahamson J, Barlow A, Daly MJ, Lincoln SE, Newburg L. 1987. MAPMAKER; an interactive computer program for constructing genetic linkage maps of experimental and natural populations. Genomics 1:174–181.

Lang NT, Subudhi PK, Virmani SS, Brar DS, Khush GS, Li ZK, Huang N. 1999. Development of PCR-based markers for thermosensitive genetic male sterility gene tms3(t) in rice (Oryza sativa L.). Hereditas 131:121–127.

Larkin PJ, Scowcroft WR. 1981. Somaclonal variation—a novel source of variability from cell cultures for plant improvement. Theor Appl Genet 60:190–214.

Lemieux B, Aharoni A, Schena M. 1998. Overview of DNA chip technology. Mol Breed 4:277–289.

Litt M, Luty JA. 1989. A hypervariable microsatellite revealed by in vitro amplification of a dinucleotide repeat within the cardiac muscle actin gene. Am J Hum Genet 44:397–401.

Liu KD, Wang J, Li HB, Xu CG, Liu AM, Li XH, Zhang Q. 1997. A genome-wide analysis of wide compatibility in rice and the precise location of the S5 locus in the molecular map. Theor Appl Genet 95:809–814.

Lu C, Shen L, Tan Z, Xu Y, He P, Chen Y, Zhu L. 1996. Comparative mapping of QTLs for agronomic traits of rice across environments using a doubled haploid population. Theor Appl Genet 93:1211–1217.

Lucca P, Hurrell R, Potrykus I. 2001. Genetic engineering approaches to improve the bioavailability and the level of iron in rice grains. Theor Appl Genet 102:392–397.

MacBeath G, Schreiber SL. 2000. Printing protein as microarrays for high throughput function determination. Science 289:1760–1763.

Mackill DJ, Nguyen HT, Zhang JX. 1999. Use of molecular markers in plant improvement programs for rainfed lowland rice. Field Crops Res 64:177–185.

Manly KP, Cudmore RH. 1998. MapManager XP. Plant and animal genome VI (San Diego, CA, January, 1998). p. 75.

Maraschin SF, de Priester W, Spaink HP, Wang M. 2005. Androgenic switch: An example of plant embryogenesis from the male gametophyte perspective. J Exp Bot 56:1711–1726.

McClintock B. 1978. Mechanisms that rapidly reorganize genome. Stadler Get Symp 10:25–48.

Meredith WR Jr. 1984. Quantitative genetics. In: Kohel RJ, Lewis CW, editors. Cotton and cotton improvement. Madison, WI: ASA and CSSA. p. 131–150.

Michelmore RW, Paran I, Kesseli RV. 1991. Identification of markers linked to disease-resistance genes by bulked segregant analysis: A rapid method to detect markers in specific genomic regions by using segregating populations. Proc Natl Acad Sci USA 88:9828–9832.

Mu P, Li ZC, Li CP, Zhang HL, Wang XK. 2004. QTL analysis for lodging resistance in rice using a DH population under lowland and upland ecosystems. Yi Chuan Xue Bao 31:717–723.

Murai H, Hashimoto Z, Sharma PN, Shimizu T, Murata K, Takumi S, Mori N, Kawasaki S, Nakamura C. 2001. Construction of a high-resolution linkage map of a rice brown plant hopper (Nilaparvata lugens Stål) resistance gene bph2. Theor Appl Genet 103:526–532.

Murai N, Sutton DV, Murray MG, Slightom JL, Merlo DJ, Reichert NA, Sengupta-Gopalan C, Stock CA, Barker RF, Kemp JD, Hall TC. 1983. Phaseolin gene from bean is expressed after transfer to sunflower via tumor-inducing plasmid vectors. Science 222:476–482.

Narasimhamoorthy B, Gill BS, Fritz AK, Nelson JC, Brown-Guedira GL. 2006. Advanced backcross QTL analysis of a hard winter wheat × synthetic wheat population. Theor Appl Genet 112:787–796.

Narayanan NN, Baisakh N, Oliva NP, Vera Cruz CM, Gnanamanickam S, Datta K, Datta SK. 2004. Molecular breeding: Marker-assisted selection combined with biolistic transformation for blast and bacterial blight resistance in Indica rice (cv. CO39). Mol Breed 14:61–71.

Nelson JC. 1997. QGENE: Software for marker-based genomic analysis and breeding. Mol Breed 3:239–245.

Neuhaus G, Spangenberg G, Mittlestein-Scheid O, Schweiger HG. 1987. Transgenic rapeseed plants obtained by the microinjection of DNA into microspore derived embryoids. Theor Appl Genet 75:30–36.

Olson M, Hood L, Cantor C, Botstein D. 1989. A common language for physical mapping of the human genome. Science 245:1434–1435.

Paine JA, Shipton CA, Chaggar S, Howells RM, Kennedy MJ, Vernon G, Wright SY, Hinchliffe E, Adams JL, Silverstone AL, Drake R. 2005. Improving the nutritional value of Golden Rice through increased pro-vitamin A content. Nat Biotechnol 23:482–487.

Parkhi V, Rai M, Tan J, Oliva N, Rehana S, Bandyopadhyay A, Torrizo L, Ghole V, Datta K, Datta SK. 2005. Molecular characterization of marker-free transgenic lines of *indica* rice that accumulate carotenoids in seed endosperm. Mol Genet Genomics 274:325–336.

Paterson AH. 1998. Prospect for cloning the genetic determinants of QTLs. In: Paterson AH, editor. Molecular dissection of complex traits. Boca Raton, FL: CRC Press. p. 289–293.

Paterson AH, Tanksley SD, Sorrells ME. 1991. DNA markers in plant improvement. Adv Agron 46:39–90.

Poehlman JM. 1987. Breeding field crops. Westport, CT: AVI Publishing Co., Inc.

Pot J, Pouwels D. 2002. Hands-on-experience AFLP-QUANTAR(pro) software. Plant, Animal & Microbe Genomes X Conference, January 12–16, 2002, Town & Country Convention Center, San Diego, CA. Available at: http://www.intl-pag.org/pag/10/abstracts/PAGX_W320.html.

Prasad SR, Bagali P, Hittalmani S, Sashidhar HE. 2000. Molecular mapping of quantitative trait loci associated with seedling tolerance to salt stress in rice (*Oryza sativa* L.). Curr Sci 78:162–164.

Qian Q, He P, Teng S, Zeng D, Zhu L. 1999a. QTL analysis of rice (*Oryza sativa* L.) tiller angle in a double haploid population derived from anther culture of indica/japonica. Chin Rice Res Newslett 7:1–2.

Qian Q, Zhen D, Huang F, He P, Zheng XW, Chen Y, Zhu LH. 1999b. The QTL analysis of seedling cold tolerance in a double haploid population derived from anther culture of indica/japonica. Chin Rice Res Newslett 7:1–2.

Rafalski A. 2002. Applications of single nucleotide polymorphisms in crop genetics. Curr Opin Plant Biol 5:94–100.

Rakoczy-Trojanowska M. 2002. Alternative methods of plant transformation—a short review. Cell Mol Biol Lett 7:849–858.

Ribaut JM, Bertan J. 1999. Single large-scale marker-assisted selection (SLS-MAS). Mol Breed 5:531–541.

Roberts JK. 2002. Proteomics and a future generation of plant molecular biologists. Plant Mol Biol 48:143–154.

Sanchez AC, Brar DS, Huang N, Li Z, Khush GS. 2000. Sequence tagged site marker-assisted selection for three bacterial blight resistance genes in rice. Crop Sci 40:792–797.

Schadt EE, Monks SA, Drake TA, Lusis AJ, Che N, Colinayo V, Ruff TG, Milligan SB, Lamb JR, Cavet G, Linsley PS, Mao M, Stoughton RB, Friend SH. 2003. Genetics of gene expression surveyed in maize, mouse and man. Nature 422:297–302.

Schena M, Shalon D, Davis RW, Brown PO. 1995. Quantitative monitoring of gene expression patterns with a complementary DNA microarray. Science 270:467–470.

Schlegel RH. 2003. Dictionary of plant breeding. Binghamton, NY: Food Products Press, Haworth Press.

Schoen DJ, Brown AH. 1993. Conservation of allelic richness in wild crop relatives is aided by assessment of genetic markers. Proc Natl Acad Sci USA 90:10623–10627.

Scowcroft WR. 1988. Genetic manipulation in crops: A symposium review. In: Genetic manipulation in crops. Philadelphia: International Rice Research Institute. p. 13–17.

Selvi A, Shanmugasundaram P, Kumar SM, Raja JAJ. 2002. Molecular markers for yellow stem borer *Scirpophaga incertulas* (Walker) resistance in rice. Euphytica 124:371–377.

Shenoi H, Burnham J, Koller S, Bitner R. 2002. Automated purification of plant DNA. Plant, Animal & Microbe Genomes X Conference, January 12–16, 2002, Town & Country Convention Center, San Diego, CA, poster 124.

Silvey V. 1981. The contribution of new wheat, barley and oat varieties to increasing yield in England and Wales 1947–78. J Natl Inst Agric Bot 15:399–412.

Simmonds NW. 1981. Genotype (G), environment (E) and GE components of crop yields. Exp Agric 17:355–362.

Stam P. 1993. Construction of integrated genetic linkage maps by means of a new computer package, JoinMap. Plant J 3:739–744.

Stuber CW. 2002. Foreword. In: Kang MS, editor. Quantitative genetics, genomics, and plant breeding. London: CABI Publishing.

Stuber CW, Polacco M, Senior ML. 1999. Synergy of empirical breeding, marker-assisted selection, and genomics to increase crop yield potential. Crop Sci 39:1571–1583.

Subudhi PK, Huang N. 1995. Identification of genes responsible for segregation distortion in a doubled haploid population of rice by using molecular markers. Rice Genet Newslett 12:239–241.

Subudhi PK, Nguyen HT. 2004. Genome mapping and genomic strategies for crop improvement. In: Nguyen HT, Blum A, editors Physiology and biotechnology integration for plant breeding. New York: Marcel Dekker, Inc. p. 403–451.

Takahashi Y, Shomura A, Sasaki T, Yano M. 2001. Hd6, a rice quantitative trait locus involved in photoperiod sensitivity, encodes the α subunit of protein kinase CK2. Proc Natl Acad Sci USA 98:7922–7927.

Tan ZB, Shen LS, Kuang HC, Lu CF, Chen Y, Zhou KD, Zhu LH. 1997a. Identification of QTLs for lengths of the top internodes and other traits in rice and analysis of their genetic effects. Chin J Genet 24:15–22.

Tan ZB, Shen LS, Yuan ZL, Lu CF, Chen Y, Zhou KD, Zhu LH. 1997b. Identification of QTLs for rationing ability and grain yield traits of rice and analysis of their genetic effects. Acta Agron Sin 23:289–295.

Tanksley SD, Young ND, Paterson AH, Bonierbale MW. 1989. RFLP mapping in plant breeding: New tools for an old science. Biotechnology 7:257–264.

Tanksley SD. 1993. Mapping polygenes. Ann Rev Genet 27:205–233.

Tanksley SD, Nelson JC. 1996. Advanced backcross QTL analysis: A method for simultaneous discovery and transfer of valuable QTL from unadapted germplasm into elite breeding lines. Theor Appl Genet 92:191–203.

Tanksley SD, McCouch SR. 1997. Seed banks and molecular maps: Unlocking genetic potential from the wild. Science 277:1063–1066.

Teng S, Zeng D, Zheng XW, Yasufumi K, Qian Q, Zhu LH. 2000. QTL analysis of rice (O. sativa L.) root vitality in a double haploid population derived from anther culture of an indica/japonica cross. Chin Rice Res Newslett 8:3–4.

Tu J, Datta K, Khush GS, Zhang Q, Datta SK. 2000. Field performance of transgenic indica rice. Theor Appl Genet 101:15–20.

Tu J, Datta K, Oliva N, Zhang G, Xu C, Khush GS, Zhang Q, Datta SK. 2003. Site-independently integrated transgenes in the elite restorer rice line Minghui 63 allow removal of a selectable marker from the gene of interest by self-segregation. Plant Biotechnol J 1:155–165.

Twyman RM, Christou P, Stoger E. 2002. Genetic transformation of plants and their cells. In: Oksman-Caldentey KM, Barz WM, editors. Plant biotechnology and transgenic plants. New York: Mercel Dekker Inc. p. 111–141.

Tyagi S, Kramer FR. 1996. Molecular beacons: Probes that fluoresce upon hybridization. Nat Biotechnol 14:303–308.

Vavilov, NI. 1926. Studies on the origin of cultivated plants. Bulletin of Applied Botany and Plant Breeding 26:1–248.

Veilleaux RE, Johnson AA. 1998. Somaclonal variation: molecular analysis, transformation, interaction, and utilization. Plant Breed Rev 16:229–268.

von Korff M, Wang H, Leon J, Pillen K. 2006. AB-QTL analysis in spring barley: II. Detection of favorable exotic alleles for agronomic traits introgressed from wild barley (H vulgare ssp. spontaneum). Theor Appl Genet 112:1221–1231.

Vos P, Hogers R, Bleeker M, Reijans M, van de Lee T, Hornes M, Frijters A, Pot J, Peleman J, Kuiper M, Zabeau M. 1995. AFLP: A new technique for DNA fingerprinting. Nucleic Acids Res 23:4407–4414.

Wang GL, Mackill DJ, Bonman JM, McCouch SR, Champoux MC, Nelson RJ. 1994. RFLP mapping of genes conferring complete and partial resistance to blast in a durably resistant rice cultivar. Genetics 136:1421–1434.

Wang M, van Bergen S, van Duijn B. 2000. Insights into a key developmental switch and its importance for efficient plant breeding. Plant Physiol 124:523–530.

Wang S, Basten CJ, Zeng ZB. 2006. Windows QTL Cartographer 2.5. Department of Statistics, North Carolina State University, Raleigh, NC. Available at: http://statgen.ncsu.edu/qtlcart/WQTLCart.htm.

Wang ZX, Iwata N. 1991. Production of $n + 1$ plants and tetrasomics by means of anther culture of trisomic plants in rice (Oryza sativa L.). Theor Appl Genet 83:12–16.

Watanabe Y. 1974. Meiotic chromosome behaviours of auto-pentaploid rice plant derived from anther culture. Cytologia 39:283–288.

Webel OD, Lonnquist JH. 1967. An evaluation of modified ear-to-row selection in a population of corn (Zea mays L.). Crop Sci 7:651–655.

Wehner T. 2006. History of plant breeding. Available at: http://cuke.hort.ncsu.edu/cucurbit/wehner/741/hs741hist.html. Accessed 5 Nov 2006.

Williams JG, Kubelik AR, Livak KJ, Rafalski JA, Tingey SV. 1990. DNA polymorphisms amplified by arbitrary primers are useful as genetic markers. Nucleic Acids Res 18:6531–6535.

Wilson ID, Barker GI, Edwards KJ. 2003. Genotype to phenotype: A technological challenge. Ann Appl Biol 142:33–39.

Xiao J, Grandillo S, Ahn SN, McCouch SR, Tanksley SD, Li JM, Yuan LP. 1996. Genes from wild rice improve yield. Nature 384:223–224.

Yan W, Kang MS. 2003. GGE biplot analysis: A graphical tool for breeders, geneticists, and agronomists. Boca Raton, FL: CRC Press.

Yano M, Katayose Y, Ashikari M, Yamanouchi U, Monna L, Fuse T, Baba T, Yamamoto K, Umehara Y, Nagamura Y, Sasaki T. 2000. *Hd1*, a major photoperiod sensitivity quantitative trait locus in rice, is closely related to the *Arabidopsis* flowering time gene CONSTANS. Plant Cell 12:2473–2484.

Yao FY, Xu CG, Yu SB, Li JX, Gao YJ, Li XH, Zhang QF. 1997. Mapping and genetic analysis of two fertility restorer loci in the wild-abortive cytoplasmic male sterility system of rice (*Oryza sativa* L.). Euphytica 98:183–187.

Ye S, Dhillon S, Ke X, Collins AR, Day IN. 2001. An efficient procedure for genotyping single nucleotide polymorphisms. Nucleic Acids Res 29: e88.

Ye X, Al-Babili S, Kloti A, Zhang J, Lucca P, Beyer P, Potrykus I. 2000. Engineering the provitamin A (β-carotene) biosynthetic pathway into (carotenoid-free) rice endosperm. Science 287:303–305.

Yoon DB, Kang KH, Kim HJ, Ju HG, Kwon SJ, Suh JP, Jeong OY, Ahn SN. 2006. Mapping quantitative trait loci for yield components and morphological traits in an advanced backcross population between *Oryza grandiglumis* and the *O. sativa.* japonica cultivar Hwaseongbyeo. Theor Appl Genet 112:1052–1062.

Young ND, Zamir D, Ganal MW, Tanksley SD. 1988. Use of isogenic lines and simultaneous probing to identify DNA markers tightly linked to the *Tm-2a* gene in tomato. Genetics 120:579–585.

Young ND. 1999. A cautiously optimistic vision for marker-assisted breeding. Mol Breed 5:505–510.

Zabeau M, Vos P. 1993. Selective restriction fragment amplification: A general method for DNA fingerprinting, European patent application number: 92402629.7. Publication no. 534 858 A1.

Zamir D. 2001. Improving plant breeding with exotic genetic libraries. Nat Rev Genet 2:983–989.

Zhu H, Briceno G, Dovel R, Hayes PM, Liu BH, Liu CT, Ullrich SE. 1999. Molecular breeding for grain yield in barley: An evaluation of QTL effects in a spring barley cross. Theor Appl Genet 98:772–779.

Chapter 2

Genetic Enhancement of Polyploid Crops Using Tools of Classical Cytogenetics and Modern Biotechnology

Prem P. Jauhar

Introduction

Traditional plant breeding has been mainly responsible for the genetic improvement of crop plants. It not only strives to exploit existing genetic variability but also to generate, manipulate, and combine new variability into cultivars most useful to humans. Sustained improvement of crop plants has been achieved through hybridization with landraces and allied species resulting in high-yielding, superior cultivars of food, fiber, oilseed, and forage crops. Although plant breeding started essentially as an art in the pre-Mendelian era and the early "breeders" produced some useful results, it became a sound, science-based technology only when supported by the principles of genetics and cytogenetics at the turn of the 20th century. During the period from about 1930 to 1970, yields of major food crops registered a substantial and steady increase.

The availability of cytoplasmic male sterility (CMS) in maize, pearl millet, and other crops facilitated the exploitation of hybrid vigor (Jauhar and Hanna, 1998; Vasal, 2002; Vasal et al., 2006; Jauhar et al., 2006). The remarkable speed with which breeders in India developed high-yielding grain hybrids of pearl millet is considered one of the most outstanding success stories of all time (Burton and Powell, 1968; Jauhar, 1981). The period from 1960 to 1980 witnessed a phenomenal increase in crop yields, especially of cereal crops (Fig. 2.1), and culminated in the most welcome Green Revolution (Khush, 2001; Evenson and Gollin, 2003; Baenziger et al., 2006; Swaminathan, 2006) that saved so many lives in Asia.

Wild relatives of crop species generally harbor desirable genes, especially for disease resistance, that have been and are being used for genetic enrichment of crops, particularly cereals. The tools of cytogenetics facilitated wide hybridization and thereby chromosome-mediated gene transfers from wild species into crop plants (Friebe et al., 1996; Fedak, 1999; Jauhar and Chibbar, 1999; Baenziger et al., 2006; Jauhar 2006a). Thus, classical cytogenetic tools played an important role in crop improvement.

Subsequent development, in the last couple of decades, of novel tools of direct gene transfer added new dimensions to breeding efforts. These asexual tools of genetic engineering help incorporate into plants new traits that are otherwise not possible to introduce by conventional breeding. Value-added traits successfully engineered into crop cultivars include resistance to diseases (fungal and viral) and insect pests and enhancement of their nutritional status (Vasil, 1998, 2003; Jauhar and Khush, 2002; Sharma et al., 2004; Jauhar 2006b). Thus, genetic engineering has accelerated crop improvement, yielding encouraging results.

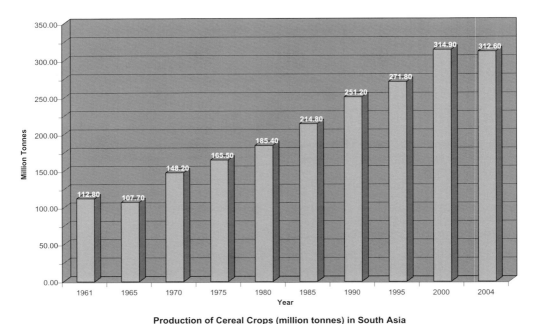

Production of Cereal Crops (million tonnes) in South Asia

Fig. 2.1.   Production of cereal crops (in million tonnes) in South Asia from 1961 to 2004. Source: faostat.fao.org.

The objective of this chapter is to discuss the cytogenetic architecture of polyploid cereals and to highlight how a multi-faceted approach to crop improvement has and can lead to genetic amelioration of these crop plants.

Incidence of polyploidy: origin and success of crop plants

Polyploidy—involving multiple sets of chromosomes, called genomes—has played a pivotal role in the evolution of plant species useful to the human race. Between 50–70 percent of angiosperms have had one or more events of chromosome increase sometime in their ancestry (Stebbins, 1971; Lewis, 1980; Masterson, 1994).

Polyploid crop plants

Plants with multiple copies of the same genome, e.g., AAAA and BBBB, are called autopolyploids or polysomic polyploids.

The potato (*Solanum tuberosum* L., 2n = 4x = 48) and alfalfa (*Medicago sativa* L., 2n = 4x = 32) are excellent examples of successful autopolyploids. Plants with multiple copies of different genomes, e.g., AABB and AABBDD, are termed allopolyploids or disomic polyploids. Hexaploid wheat (*Triticum aestivum* L.) and hexaploid oat (*Avena sativa* L.) are typical examples of successful allopolyploids.

Some diploid cereals are diploidized polyploids

Several of our crop plants, like maize (*Zea mays* L. ssp. *mays*), rice (*Oryza sativa* L.), pearl millet (*Pennisetum glaucum* [L.] R. Brown = *Pennisetum typhoides* [Burm.] Stapf et Hubb.), and sorghum (*Sorghum bicolor* L. Moench), are traditionally considered to be diploids. However, genetic mapping studies have shown that several or at least some of these crop species are in fact diploidized polyploids, having resulted from ancient polyploids (perhaps allopolyploids) through some sort of structural repatterning and differentiation of chromosomes. The formation of up to two bivalents in haploids ($2n = 2x = 7$) of pearl millet (Jauhar, 1970a) and intergenomic and intragenomic chromosome pairing in its interspecific hybrids (Jauhar, 1968, 1981) indicated that the pearl millet complement was derived from an ancestral chromosome number of $\times = 5$ as a result of duplication during the evolutionary course. Thus, it was considered to be a secondarily balanced species (Jauhar, 1968). Study of restriction fragment length polymorphism (RFLP) linkage maps provided substantiating evidence of the presence of duplicate loci in pearl millet (Liu et al., 1994).

Maize is not a truly diploid species as earlier thought. It is clearly a diploidized polyploid with extensive chromosome duplications—so much so that it tolerates chromosome deficiency, such as monosomy for one or more chromosomes (Weber, 1970, 1994). This clearly suggests that maize is an "ancient, secondarily balanced" species with extensive duplication, and perhaps redundancy, of genetic information in the form of whole chromosomes. Thus, surprisingly, maize seems to have more genetic buffering than tetraploid durum wheat in the sense that the latter does not generally tolerate even a simple monosomy for a chromosome. Maize has clearly resulted from an ancient polyploidization event(s) (Gaut, 2001).

Sorghum ($2n = 2x = 20$) is also considered to be an ancient polyploid (Gómez et al., 1998); and the diploid nature of rice (*Oryza sativa* L.; $2n = 24$) has long been questioned as well (Nandi, 1936; Vandepoele et al., 2003; Guyot and Keller, 2004).

All these supposedly diploid cereal crops have clearly resulted from ancient polyploidization events. However, they show disomic inheritance, and the breeding methods applied to true diploids are applicable to these cereals as well.

Our allopolyploid crop plants: sexual polyploidization

Among the polyploids, allopolyploids—incorporating more than two different genomes—are preponderant in nature. Sexual polyploidization, resulting from the functioning of meiotically unreduced (2n) gametes, is a major source of allopolyploids (Harlan and de Wet, 1975; Jauhar et al., 2000; Jauhar, 2003a). Chromosome doubling resulting from fusion of 2n male and female gametes in an interspecific hybrid (amphihaploid) could instantly lead to a fertile amphidiploid that, in turn, may become a new species (Jauhar, 2003a, 2007). The role of 2n gametes in rapid production of polyploids is well documented (Jauhar, 2007).

Sexuality and allopolyploidy

Sexuality and allopolyploidy seem to go together. Because allopolyploidy involves hybridization at the species or genus level and is associated with chromosome doubling, allopolyploidy occurs only in sexually reproducing plants. Thus, sexuality and allopolyploidy form an important, successful linkage. The cataclysmic effect of hybridization on evolution is well known. The first man-made cereal, triticale (AABBRR)—a successful hybrid between *Triticum* species and *Secale cereale* L.—provides an excellent example of rapid evolution by allopolyploidy. Such an accelerated evolution inspired Haldane (1958) to modify William Blake's 1792 famous dictum to read: "To create a little flower is the labour of ages, *except by allopolyploidy.*" The italicized portion was added by Haldane. A "little flower" in this quote means a "new species."

Allopolyploidy and genetic control of chromosome pairing

Since 1958, we have come to understand that successful establishment of a sexually reproducing allopolyploid depends essentially upon the integration of constituent genomes into a meiotically and, hence, reproductively stable form that is effectively achieved by a precise genetic control of chromosome pairing. Such integration has been demonstrated in hexaploid wheat (Riley and Chapman, 1958; Sears and Okamoto, 1958), hexaploid tall fescue, *Festuca arundincea* Schreb (Jauhar, 1975a,b), and hexaploid oat (Rajhathy and Thomas, 1972; Jauhar, 1977). And genetic regulation ensures diploid-like chromosome pairing and hence disomic inheritance and would, therefore, be necessary for allopolyploidy to become a powerful force in evolution. This may lead to the conclusion that genetic control of pairing impacts not only fertility in a newly formed (or an existing) allopolyploid, but also its propensity to successfully form and perpetuate higher forms of allopolyploidy. Jauhar (1975c) argued, therefore, that "polyploidy and sexuality cannot coexist in nature without such a regulatory mechanism" (see also Jauhar, 2003a, 2007).

Jauhar (2003a) stated: "Most, if not all, natural disomic polyploids have some form of control of chromosome pairing. Without such a control, precise bivalent formation in natural polyploids (with several sets of related chromosomes) would not be achieved." It is this fact that led Jauhar (1975c) to further modify the Blake-Haldane statement to read: "To create a little flower is the labour of ages except by allopolyploidy *coupled with genetic control of chromosome pairing*" (the italicized portion added by Jauhar). Thus, Jauhar (2003a) made a strong case that three factors—sexuality, allopolyploidy, and genetic control of chromosome pairing—jointly constitute a perfect recipe for rapid or cataclysmic evolution and speciation in nature.

Allopolyploid crop plants: benefits of polyploidy and hybridity

Among the sexual polyploids, allopolyploids are much more successful and hence they overwhelmingly outnumber auto-polyploids in nature. Autopolyploidy does not add any new genes and hence confers little adaptive advantage, although the dosage effect of genes may be somewhat helpful. Moreover, autopolyploidy is accompanied by sterility unless there is cytological diploidization (Jauhar, 1970b) or some other, possibly genetic, control ensuring regular and equal disjunction of multivalent associations of chromosomes (Crowley and Rees, 1968). Evolution by straight multiplication of chromosomes is therefore inefficient and slow, which accounts for the rare occurrence of autopolyploids in nature.

Allopolyploidy, resulting from interspecific or intergeneric hybridization between diploid progenitors, coupled with concomitant or subsequent chromosome doubling, leads to sexual polyploidization that, in turn, results in new species. Allopolyploid species are natural, stable hybrids that enjoy the benefits of both perpetual hybridity as well as polyploidy-endowed genetic redundancy. Thus, allopolyploidy has been instrumental in the production of many of our most important grain, fiber, oilseed, and forage crops that sustain humankind. Polyploid wheats—bread wheat (*Triticum aestivum* L., $2n = 6x = 42$) and durum wheat (*Triticum turgidum* L., $2n = 4x = 28$) are very important cereals.

Durum wheat: a model for allopolyploid evolution

Durum wheat (AABB genomes) and bread wheat (AABBDD genomes) are the most important cereal crops that arose by allopolyploidy. In fact, they offer a nice model for evolution by allopolyploidy. Both cereals are stable natural hybrids of wild diploid species (Fig. 2.2).

Evolution of tetraploid (AABB) wheat

Durum wheat's genomes A and B came from two diploid grasses. The donor of the A genome is *Triticum urartu* Tum. (Nishikawa, 1983; Dvořák et al., 1993), which is closely related to einkorn wheat (*Triticum monococcum* L.) domesticated in southeastern Italy about 12,000 years ago (Heun et al., 1997). The B genome was derived from *Aegilops speltoides* Tausch (Sarkar and

Stebbins, 1956; Wang et al., 1997; Dvořák, 1998). The two progenitors, native to the Middle East, crossed in nature about half a million years ago (Huang et al., 2002) and produced tetraploid emmer wheat (*T. turgidum* var. *dicoccoides* Körn), presumably as a result of fusion of unreduced male and female gametes in the BA hybrid (amphihaploid). This step of evolution can be recreated by inducing haploids (BA) of durum wheat by crossing it with maize (Jauhar, 2003b). We have shown (see Fig. 2.2) that durum haploids (BA) so derived produce unreduced gametes by first-division restitution (FDR), resulting in viable seed which then produce normal, disomic durum wheat

Steps in the Evolution of Bread Wheat

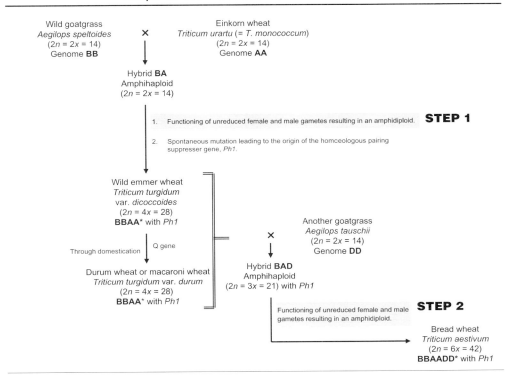

*Because *Ae. speltoides* (BB) is also the donor of cytoplasm to cultivated wheats, the correct designation for durum wheat is BBAA and that for bread wheat BBAADD. However, they are generally designated as AABB and AABBDD, respectively.

Fig. 2.2.   Diagram showing the two major steps that led to the evolution of bread wheat. The first cycle of hybridization between two grasses resulted in durum wheat, and then the second hybridization with a third grass led to the production of the bread wheat that sustains humankind. Source: Jauhar, P.P. 2007. J. Hered. 98:188–193.

(2n = 4x = 28) (Jauhar et al., 2000; Matsuoka and Nasuda, 2004).

Tetraploid emmer wheat (the wild form of durum wheat) may be the predecessor of cultivated durum as well as bread wheats. Emmer wheat was domesticated in the Fertile Crescent, where it is believed to have acquired the *Q* gene for free-threshing (Muramatsu, 1986) and gave rise to durum wheat that became one of the earliest domesticated crops. Thus, the acquisition of *Q* gene signaled the dawn of human civilization in the Near East. For the newly established farming settlements in this region, the tetraploid wheat was the principal wheat crop, which then spread to Egypt, India, and Europe (Nevo et al., 2002; Weiss et al., 2004).

Establishment of genetic regulation of chromosome pairing

The corresponding chromosomes of the A and B genomes are genetically related and are termed as homoeologous chromosomes, their own, identical partners being homologous. Because homoeologous chromosomes, such as 1A and 1B, are genetically similar and hence capable of pairing with one another (Jauhar et al., 1999), a stringent regulation of pairing would be necessary for meiotic regularity and, hence, reproductive stability. Therefore, the *Pairing homoeologous (Ph1)* gene in the long arm of chromosome 5B was somehow acquired, and it ensured diploid-like pairing, i.e., pairing between homologous partners only (Riley and Chapman, 1958; Sears and Okamoto, 1958; Jauhar et al., 1991; Jauhar and Joppa, 1996). The acquisition of *Ph1* was imperative for the evolution of first the durum wheat, and later the bread wheat, and it happened either at the BA hybrid level or, more plausibly, at the origin of tetraploid wild emmer. A spontaneous mutation could have given rise to *Ph1* that permitted only homologous chromosome pairing, ensuring

disomic inheritance. Thus, it is because of *Ph1* that the meiotic integrity of the A and B genomes has been largely maintained across centuries.

Evolution of bread wheat (AABBDD): a boon to humankind

A spontaneous hybridization, followed or accompanied by acquisition of *Ph1*, gave rise to a meiotically regular and reproductively stable emmer wheat (Step 1 in Fig. 2.2). Another cycle of spontaneous hybridization occurred between tetraploid wheat and a diploid goatgrass (*Aegilops tauschii* Coss., 2n = 2x = 14, DD genome) (McFadden and Sears, 1946) some 8,000 years ago (Huang et al., 2002) and produced bread wheat (Step 2 in Fig. 2.2). Thus, three wild grasses hybridized in nature, in two steps, and produced such a useful crop as wheat that sustains humankind.

It is likely that hybridization between emmer or durum wheat and *Ae. tauschii* gave rise to triploid hybrid BAD, which produced unreduced male and female gametes that fused to set viable seed, which then produced hexaploid bread wheat (Jauhar, 2007). This parallels the events in durum haploids (BA) that set seed to give rise to disomic durum (Jauhar et al., 2000). The *Ph1*-enforced diploid-like pairing, which was established in its predecessor tetraploid, emmer wheat, confers meiotic normality and reproductive fertility to bread wheat also.

Hexaploid oat

Like common wheat, common oat (*Avena sativa* L., 2n = 6x = 42; AACCDD genomes) is also an allohexaploid species with three genomes. The donors of the three genomes are not well characterized, but the allotetraploid progenitors AABB and AACC have been identified (see Jellen and Leggett, 2006 for references). Although the three genomes of wheat can be characterized, discrimination between two of the oat genomes, A and D, has been difficult, and a satellite DNA sequence was used to distinguish these closely related genomes (Linares et al., 1998).

Drossou et al. (2004) employed AFLP (amplified fragment length polymorphisms) and RAPD (random amplified polymorphic DNA) markers to decipher intergenomic and species relationships in *Avena,* but the full picture is not clear as yet. The diploid-like pairing in hexaploid oat is, nevertheless, under genetic control (Rajhathy and Thomas, 1972; Jauhar, 1977). Moreover, as in wheat, this genetic control is hemizygous-effective as evidenced by lack of pairing in oat poly-haploids (Nishiyama and Tabata, 1964) with one dose of the diploidizing gene(s) and is genetically repressible by an appropriate genotype of its diploid wild relative, leading to homoeologous chromosome pairing (Jauhar, 1977). Although the pairing control mechanism in hexaploid oat is not fully elucidated so far, genetic suppression of homoeologous pairing is an important approach to alien gene transfers into this cereal.

Genetic enhancement of non-cereal polyploid crops: the potato

The potato is an important staple food widely grown in some 149 countries (Hijmans, 2001). Being the world's fourth most important food crop, it has been subjected to numerous investigations for its genetic improvement using both conventional and modern tools (Solomon-Blackburn and Barker, 2001; see also chapter 10). It is an autotetraploid ($2n = 4x = 48$) that shows tetrasomic inheritance. Moreover, being vegetatively propagated, potato is highly heterozygous, which further complicates its improvement by conventional breeding. Some of the wild and cultivated species of potato have been utilized for its genetic

improvement (Bradshaw et al., 2006). For making appropriate crosses, however, appropriate ploidy-level manipulations (scaling up or down the whole chromosome sets) need to be made, for which several of the *Solanum* species are suitable (Ramanna and Jacobsen, 2003; Carputo and Barone, 2005).

Breeding at haploid level

Dihaploids—diploids derived from tetraploid potato—have been effectively used for transferring genes from diploid species. Breeding at the dihaploid ($2n = 2x = 24$) level offers promise for making appropriate crosses and producing improved cultivars (De Maine, 1982; see also Ramanna and Jacobsen, 2003). Several breeding approaches to potato improvement for various traits are discussed in detail in chapter 10.

Mutation breeding

Because of clonal propagation, the potato lends itself to improvement by mutation breeding. Desirable mutants produced by physical or chemical mutagenesis can be vegetatively propagated (see for example, Jauhar and Swaminathan, 1967; Jauhar, 1969; Solomon-Blackburn and Barker, 2001). Somaclonal variation generated during plant regeneration in cultures can lead to some novel phenotypes in potato and improved clones have thus been derived (Kawchuk et al., 1997)

Transgenic approaches to genetic improvement

Genetic engineering with cloned genes has proved useful in introducing desirable traits in the potato. And transgenes have been widely used for its improvement, particularly resistance breeding. Potato has been successfully transformed using *Agrobacterium* (Visser et al., 1989; Newell et al., 1991) and particle bombardment (Romano et al., 2001). However, genetic transformation of potato is genotype-dependent (Heeres et al., 2002). Transformation with multiple genes can also be achieved (Romano et al., 2001).

Genes for desirable traits, such as resistance to potato virus X (Hoekema et al., 1989), tuber moth (Douches et al., 1998), and the fungus *Verticillium dahliae* Kleb. (Gao et al., 2000), have been introduced in potato cultivars. The technology of pathogen-derived-resistance (PDR) (Powell-Abel et al., 1986; Beachy et al., 1990) offers promise for introducing virus resistance in potato (Schubert et al., 2004; see Solomon-Blackburn and Barker, 2001, for earlier references).

The potato is the most important non-cereal food staple in the world. The need to improve its nutrition value cannot be over-emphasized, as a large number of the poor people in Asia and Africa are malnourished. Transgenic approaches offer an efficient means for the nutritional enhancement of potato. The expression of the *AmAl* gene from amaranth (*Amaranthus hypochondriacus* L.) in transgenic tubers produced a significant increase in most essential amino acids and protein compared to non-transgenic tubers (Chakraborty et al., 2000). Indian scientists in New Delhi have produced a genetically transformed potato called "protato" that is rich in protein (Arthur, 2003). Employing metabolic engineering, Ducreux et al. (2005) produced high-carotenoid potato tubers with higher content of β-carotene and lutein.

Transgenic approaches to production of edible or oral vaccines

Vaccines have played an important role in maintenance of human health and saving human lives. Modern biotechnology could help or facilitate production of inexpensive oral vaccines (Alvarez et al., 2006). Plant-based vaccines that can be eaten uncooked in fruits or vegetables, such as melons, banana, or a tomato, could help immunize against some diseases. Thus, edible vaccines against hepatitis B have been developed in transgenic banana (Kumar et al., 2005). Scientists at Arizona State University have genetically engineered potatoes to produce a vaccine against hepatitis B virus, which kills almost a million people every year. In a trial of an edible vaccine, up to 60 percent of volunteers who ate chunks of the raw potato were reported to develop antibodies against the virus (Ariza, 2005; Charles Arntzen, personal communication, February 2006).

Genetic enhancement of polyploid cereals

Cereal crops have an extremely important place in human diet and healthy life. Diet of whole grain cereals has been linked to cardiovascular health. According to researchers at Harvard Medical School, eating a bowl of whole grain cereals could reduce the risk of heart attack by 27 percent (Djousse and Gaziano, 2007). In order to feed the world wholesome food, it is important to bring about genetic improvement in cereal crops using classical and modern tools.

As discussed above, the allopolyploid cereals—durum wheat ($2n = 4x = 28$; AABB), bread wheat ($2n = 6x = 42$; AABBDD), and oat ($2n = 6x = 42$; AACCDD)—show a highly disciplined, diploid-like chromosome pairing because of precise pairing control. Because the *Ph1* in polyploid wheats and a *Ph1*-like gene(s) in oat do not allow pairing (adultery, so to speak) among related (or homoeologous) chromosomes, this poses limitations to the transfer of desirable genes across species. However, homoeologous pairing is the key to alien gene transfers. Manipulation of this pairing-control mechanism would, therefore, be important for alien gene transfers into the cereal crops.

Wild relatives as reservoirs of desirable genes

Several wild relatives of wheat are rich reservoirs of genes that can be transferred to durum and bread wheat cultivars via sexual hybridization. Thus, wild emmer (AABB) shares the A and B genomes with durum wheat and is in the primary gene pool of the latter. Wild emmer has genes for resistance to several diseases, including stem rust (Rajaram et al., 2001), Fusarium head blight (FHB) (Stack et al., 2002), stripe rust (Sun et al., 1997), and powdery mildew (Liu et al., 2002; Xie et al., 2003; Mohler et al., 2005). Perennial wild grasses, in the secondary gene pool of wheat, also harbor desirable genes for resistance to several diseases, including FHB (Jauhar and Peterson, 1996; Jauhar and Xu, 2004; Cai et al., 2005). *Elymus humidus* Osada, in the tertiary gene pool of wheat, is reported to be immune to FHB (Fedak, 2000). Desirable genes from these wild relatives can be transferred to cultivated wheats by hybridization and promoting, in the hybrids, pairing between wheat chromosomes and alien chromosomes leading to genomic reconstruction.

Chromosome engineering in durum wheat and bread wheat

Cytogenetic manipulation for suppressing the *Ph1* system for incorporating desirable alien chromatin into wheat was termed "chromosome engineering" (Sears, 1972), which leads to genomic reconstruction in wheat. There are several means of circumventing the *Ph1*-control to promote homoeologous pairing and alien gene transfers.

Chromosome pairing in the absence of Ph1

The use of *ph1bph1b* mutant, involving a small intercalary deficiency for *Ph1* (Sears, 1977), in crosses with alien species promotes homoeologous pairing in hybrids. Use of 5B-deficient stocks such as Langdon 5D(5B) disomic substitution line also promotes intergenomic chromosome pairing and hence interspecific or intergeneric gene transfers (Sears, 1981; Jauhar and Chibbar, 1999; see also Jauhar, 2006a, for recent references).

Using 5D(5B) disomic substitution line as female parent in crosses with a diploid wheatgrass, *Lophopyrum elongatum* (Host) Á. Löve (2n = 2x = 14; EE genome), we promoted pairing among parental chromosomes and transferred alien chromatin into the durum genome. Thus, we produced scab-resistant durum germplasm by transferring chromosome segments from diploid wheatgrass, *L. elongatum* (Jauhar and Peterson, 2000; Jauhar and Xu, 2004). We adopted the same approach for transferring chromatin from another diploid wheatgrass, *Thinopyrum bessarabicum* (Savul. and Rayss) Á. Löve (2n = 2x = 14; JJ genome), into durum.

Promotion of chromosome pairing by suppressing Ph1

Certain genotypes of alien donors have the inherent ability to suppress the activity of *Ph1* in their hybrids with wheat, thereby raising homoeologous pairing. This is perhaps a better approach than the one described above for alien gene transfers into wheat. We have used this strategy in our wheat germplasm enhancement program. Tetraploid crested wheatgrass, *Agropyron cristatum* (L.) Gaertner (2n = 4x = 28; PPPP), is a useful source of genes for cold tolerance. By crossing this alien donor with bread wheat (as female parent), we produced pentaploid hybrids (2n = 5x = 35; ABDPP) that showed extensive pairing, some between parental chromosomes, offering the possibility of alien chromatin integration into the wheat genome (Jauhar, 1992).

Alien genotype-induced homoeologous pairing was also observed in hybrids between bread wheat and decaploid (2n = 10x = 70) tall wheatgrass, *Thinopyrum ponticum* (Podp.) Barkworth and Dewey (Jauhar, 1995). The F_1 hybrids showed considerable pairing among the parental chromosomes. It is remarkable that these hybrids show substantial pollen and seed fertility and therefore can be easily backcrossed to the wheat parent (Jauhar, 1995). The seed set on the perennial hybrid derivatives resembles the wheat seed, offering prospects for breeding perennial wheat.

We produced some scab-resistant durum germplasm by crossing durum cultivars (as female parent) with tetraploid wheatgrass, *Thinopyrum junceiforme* (Löve & Löve) Löve (2n = 4x = 28; $J_1J_1J_2J_2$ genomes) (Jauhar and Peterson, 2001). Hybridization with perennial grasses in the genera *Agropyron, Thinopyrum,* and *Lophopyrum,* among others, has contributed to the genetic enrichment of wheat (Feldman and Sears, 1981; McIntosh, 1991; Jauhar, 1993, 2006a; Friebe et al., 1996; Cai et al., 2005).

Using chromosome engineering, considerable progress has been made in transferring alien chromosome segments carrying desirable genes into bread wheat (Sears, 1981, 1983; Feldman, 1988; Mujeeb-Kazi and Rajaram, 2002; Yang et al., 2005; Mujeeb-Kazi, 2006; Oliver et al., 2006) and durum wheat (Ceoloni et al., 2005a,b; Ceoloni and Jauhar, 2006; Jauhar, 2006a).

Addition of full alien chromosomes or chromosome segments

Wheat alien addition lines involving diploid wheatgrass chromosomes have also been synthesized (Jauhar and Peterson, 2007) and they may serve as useful material for incorporation of alien segments into bread wheat or durum wheat. Such a progressive reduction of unwanted chromatin from the alien donor would help retain the chromosome segment with the desirable gene(s). Thus, this constitutes elegant and targeted chromosome engineering. We are in the process of producing a series of alien addition lines in the durum wheat background. In view of the fact that durum wheat has less genetic buffering than bread wheat, alien additions are more stable and are better tolerated in the latter (McIntosh et al., 1995; Ceoloni et al., 2005b).

Other methods of transferring alien chromatin into the wheat complement are known. Thus, X-irradiation has been successfully tried (Sears, 1993), although it may induce undesirable translocations also. Promotion of homoeologous pairing described above offers a more precise technique of "chromosome surgery" to recombine desirable chromosome segments and thereby desirable genes.

Genomic reconstruction of wheat

Both annual and perennial species have helped in the genomic reconstruction of wheat and its genetic improvement. Superior traits, such as resistance to biotic stress, in wheat cultivars have resulted from translocation between wheat chromosomes and alien chromosomes. Such translocations have been shown to confer resistance to diseases and pests and tolerance to abiotic stresses (Gill and Friebe, 2002). Rye (*Secale cereale* L.) chromatin has contributed to genomic reconstruction of wheat, resulting in superior agronomic performance. Thus, gain in wheat yield has been obtained via wheat-rye translocation 1B-1R, where the short arm of chromosome 1B of wheat is replaced by the short arm of 1R of rye (Rajaram et al., 2002).

A wheat-alien translocation involving chromatin from tall wheat grass, *Thinopyrum ponticum* (= *Agropyron elongatum* ssp. *ruthenicum* Beldie) that carries leaf and stem resistance genes, *Lr19* and *Sr25,* respectively, is known to boost the yield potential of wheat to varying degrees (Singh et al., 1998; Reynolds et al., 2001). Wheat—

Leymus racemosus (Lam.) Tzvelev translocation lines with resistance to FHB were produced by Chen and Liu (2000). Durum wheat—*Haynaldia villosa* (L.) Shur. translocation lines, T6DL·6VS, with resistance to powdery mildew caused by *Erysiphe graminis* have also been developed (Li et al., 2005).

Characterization of alien chromatin in the wheat genome

Once alien chromatin is incorporated into a wheat genome as a result of homoeologous pairing, it would be useful to characterize this chromatin as a terminal or interstitial segment. Fluorescent genomic in situ hybridization (fl-GISH) provides an excellent tool for visualizing such alien chromatin—its size and location. Fl-GISH has been successfully employed to characterize introgression products in various interspecific and intergeneric hybrids (Jauhar et al., 2004; Ceoloni and Jauhar, 2006). We have also applied multicolor fl-GISH to decipher intergenomic pairing among more than two genomes of different species (Jauhar and Peterson, 2006).

Chromosome engineering in hexaploid oat

Hexaploid oat (2n = 6x = 42; AACCDD) forms only 21 bivalents at meiosis. This diploid-like pairing is under genetic control, which that can be genetically suppressed. A genotype of diploid grass, *Avena longiglumis* L. brings about homoeologous pairing in its hybrids with hexaploid oat.

Using wide hybridization, coupled with manipulation of chromosome pairing, some desirable genes were transferred into cultivated oat. Thus, a mildew resistance gene was transferred from *Avena barbata* L. into oat (Thomas et al., 1980). Several other examples of alien gene transfer into oat are given in Jellen and Leggett (2006).

Transgenic approaches to cereal crop improvement: biotic and abiotic stresses

Traditional breeding, sometimes aided by marker-assisted selection (Dubcovsky, 2004; Lapitan and Jauhar, 2006) and tools of cytogenetics (Friebe et al., 1996; Jauhar and Chibbar, 1999; Jauhar, 2006a), has produced crop cultivars with superior traits. However, these techniques are notoriously slow and have several limitations. Unwanted chromosomes or chromosome segments and, hence, undesirable characters of wild relatives get incorporated into cereal cultivars, following their sexual hybridization with the wild donors. Sexual incompatibility between parental taxa may pose another serious problem. Newer biotechnological tools of direct gene transfer (Vasil et al., 1993; Vasil, 2003; Jauhar and Chibbar, 1999; Altpeter et al., 2005; Jauhar 2006b) have accelerated the pace of genetic improvement. A gene from a soil-borne bacterium, *Bacillus thuringiensis* (Bt), when bioengineered into a crop genome, confers almost complete resistance to some pests (Palevitz, 2001). This Bt technology has been successfully employed in several crop plants (See Jauhar, 2006b for references).

Genetic engineering offers a rapid means of asexually inserting a gene(s) of unrelated organisms into plant cells, which may then be regenerated into full plants with the inserted gene(s) (or transgene[s]) in their genome. This process is not only rapid, but it also allows access to an unlimited gene pool without the constraint of sexual hybridization.

Important factors for successful genetic transformation

For successful genetic transformation, certain prerequisites must be met. These are: (1) an efficient in vitro regeneration protocol to produce complete plants from single cells;

(2) a DNA delivery system to introduce the desired genes into cells; and (3) a functional transgene(s) that is/are expressed in the recipient plant. Several crop plants, including polyploid cereals with these attributes, have been genetically transformed. Thus, when all these prerequisites were met, it became possible to transform bread wheat by several groups, in the United States, Canada, and Germany (Vasil et al., 1992; Weeks et al., 1993; Becker et al., 1994; Nehra et al., 1994).

However, early attempts to transform durum wheat were not successful, primarily because the in vitro regeneration system was not standardized for durum. Therefore, we developed an effective in vitro regeneration protocol for durum cultivars (Bommineni and Jauhar, 1996) and, using that protocol, we produced the first transgenic durum

wheat carrying glufosinate ammonium resistance (Bommineni et al., 1997) (Fig. 2.3). Thus, we standardized the transgenic technology for durum wheat that opened up new avenues for its genetic enhancement.

Resistance to diseases in wheat

Transgenic technology has been applied to combat several diseases and pests. Scab of FHB caused by the fungus *Fusarium graminearum* Schwabe, for example, is a ravaging disease of durum and bread wheat. Transgenic approaches to combating FHB have been tried. Genetic engineering with antifungal and antitoxin genes has been applied to contain FHB in wheat (Anand et al., 2003, 2004).

Transgenic technology offers an effective tool for protecting crop plants against

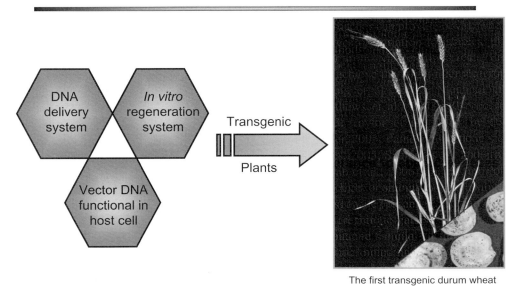

Interaction of Technologies to Produce Transgenic Plants

The first transgenic durum wheat produced in Jauhar's Laboratory in 1996

Fig. 2.3.   Three prerequisites for genetic transformation of a crop plant. Using the three (after standardizing the in vitro regeneration protocol), we produced the first transgenic durum wheat. Also shown is the transient expression of *gus* gene in scutellar cells that initially showed the incorporation of the desired gene.

devastating viral pathogens. Engineering plants with nucleotide sequences derived from viral genomes provides protection against the virus from which the sequences were derived. Such a pathogen-derived resistance (Powell-Abel et al., 1986) has helped control some viral pathogens. It has been suggested that expression of a virus coat protein as a transgene in a plant confers resistance to the virus in direct proportion to the quantity of coat protein produced by the transformed plant (Beachy et al., 1990).

The PDR technology has been applied in cereal crops. Sivamani et al. (2002) found that transgenic wheat plants engineered with the coat-protein gene of wheat streak mosaic virus (WSMV) conferred protection against some WSMV strains.

Tolerance to abiotic stresses in wheat

Abiotic stresses like drought and salinity cause an estimated yield loss of over 50% worldwide (Bray et al., 2000; Roy and Wu, 2002). Thus, drought is a major cause of yield loss in wheat (Trethowan et al., 2001), and it is difficult to breed drought tolerance by traditional breeding methods. Abebe et al. (2003) showed that wheat engineered with the *mt1D* gene from *Escherichia coli* improved tolerance to drought stress and salinity.

Tolerance to abiotic stresses in oat

Osmotic stress caused by drought or salinity leads to yield loss in oat worldwide (Tamm, 2003). Because of the lack of genetic variability for salt tolerance in oat and the complexity of this trait, conventional breeding is not effective in producing salt-tolerant oat cultivars (Cushman and Bohnert, 2000). Transgenic oat expressing barley *HVA1* gene has been shown to have salt tolerance (Oraby et al., 2005).

Microprojection and haploidy

Haploidy and transgenic technology can be combined to produce stable transgenics (Jauhar and Chibbar, 1999). During microprojectile bombardment, the target gene can land on any of the chromosomes in the cells being bombarded, so it would be difficult to stabilize the gene in question. Haploid cultures offer useful material for microprojection because subsequent chromosome doubling would lead to homozygosity and hence stability of the transgene(s).

Biofortification and nutritional enhancement of cereal crops

Micronutrient deficiency affects a large proportion of the poor populace. About 842 million people in Asia and Africa are malnourished (http://www.fao.org/english/newsroom/news/2003/26659-en.html). Improvement of the nutritional status of food crops should be a high priority to combat malnutrition and its associated problems. Biofortification is the enrichment of staple crops with micronutrients and vitamins. The need for biofortification of food staples like wheat cannot be overemphasized. Production of micronutrient-rich staple crops, using traditional and modern technologies, should be a high priority for all concerned (Nestel et al., 2006).

The development of Golden Rice with genetic instructions to produce both vitamin A and iron (Ye et al., 2000; Beyer et al., 2002; Potrykus, 2001, 2003; Bajaj and Mohanty, 2005; Paine et al., 2005) is a most remarkable development because it offers "grains of hope" for some 400 million children threatened with night blindness and for anemic pregnant women suffering from iron deficiency. These traits are being introduced in indica rice cultivars, IR64 and BR29 (Datta et al., 2006). It would be unimaginable to engineer those traits into rice or any other crops by traditional means. Similar approaches to biofortification of other cereals such as wheat will have to be adopted.

Conclusion: prospects and challenges

Crop improvement has been achieved mainly by traditional plant breeding that was initially practiced as an art and, to a certain extent, helped transform some weedy species into our crop plants. However, with the discovery of the principles of genetics and cytogenetics, plant breeding was strengthened as a science-based technology. Sexual hybridization with suitable gene donors in the primary or secondary gene pool, in conjunction with manipulation of chromosome pairing, led to enrichment of crop cultivars with superior genes of alien origin (Friebe et al., 1996; Ceoloni and Jauhar, 2006; Jauhar, 2006a). Thus, plant breeding kept evolving and the breeders became better equipped to tackle problems of disease and pest resistance, for example. Integrations of alien chromosome segments into the crop genomes led to genomic reconstruction and acquisition of newer traits. Important advances in increasing yield, disease and pest resistance, and nutritional quality of crop cultivars became possible.

More recently, the advent of the techniques of molecular cytogenetics and biotechnology has put new tools in the tool-kit of breeders and other professionals engaged in crop improvement. It became possible to engineer new traits into crop plants that otherwise would be very difficult or impossible to introduce by traditional means. The successful application of transgenic approaches to engineer resistance to viral diseases is a remarkable development. And nutritional enhancement of crop cultivars made possible by genetic engineering, would be unimaginable by conventional means. Biofortification of staple crops by biotechnological approaches offers one of the best means of tackling global malnutrition.

Acknowledgments

Mention of tradenames or commercial products in this publication is solely to provide specific information and does not imply recommendation or endorsement by the U.S. Department of Agriculture.

I acknowledge the excellent technical help of Terrance Peterson in the preparation of the manuscript.

References

Abebe, T., Guenzi, A.C., Martin, B. and Cushman, J.C. 2003. Tolerance of mannitol-accumulating transgenic wheat to water stress and salinity. Plant Physiol. 131:1748–1755.

Altpeter, F., Varshney, A., Abderhalden, O., Douchkov, D., Sautter, C., Kimlehn, J., Dudler, R. and Schweizer, P. 2005. Stable expression of a defense-related gene in wheat epidermis under transcriptional control of a novel promoter confers pathogen resistance. Plant Mol. Biol. 52:271–283.

Alvarez, M.L., Pinyerd, H.L., Crisantes, J.D., Rigano, M.M., Pinkhasov, J., Walmsley, A.M., Mason, H.S. and Cardineau, G.A. 2006. Plant-made subunit vaccine against pneumonia and bubonic plague is orally immunogenic in mice. Vaccine 24:2477–2490.

Anand, A., Zhou, T., Trick, H.N., Gill, B.S. and Muthukrishnan, S. 2003. Greenhouse and field testing of transgenic wheat plants stably expressing genes for thaumatin-like protein, chitinase and glucanase against *Fusarium graminearum*. J. Exp. Bot. 54:1101–1111.

Anand, A., Lei, Z., Sumner, L.W., Mysore, K.S., Arakane, Y., Bockus, W.W. and Muthukrishnan, S. 2004. Apoplastic extracts from a transgenic wheat line exhibiting lesion-mimic phenotype have multiple pathogenesis-related proteins that are antifungal. Mol. Plant Microbe Interact. 17:1306–1317.

Ariza, L.M. 2005. Defensive eating. Sci. Am. 292 (5):25.

Arthur, C. 2003. Creation of GM potato to fight hunger sets India's scientists against green groups. The Independent London, England, Jan 3, 2003.

Baenziger, P.S., Russell, W.K., Graef, G.L. and Campbell, B.T. 2006. Improving lives: 50 years of crop breeding, genetics and cytology (C-1). Crop Sci. 46:2230–2244.

Bajaj, S. and Mohanty, A. 2005. Recent advances in rice biotechnology—Towards genetically superior transgenic rice. Plant Biotech. J. 3:275–307.

Beachy, R.N., Loesch-Fries, S. and Turner, N.E. 1990. Coat-protein-mediated resistance against virus infection. Annu. Rev. Phytopathol. 28:451–474.

Becker, D., Brettscneider, R. and Lörz, H. 1994. Fertile transgenic wheat from microprojectile bombardment of scutellar tissue. Plant J. 5:299–307.

Beyer, P., Al-Babili, S., Ye, X.D., Lucca, P., Schuab, P., Welsch, R. and Potrykus, I. 2002. Golden rice, introducing the β-carotene biosynthesis pathway into rice endosperm by genetic engineering to defeat vitamin A deficiency. J. Nutr. 132: 506S–510S.

Blake, W. 1792. The Marriage of Heaven and Hell. (Privately printed).

Bommineni, V.R. and Jauhar, P.P. 1996. Regeneration of plantlets through isolated scutellum culture of durum wheat. Plant Sci. 116:197–203.

Bommineni, V.R., Jauhar, P.P. and Peterson, T.S. 1997. Transgenic durum wheat by microprojectile bombardment of isolated scutella. J. Hered. 88:475–481.

Bradshaw, J.E., Bryan, G.J. and Ramsay, G. 2006. Genetic resources (including wild and cultivated Solanum species) and progress in their utilization in potato breeding. Potato Res. 49:49–65.

Bray, E.A., Bailey-Serres, J. and Weretilnyk, E. 2000. Responses to abiotic stresses. pp. 1158–1249. In Gruissem W, Buchanan B, and Jones R (ed.) Biochemistry and Molecular Biology of Plants. Am. Soc. of Plant Physiol., Rockville, MD, USA.

Burton, G.W. and Powell, J.B. 1968. Pearl millet breeding and cytogenetics. Adv. Agron. 20:49–89.

Cai, X., Chen, P.D., Xu, S.S., Oliver, R.E. and Chen, X. 2005. Utilization of alien genes to enhance Fusarium head blight resistance in wheat—a review. Euphytica 142:309–318.

Carputo, D. and Barone, A. 2005. Ploidy level manipulation in potato through sexual hybridization. Annals Appl. Biol. 146:71–79.

Ceoloni, C. and Jauhar, P.P. 2006. Chromosome engineering of the durum wheat genome: Strategies and applications of potential breeding value. In Singh RJ, and Jauhar PP (eds.) Genetic Resources, Chromosome Engineering, and Crop Improvement, Volume 2: Cereals. CRC Taylor & Francis Press, Boca Raton, FL, USA. pp. 27–59.

Ceoloni, C., Forte, P., Gennaro, A., Micali, S., Carozza, R. and Bitti, A. 2005a. Recent developments in durum wheat chromosome engineering. Cytogenet. Genome Res. 109:328–334.

Ceoloni, C., Pasquini, M. and Simeone, R. 2005b. The cytogenetic contribution to the analysis and manipulation of the durum wheat genome. pp. 165–208. In Royo C, Nachit MM, Di Fonzo N, Araus JL, Pfeiffer WH, and Slafer GA (eds.) Durum Wheat Breeding: Current Approaches and Future Strategies. Volume 1. The Haworth Press, New York, USA.

Chakraborty, S., Chakraborty, N. and Datta, A. 2000. Increased nutritive value of transgenic potato by expressing a non-allergenic seed albumin gene from Amaranthus hypochondriacus. Proc. Natl. Acad. Sci. USA 97:3724–3729.

Chen, P.D. and Liu, D.J. 2000. Transfer of scab resistance from Leymus racemosus, Roegneria ciliaris, and Roegneria kamoji into common wheat. In Raupp WJ, Ma Z, Chen PD, and Liu DJ (eds.) Proc. International Symp. Wheat Improvement for Scab Resistance, Suzhou and Nanjing, China, pp. 62–67.

Crowley, J.G. and Rees, H. 1968. Fertility and selection in tetraploid Lolium. Chromosoma 24:300–308.

Cushman, J.C. and Bohnert, H.J. 2000. Genomic approach to plant stress tolerance. Curr. Opin. Plant Biol. 3:117–124.

Datta, K., Rai, M., Parkhi, V., Oliva, N., Tan, J. and Datta, S.K. 2006. Improved 'golden' indica rice and post-transgeneration enhancement of metabolic target products of carotenoids (β-carotene) in transgenic elite cultivars (IR64 and BR29). Current Science 91:935–939.

De Maine, M.J. 1982. An evaluation of the use of dihaploids and unreduced gametes in breeding for quantitative resistance to potato pathogens. J. Agric Sci. Camb. 99:175–180.

Djousse, L. and Gaziano, J.M. 2007. Consumption of breakfast cereals and risk of heart failure: The physicians health story. American Heart Association's 47th Annual Conference on Cardiovascular Disease Epidemiology and Prevention, Friday March 2, 2007.

Douches, D.S., Westedt, A.L., Zarka, D., Schroeter, B. and Grafuis, E.J. 1998. Potato transformation to combine natural and engineered resistance for controlling tuber moth. HortScience 33:1053–1056.

Drossou, A., Katsiotis, A., Leggett, J.M., Loukas, M., and Tsakas, S. 2004. Genome and species relationships in genus Avena based on RAPD and AFLP molecular markers. Theor. Appl. Genet. 109: 48–54.

Dubcovsky, J. 2004. Marker-assisted selection in public breeding programs: The wheat experience. Crop Sci. 44:1895–1898.

Ducreux, L.J.M., Morris, W.L., Hedley, P.E., Shepherd, T., Davies, H.V., Millam, S. and Taylor, M.A. 2005. Metabolic engineering of high carotenoid potato tubers containing enhanced levels of β-carotene and lutein. J. Exp. Bot. 56:81–89.

Dvořák, J. 1998. Genome analysis in the Triticum-Aegilops alliance. Proceedings of the 9th International Wheat Genetics Symposium, University of Saskatchewan, Saskatoon, Canada 1:8–11.

Dvořák, J., DiTerlizzi, P., Zhang, H-B., and Resta, P. 1993. The evolution of polyploid wheats: identification of the A genome donor species. Genome 36:21–31.

Evenson, R.E. and Gollin, D. 2003. Assessing the impact of the green revolution, 1960–2000. Science 300:758–762.

Fedak, G. 1999. Molecular aids for integration of alien chromatin through wide crosses. Genome 42: 584–591.

Fedak, G. 2000. Sources of resistance to Fusarium head blight. *In* Proc. of the International Symposium on Wheat Improvement for Scab Resistance, Suzhou and Nanjing, China, May 5–7, 2000. Nanjing Agricultural University, Jiangsu Academy of Agricultural Science, Nanjing, China, p. 4.

Feldman, M. 1988. Cytogenetic and molecular approaches to alien gene transfer in wheat. *In* Miller TE and Koebner RMD (eds.) Proc. 7th Int. Wheat Genet. Symp., Cambridge, UK. 13–19 July 1988. Inst. Plant Sci. Res., Cambridge, pp. 23–32.

Feldman, M. and Sears, E.R. 1981. The wild gene resources of wheat. Sci. Am. 244:98–109.

Friebe, B., Jiang, J., Raupp, W.J., McIntosh, R.A. and Gill, B.S. 1996. Characterization of wheat-alien translocations conferring resistance to diseases and pests: Current status. Euphytica 91:59–87.

Gao, A-G., Hakin, S.M., Mittanck, C.A., Wu, Y., Woerner, B.M., Stark, D.M., Shah, D.M., Liang, J. and Rommens, C.M.T. 2000. Fungal pathogen protection in potato by expression of a plant defensin peptide. Nat. Biotechnol. 18:1307–1310.

Gaut, B.S. 2001. Patterns of chromosomal duplication in maize and their implications for comparative maps of the grasses. Genome Res. 11:55–66.

Gill, B.S. and Feiebe, B. 2002. Cytogenetics, phylogeny and evolution of cultivated wheats. *In* Curtis BC, Rajaram S, and Gomez Macpherson H (eds.) Bread Wheat Improvement and Production. FAO, Rome, pp. 71–88.

Gómez, M.I., Islam-Faridi, M.N., Zwick, M.S., Czeschin, D.G., Hart, G.E., Wing, R.A., Stelly, D.M. and Price, H.J. 1998. Tetraploid nature of *Sorghum bicolor* (L.) Moench. J. Hered. 89:188–190.

Guyot, R. and Keller, B. 2004. Ancestral genome duplication in rice. Genome. 47:610–614.

Haldane, J.B.S. 1958. The theory of evolution, before and after Bateson. J. Genet 56:1–17.

Harlan, J.R. and de Wet, J.M.J. 1975. On Ö. Winge and a prayer: The origins of polyploidy. Bot. Rev. 41:361–390.

Heeres, P., Chippers-Rozenboom, M., Jacobsen, E. and Visser, R.G.F. 2002. Transformation of a large number of potato varieties: Genotype-dependent variation in efficiency and somaclonal variability. Euphytica 124:13–22.

Heun, M., Schäfer-Pregl, R., Klawan, R.C., Accerbi, M., Borghi, B. and Salamini, F. 1997. Site of einkorn wheat domestication identified by DNA fingerprinting. Science 278:1312–1314.

Hijmans, R.J. 2001. Global distribution of the potato crop. Am. J. Potato Res. 78:403–412.

Hoekema, A., Huisman, M., Molendijk, L., Elzen van den, P. and Cornelissen, B. 1989. The genetic engineer of two commercial potato cultivars for resistance to potato virus X. Bio/Technol. 7:273–278.

Huang, S., Sirikhachornkit, A., Su, X., Faris, J., Gill, B., Haselkorn, R. and Gornicki, P. 2002. Genes encoding plastid acetyl-CoA carboxylase and 3-phosphoglycerate kinase of the *Triticum/Aegilops* complex and the evolutionary history of polyploid wheat. Proc. Natl. Acad. Sci. USA 99:8133–8138.

Jauhar, P.P. 1968. Inter- and intra-genomic chromosome pairing in an interspecific hybrid and its bearing on basic chromosome number in *Pennisetum*. Genetica 39:360–370.

Jauhar, P.P. 1969. Induction of some rare somatic mutations in *Solanum tuberosum* L. by ionizing radiations and radio-phosphorus. Eur. Potato J. 12:8–12.

Jauhar, P.P. 1970a. Haploid meiosis and its bearing on phylogeny of pearl millet, *Pennisetum typhoides* Stapf et Hubb. Genetica 41:532–540.

Jauhar, P.P. 1970b. Chromosome behaviour and fertility of the raw and evolved synthetic tetraploids of pearl millet, *Pennisetum typhoides* Stapf et Hubb. Genetica 41:407–424.

Jauhar, P.P. 1975a. Genetic control of diploid-like meiosis in hexaploid tall fescue. Nature 254:595–597.

Jauhar, P.P. 1975b. Genetic regulation of diploid-like chromosome pairing in the hexaploid species, *Festuca arundinacea* Schreb. and *F. rubra* L. (Gramineae). Chromosoma 52:363–382.

Jauhar, P.P. 1975c. Polyploidy, genetic control of chromosome pairing and evolution in the *Festuca-Lolium* complex. Proc. 178th Meeting of the British Genetical Society. John Innes Institute, Norwich, England, April 1975. Heredity 35:430.

Jauhar, P.P. 1977. Genetic regulation of diploid-like chromosome pairing in *Avena*. Theor. Appl. Genet. 49:287–295.

Jauhar, P.P. 1981. Cytogenetics of pearl millet. Adv. Agron. 34:407–479.

Jauhar, P.P. 1992. Chromosome pairing in hybrids between hexaploid bread wheat and tetraploid crested wheatgrass (*Agropyron cristatum*). Hereditas 116:107–109.

Jauhar, P.P. 1993. Alien gene transfer and genetic enrichment of wheat. *In* Damania, AB (ed.) Biodiversity and Wheat Improvement. John Wiley and Sons, Chichester, England, pp. 103–119.

Jauhar, P.P. 1995. Meiosis and fertility of F_1 hybrids between hexaploid bread wheat and decaploid tall wheatgrass (*Thinopyrum ponticum*). Theor. Appl. Genet. 90:865–871.

Jauhar, P.P. 2003a. Formation of 2n gametes in durum wheat haploids: Sexual polyploidization. Euphytica 133:81–94.

Jauhar, P.P. 2003b. Haploid and doubled haploid production in durum wheat by wide hybridization. *In* Maluszynski M, Kasha KJ, Forster BP, and

I Szarejko (eds.) Doubled Haploid Production in Crop Plants: A Manual. Kluwer Academic Publishers, Dordrecht, Netherlands. pp. 161–167.

Jauhar, P.P. 2006a. Cytogenetic architecture of cereal crops and their manipulation to fit human needs: Opportunities and challenges. *In* Singh RJ, and Jauhar PP (eds.) Genetic Resources, Chromosome Engineering, and Crop Improvement, Volume 2: Cereals. CRC Taylor & Francis Press, Boca Raton, FL, USA.

Jauhar, P.P. 2006b. Modern biotechnology as an integral supplement to conventional plant breeding: The prospects and challenges. Crop Science 46:1841–1859.

Jauhar, P.P. 2007. Meiotic restitution in wheat polyhaploids (amphihaploids): A potent evolutionary force. J. Hered. 98:188–193.

Jauhar, P.P. and Chibbar, R.N. 1999. Chromosome-mediated and direct gene transfers in wheat. Genome 42:570–583.

Jauhar, P.P. and Hanna, W.W. 1998. Cytogenetics and genetics of pearl millet. Adv. Agron. 64:1–26.

Jauhar, P.P. and Joppa, L.R. 1996. Chromosome pairing as a tool in genome analysis: Merits and limitations. *In* Jauhar PP (ed.) Methods of Genome Analysis in Plants. CRC Press, Boca Raton, FL, USA. pp. 9–37.

Jauhar, P.P. and Khush, G.S. 2002. Importance of biotechnology in global food security. *In* Lal R, Hansen DO, Uphoff N, and Slack S (eds.) Food Security and Environmental Quality in the Developing World. Lewis Publishers, CRC Press, Boca Raton, London, Tokyo. pp. 107–128.

Jauhar, P.P. and Peterson, T.S. 1996. *Thinopyrum* and *Lophopyrum* as sources of genes for wheat improvement. Cereal Res. Commun. 24:15–21.

Jauhar, P.P. and Peterson, T.S. 2000. Toward transferring scab resistance from a diploid wild grass, *Lophopyrum elongatum,* into durum wheat. *In* Proc. of the 2000 National Fusarium Head Blight Forum, Cincinnati. 10–12 Dec. 2000, pp. 201–204.

Jauhar, P.P. and Peterson, T.S. 2001. Hybrids between durum wheat and *Thinopyrum junceiforme:* Prospects for breeding for scab resistance. Euphytica 118:127–136.

Jauhar, P.P. and Peterson, T.S. 2006. Cytological analyses of hybrids and hybrid derivatives between durum wheat and *Thinopyrum bessarabicum,* using multicolour fluorescent GISH. Plant Breeding 125:19–26.

Jauhar, P.P. and Peterson, T.S. 2007. Registration of DGE-1, a durum alien disomic addition line with resistance to Fusarium head blight. Crop Sci. (in press).

Jauhar, P.P. and Swaminathan, M.S. 1967. Mutational rectification of specific defects in some potato varieties. Curr. Sci. 36:340–342.

Jauhar, P.P. and Xu, S.S. 2004. Multidisciplinary approaches to breeding Fusarium head blight resistance into commercial wheat cultivars: Challenges ahead. *In* Proc. of the 2nd Int. Symp. on Fusarium Head Blight, Orlando, Florida. 11–15 Dec. 2004. Michigan State Univ., East Lansing, MI, USA, pp. 77–81.

Jauhar, P.P., Riera-Lizarazu, O., Dewey, W.G., Gill, B.S., Crane, C.F. and Bennett, J.H. 1991. Chromosome pairing relationships among the A, B, and D genomes of bread wheat. Theor. Appl. Genet. 82:441–449.

Jauhar, P.P., Almouslem, A.B., Peterson, T.S. and Joppa, L.R. 1999. Inter- and intragenomic chromosome pairing relationships in synthetic haploids of durum wheat. J. Hered. 90:437–445.

Jauhar, P.P., Doğramacı-Altuntepe, M., Peterson, T.S. and Almouslem, A.B. 2000. Seedset on synthetic haploids of durum wheat: Cytological and molecular investigations. Crop Sci. 40:1742–1749.

Jauhar, P.P., Doğramacı, M. and Peterson, T.S. 2004. Synthesis and cytological characterization of trigeneric hybrids involving durum wheat with and without *Ph1.* Genome 47:1173–1181.

Jauhar, P.P., Rai, K.N., Ozias-Akins, P., Chen, Z. and Hanna, W.W. 2006. Genetic improvement of pearl millet for fodder and grain: Cytogenetic manipulation and heterosis breeding. *In* Singh RJ, and Jauhar PP (eds.) Genetic Resources, Chromosome Engineering, and Crop Improvement, Volume 2: Cereals. CRC Taylor & Francis Press, Boca Raton, FL, USA, pp. 281–307.

Jellen, E.N. and Leggett, J.M. 2006. Cytogenetic manipulation in oat improvement. *In* Singh RJ, and Jauhar PP (eds.) Genetic Resources, Chromosome Engineering, and Crop Improvement, Volume 2: Cereals. CRC Taylor & Francis Press, Boca Raton, FL, USA, pp. 199–231.

Kawchuk, L.M., Lynch, D.R., Martin, R.R., Kozub, G.C. and Farries, B. 1997. Field resistance to the potato leafroll luteovirus in transgenic and somaclone potato plants reduce tuber disease symptoms. Can. J. Plant Pathol. 19:260–266.

Khush, G.S. 2001. Green revolution: The way forward. Nature Rev. Genet. 2:815–822.

Kumar, G.B., Ganapathi, T.R., Revathi, C.J., Srinivas, L. and Bapat, V.A. 2005. Expression of hepatitis B surface antigen in transgenic banana plans. Planta 222:484–493.

Lapitan, N. and Jauhar, P.P. 2006. Molecular markers, genomics and genetic engineering in wheat. *In* Singh RJ, and Jauhar PP (eds.) Genetic Resources, Chromosome Engineering, and Crop Improvement, Volume 2: Cereals. CRC Taylor & Francis Press, Boca Raton, FL, USA, pp. 99–114.

Lewis, W.H. (ed.) 1980. Polyploidy: Biological Relevance. Plenum Press, New York and London.

Li, H., Chen, S., Xin, Z.Y., Ma, Z., Xu, H.J., Chen, X.Y. and Jia, X. 2005. Development and identification of wheat—*Haynaldia villosa* T6DL.6VS chromosome translocation lines conferring resistance to powdery mildew. Plant Breed. 124:203–205.

Linares, C., Ferrer, E. and Fominaya, A. 1998. Discrimination of the closely related A and D genomes of the hexaploid oat, *Avena sativa* L. Proc. Natl. Acad. Sci. USA 95:12450–12455.

Liu, C.J., Witcombe, J.K., Pittaway, T.S., Nash, M., Hash, C.T., Busso, C.S. and Gale, M.D. 1994. An RFLP-based genetic map of pearl millet (*Pennisetum glaucum*). Theor. Appl. Genet. 89:481–487.

Liu, Z.Y., Sun, Q.X., Ni, Z.F., Nevo, E. and Yang, T.M. 2002. Molecular characterization of a novel powdery mildew resistance gene *Pm30* in wheat originating from wild emmer. Euphytica 123:21–29.

Masterson, J. 1994. Stomatal size in fossil plants: Evidence for polyploidy in majority of angiosperms. Science 264:421–424.

Matsuoka, Y. and Nasuda, S. 2004. Durum wheat as a candidate for the unknown female progenitor of bread wheat: An empirical study with a highly fertile F$_1$ hybrid with *Aegilops tauschii* Coss. Theor. Appl. Genet. 109:1710–1717.

McFadden, E.S. and Sears, E.R. 1946. The origin of *Triticum spelta* and its free threshing hexaploid relatives. J. Hered. 37:81–89.

McIntosh, R.A. 1991. Alien sources of disease resistance in bread wheats. *In* Sasakuma T, and Kinoshita T (eds.) Nuclear and Organellar Genomes of Wheat Species: Proceedings of the Dr. H. Kihara Memorial International Symposium on Cytoplasmic Engineering in Wheat, July 1991, Hokkaido University, Japan, pp. 321–332.

McIntosh, R.A., Wellings, C.R. and Park, R.F. 1995. Wheat rusts: An Atlas of Resistance Genes, CSIRO, Australia.

Mohler, V., Zeller, F.J., Wenzel, G. and Hsam, S.L.K. 2005. Chromosomal location of genes for resistance to powdery mildew in common wheat (*Triticum aestivum* L. em Thell.). 9. Gene *MlZec1* from the *Triticum dicoccoides*-derived wheat line Zecoi-1. Euphytica 142:161–167.

Mujeeb-Kazi, A. 2006. Utilization of genetic resources for bread wheat improvement. *In* Singh RJ, and Jauhar PP (eds.) Genetic Resources, Chromosome Engineering, and Crop Improvement, Volume 2: Cereals. CRC Taylor & Francis Press, Boca Raton, FL, USA, pp. 61–97.

Mujeeb-Kazi, A. and Rajaram, S. 2002. Transferring alien genes from related species and genera for wheat improvement. *In* Bread Wheat Improvement and Production. FAO, pp. 199–215.

Muramatsu, M. 1986. The *vulgare* super gene, *Q:* Its universality in *durum* wheat and its phenotypic effects in tetraploid and hexaploid wheats. Can. J. Genet. Cytol. 28:30–41.

Nandi, H.K. 1936. The chromosome morphology, secondary association and origin of cultivated rice. J. Genet. 33:315–336.

Nehra, N.S., Chibbar, R.N., Leung, N., Caswel, K., Mallard, C.S., Steinhauer, L., Båga, M. and Kartha, K.K. 1994. Self-fertile transgenic wheat plants regenerated from isolated scutellar tissues following micropojectile bombardment with two distinct gene constructs. Plant J. 5:285–297.

Nestel, P., Bouis, H.E., Meenakshi, J.V. and Pfeiffer, W. 2006. Biofortification of staple food crops. Symposium: Food Fortification in Developing Countries. J. Nutr. 136:1064–1067.

Nevo, E., Korol, A.B., Beiles, A. and Fahima, T. 2002. Evolution of Wild Emmer and Wheat Improvement—Population Genetics, Genetic Resources, and Genome Organization of Wheat's Progenitor, *Triticum dicoccoides.* Springer, Berlin.

Newell, C.A., Rozman, R., Hinchee, M.A., Lawson, E.C., Haley, L., Sanders, P., Kaniewski, W., Tumer, N.E., Horsch, R.B. and Fraley, R.T. 1991. *Agrobacterium*-mediated transformation of *Solanum tuberosum* L. cv. 'Russet Burbank'. Plant Cell Rep. 10:30–34.

Nishikawa, K. 1983. Species relationship of wheat and its putative ancestors as viewed from isozyme variation. Proceedings of the 6th International Wheat Genetics Symposium, Kyoto, Japan, pp. 59–63.

Nishiyama, I. and Tabata, M. 1964 Cytogenetic studies in *Avena*-XII. Meiotic chromosome behaviour in a haploid cultivated oat. Jap. J. Genet. 38:311–316.

Oliver, R.E., Xu, S.S., Stack, R.W., Friesen, T.L., Jin, Y., Cai, X. 2006. Molecular cytogenetics characterization of four partial wheat- *Thinopyrum ponticum* amphiploids and their reactions to *Fusarium* head blight, tan spot and *Stagonospora nodorum* blotch. Theor. Appl. Genet. 112:1473—1479.

Oraby, H.F., Ransom, C.B., Kravchenko, A.N. and Sticklen, M.B. 2005. Barley *HVA1* gene confers salt tolerance in R3 transgenic oat. Crop Sci. 45:2218–2227.

Paine, J.A., Shipton, C.A., Chaggar, S., Howells, R.M., Kennedy, M.J., Vernon, G., Wright, S.Y., Hinchliffe, E., Adams, J.L., Silverstone, A.L. and Drake, R. 2005. Improving the nutritional value of Golden Rice through increased pro-vitamin A content. Nat. Biotechnol. 23:482–487.

Palevitz, B.A. 2001. EPA reauthorizes Bt corn. Scientist 15:11.

Potrykus, I. 2001. Golden rice and beyond. Plant Physiol. 125:1157–1161.

Potrykus, I. 2003. Nutritionally enhanced rice to combat malnutrition disorders for the poor. Nutrition Rev. 61:S101–S104.

Powell-Abel, P., Nelson, R.S., De, B., Hoffmann, N., Rogers, S.G., Fraley, R.T. and Beachy, R.N. 1986.

Delay of disease development in transgenic plants that express the tobacco mosaic virus coat protein. Science 232:738–743.

Rajaram, S., Peña, R.J., Villareal, R.L., Mujeeb-Kazi, A., Singh, R. and Gilchrest, L. 2001. Utilization of wild and cultivated emmer and of diploid wheat relatives in breeding. Israel J. Plant Sci, 49(1): 93–104.

Rajaram, S., Borlaug, N.E. and Van Ginkel, M. 2002. CIMMYT international wheat breeding. *In* Curtis BC, Rajaram S, and Gomez Macpherson H (eds.) Bread Wheat Improvement and Production. FAO, Rome, pp. 103–117.

Rajhathy, T. and Thomas, H. 1972. Genetic control of chromosome pairing in hexaploid oats. Nature New Biol. 239:217–219.

Ramanna, M.S. and Jacobsen, E. 2003. Relevence of sexual polyploidization for crop improvement—a review. Euphytica 133:3–18.

Reynolds, M.P., Calderini, D.F., Condon, A.G. and Rajaram, S. 2001. Physiological basis of yield gains in wheat associated with the *Lr19* translocation from *Agropyron elongatum*. *In* Bedo Z, and Lang L (eds.) Wheat in a Global Environment. Kluwer Academic Publishers, the Netherlands, pp. 345–351.

Riley, R. and Chapman, V. 1958. Genetic control of the cytologically diploid behaviour of hexaploid wheat. Nature 182:713–715.

Romano, A., Raemakers, K., Visser, R. and Mooibroek, H. 2001. Transformation of potato (*Solanum tuberosum*) using particle bombardment. Plant Cell Rep. 20:198–204.

Roy, M. and Wu, R. 2002. Overexpression of S-adenosylmethionine decarboxylase gene in rice increases polyamine level and enhances sodium chloride-stress tolerance. Plant Sci. 163:987–992.

Sarkar, P. and Stebbins, G.L. 1956. Morphological evidence concerning the origin of the B genome in wheat. Am. J. Bot. 43:297–304.

Schubert, J., Matoušek, J. and Mattern, D. 2004. Pathogen-derived resistance in potato to *Potato virus Y*: Aspects of stability and biosafety under field conditions. Virus Res. 100:41–50.

Sears, E.R. 1972. Chromosome engineering in wheat. Stadler Genetics Symp. 4:23–38.

Sears, E.R. 1977. An induced mutant with homoeologous pairing in common wheat. Can. J. Genet. Cytol. 19:585–593.

Sears, E.R. 1981. Transfer of alien genetic material to wheat. *In* Evans LT, and Peacock WJ (eds.) Wheat Science—Today and Tomorrow. Cambridge Univ. Press. pp. 75–89.

Sears, E.R. 1983. The transfer to wheat of interstitial segments of alien chromosomes. *In* Sakamoto S (ed.) Proc. 6th Int. Wheat Genet. Symp., Kyoto, Japan. 28 Nov.–3 Dec. 1983. Plant Germplasm Inst., Kyoto, Japan, pp. 5–12.

Sears, E.R. 1993. Use of radiation to transfer alien segments to wheat. Crop Sci. 33:897–901.

Sears, E.R. and Okamoto, M. 1958. Intergenomic chromosome relationships in hexaploid wheat. Proc. X Intern. Cong. Genet. 2:258–259.

Sharma, H.C., Sharma, K. and Crouch, J. 2004. Genetic transformation of crops for insect resistance: Potential and limitations. Crit. Rev. Plant Sci. 23:47–72.

Singh, R.P., Huerta-Espino, J., Rajaram, S. and Crossa, J. 1998. Agronomic effects from chromosome translocations 7DL.7Ag and 1BL.1RS in spring wheat. Crop Sci 38:27–33.

Sivamani, E., Brey, C.W., Talbert, L.E., Young, M.A., Dyer, W.E., Kaniewski, W.K. and Qu, R. 2002. Resistance to wheat streak mosaic virus in transgenic wheat engineered with the viral coat protein gene. Transgenic Res. 11:31–41.

Solomon-Blackburn, R.M. and Barker, H. 2001. Breeding virus resistant potatoes (*Solanum tuberosum*): A review of traditional and molecular approaches. Heredity 86:17–35.

Stebbins, G.L. 1971. Chromosomal Evolution in Higher Plants. Edward Arnold, London.

Sun, G.L., Fahima, T., Korol, A.B., Turpeinen, T., Grama, A., Ronin, Y.I. and Nevo, E. 1997. Identification of molecular markers linked to the Yr15 stripe rust resistance gene of wheat originated in wild emmer wheat, *Triticum dicoccoides*. Theor. Appl. Genet. 95:622–628.

Stack, R.W., Elias, E.M., Mitchell, J.F., Miller, J.D. and Joppa, L.R. 2002. Fusarium head blight reaction of Langdon Durum –*Triticum dicoccoides* chromosome substitution lines. Crop Sci. 42:637–642.

Swaminathan, M.S. 2006. An evergreen revolution. Crop Sci. 46:2293–2303.

Tamm, I. 2003. Genetic and environmental variation of kernel yield of oat varieties. Agron. Res. 1:93–97.

Thomas, H., Powell, W. and Aung, T. 1980. Interfering with regular meiotic behaviour in *Avena sativa* as a method of incorporating the gene for mildew resistance from *A. barbata*. Euphytica 29:635–640.

Trethowan, R.M., Crossa, J., van Ginkel, M. and Rajaram, S. 2001. Relationships among bread wheat international yield testing locations in dry areas. Crop Sci. 41:1461–1469.

Vandepoele, K., Simillion, C. and Van der Peer, Y. 2003. Evidence that rice and other cereals are ancient aneuploids. Plant Cell 15:2192–2202.

Vasal, S.K. 2002. Quality protein maize: Overcoming the hurdles. J. Crop Prod. 6:193–227.

Vasal, S.K., Riera-Lizarazu, O. and Jauhar, P.P. 2006. Genetic enhancement of maize by cytogenetic manipulation, and breeding for yield, stress tolerance, and high protein quality. *In* Singh RJ, and Jauhar PP (eds.) Genetic Resources, Chromosome Engineering, and Crop Improvement, Volume 2:

Cereals. CRC Taylor & Francis Press, Boca Raton, FL, USA, pp. 159–197.

Vasil, I.K. 1998. Biotechnology and food security for the 21st century: A real-world perspective. Nature Biotech. 16:399–400.

Vasil, I.K. 2003. The science of politics of plant biotechnology 2002 and beyond. In Vasil IK (ed.) Plant Biotechnology 2002 and Beyond. Kluwer Academic Publ., Dordrecht, the Netherlands, pp. 1–9.

Vasil, V., Castillo, A.M., Fromm, M.E. and Vasil, I.K. 1992. Herbicide resistant fertile transgenic wheat plants obtained by microprojectile bombardment of regenerable embryonic callus. Bio/Technol. 10:667–674.

Vasil, V., Srivastava, V., Castillo, A.M., Fromm, M.E. and Vasil, I.K. 1993. Rapid production of transgenic wheat plants by direct bombardment of cultured immature embryos. Bio/Technol. 11:1553–1558.

Visser, R.G.F., Jacobsen, E. and Feenstra, W.J. 1989. Efficient transformation of potato (Solanum tuberosum L.) using a binary vector in Agrobacterium rhizogenes. Theor. Appl. Genet. 78:594–600.

Wang, G.Z., Miyashita, N.T. and Tsunewaki, K. 1997. Plasmon analyses of Triticum (wheat) and Aegilops: PCR single-strand conformational polymorphism (PRC-SSCP) analyses of organellar DNAs. Proc. Natl. Acad. Sci. USA 94:14570–14577.

Weber, D.F. 1970. Doubly and triply monosomic Zea mays. Maize Genet. Coop. Newsl. 44:203.

Weber, D.F. 1994. Use of maize monosomics for gene localization and dosage studies. In Freeling M, and Walbot V (eds.) The Maize Handbook. Springer-Verlag, New York, pp. 350–358.

Weeks, J.T., Anderson, O.D. and Blechl, A.E. 1993. Rapid production of multiple independent lines of fertile transgenic wheat (Triticum aestivum L.). Plant Physiol. 102:1077–1084.

Weiss, E., Wetterstrom, W., Nadel, D. and Bar-Yosef, O. 2004. The broad spectrum revisited: Evidence from plant remains. Proc. Natl. Acad. Sci. USA 101:9551–9555.

Xie, C.J., Sun, Q.X., Ni, Z.F., Yang, T.M., Nevo, E. and Fahima, T. 2003. Chromosomal location of a Triticum dicoccoides-derived powdery mildew resistance gene in common wheat by using microsatellite markers. Theor. Appl. Genet. 106:341–345.

Yang, Z.J., Li, G.R., Feng, J., Jiang, H.R. and Ren, Z. L. 2005. Molecular cytogenetics characterization and disease resistance observation of wheat-Dasypyrum breviaristatum partial amphiploid and its derivatives. Hereditas 142:80–85.

Ye, X., Al-Babill, S., Klötl, A., Zhang, J., Lucca, P., Beyer, P. and Potrykus, I. 2000. Engineering the provitamin A (β-carotene) biosynthetic pathway into (carotenoid-free) rice endosperm. Science 287:303–305.

Chapter 3

Biofortification: Breeding Micronutrient-Dense Crops

Wolfgang H. Pfeiffer and Bonnie McClafferty

Introduction

In life, minute things can be enormously important. Mineral micronutrients make up a minuscule fraction of the physical mass of a grain, tuber, or fruit; nonetheless, they are crucial to human health. The wide array of micronutrients—more than 20 mineral elements and more than 40 nutrients—necessary for human health, can all be provided by a well-balanced diet. However, the daily diets of large portions of urban and rural populations in the developing world consist mainly of staple foods, such as rice (*Oryza sativa* L.), wheat (*Triticum aestivum* L.), maize (*Zea mays* L.), and cassava (*Manihot esculenta* Crantz), which are good calorie sources, but supply insufficient amounts of the basic micronutrients. Low in minerals, vitamins, and protein from animal and plant sources, poor-quality diets cause micronutrient malnutrition, a burden that afflicts more than one-half of the world's population (UN SCN, 2004).

Micronutrient malnutrition can have disastrous consequences for the more vulnerable members of the human family, especially poor women and preschool children in developing countries. Vitamin A deficiency is the single most important cause of total blindness in developing countries; iron deficiency anemia dramatically reduces the likelihood that mothers will survive childbirth; and even mild levels of micronutrient deficiency can affect physical and cognitive development and lower disease resistance in children. The costs of these deficiencies in terms of diminished quality of life and lives lost are staggering (www.harvestplus.org).

Despite past progress in controlling micronutrient deficiencies through supplementation and fortification, new approaches are needed to expand the reach of these interventions to the rural poor and contribute to sustainable micronutrient deficiency alleviation in burgeoning urban populations. In recent years, an alternative solution is being brought to bear on the problem of micronutrient malnutrition: biofortification of staple crops (Graham and Welch, 1996; Graham et al., 1999, 2001; Bouis et al., 2000; Pinstrup-Andersen, 2000; Underwood, 2000; Bouis, 2003). Biofortification generates nutritionally improved crop varieties through conventional plant breeding and modern biotechnology. Haas et al. (2005) and Van Jaarsveld et al. (2005) demonstrated the feasibility of the biofortification concept from a nutrition perspective. In large-scale human efficacy trials, consumption of diets based on biofortified rice high in iron and orange-fleshed sweetpotato (*Ipomoea batatas* [L.] Lam) high in provitamin A significantly improved human micronutrient status. Hence, biofortified varieties offer the hope that poor at-risk populations in developing countries will be able to meet their micronutrient requirements by consuming the staple crops in their typical diet, at no additional cost.

Plant breeding for micronutrient density began to gain legitimacy when deficiencies in micronutrients, such as iron, iodine, zinc, and vitamins, were recognized as an issue of overwhelming global public health significance and one of the major development challenges of the 21st century. In July of 2003, the Consultative Group on International Agricultural Research (CGIAR) established HarvestPlus: the Biofortification Challenge Program[1] to add food nutritional quality to its agricultural production research paradigm and capitalize on agricultural research as a tool for public health interventions.

The goal of HarvestPlus is to reduce micronutrient malnutrition among poor at-risk populations in Africa, Asia, and Latin America, thereby improving food security and enhancing the quality of life. Harvest-Plus seeks to bring the full potential of agricultural and nutrition sciences to bear on the persistent problem of micronutrient malnutrition. To accomplish this task, HarvestPlus has assembled a multidisciplinary global alliance of more than 150 scientists from CGIAR research centers, private agricultural research institutions, national agricultural research and extension systems (NARES), and non-government organizations (NGOs). Ten CGIAR research centers form the nexus of development of biofortified crops and their NARES partners make up a research alliance that conducts adaptive and participatory breeding as well as participatory variety selection of promising candidate varieties in target zones.

Iron, zinc, and vitamin A, three micronutrients recognized by the World Health Organization as limiting to human health, are the target micronutrients of HarvestPlus. Biofortification research conducted under the auspices of HarvestPlus focuses on a multitude of crops that are a regular part of the staple-based diets of the poor and thus indispensable nutrient sources. Full-fledged plant breeding programs are already under-

way for six "phase I" crops: the mega-crops—rice, wheat, and maize, plus cassava, common beans, and orange-fleshed sweet-potato. For 10 additional "phase II" crops (bananas/plantains, barley, cowpeas, groundnuts, lentils, millet, pigeon peas, potatoes, sorghum, and yams), pre-breeding feasibility studies have been completed and population development has begun.

Given that discussing the latest research and progress achieved in all these crops is beyond the scope of this chapter, we will give an overview of interdisciplinary research activities currently underway, introduce the underlying principles of breeding micronutrient-dense crops, and address the key issues along the HarvestPlus impact pathway (Fig. 3.1). For the purposes of this chapter, dedicated to highlighting new insights related to breeding micronutrient-dense crops, a clear emphasis will be placed on the current objectives, strategies, results, and activities of the plant-breeding component of this multidisciplinary program.[2]

Biofortification: a new process, a new concept

Biofortification is the process of increasing the bioavailable micronutrient density of staple crops through conventional plant breeding and modern biotechnology to achieve a measurable and positive impact on human health. There are marked differences between traditional plant breeding and crop biofortification. Traditional breeding focuses on improving traits of known economic value and developing product concepts for existing markets. Traits are targeted for selection based on whether they can provide better crop and/or utilization options to farmers, but nutritional value as a trait for selection has been largely ignored. Biofortification breeding, on the other hand, seeks to make an impact on human micronutrient status, an endeavor that entails merging breeding with nutrition and socio-

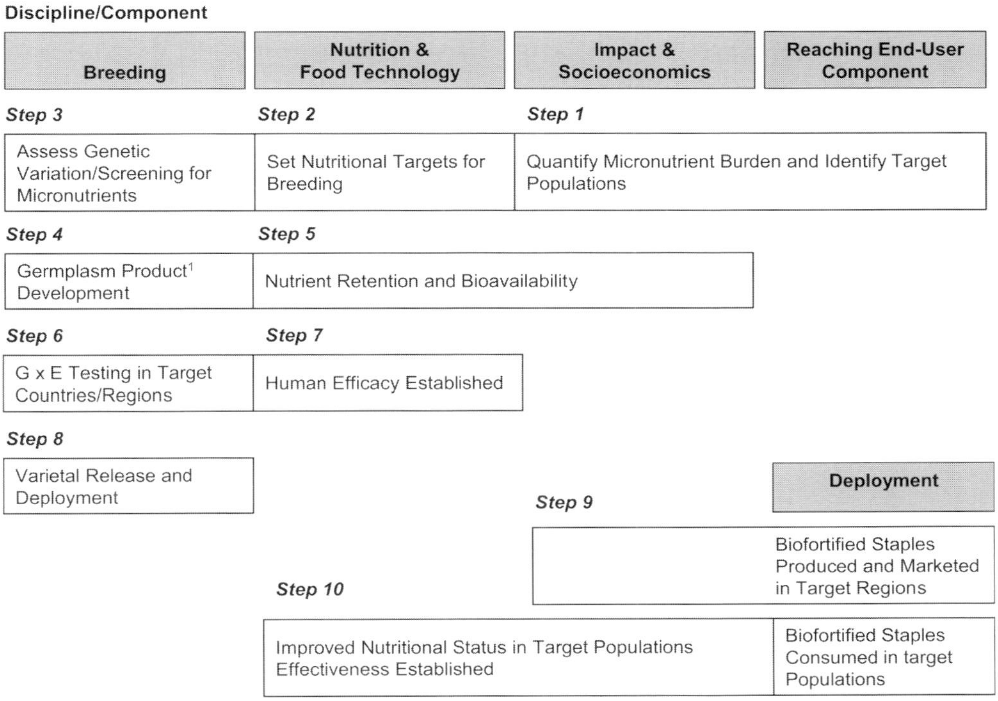

Discipline/Component

Breeding	Nutrition & Food Technology	Impact & Socioeconomics	Reaching End-User Component
Step 3	**Step 2**	**Step 1**	
Assess Genetic Variation/Screening for Micronutrients	Set Nutritional Targets for Breeding	Quantify Micronutrient Burden and Identify Target Populations	
Step 4	**Step 5**		
Germplasm Product[1] Development	Nutrient Retention and Bioavailability		
Step 6	**Step 7**		
G x E Testing in Target Countries/Regions	Human Efficacy Established		
Step 8			
Varietal Release and Deployment		**Step 9**	**Deployment**
			Biofortified Staples Produced and Marketed in Target Regions
	Step 10		
	Improved Nutritional Status in Target Populations Effectiveness Established		Biofortified Staples Consumed in target Populations

[1] Includes nutrition genomics.
Source: HarvetsPlus Impact Pathway, 2006.

Fig. 3.1. HarvestPlus Impact Pathway.

economics research to enhance traits that have measurable value in health outcomes. Biofortification breeding accesses the information it needs to identify such traits by linking directly to the human health and nutrition sectors (Nestel et al., 2006), which have become an integral part of crop improvement and product concept development.

At the core of any biofortification breeding program is a product pathway driven by potential impacts of research and nutrition (see Fig. 3.1). Collaboration between plant breeding and socioeconomics allows the exchange of information to identify target populations that consume target crops based on their micronutrient burden (see Fig. 3.1, step 1). The micronutrient burden is the burden imposed on individuals by micronutrient deficiencies and related diseases; it can

be quantified using the disability-adjusted-life-years (DALYs) approach (Stein et al., 2005). Measuring the effectiveness of biofortified crops in improving human health provides a benchmark for quantifying the ultimate success of biofortification as a cost-effective public health intervention (see Fig. 3.1, step 10).

For biofortification to be successful, micronutrient levels targeted by breeding programs must be derived from nutrition goals set by nutritionists who understand the complexities of making a measurable impact on human health (see Fig. 3.1, step 2). To set target levels and determine the likely contribution to nutritional status, critical information is needed on the bioconversion/bioavailability of ingested nutrients; retention of the micronutrient after storage, processing, and cooking; human micronutrient requirements; and potential levels of consumption by the target population. Genotypic differences in retention, post-harvest micronutrient deterioration, and concentrations of antinutrients and promoters that inhibit or enhance micronutrient bioavailability have been established. Throughout crop development, nutrition and food technology (see Fig. 3.1, steps 4–6) are engaged in assessing the magnitude of genetic variation and genotypic differences for these traits; this allows breeders to increase bioavailability (see Fig. 3.1, step 5) and determine the effect of micronutrient-dense crops or candidate varieties on micronutrient status via human efficacy trials (see Fig. 3.1, step 7).

A conceptual framework for breeding biofortified germplasm

Figure 3.2 outlines the key HarvestPlus biofortified germplasm development activities. Different research categories reflect sequentially arranged stages and milestones, and are superimposed upon a decision-tree that allows monitoring progress and making strategic and "go/no-go" decisions when goals and targets cannot be achieved. The role of nutrition, food technology, and socioeconomics in product development is illustrated in Figure 3.2.

Crop improvement activities of Harvest-Plus focus, first, on exploring the available

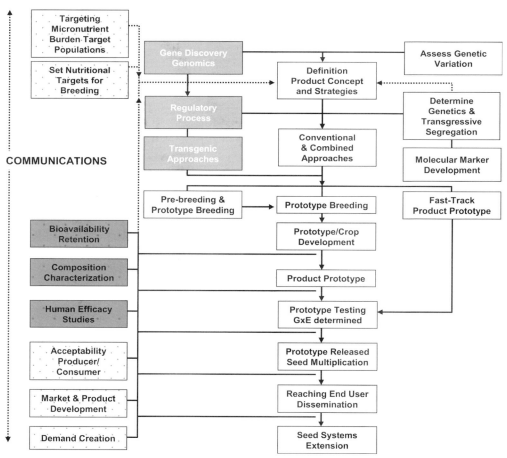

Fig. 3.2. HarvestPlus Breeding Framework.

genetic diversity for iron, zinc, and provitamin A carotenoids. At the same time (or during subsequent screening), agronomic and end-use features are characterized. The objectives when exploring the available genetic diversity are to identify: (1) parental genotypes that can be used in crosses, genetic studies, molecular-marker development, and parent-building, and (2) existing varieties, pre-varieties in the release pipeline, or finished germplasm products for "fast-tracking." Fast-tracking refers to releasing, commercializing, or introducing genotypes that combine the target micronutrient density with the required agronomic and end-use traits so they can be quickly delivered to producers and have an immediate impact on micronutrient-deficient populations.

Identifying the source of genetic variation is essential for the next breeding steps. If variation is present in the strategic gene pool (unadapted trait sources), pre-breeding is necessary prior to using the trait in final product development; if variation is present in the tactical gene pool, by definition, the materials can be used directly to develop competitive varieties. Most breeding programs simultaneously conduct pre-breeding and product enhancement activities to develop germplasm combining high levels of one or more micronutrients. If the available genetic variation suggests that target increments are unlikely to be reached, it is possible to find additional genetic variation through transgressive segregation or by exploiting heterosis. When genetic variation is absent, or micronutrient levels are insufficient to have an impact on human health, a transgenic approach may be one remaining option, e.g., for provitamin A in rice (Khush, 2002; Bouis et al., 2003; Al-Babili and Beyer, 2005). The next breeding steps involve developing and testing micronutrient-dense germplasm, conducting genetic studies, and developing molecular markers to facilitate breeding. Also, genotype × environment interaction (G × E)—the influence of the growing environment on micronutrient expression—needs to be determined at experiment stations and in farmers' fields in the target countries.

Factors related to adoption, commercialization, and product concepts

Activities aimed at reaching the end-users of biofortified crops include delivering seed to farmers and dissemination partners, and, more specifically for HarvestPlus, establishing productive research networks with national program partners and advanced research institutes, and building research capacity to enable sustainable development of biofortified crops within national research systems. However, for biofortified products that contain novel traits and, at times, altered features, product acceptance and marketability need to be assessed prior to deployment throughout an end-user value chain (seed producer, producer/grower, primary processor, manufacturer, distributor, wholesaler/retailer or retail consumer, and consumer) to develop relevant product concepts that will achieve the commercial goal. This information is generally not available for biofortified crops, and, for example, test marketing is not possible without biofortified germplasm products at hand. Hence, product concepts need to be modified and adjusted as part of an iterative process that builds on results generated by crop enhancement, nutrition, food technology, and socio-economics, as breeding advances through recurrent crop development cycles.

The acceptance of biofortified crops by producers and consumers hinges on developing attractive trait packages without compromising agronomic and end-use characteristics. Crucial to developing such trait packages is research to understand the value farmers place on these traits and, hence, the social, economic, and cultural factors that determine crop adoption. Likewise, consumer acceptance studies among the undernourished are essential to accurately gauge demand and identify targeted messages and marketing strategies. In many cases, trait packages must take into account the needs and demands of women and mothers as both consumer targets for nutrient-dense food and key guardians of the most nutritionally vulnerable target population, undernourished children. Developing relevant product concepts for biofortified crops relies on feedback and a continuous flow of information from socioeconomics, impact-assessment studies, and marketing and consumer behavior research (see Fig. 3.2). The outcomes of these diagnostic studies contribute to designing communication strategies that are crucial to the diffusion of innovations and product adoption (see Rogers, 1983).

Issues related to trait visibility in micronutrient-dense biofortified crops are crucial for developing product concepts. Higher levels of provitamin A carotenoids will turn endosperm, seed, or tuber color from white or light yellow to dark yellow and orange. Such color changes are important because consumers often prefer products, for example, from white-seeded or white-fleshed varieties. For consumers and producers to accept biofortified (yellow or orange) versions of their staple crops, they would have to be convinced of their health benefits. Therefore, diagnostic studies on the feasibility of achieving acceptability and behavioral changes in a target population are crucial to developing product concepts for biofortified crops, particularly transgenic crops (Chong, 2003).

In contrast to carotenoids, high mineral concentrations are not visible and do not affect sensory traits. However, trait invisibility impedes distinguishing biofortified varieties from regular varieties and raises issues associated with product identity, branding, and procurement. Thus, Harvest-Plus has to consider effective formal and informal seed systems, development of

markets and food products, effective communications to reach end-users, and demand creation.

Ultimately, biofortified foods and the HarvestPlus Alliance, as their architect, are expanding the existing production-driven agricultural research paradigm to include food quality. This transformation is no small task, for it is forcing traditional agricultural research institutions and scientists to cross disciplines, and human nutritionists to work with crop scientists to develop effective methods, define new protocols, and even create a new lexicon. It is working to re-cast agricultural research as a public health intervention that will reach the most vulnerable and undernourished.

Crop improvement

Existing genetic variation, trait heritability, gene action, associations among traits, available screening techniques, and diagnostic tools are criteria commonly used to identify selectable traits and estimate potential genetic gains. However, for novel traits such as micronutrients, biofortified product concepts have to consider factors associated with probability of success. These factors encompass: (1) technological goals to identify a trait that enables the desired phenotypic and nutritional performance under all production conditions; (2) legal goals to facilitate the development of a final biofortified product unencumbered by intellectual property rights or other legal barriers to development, manufacture, or sale (freedom to operate); (3) production (breeding) goals to generate a plant product containing the trait that enables the desired performance in target populations, target areas, and in all biofortified varieties without compromising agronomic performance, nutrition, or end-use quality; (4) regulatory goals to ensure qualitative traits (in this case, assured seed nutrient density level) and to facilitate the development of a "transgenic event" embodying the trait or technology in the plant genome that meets all domestic and/or international regulatory requirements for food and feed; and (5) commercial goals to guide the design and delivery of a technology (modified from McElroy, 2004). Achieving each of these product development goals presents their own set of process-specific challenges, costs, and risks (McElroy, 2004). In the following sections, we will focus on: (1) technological and product goals in the context of biofortification; (2) factors relating to genetic advance and the likelihood of success, and their contribution to creating feasible product concepts; and (3) the methodologies used to implement them.

Product concepts

Crop enhancement methodologies and procedures for breeding micronutrient-dense crops follow the standard principles applied to traits with equivalent characteristics (e.g., number of genes and mode of inheritance). In biofortification breeding, as enabling technologies are developed, methodologies have to be tailored according to available trait diagnostics and breeding objectives. Furthermore, product concepts are specific to targeted countries/zones. For example, product concepts for HarvestPlus maize embrace a range of biofortified germplasm products for various target countries or maize-producing areas in Africa, Asia, and Latin America. These include germplasm products enriched with individual micronutrients or combinations of iron, zinc, and provitamin A (according to the micronutrient burden and target countries identified in step 1, Figure 3.1). Elite adapted genetic backgrounds of both conventional and quality protein maize (QPM) are being used as platforms to add micronutrient density. Breeding efforts at CGIAR centers and NARES are focusing on developing hybrids, but they also consider synthetics and open-pollinated varieties on a smaller scale, during a transitional period, until formal and informal hybrid seed systems are established. Agronomic superiority can more easily be achieved by replacing open-pollinated varieties with hybrids. Launching hybrids also overcomes the problem of degeneration of micronutrient density in open-pollinated varieties due to outcrossing. Depending on the target micronutrient(s), product concepts embrace white, mineral-enriched maize, yellow/orange-colored provitamin-A-dense maize, and yellow/orange maize biofortified with both provitamin A and minerals.

Due to a strong cultural preference for white maize for human consumption in Africa, Asia, and Latin America, product concepts for provitamin-A-dense maize have to consider its acceptability to producers and consumers, and feasibility studies are required to guide breeding decisions. If crops are new or non-traditional, color preferences have not usually been established. Provitamin-A-dense maize with orange-colored grain may be perceived as a new product and accepted. Visible traits confer product identity, and added nutritional value may constitute an incentive for adopting biofortified varieties.

For crops biofortified with iron and zinc, the scenario is different because the trait is invisible and does not affect sensory characteristics. Product concepts must consider farmers' criteria for changing varieties, including factors related to food and income security. Farmers usually weigh risk factors against the higher income they would obtain from increased production or improved production efficiency as a result of adopting the new technology.

Because technically genotypes and germplasm can be distinguished on the basis of morphological, biochemical, or molecular characteristics, breeders could incorporate marker traits (e.g., a morphological trait) to distinguish micronutrient-dense genotypes. However, these markers are impractical for identifying, procuring, and labeling or branding a product, since they are not directly linked to a mineral and not apparent to growers and consumers. In addition, breeding for these types of markers is not viable, given the added costs and negative impact on genetic progress that can result from breeding for additional traits. Hence, breeding for micronutrient density must consider strategies to keep pace with rates of progress for value-added traits, particularly yield, in non-biofortified germplasm, while simultaneously incorporating additional traits for micronutrient density.

Assessing genetic variation

Developing enabling technologies (e.g., analytical methods and high throughput screening methods to assay micronutrients) and establishing germplasm screening are prerequisites for effectively assessing genetic variation. Inexpensive rapid screening methods boost breeding effectiveness and are crucial for assessing the large number of genotypes in plant population development and coping with screening and sample turnaround requirements for crops with two or more cycles per year. A factor that poses a challenge to sampling and trait diagnostics is rapid post-harvest deterioration, particularly of tuber crops or fruits, which are harvested with high moisture content. In contrast to minerals, provitamin A carotenoids experience greater degradation during storage, drying, milling, and processing.

Initially, the lack of rapid techniques for screening cereals, legumes, and tubers for minerals and provitamin A negatively affected progress in breeding for micronutrient-dense crops. In addition, crop sampling protocols and protocols for conventional analytical methods, including sample preparation, digestion, extraction, and milling procedures, had to be developed and standardized across laboratories. HarvestPlus has made considerable investment in assessing the analytical accuracy of participating laboratories by using external quality assurance programs to allow comparison of results. Tremendous progress has been achieved in this area, and these enabling technologies are currently being validated and implemented at various CGIAR centers and national research institutes. Furthermore, recent studies have found that iron contamination (e.g., from soil), degree of milling/polishing, and seed size/seed shriveling have significant effects on mineral concentrations; earlier research considered these effects only rarely, if at all. In view of the lack of published information on micronutrient analysis and its critical role in breeding, we have included a section that describes in greater detail micronutrient analysis and related research conducted to date.

Micronutrient analysis

Mineral micronutrients make up a minuscule fraction of the physical mass of a grain, tuber, or fruit, with concentrations in parts per million (ppm) or even parts per billion range ($\mu gg^{-1} = mg\,kg^{-1} = ppm$). Typical iron and zinc concentrations in major crops range from $5\,\mu gg^{-1}$ to $150\,\mu gg^{-1}$. In crops that show genetic variation for provitamin A or β-carotene, typical concentrations range from $>1\,\mu gg^{-1}$ to $>400\,\mu gg^{-1}$ (for example, in orange-fleshed sweetpotato). Although sensitive analytical methods are required to accurately determine micronutrients, the sensitivity requirements in applied breeding may vary greatly for pre-screening large segregating populations and characterizing progenitors for crosses or candidate varieties.

Minerals

Table 3.1 contains a summary of precision analysis methods and state-of-the-art high-throughput screening methods (HTMs) that are applied and/or being tested for use in

Table 3.1. Analytical methods for micronutrients.

	ICP-OES: Inductively Coupled Plasma Spectrometer	AAS: Atomic Absorption Spectrometer	XRF: X-Ray Fluorescence Spectrometer	NIRS: Near Infrared Reflectance Spectrophotometer	Modified Perl's Prussian Blue	Modified 2,2 Dipyridal	Modified Zincon
Principle, Throughput, and Practical Considerations							
Principle	Excitation and emissions at various wavelengths	Absorption	X-ray fluorescence	Absorption at wavelengths in the near infra-red	Color reaction	Color reaction	Color reaction
Digestion required	Yes	Yes	No	No	No	Yes	Yes
Sample destructive	Yes	Yes	No	No	Yes	Yes	Yes
Throughput	Up to 2.5 min per sample regardless of number of elements analyzed	~2.5 min per element	5–10 min per sample depending on the number of elements analyzed	~2 min per sample	~4 min per sample	~4 min per sample	~4 min per sample
Pros/comments	Total recovery of nutrient achieved but subject to type of digestion procedure used; good sensitivity	Total recovery of nutrient achieved but subject to type of digestion used; good sensitivity; low cost of purchasing equipment	Total recovery of nutrient achieved; good sensitivity with more expensive models; low cost of bench-type equipment; no digestion step	Low cost	Low cost	Low cost	Low cost
Cons/comments	Gas required; high start-up cost; digestion step; destructive	Gas required; digestion step needed; destructive analysis; samples require milling	Problems with Al and Ti sensitivity; samples require milling	Calibration cost can be high; requires ongoing addition of calibrations	Labor-intensive; semi-quantitative	Semi-quantitative	Semi-quantitative
Analytical Capability							
Iron	Yes	Yes	Yes	Yes	Yes; separation into high and low groups	Yes; separation into high and low groups	Yes; separation into high and low groups
Zinc	Yes	Yes	Yes	Yes	No	No	Yes; separation into high and low groups
Contamination indicators	Yes	No	Yes, but sensitivity for Al is inadequate as contaminant indicator	Not tested	No	No	No
Other elements	Yes	Yes	Yes	Yes, although data limited in this area	No	No	No
Other relevant compounds—promoters & inhibitors	No	No	No	For example, protein, carotenoids, antinutrients	No	No	No

69

Table 3.1. *Continued.*

	ICP-OES: Inductively Coupled Plasma Spectrometer	AAS: Atomic Absorption Spectrometer	XRF: X-Ray Fluorescence Spectrometer	NIRS: Near Infrared Reflectance Spectrophotometer	Modified Perl's Prussian Blue	Modified 2,2 Dipyridal	Modified Zincon
Precision							
Application	Plant and soil material	Plant and soil material	Plant and soil material	Plant material	Plant material	Plant material	Plant material
Accuracy for Fe	High	High	High	High	High	High	No
Accuracy for Zn	High	High	High	High	No	No	High
Accuracy for carotenoids	No	No	No	High to medium-high confirmed for different crops	No	No	No
Accuracy for total carotenoids	No	No	No	High to medium-high confirmed for different crops	No	No	No
Accuracy for β-carotene	No	No	No	High to medium-high confirmed for different crops	No	No	No
Accuracy for minerals	Very accurate	Very accurate	Very accurate	Separates out only high- and low-nutrient genotypes	Separates out only high- and low-Fe genotypes	Separates out only high- and low-Fe genotypes	Separates out only high- and low-Zn genotypes
Economics							
Start-up costs (equipment)	$50,000–$300,000 depending on make and model	$10,000–$40,000 depending on make and model	$50,000–$350,000 depending on make and model	$60,000–$90,000	minimal	minimal	minimal
Running costs	Varies from lab to lab; between $4.00 and $7.00/sample	Varies from lab to lab; e.g., Argon between $0.22 and $2.00/sample; labor the greatest cost in analysis	$15–$25 AUD/sample for XRF analysis; no cost for gas, just instrument upkeep and labor	$0.5–$2.00/sample; dramatically decreasing costs/measurement as more components are measured	0.5–$1.00/sample	$0.5–$1.00/sample	$0.5–$1.00/sample
Application in Breeding							
Recommended application in breeding	Pre-screening in population development and validation of Fe- and Zn-dense genotypes identified by rapid screening techniques	Pre-screening in population development and validation of Fe- and Zn-dense genotypes identified by rapid screening techniques	Pre-screening in population development and validation of Fe- and Zn-dense genotypes identified by rapid screening techniques	Pre-screening in population development for minerals and provitamins A; precison analysis for carotenoids and β-carotene for certain crops	Pre-screening in population development	Pre-screening in population development	Pre-screening in population development

Source: James Stangoulis, School of Agriculture and Wine, University of Adelaide, Waite Campus.

breeding different crops. For precision analysis, the Inductively Coupled Plasma Argon Optical Emission Spectrometer (ICP), Atomic Absorption Spectrometer (AAS), and X-Ray Fluorescence Spectrometer (XRF) allow identification of a wide range of micronutrients, including elements, such as phosphorus, which is indicative of the antinutrient phytate. The ICP is the current method of choice for quantifying elements, such as aluminum, which has been proposed as an indicator of contaminant iron. Various HTMs are applied in pre-screening to reduce the large number of samples in populations segregating for micronutrient concentration to a more practical number for subsequent, more expensive high precision analyses. Depending on the method used, an approximate 66% proportion for more qualitative colorimetric methods (Modified Perl's Prussian Blue) to a 75–85% proportion for semi-quantitative methods (2,2 Dipyridal) of "lows" can be discarded. The accuracy of semi-quantitative methods can be increased by using computerized systems with image analyzers. Correlations between iron determined by ICP and the 2,2 Dipyridal colorimetric method in rice, wheat, maize, sweetpotato, and cassava ranged between 0.88 and 0.98 (James Stangoulis, personal communication). Colorimetric techniques are simple, fast, and low-cost, but destructive because samples must be milled. Elements indicative of contamination are not determined with color-staining techniques; they need to be quantified during subsequent precision mineral analysis.

The Near-Infrared Reflectance Spectrophotometry (NIRS) method relates a sample's reflectance of near-infrared light to its chemical composition and covers wavelengths between 730 and 2500 nm emitted by major plant compounds, such as oil, starch, cellulose, water, and protein. In breeding, NIRS is routinely used to determine, for example, grain protein, which it can measure with high accuracy. The latest research has shown that NIRS has potential to predict iron and zinc with sufficient precision for pre-screening, although the causality of the association is not yet understood. Correlations between iron and zinc determined by ICP and NIRS in potato, sweetpotato, and beans range between 0.77 and 0.85 (Wolfgang Grüneberg, CIP [International Potato Center], and Steve Beebe, CIAT [International Center for Tropical Agriculture], personal communications), and NIRS has been used to separate high-, medium-, and low-iron barley genotypes (James Stangoulis, personal communication). NIRS is environmentally friendly (no reagents are involved) and easy to operate, but it requires continuing calibration to make sure the calibration set includes samples representative of the genetic variation used in breeding, and it must also account for environmental impact on readings. The section "Provitamin A Carotenoids" later in this chapter elaborates further on NIRS in the context of carotenoid analysis.

Contamination in mineral analyses

Detecting contamination while assaying micronutrient concentration is complicated, and references are not yet available to guide researchers. In the past, contamination has, in general, not been addressed in the literature; extremely high iron concentrations that have been reported are likely due to contamination.

Contamination—for example, by iron—can result from soil, dust, metal parts or paint in threshing equipment, rubber products (particularly silicon and neoprene), sample preparation, or seed handling. Zinc is less subject to be a contaminat than iron. Research to establish protocols and guidelines for determining approximate thresholds and, in particular, for developing corrective measures, is currently underway. However, diagnostics for contamination cannot substitute for validating micronutrient concentrations of selected genotypes through additional screening. Procedures that can eliminate sources of contamination in experimentation—for example, preventing lodging in cereals or washing tubers before sampling—should be routine. Washing maize grain samples to eliminate iron contamination has been investigated with varying degrees of success (Kevin Pixley, CIMMYT; James Stangoulis, personal communications). Using appropriate spatial experimental designs with replicated standards or checks to estimate error is also warranted, particularly in unreplicated nurseries.

One approach for detecting contamination entails using indicator elements that are: (1) abundantly found in soil, dust, or equipment; (2) uniform in concentration of contaminating fractions (e.g., of soil); and (3) reproducibly released and easily measured. However, contamination-indicator elements must be absent in plants or seed, or present only in trace amounts. Earlier research investigated Al, Ti, and Cr (or a combination of these) for their potential to act as indicator elements, and attempted to establish threshold levels/bands. Other elements may also be suitable, but good indicator elements for soil are, by definition, very hard to determine accurately in plant tissues, if any soil at all is present. To date, research in this area has identified Al as the most suitable indicator element. Effectively correcting for soil/dust contamination is complicated by the large variety of soil types and the varied ratios of Al, Ti, Cr, and Fe and Zn in plant sample analyses, which result, for example, in different recovery rates, depending on particle size/surface area; hence, confidence in such corrections is not high.

Statistical approaches use population parameters to identify values that fall outside the range expected for an assumed normal distribution and are probably erroneously high. The accuracy of detection increases when data from replicated trials or replicated check varieties are available. These two approaches are complementary and should be used in combination.

Data from mineral analyses conducted at Waite Analytical Laboratories (Adelaide, Australia) and from micronutrient screening of cereal, legume, and tuber crops provided by HarvestPlus crop leaders have been studied to establish tentative Al thresholds for iron contamination. The criteria used to examine contamination consisted of analyzing and comparing data subsets with different Al ranges and using replicated check data to validate results in combination with statistical outlier tests and correlations among elements. Results suggest that Al concentrations of more than 5 to $10 \mu g g^{-1}$ are frequently associated with contaminant Fe. Analyses also revealed that significant correlations between Fe and Al levels in adapted genotypes generally indicate Fe contamination. Eliminating data for samples with Al values $>10 \mu g g^{-1}$ reduced the average correlation between Fe and Al across datasets/crops from $r = 0.35$ to 0.18; including only data for samples with Al $< 5 \mu g g^{-1}$ further reduced the correlation to $r = 0.11$. These findings coincided with results from statistical analyses.

Effects of milling/polishing

Minerals in rice, wheat, maize, and other cereals are concentrated in the aleurone and embryo, as shown in Figure 3.3 (Ozturk et al., 2006); mineral concentration in the endosperm is much lower and decreases sharply toward the center of gravity of the kernel. During polishing/milling, the mineral-containing aleurone layer and embryo are completely or partially removed; small differences in wheat flour extraction rates or in rice polishing can, thus, have an over-proportional effect on micronutrient concentration. In wheat, significant portions of non-endosperm particles are retained for flour extraction rates >80%; similarly, mineral concentrations are significantly

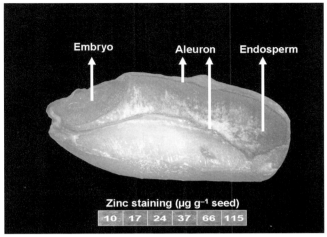

Fig. 3.3. Localization and staining of zinc in wheat seed (*Triticum aestivum* L., cv. Balatilla). Seed was stained with Dithizone (DTZ) 52 days after anthesis and Zn content analyzed with ICP-OES (source: Ozturk et al., 2006).

higher in under-polished rice. Mineral concentrations in wheat grain and endosperm are closely associated, which contrasts with the findings for rice: extensive experience at the International Rice Research Institute (IRRI) revealed a poor association between mineral concentrations in brown rice and polished rice (Gerard Barry, personal communication), although a recent study by Sison et al. (2006) found a close correlation. Research on how the degree of polishing/milling affects micronutrient concentration is complicated, and researchers are simultaneously testing and modifying non-contaminating equipment suitable for milling small samples (such as seed from individual plants). Standardized screening protocols have been developed and are now being validated and implemented in breeding projects to achieve a level of standardization that would permit data comparison.

Micronutrient concentration versus content

Special care in milling is warranted when assessing seed of, for example, wild relatives of crop species, genetic stocks, inbred lines, and unadapted germplasm or genotypes that may have small, shriveled seed and/or incomplete seed set. Biotic and abiotic stress or other production constraints may have the same effect on seed and seed set in adapted genotypes. If the remaining portions of a seed fraction—for example, the embryo—have high micronutrient concentration, concentration levels can be inflated. Furthermore, shriveled seed may require a disproportionate degree of milling or polishing to make processed products that satisfy commercial or laboratory standards.

Seed shriveling, wrinkling, and weathering can have dramatic effects on grain micronutrient density, given that micronutrient concentration in the embryo and seed coat is much higher than micronutrient levels in the endosperm. The seed coat to endosperm ratio is high, which can result in elevated micronutrient concentrations, i.e., the "concentration" effect. The concentration effect can result when fewer grains act as the micronutrient sink because few grains per spike have been produced due to sterility or poor seed set. In plump seed, the seed coat to endosperm ratio is much lower, causing a "dilution" effect. Because micronutrient concentration is generally determined on whole grain, concentration levels in shriveled seed can be overestimated (Cakmak et al., 2000; Imtiaz et al., 2003).

Figure 3.4 displays iron concentration versus content in 438 synthetic wheat accessions. The variation in content for narrow concentration ranges (e.g., 48–52 μgg^{-1}) can be crucial in determining micronutrient content (μg seed^{-1} or, in certain cases, μg plant^{-1}) rather than micronutrient concentration (μgg^{-1}) when characterizing germplasm. In hybrid crops, correlations between mineral content and mineral concentration can be significantly affected by the concentration effect (e.g., in inbred or sibbed lines) and warrant consideration in breeding. There are few reports in the literature regarding content within the context of mineral accumulation in conventional germplasm and transgenic materials.

Grain yield, agronomic performance, and end-use quality attributes (e.g., protein concentration) of large- or plump-seeded adapted genotypes are often compared with those of non-adapted genotypes with small, shriveled grain. Since non-adapted genotypes have higher grain protein concentration due to the concentration effect and regularly produce lower grain yields, correlations between these traits and micronutrient concentration can be overrated. Correlations based on content are usually lower and can partially remove the masking effect of seed size and shriveling. These factors must be taken into account when comparing different types of germplasm or in germplasm selection.

Provitamin A carotenoids

Spectrometric Measurement The quantification of major carotenoids is a challenging task; one of the difficulties results from the

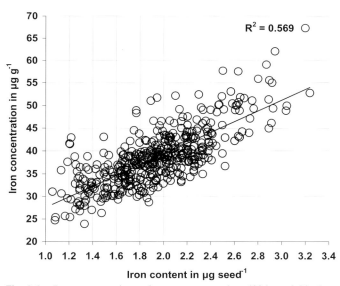

Fig. 3.4. Iron concentration and content measured on 438 hexaploid wheat synthetics accessions (reconstituted wheat from crosses between 4X wheat and *T. tauschii*; AABBDD genome) grown in field experiments at Cd. Obregón, Sonora, Mexico (data source: I. Ortiz-Monasterio, CIMMYT, Mexico).

variation in the carotenoid composition of different crops (Rodriguez-Amaya and Kimura, 2004; Kimura et al., 2007). Knowing a crop's carotenoid composition is essential for determining which measuring method to use. Due to its sensitivity and selectivity, high performance liquid chromatography (HPLC) is the method of choice to quantify individual carotenoids and their isomers. Other methods may not distinguish between individual carotenoids that differ in their provitamin A activity. However, HPLC equipment is more specialized and more expensive to run. For example, cost per sample can range between US$50 and US$70. The low throughput of 15–45 samples per day makes it unsuitable for rapid screening purposes.

Spectrophotometric methods are generally simpler to use. However, they have less capacity to distinguish between different carotenoids and their isomers. For example, a thin-layer chromatography (TLC) method only separates the three different carotenoid groups (β-carotene and α-carotene; β-cryptoxanthin; and lutein and zeaxanthin) but does not detect the presence of trans and cis isomers of β-carotene; cis isomers are known to have lower provitamin A activity and can be present in significant amounts compared with trans isomers.

It may be adequate to use a spectrophotometric method, for example, on cassava or orange-fleshed sweetpotato, where there is typically one major carotenoid—all trans β-carotene. In maize, however, the major carotenoids are zeaxanthin and lutein, which do not show provitamin A activity, but have much smaller amounts of β-carotene and β-cryptoxanthin. With TLC, lutein and zeaxanthin interfere with the detection of other carotenoids, for example, β-carotene. Hence, for maize, use of HPLC is necessary.

Near-Infrared Reflectance Spectrophotometry Carotenoids show absorption in the visible and infrared regions of the electro-magnetic spectrum, and NIRS has revealed potential for screening a range of crops for total carotenoids and provitamin A. Brenna and Bernardo (2004) applied NIRS to determine carotenoids and the vitamin A precursors β-carotene and β-cryptoxanthin, which are relevant to breeding biofortified maize. Cross-validation procedures indicated close associations between HPLC values and NIRS estimates for major carotenoids. In sweetpotato and cassava, β-carotene predominates among total carotenoids, and NIRS screening of the two crops revealed correlations between total carotenoids/β-carotene, as determined by HPLC and NIRS estimates, ranging from >0.80 to >0.90 (Wolfgang Grüneberg, CIP; Thomas zum Felde, personal communications).

Based on experience to date, NIRS shows good potential for estimating carotenoids and provitamin A carotenoids with high or medium-to-high precision; determine iron and zinc with the sensitivity required for pre-screening; and predict with high accuracy antinutrients, such as phenolic compounds and promoters, and value-added traits, such as protein, oil, or kernel hardness, which affect flour yield in milling. Since multiple traits can be determined on the same sample, the costs per compound will be proportionately lower. Hence, NIRS may provide a very inexpensive and rapid method for screening large numbers of genotypes for a wide array of traits.

Visual screening of provitamin A carotenoids

The crop-specific variation in the carotenoids profile reflects differences in the association between provitamin A concentration and visual color intensity; it also determines the suitability of using color intensity for visual grading or selection with color charts when pre-screening for provitamin A. Color charts can be used for cassava (Chavez et al., 2005) and orange-fleshed sweetpotato (Zhang and

Xie, 1988; Simonne et al., 1993), and for crops in which β-carotene or provitamin A constitutes the major portion of total carotenoids. For maize, visual color is dominated by the non-provitamin A precursors zeaxanthin and lutein, and inexpensive high-throughput visual selection can only be applied to separate white or light yellow maize grains from dark yellow and orange color types.

Genetic variation—germplasm screening

The availability of genetic variation for micronutrient density is essential for determining the feasibility of achieving meaningful increments through conventional breeding and high rates of genetic progress under selection (G_s) ($G_s = i\sigma_p h^2$, where i is selection intensity, σ_p is phenotypic standard deviation, and h^2 is heritability). Breeders can capitalize on additive gene effects, transgressive segregation, heterosis, and maternal effects to improve micronutrient density. When the required genetic variation is not available, transgenic approaches can provide novel and additional sources of variation to increase provitamin A or iron—via ferritin—in the endosperm and achieve the target micronutrient density (Bouis et al., 2002; Nandi et al., 2002; Matthews et al., 2003; Vasconcelos et al., 2003; Taylor et al., 2004; Al-Babili and Beyer, 2005; Paine et al., 2005; Shewry and Jones, 2005; Sautter et al., 2006; Khalekuzzaman et al., 2006). In the future, breeding will likely combine both conventional and transgenic approaches.

Screening objectives entail assaying representative samples of the genetic diversity for micronutrient density contained in the tactical and strategic gene pools, along with agronomic and end-use quality features of trait-source genotypes. To date, only a relatively small portion of the existing genetic diversity for micronutrients has been evaluated, and an even lower proportion of the genetic diversity for antinutrients and promoters has been assessed. Evaluating all accessions of each relevant species conserved in gene banks is beyond the scope of this chapter, as gene banks at CGIAR centers alone preserve more than 530,000 accessions in-trust (see http://www.cgiar.org/impact/accessions.html). Because of these large numbers of accessions, current screening by breeding programs under HarvestPlus is focusing on core collections. Screening and research on screening methodology need to be expanded to include generic and crop-specific inhibitors and promoters for bioavailability. Future searches for variation outside core collections may employ state-of-the-art geographical information system (GIS) and molecular tools to enhance screening effectiveness. These modern techniques allow targeting accessions that are most likely to possess untapped genetic variation based on, for example, phytogeography and ancestral, genetic, or functional genomic relationships.

Micronutrient concentrations are affected by micro-environmental variation, $G \times E$ interaction, germplasm type, and numerous other factors. Consequently, published data on micronutrient genetic diversity reveal significant variation in average values and genotypic variation per se among crops and within crop species. Thus, when interpreting these data, one must consider differences that can produce error such as sampling, milling, analytical protocols, and the type of experimental screening design used.

Ranges in micronutrient concentrations reported in the literature reveal there is significant genetic variation in barley (Ma et al., 2004), beans (Beebe et al., 2000; Nunez-Gonzalez et al., 2002; Wissuwa, 2005), cassava (Simonne et al., 1993; Maziya-Dixon et al., 2000; Chavez et al., 2000), cowpea (Farinu and Ingrao, 1991), maize (Bänziger and Long, 2000; Mi et al., 2004), rice (Gregorio et al., 2000), sorghum

(Reddy et al., 2005; Kayodé et al., 2006) and wheat (Feil and Fossati, 1995; Cakmak et al., 2000; Ortiz-Monasterio and Graham, 2000). In transgenics, there is genetic variation for provitamin A in potato (Taylor et al., 2004; Ducreux et al., 2005), iron in rice (Lucca et al., 2000; Qu et al., 2005; Khalekuzzaman et al., 2006), and β-carotene in rice (Paine et al., 2005; Parkhi et al., 2005). White and Broadley (2005) provided a recent review of genetic variation for minerals.

Maximum micronutrient levels are frequently present in the strategic gene pool, in genetically distant sources, such as wild relative species, landraces, or germplasm unadapted to the agroecology of the target environment (Cakmak et al., 2000; Zeng et al., 2004). The difficulty is accessing the genetic variation in unadapted sources, and the extent to which variation in the tactical gene pool can be recovered depends on factors such as genetic distance, differences in ploidy levels, and trait linkages. In prebreeding, eliminating unfavorable traits associated with the target trait causes "linkage drag," which, depending on its magnitude, adds to product development time and costs. Since trait recovery from gene sources in the strategic gene pool varies, we used the genetic variation present in the tactical gene pool (Fig. 3.5) to predict progress in the shorter term.

Figure 3.5 displays average (baseline) and maximum values for iron (Fig. 3.5a) and zinc (Fig. 3.5b) in adapted germplasm. For rice, HarvestPlus used data for polished rice because brown rice is rarely consumed. For both iron and zinc, the short-term exploitable variation in adapted germplasm is of similar magnitude for cereals and legumes, and lower for tuber crops and rice, regardless of the baseline level. The variation for iron in beans is higher than in other crops; HarvestPlus has found an approximate increase in the maximum of $20\,\mu gg^{-1}$ resulted from a breeding cycle for high-iron beans (Steve Beebe, CIAT, personal communica-

tion). Maximum Fe values reported in the literature for certain crops could be up to 10 times higher than those encountered in later studies and likely coincide with the maximum Al values reported. HarvestPlus has found that maximum values for β-carotene (fresh weight basis) are about $9\,\mu gg^{-1}$ in cassava and $>300\,\mu gg^{-1}$ in orange-fleshed sweetpotato, whereas for provitamin A in maize (dry matter basis), they are about $15\,\mu gg^{-1}$.

Setting nutritional target levels

The available genetic variation allows the prediction of the magnitude of micronutrient increments that can be added through breeding. However, only a portion of an increment contributes to human micronutrient status; this portion, the bioavailable amount, largely depends on how much nutrient is lost from crop harvest until ingestion and on the bioavailability of a nutrient once ingested. Critical information needed to set nutritional target levels for breeding (for a target country) includes the amount of nutrient retained after storage, processing, and cooking; micronutrient bioconversion/bioavailability in a typical diet once the nutrient is ingested; and nutrient requirements of a target population (Institute of Medicine, 2001; Nestel et al., 2006; White and Broadley, 2005). The daily micronutrient intake supplied by a crop must also be considered when setting target levels. Many of these parameters are interrelated in a highly complex manner, since human micronutrient status, dietary composition, and health status affect bioavailability (for example, for β-carotene: β-carotene absorbed/β-carotene in food) and its components' bioaccessibility (β-carotene freed/micronutrient in food), bioconversion (retinol formed/β-carotene absorbed), and bioefficacy (retinol formed/β-carotene in food). A more detailed discussion of these factors is beyond the scope of this chapter.

As a starting point, when initially setting tentative target levels without detailed

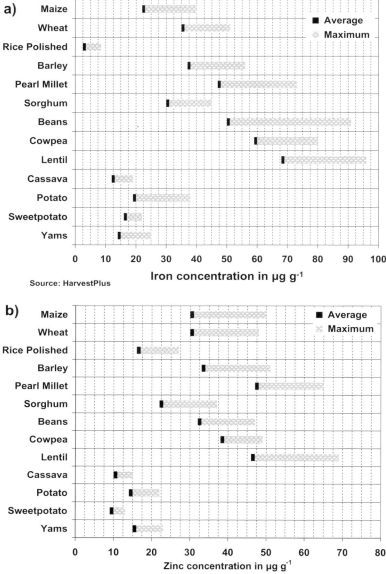

Fig. 3.5. Typical average and maximum concentrations for iron (a) and zinc (b) for adapted genotypes evaluated in field experiments for major cereals, legumes, and tubers (data source: HarvestPlus database; data provided by HarvestPlus crop leaders and published/unpublished sources).

information at hand for accurate assessment, bioavailability of zinc can be assumed to be 25% and bioavailability of iron to be 5% for legumes (e.g., beans, lentil, cowpea) and cereals with significant phytate concentrations (e.g., wheat, maize, sorghum, pearl millet, barley). For tubers (e.g., cassava, potato, sweetpotato, yams) and low-phytate rice, 10% bioavailability can be assumed.

Within a context of general assumptions, Figure 3.6 (a, b, c) illustrates the relationship between micronutrient intake and the micro-a)

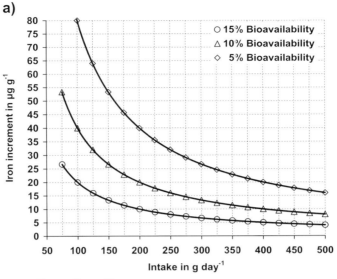

Source: HarvestPlus

b)

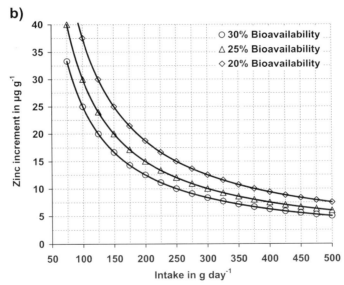

Fig. 3.6. Micronutrient increments from a baseline concentration for a measurable biological impact for women from a public health perspective for various intake levels assuming 100% retention. For iron (**a**), assumed bioavailability is 5% and 10% based on an 8 mg/day^{-1} requirement. For zinc (**b**). assumed bioavailability is 25% based on a requirement of 3 mg/day^{-1}. For β-carotene/provitamins A (**c**), assumed β-carotene/provitamins A:retinol bioconversion rates are 3:1, 6:1, and 12:1, assuming the crop provides 50% of the Estimated Average Requirement (EAR) based on a requirement of 500 mg Retinol Activity Equivalents (RAE)/day^{-1}.

nutrient increment from the baseline concentration needed to make a measurable biological impact on women of childbearing age from a public health perspective (for iron [3.6a], zinc [3.6b], and β-carotene/provitamin A [3.6c]). In Figure 3.6, 100% retention has been assumed to allow generalizations across crops. In Figure 3.6, target

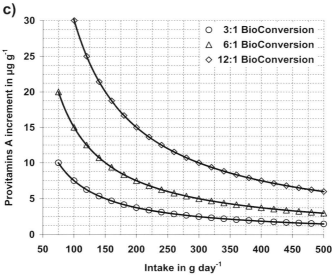

Fig. 3.6. *Continued.*

increment functions are given for iron for 5%, 10%, and 15% bioavailability (Fig. 3.6a), for 20%, 25%, and 30% biovailability for zinc (Fig. 3.6b), and for β-carotene/ provitamin A, for bioconversion rates for β-carotene/provitamin A to retinol of 12:1, 6:1, and 3:1 (Fig. 3.6c).

The feasibility of reaching nutritional target increments through conventional breeding largely depends on bioavailability (provitamin A: retinol bioconversion) and intake (Fig. 3.6). There is scope for increasing micronutrient density in the shorter term via transgressive segregation, heterosis, and maternal effects beyond the variation displayed in Figure 3.5, and for discovering higher levels in the tactical gene pool. Figure 3.6 implies, in combination with the variation displayed in Figure 3.5, that in the shorter term, target increments of zinc and provitamin A can likely be reached in most crops. The genetic variation for zinc in varieties, germplasm lines, and parental stocks, especially of cereal and legume crops, is high (see Fig. 3.5b; White and Broadley, 2005). However, due to the substantially

lower bioavailability of iron when compared with zinc, significantly higher micronutrient increments have to be added to reach nutritional target levels and achieve a measurable impact on human health.

The bioavailability of iron and zinc is associated with the presence of antinutrients and/or the lack of promoter substances for micronutrients (White and Broadley, 2005). Since an increase in bioavailability translates into a proportional decrease in the nutritional target increment (increasing iron bioavailability from 5% to 10% reduces the target increment by 50%), strategies for breeding micronutrient-dense crops should consider indirect breeding for increased bioavailability, increased retention, or reduced post-harvest micronutrient deterioration, in addition to direct breeding for increased concentration. Although not yet well understood, breeding for increased bioavailability via conventional (Welch et al., 2000; Oikeh et al., 2003) or transgenic approaches (Lucca et al., 2000; 2001; Drakakaki et al., 2005) offers tremendous potential (Hambidge et al., 2004).

Breeding for increased bioavailability

Breeding for micronutrient bioavailability per se is greatly limited by the lack of diagnostic tools for large-scale, rapid germplasm evaluation, such as adequate in vitro and/or animal bioavailability models. Current studies are exploring the feasibility of dissecting overall bioavailability into its causal components, such as antinutrients and promoters, which can be addressed by breeding. Ongoing exploratory research is investigating the feasibility of breeding for inhibitors/enhancers from both the crop improvement and human nutrition perspectives. Breeding studies entail determining the genetic variation for antinutrients and promoters, the magnitude of trait expression/stability through G × E studies, trait heritability, and associations with agronomic and end-use quality traits. Screening methods are being evaluated in a parallel effort, whereas nutrition research and food science are investigating bioavailability and nutritional impact using in vitro and animal bioavailability models and, subsequently, efficacy and retention studies involving human subjects.

Phytate occupied center stage in past and current research on antinutrients (Raboy, 2000). Genetic variation for phytate has been reported in numerous crops: cowpea (Farinu and Ingrao, 1991), beans (Muzquiz et al., 1999), and sorghum (Kayodé, et al., 2006). Low-phytate mutants are available and have been found in barley (Dorsch et al., 2003; Bregitzer and Raboy, 2006), maize (Raboy et al., 2000; Shi et al., 2003; Shukla et al., 2004), and wheat (Guttierie et al., 2004). Transgenic research focuses on the enzyme phytase (potato, Ulla et al., 2003; rice, Hong et al., 2004; maize, Drakakaki et al., 2005). Introduction of phytase in rice (Lucca et al., 2000), in combination with ferritin in maize (Drakakaki et al., 2005), and co-expression of phytase, ferritin, and a metallothionein-like gene in rice (Lucca et al., 2001a,b) significantly increased iron absorption/bioavailability and demonstrated the potential of transgenic approaches to capitalize on higher micronutrient concentration and increased bioavailability.

Lowering phytate concentration can have adverse effects on human health and must be researched by nutrition experts. Safe threshold levels have to be established before addressing the trait in breeding crops for human consumption. The viability of addressing phytate (or heat-stable phytase) in crop improvement via selecting for phytate:zinc or phytate:iron molar ratios also depends on how the crop is consumed. For example, when wheat is milled, there is an over-proportional reduction of phytate (compared with iron and zinc) with decreasing flour extraction rates; in white flour (72% extraction according to international trade standards), iron and zinc concentrations are reduced by approximately 50–60% compared with concentrations in the whole grain, whereas phytate concentration is reduced by about 90%. This over-proportional reduction causes a concomitant decrease in the phytate:mineral molar ratio and, hence, in increased bioavailability. Thus, lowering phytate via breeding, may not have the desired effect if wheat is predominantly consumed as white flour products. Phytate can also be significantly reduced by processing methods (Mamta and Darshan, 2000).

Selecting for increased nutrient retention can increase the micronutrient concentration "on the plate" and lower target increments. Significant genotypic differences in retention that could be exploited in breeding have been found, for example, in cassava and yams for provitamin A and, to a lesser extent, for minerals; evaluation of the genetic variation for micronutrient retention in other crops is also warranted. Micronutrient retention has been related to factors associated with flour extraction rate in wheat (e.g., grain hardness, texture, grain shape) and degree of polishing in rice.

Genetics

Knowledge of heritability as it relates to genetic progress ($G_s = i \sigma_p h^2$) and associated genetics is crucial for establishing screening, population development, and $G \times E$ testing strategies, and hence, for effective breeding. All plant-breeding components (such as crossing strategies, breeding methodologies, the operational scale with plot size and number of sites and years required for testing) are based on genetic parameters. Furthermore, the potential for developing molecular markers is closely associated with factors such as the number of genes and their individual contributions. Molecular markers and marker-assisted protocols for applied breeding that can be used at early plant stages to select for micronutrient density can greatly increase breeding efficiency; quantitative trait loci (QTL) for micronutrients have been identified in several crops including beans, maize, and rice (Guzmán-Maldonado et al., 2003; Wong et al., 2004; Gregorio and Htut, 2003; Wissuwa, 2005).

Growing evidence from HarvestPlus research supports findings that iron and zinc concentration is controlled by several (2–5) relevant genes, and that mineral heritability is of intermediate magnitude (Maloo et al., 1998; Philip and Maloo, 1996; Gregorio and Htut, 2003; Long et al., 2004; Cichy et al., 2005). Provitamin A appears to be controlled by a few (~2) major genes, and trait heritability is high (Brown et al., 1993; Egesel et al., 2003b; Menkir and Maziya-Dixon, 2004; Grüneberg et al., 2005). However, heritability can be overestimated if studies contrast non-provitamin A and high provitamin A genotypes. For both minerals and provitamin A, additive gene action and general combining ability predominate (Maloo et al., 1998; Egesel et al., 2003a; Egesel et al., 2003b; Gregorio and Htut, 2003; Long et al., 2004). Heterosis has been found for maize and cassava. Also, studies on temperate maize have revealed signifi-cant reciprocal (maternal) effects for provitamin A (Egesel et al., 2003a). Transgressive segregation for provitamin A has been encountered, for example, in cassava crossed with wild relatives, for iron in beans, and for zinc in wheat. Although information on transgressive segregation or heterosis is still incomplete, there is growing evidence that complementary genes are present, particularly in genetically distant sources, such as wild relative species. This is not unexpected, given that in the past, breeders did not select for micronutrients, and latent variation may have been lost. Furthermore, in the past, breeders would often select for white endosperm and, hence, against carotenoids (e.g., in maize and wheat), or for lower ash content, which is associated with lower mineral concentration (e.g., in wheat).

Correlations among minerals and value-added traits

Figure 3.7 displays correlations among minerals in cereals (maize and wheat), a legume (lentil), and tubers (potato and yams). The data reveal a generic positive correlation between iron and zinc concentrations, and between iron and zinc and all other nutritionally important minerals and trace elements (Ca, Cu, K, Mg, Mn, P, and S). The substantial positive association among minerals suggests an opportunity for raising levels of a number of micronutrients simultaneously by direct selection for numerous micronutrients or by capitalizing on the indirect selection response (i.e., a parallel increase in micronutrients not targeted in selection). Strain and Cashman (2002) provide an overview of the importance of minerals and trace elements in human nutrition.

Several publications report significant associations between mineral concentrations (in particular zinc) and grain protein concentration in wheat (Peterson et al., 1986; Zebarath et al., 1992; Feil and Fossati, 1995;

Fe : Cu	
Maize	0.14
Wheat	0.24
Lentil	0.41
Potato	0.13
Yams	0.28

Fe : Mg	
Maize	0.44
Wheat	0.30
Lentil	0.46
Potato	0.11
Yams	0.40

Fe : P	
Maize	0.44
Wheat	0.32
Lentil	0.38
Potato	0.25
Yams	0.35

Fe : Ca	
Maize	0.10
Wheat	0.21
Potato	0.36
Yams	0.48
Lentil	0.20

Fe : K	
Maize	0.42
Wheat	-0.02
Lentil	0.43
Potato	0.26
Yams	0.47

Fe : Mn	
Maize	0.27
Wheat	0.57
Lentil	0.47
Potato	0.36
Yams	0.32

Fe : S	
Maize	0.32
Wheat	0.42
Lentil	0.08
Potato	0.39
Yams	0.32

Iron

Fe : Zn	
Maize	0.49
Wheat	0.46
Lentil	0.52
Potato	0.44
Yams	0.61

Zn : Mn	
Maize	0.35
Wheat	0.51
Lentil	0.45
Potato	0.47
Yams	0.26

Zinc

Zn : S	
Maize	0.33
Wheat	0.52
Lentil	0.32
Potato	0.54
Yams	0.54

Zn : Cu	
Maize	0.28
Wheat	0.69
Lentil	0.40
Potato	0.38
Yams	0.55

Zn : Mg	
Maize	0.61
Wheat	0.50
Lentil	0.57
Potato	0.07
Yams	0.46

Zn : P	
Maize	0.57
Wheat	0.52
Lentil	0.61
Potato	0.36
Yams	0.50

Zn : Ca	
Maize	0.03
Wheat	0.12
Lentil	0.20
Potato	0.19
Yams	0.37

Zn : K	
Maize	0.48
Wheat	0.13
Lentil	0.63
Potato	0.16
Yams	0.64

Source: HarvestPlus

Fig. 3.7. Average correlations among micronutrient concentrations for maize, wheat, lentil, potato and yams. The average for each crop was calculated as arithmetic mean across different datasets. Data for elements with Al values >10 µg g^{-1} were excluded from the analyses. Data for crops comprised between 500 and 2500 micronutrient analyses per crop. Micronutrient analyses were conducted at Waite Analytical Services, Adelaide (data source: HarvestPlus database; data provided by HarvestPlus crop leaders and published/unpublished sources).

Cakmak et al., 2000; Alex Morgounov, CIMMYT, personal communication). A lower magnitude of association has been found in maize (Arnold and Bauman, 1976; Arnold et al., 1977; Bänziger et al., 2000). Due to a negative association between grain protein and grain yield, particularly in wheat, HarvestPlus researchers are now investigating whether these correlations could result in a significant negative association with grain yield and other traits positively correlated with grain yield. However, correlations can be overestimated (see section on micro-

nutrient concentration versus content). Data on HarvestPlus crops (except wheat) have not revealed relevant negative associations between micronutrients and productivity traits (Menkir and Maziya-Dixon, 2004; Mi et al., 2004), but knowledge is still incomplete. Associations between micronutrients and sensory characteristics can be relevant. For example, in sweetpotato, dry matter content and β-carotene concentration are negatively associated (Zhang and Xie, 1988; Grüneberg et al., 2005); this complicates breeding because adult African consumers prefer sweetpotato with high dry matter content (Tomlins et al., 2004).

Genotype × environment interaction

Our genetic understanding of micronutrient trait expression and trait stability (spatial, temporal, and system-dependent) is limited, and G × E interaction can greatly influence genotypic performance across different crop growing scenarios. Micronutrient trait expression and the extent of G × E interactions in different environments largely determine the screening and breeding methodologies used, as well as the genetic gains achieved from selection. Early biofortification efforts were challenged by knowledge gaps regarding site suitability for trait assessment and the effect of permanent and variable environmental factors, production constraints, and crop management practices on micronutrient concentration. Mineral traits were frequently perceived as qualitative traits until results from multi-environment experiments revealed significant G × E interactions and substantial differences in the suitability of test sites for micronutrient selection in expressing variation and discriminating among genotypes, and, hence, their quantitative nature (Reynolds et al., 2005).

An increasing body of evidence suggests that the expression of provitamin A is relatively stable under different growing conditions (Egesel et al., 2003b; Menkir and Maziya-Dixon, 2004). HarvestPlus researchers have identified cassava, maize, and sweetpotato genotypes with high and stable expression across environments, with G × E interactions predominantly of the noncrossover type. These results agree with findings that β-carotene/provitamin A are controlled by relatively few genes and more simply inherited.

The expression of zinc (and, to a lesser extent, iron) concentration is related to and affected by permanent and variable environmental factors; the higher variation due to G × E when compared with provitamin A reflects the more complex inheritance of iron and zinc, particularly in cereals. However, results from multi-environment trials revealed micronutrient-dense genotypes of cereals, legumes, and tubers with high, stable trait expression in the presence of high G × E interaction.

Differences in genotypic variation for minerals in different environments can be large. In cassava multi-location experiments conducted in Colombia, site means at proximate test locations varied 2–3 fold for zinc, and zinc standard deviations at sites varied 2–4 fold. In contrast, site mean values for iron were comparable to the respective standard deviations. Similar results for zinc have been obtained for wheat in multi-location trials in Kazakhstan (Alex Morgounov, CIMMYT, personal communication). Since soil zinc deficiency is a common problem in major agricultural areas (Cakmak et al., 1990), such results are not unexpected. Due to the complexity of soil mineral dynamics and the interaction with environmental factors, soil mineral status often explains only part of poor iron or zinc expression in the plant. Research aimed at understanding the underlying factors of G × E interactions and micronutrient trait expression by analyzing soil and plant samples is warranted. Such research would also help in understanding the association between soil micronutrient status and crop mineral concentration.

Micronutrient fertilizer can be used to generate repeatable screening and testing environments for minerals, increase breeding effectiveness, and overcome soil mineral deficiency. HarvestPlus is conducting research on the synergistic effects of zinc fertilizer on crop zinc levels in target areas. This research will provide farmers with crop-management recommendations to increase mineral density and reduce spatial and temporal fluctuations due to G × E interactions.

Microenvironment variation for minerals can be highly significant and, if not adequately sampled, may cause false high positives in mineral screening. Plot size needs to be adjusted to sample microenvironment variation. Standards, repeated checks, replications, and spatial experimental designs are used to take on-site spatial variation into account. Using common checks or standards across experiments allows results from different environments and for different types of germplasm to be compared. Among factors that can influence micronutrient expression are planting date and season: HarvestPlus G × E interaction trials revealed highly significant differences in average mineral concentration and genetic variation between planting seasons for rice and pearl millet, and between different planting dates for wheat. Hence, next to spatial and temporal variation, systems variation caused by differential crop-management practices can have significant effects.

Strategies and approaches for breeding competitive biofortified crops

Malnutrition and micronutrient deficiency frequently coincide in major target areas for biofortified crops in developing countries. In the coming decades, crop production will have to increase to compensate for diminishing per capita land and water resources and keep pace with rising global food demand. Increasing and stabilizing crop production under these circumstances pose one of the greatest challenges for agricultural research of the 21st century, given the fragile and highly variable nature of biofortification target areas and the continued deterioration of natural resources. Thus, at the same time as crop micronutrient concentration is being improved, production efficiency in the different agroecological cropping systems must

be maximized while protecting the natural resource base. Within this context, environmental, cultural, and political sustainability is what defines the focus of the research agenda (Pfeiffer et al., 2005a,b).

Breeding for high yield and micronutrient density

Breeding for additional traits not associated with crop productivity or economic yield and, in particular, for novel traits, causes lower rates of progress for productivity traits, if additional resources are not invested. Increasing the operational scale and scope of breeding activities can substantially increase genetic progress, via both genetic variation and higher selection intensity, and avoid compromising yield by breeding for micronutrient density. Other factors that increase breeding efficiency and enhance G_s are breeding and testing in target environments and/or controlled environments that reliably simulate target environments to: (1) increase heritability or the correlation between selection and target environments; (2) intensify testing of experimental germplasm in target environments; and (3) facilitate the use of molecular markers in selection and of molecular marker-assisted background selection to accelerate the introgression of micronutrient density from the strategic gene pool to locally adapted elite germplasm (Guzmán-Maldonado et al., 2003; Wong et al., 2004). The development and implementation of enabling technologies, such as inexpensive high-throughput diagnostic tools, can dramatically increase breeding efficiency.

Introgressing novel traits into the tactical gene pool initially requires additional resources. Novel traits, such as micronutrient density, are therefore being addressed as "specific traits" to accelerate developing micronutrient-rich products for rapid commercialization and/or for immediate impact on micronutrient deficiency alleviation.

Micronutrient traits are presumably not subject to genetic erosion (such as that caused by the evolution of pathogenic races) and require little maintenance breeding once genes have been incorporated. Hence, the cost of breeding for micronutrients decreases across time, and micronutrient density built into the gene pool will not affect future breeding for productivity traits. If micronutrient traits are incorporated into the tactical gene pool, micronutrient concentration can be taken up as a generic trait present in all germplasm products, which is a requirement for biofortification to be sustainable.

Strategies for achieving agronomic superiority

Agronomic superiority and farmer adoption are critical to the success of biofortification. Production increases can originate from various sources (Pfeiffer et al., 2005b): (1) genetic gains in yield potential; (2) genetic gains in tolerance/resistance to abiotic and biotic stresses; (3) improved, sustainable crop-management techniques; and (4) the synergistic effects of all these factors within the context of production economics. Circumstantial evidence indicates that progress in any of these factors will, directly or indirectly, enhance yield.

In reality, there is no environment that is completely free of stress, and breeding for stress tolerance occupies center stage in crop improvement. In practice, indirect selection for tolerance/resistance to key constraints is frequently more efficient for raising genotypic production potential (and, eventually, triggering farmer adoption) than selecting for yield or a specific abiotic stress per se. Breeders can raise productivity by concentrating on improving resistance/tolerance to biotic/abiotic factors and, particularly, resistance to diseases for which they have known and repeatable variation (Pfeiffer et al., 2005a).

Protecting yield through the incorporation of traits that buffer production vagaries and result in higher yield stability can be the key to adoption, given that, in developing countries, farmers' criteria for changing varieties are often based on risk factors related to food and income security.

The agricultural production paradigm that focuses on higher yield to maximize profit has changed along with declining commodity prices and higher proportional costs of crop management and biocides. Farmers have to maximize their revenue by reducing production costs, i.e., by aiming at higher economic yields rather than by increasing crop yields. The possibility of obtaining higher economic returns from varieties with increased input efficiency and/or responsiveness provides an incentive for adoption. Input-use efficiency at low levels of input availability and input responsiveness (i.e., the plant's effectiveness in transforming additional inputs into yield) are under genetic control and can be improved by breeding. Pleiotropic effects associated with high micronutrient content can affect agronomic performance and, hence, agronomic options. For example, seed zinc concentration and micronutrient-dense seeds in wheat are closely associated with greater seedling vigor, increased stand establishment, and higher grain yield, particularly in zinc-deficient soils (Cakmak et al., 1990). In a zinc-deficient soil, seed zinc concentrations of $355 \, \text{ng seed}^{-1}$, $800 \, \text{ng seed}^{-1}$, and $1465 \, \text{ng seed}^{-1}$ resulted in grain yields of $480 \, \text{kg ha}^{-1}$, $920 \, \text{kg ha}^{-1}$, and $1240 \, \text{kg ha}^{-1}$, respectively. Product concepts can capitalize on these effects.

The end-use quality of a variety used to produce local and processed food products can be a criterion for variety adoption in subsistence farming systems and market economies. In market economies, higher returns from premium prices for end-use quality traits, such as percent protein, grade requirement, or sensory traits (size, shape, color, taste), can compensate for any reduction in income due to a yield penalty.

In most developing countries where micronutrient deficiency is prevalent, agronomic superiority can be achieved more easily by replacing open-pollinated varieties (for example, of maize, sorghum, and pearl millet) with hybrids or synthetics. However, if a product concept entails deploying hybrid technologies, it must consider the feasibility of having sustainable seed systems in place.

Adoption often entails implementing a technological package. Improved agronomic practices, such as direct seeding under reduced or zero tillage and stubble retention, along with germplasm adapted to these practices, capitalize on synergies between genetic and agronomic solutions to achieve production and end-use quality objectives. Breeding for adaptation to direct seeding under reduced or zero tillage is feasible (Trethowan et al., 2005). The arsenal of modern crop-management techniques can provide an agronomic platform for successfully producing biofortified varieties and achieving both nutritional and commercial goals.

Acknowledgments

The authors recognize the intellectual contributions made by research partners within the CGIAR, NARES, and nutrition community throughout the world. They also acknowledge the support received from key members of the donor community, including the Asian Development Bank, the Danish Agency for International Development Assistance, the U.K. Department of International Development, the Bill and Melinda Gates Foundation, the United States Agency for International Development, and the World Bank. Special thanks are extended to Alma McNab for her editorial expertise.

Endnotes

1. The HarvestPlus program is funded by the Bill and Melinda Gates Foundation, the World Bank, US Agency for International Development, UK Department for International Development, and the Danish International Development Agency.

2. Challenge programs are time-bound, independently governed programs of high-impact research that target CGIAR goals in relation to complex issues of overwhelming global and/or regional significance, and rely on partnerships among a wide range of institutions to deliver their products. In the case of HarvestPlus, biofortification research is conducted by a global alliance of research institutions and implementing agencies in developed and developing countries, and coordinated by two CGIAR centers, the International Center for Tropical Agriculture (CIAT) and the International Food Policy Research Institute (IFPRI).

References

Al-Babili, S., and P. Beyer. 2005. Golden Rice—five years on the road—five years to go? Trends in Plant Science 10(12):565–573.

Arnold, J.M., and L.F. Bauman. 1976. Inheritance of and interrelationships among maize kernel traits and elemental contents. Crop Sci. 16(3):439–440.

Arnold, J.M., L.F. Bauman, and H.S. Aycock. 1977. Interrelations among protein, lysine, oil, certain mineral element concentrations, and physical kernel characteristics in two maize populations. Crop Sci. 17(3):421–425.

Bänziger, M., and J. Long. 2000. The potential of increasing the iron and zinc density of maize through plant breeding. Food Nutr. Bull. 21:397–400.

Beebe, S., A.V. Gomez, and J. Renfigo. 2000. Research on trace minerals in the common bean. Food Nutr. Bull. 21:387–391.

Bouis, H.E. 2003. Micronutrient fortification of plants through plant breeding: Can it improve nutrition in man at low cost? Proc. Nutr. Soc. 62:403–11.

Bouis, H.E., R.D. Graham, and R.M. Welch. 2000. The Consultative Group on International Agricultural Research (CGIAR) Micronutrient Project: Justification and objectives. Food Nutr. Bull. 21:374–381.

Bouis, H.E., S. Harris, and D. Lineback. 2002. Biotechnology-derived nutritious foods for developing countries: Needs, opportunities, and barriers. Food and Nutrition Bulletin 23(4):342–383.

Bregitzer, P., and V. Raboy. 2006. Effect of four independent low phytate mutations on barley agronomic performance. Crop Sci. 46:1318–1322.

Brenna, O.V., and N. Bernardo. 2004. Application of near-infrared spectroscopy (NIRS) to the evaluation of carotenoids in maize. J. Agric. Food Chem. 52:5577–5582.

Brown, C.R., C.G. Edwards, C.P. Yang, and B.B. Dean. 1993. Orange flesh trait in potato: Inheritance and carotenoid content. Journal of the American Society for Horticultural Science 118(1):145–150.

Cakmak, I., H. Ozkan, H.J. Braun, R.M. Welch, and V. Römheld. 2000. Zinc and iron concentrations in seeds of wild, primitive, and modern wheats. Food Nutr. Bull. 21:401–403.

Cakmak, I., M. Kalayc, H. Ekiz, H.J. Braun, Y. Kılınc, and Y. Yılmaz. 1990. Zinc deficiency as a practical problem in plant and human nutrition in Turkey: A NATO science-for-stability project. Field Crops Res. 60:175–188.

Chavez, A.L., J.M. Bedoya, T. Sanchez, C. Iglesias, H. Ceballos, and W. Roca. 2000. Iron, carotene, and ascorbic acid in cassava roots. Food Nutr. Bull. 21:410–413.

Chavez, A.L., T. Sanchez, G. Jaramillo, J.M. Bedoya, J. Echeverry, E.A. Bolanos, H. Ceballos, and C.A. Iglesias. 2005. Variation of quality traits in cassava roots evaluated in landraces and improved clones. Euphytica 143:125–133.

Chong, M. 2003. Acceptance of Golden Rice in the Philippine "rice bowl." Nature Biotechnology 21(9):971–972.

Cichy, K.A., S. Forster, K.F. Grafton, and G.L. Hosfield. 2005. Inheritance of seed zinc accumulation in navy bean. Crop Sci. 45(3):864–870.

Dorsch, J.A., A. Cook, K.A. Young, J.M. Anderson, A.T. Bauman, C.J. Volkmann, P.P.N. Murthy, and V. Raboy. 2003. Seed phosphorus and inositol phosphate phenotype of barley low phytic acid genotypes. Phytochemistry 62(5):691–706.

Drakakaki, G., S. Marcel, R.P. Glahn, E.K. Lund, S. Pariagh, R. Fischer, P. Christou, and E. Stoger. 2005. Endosperm-specific co-expression of recombinant soybean ferritin and Aspergillus phytase in maize results in significant increases in the levels of bioavailable iron. Plant Molecular Biology 59(6):869–880.

Ducreux, L.J.M., W.L. Morris, P.E. Hedley, T. Shepherd, H.V. Davies, S. Millam, and M.A. Taylor. 2005. Metabolic engineering of high carotenoid potato tubers containing enhanced levels of β-carotene and lutein. J. Exp. Bot. 409:81–89.

Egesel, C.O., J.C. Wong, R.J. Lambert, and T.R. Rocheford. 2003a. Gene dosage effects on carotenoid concentration in maize grain. Maydica 48(3):183–190.

Egesel, C.O., J.C. Wong, R.J. Lambert, and T.R. Rocheford. 2003b. Combining ability of maize inbreds for carotenoids and tocopherols. Crop Sci. 43(3):818–823.

Farinu, G.O., and G. Ingrao. 1991. Gross composition, amino acid, phytic acid and trace element contents of thirteen cowpea cultivars and their nutritional significance. Journal of the Science of Food and Agriculture 55(3):401–410.

Feil, B., and D. Fossati. 1995. Mineral composition of triticale grain as related to grain yield and grain protein. Crop Sci. 35:1426–1431.

Graham, R.D., and R.M. Welch. 1996. Breeding for staple-food crops with high micronutrient density: Agricultural strategies for micronutrients. Working Paper No. 3; 1996. International Food Policy Research Institute, Washington, D.C.

Graham, R.D., D. Senadhira, S.E Beebe, C. Iglesias, and I. Ortiz-Monasterio. 1999. Breeding for micronutrient density inedible portions of staple food crops: Conventional approaches. Field Crops Res. 60:57–80.

Gregorio, G.B., and T. Htut. 2003. Micronutrient-dense rice: Breeding tools at IRRI. In: T.W. Mew, D.S. Brar, S. Peng, D. Dawe, and B. Hardy, editors. Rice science: Innovations and impact for livelihood. Proceedings of the International Rice Research Conference. Beijing, China, 16–19, September 2002. p. 371–378.

Gregorio, G.B., D. Senadhira, H. Htut, and R.D. Graham. 2000. Breeding for trace mineral density in rice. Food Nutr. Bull. 21:382–386.

Grüneberg, W.J., K. Manrique, Zhang DaPeng, and M. Hermann. 2005. Genotype×environment interactions for a diverse set of sweetpotato clones evaluated across varying ecogeographic conditions in Peru. Crop Sci. 45:2160–2171.

Guttierie, M., D. Bowen, J.A. Dorsch, V. Raboy, and E. Souza. 2004. Identification and characterization of a low phytic acid wheat. Crop Sci. 44:418–424.

Guzmán-Maldonado, S.H., O. Martínez, J.A. Acosta-Gallegos, F. Guevara-Lara, and O. Paredes-López. 2003. Putative quantitative trait loci for physical and chemical components of common bean. Crop Sci. 43(3):1029–1035.

Hambidge, K.M., J.W. Huffer, V. Raboy, G.K. Grunwald, J.L. Westcott, L. Sian, L.V. Miller, J.A. Dorsch, and N.F. Krebs. 2004. Zinc absorption from low-phytate hybrids of maize and their wild-type isohybrids. American Journal of Clinical Nutrition 79(6):1053–1059.

Hong, C.Y., K.J. Cheng, T.H. Tseng, C.S. Wang, L.F. Liu, and S.M. Yu. 2004. Production of two highly active bacterial phytases with broad pH optima in germinated transgenic rice seeds. Transgenic Research 13(1):29–39.

Imtiaz, M., J.B. Alloway, A.K. Shah, S.H. Siddiqui, M.Y. Memon, M. Aslam, and P. Khan. 2003. Zinc contents in the seed of some domestic and exotic wheat genotypes. Asian Journal of Plant Sciences 2(15–16):1118–1120.

Institute of Medicine. 2001. Dietary reference intakes for vitamin A, vitamin K, arsenic, boron, chromium, copper, iodine, iron, manganese, molybdenum, nickel, silicon, vanadium, and zinc. National Academy Press, Washington, D.C.

Kayodé, A.P.P., A.R. Linnemann, J.D. Hounhouigan, M.J.R. Nout, and M.A.J.S. van Boekel. 2006. Genetic and environmental impact on iron, zinc, and phytate in food sorghum grown in Benin. Journal of Agricultural and Food Chemistry 54:256–262.

Khalekuzzaman, M., K. Datta, N. Oliva, M.F. Alam, O.I. Joarder, and S.K. Datta. 2006. Stable integration, expression and inheritance of the ferritin gene in transgenic elite indica rice cultivar BR29 with enhanced iron level in the endosperm. Indian Journal of Biotechnology 5(1):26–31.

Khush, G.S. 2002. The promise of biotechnology in addressing current nutritional problems in developing countries. Food and Nutrition Bulletin 23(4): 354–357.

Kimura, M., C.N. Cobori, D.B. Rodriguez-Amaya, and P. Nestel. 2007. Screening and HPLC methods for carotenoids in sweetpotato, cassava, and maize for plant breeding trials. Food Chemistry 100(4):1734–1746.

Long, J.K., M. Bänziger, and M.E. Smith. 2004. Diallel analysis of grain iron and zinc density in Southern African-adapted maize inbreds. Crop Sci. 44(6): 2019–2026.

Lucca, P., J. Wünn, R.F. Hurrell, and I. Potrykus. 2000. Development of iron-rich rice and improvement of its absorption in humans by genetic engineering. Journal of Plant Nutrition 23(11/12):1983–1988.

Lucca, P., R. Hurrell, and I. Potrykus. 2001a. Approaches to improving the bioavailability and level of iron in rice seeds. Journal of the Science of Food and Agriculture 81(9):828–834.

Lucca, P., R. Hurrell, and I. Potrykus. 2001b. Genetic engineering approaches to improve the bioavailability and the level of iron in rice grains. Theoretical and Applied Genetics 102(2/3):392–397.

Ma, J.F., A. Higashitani, K. Sato, and K. Takeda. 2004. Genotypic variation in Fe concentration of barley grain. Soil Science and Plant Nutrition 50(7):1115–1117.

Maloo, S.R., J.S. Solanki, and S.P. Sharma. 1998. Genotypic variability for quality traits in finger millet (Eleusine coracana (L.) Gaertn.). International Sorghum and Millets Newsletter 39:126–128.

Mamta, P., and P. Darshan. 2000. Effect of pressure and solar cooking on phytic acid and polyphenol content of cowpeas. Nutrition & Food Science 2/3:133–136.

Matthews, P.D., R.B. Luo, and E.T. Wurtzel. 2003. Maize phytoene desaturase and β-carotene desaturase catalyse a poly-Z desaturation pathway: Implications for genetic engineering of carotenoid content among cereal crops. J. Exp. Bot. 54(391):2215–2230.

Maziya-Dixon, B., J.G. Kling, A. Menkir, and A. Dixon. 2000. Genetic variation in total carotene, iron, and zinc contents of maize and cassava genotypes, and ascorbic acid in cassava roots. Food Nutr. Bull. 21:419–422.

McElroy, D. 2004. Valuing the product development cycle in agricultural biotechnology—what's in a name? Nature Biotechnology 22(7):817–822.

Menkir, A., and B. Maziya-Dixon. 2004. Influence of genotype and environment on β-carotene content of tropical yellow-endosperm maize genotypes. Maydica 49(4):313–318.

Mi, G.H., F.J. Chen, X.S. Liu, L. Chun, and J.L. Song. 2004. Genotype difference in iron content in kernels of maize. Journal of Maize Sciences 12(2):13–15.

Muzquiz, M., C. Burbano, G. Ayet, M.M. Pedrosa, and C. Cuadrado. 1999. The investigation of antinutritional factors in Phaseolus vulgaris. Environmental and varietal differences. BASE: Biotechnologie, Agronomie, Société et Environnement 3(4):210–216.

Nandi, S., Y.A. Suzuki, J.M. Huang, D. Yalda, P. Pham, L.Y. Wu, G. Bartley, N. Huang, and B. Lönnerdal. 2002. Expression of human lactoferrin in transgenic rice grains for the application in infant formula. Plant Science 163(4):713–722.

Nestel, P., H.E. Bouis, J.V. Meenakshi, and W.H. Pfeiffer. 2006. Biofortification of staple food crops. J. Nutr. 136:1064–1067.

Nunez-Gonzalez, A., R.K. Maiti, J. Verde-Star, M.L. Cardenas, R. Foroughbakch, J.L. Hernandez-Pinero, S. Moreno-Limon, and G. Garcia-Diaz. 2002. Variability in mineral profile in seven varieties of bean (Phaseolus vulgaris L.) adapted in North East of Mexico. Legume Research 25(4):284–287.

Oikeh, S.O., A. Menkir, B. Maziya-Dixon, R. Welch, and R.P. Glahn. 2003. Genotypic differences in concentration and bioavailability of kernel iron in tropical maize varieties grown under field conditions. Journal of Plant Nutrition 26(10/11): 2307–2319.

Ortiz-Monasterio, I., and R.D. Graham. 2000. Breeding for trace minerals in wheat. Food Nutr. Bull. 21:392–396.

Ozturk, L., M.A. Yazici, C. Yucel, A. Torun, C. Cekic, A. Bagci, H. Ozkan, H.J. Braun, Z. Sayers, and I. Cakmak. 2006. Concentration and localization of zinc during seed development and germination in wheat. Physiologia Plantarum 128:144–152.

Paine, J.A., C.A. Shipton, S. Chaggar, R.M. Howells, M.J. Kennedy, G. Vernon, S.Y. Wright, E. Hinchliffe, J.L. Adams, A.L. Silverstone, and R. Drake. 2005. Improving the nutritional value of Golden Rice through increased pro-vitamin A content. Nature Biotechnology 23(4):482–487.

Parkhi, V., M. Rai, J. Tan, N. Oliva, S. Rehana, A. Bandyopadhyay, L. Torrizo, V. Ghole, K. Datta, and S.K. Datta. 2005. Molecular characterization of marker-free transgenic lines of indica rice that accumulate carotenoids in seed endosperm. Molecular Genetics and Genomics 274(4):325–336.

Peterson, C.J., V.A. Johnson, and P.A. Mattern. 1986. Influence of cultivar and environment on mineral and protein concentrations of wheat flour, bran, and grain. Cereal Chem. 63:118–126.

Pfeiffer, W.H., R.M. Trethowan, M. van Ginkel, I. Ortiz-Monasterio, and S. Rajaram. 2005a. Breeding for stress tolerance in wheat. In M. Ashraf and P. J.C. Harris (eds.). *Abiotic Stresses: Plant Resistance through Breeding and Molecular Approaches.* New York: The Haworth Press, Inc. p. 401–489.

Pfeiffer, W.H., R.M. Trethowan, K. Ammar, and K.D. Sayre. 2005b. Increasing yield potential and yield stability in durum wheat. In C. Royo, M.M. Nachit, N. di Fonzo, J.L. Araus, W.H. Pfeiffer, and G.A. Slafer (eds.). Vol. 2. *Durum Wheat Breeding: Current Approaches and Future Strategies.* New York: The Haworth Press, Inc. p. 531–544.

Philip, J., and S.R. Maloo. 1996. An evaluation of Setaria italica for seed iron content. International Sorghum and Millets Newsletter 37:82–83.

Pinstrup-Andersen, P. 2000. Improving human nutrition through agricultural research: Overview and objectives. Food Nutr. Bull. 21:352–355.

Qu, L.Q., T. Yoshihara, A. Ooyama, F. Goto, and F. Takaiwa. 2005. Iron accumulation does not parallel the high expression level of ferritin in transgenic rice seeds. Planta 222(2):225–233.

Raboy, V. 2000. Low-phytic-acid grains. Food Nutr. Bull. 21:423–427.

Raboy, V., P.F. Gerbasi, K.A. Young, S.D. Stoneberg, S.G. Pickett, A.T Bauman, P.P.N. Murthy, W.F. Sheridan, and D.S. Ertl. 2000. Origin and seed phenotype of maize low phytic acid 1–1 and low phytic acid 2–1. Plant Physiology 124(1):355–368.

Reddy, B.V.S., S. Ramesh, and T. Longvah. 2005. Prospects of breeding for micronutrients and β-carotene-dense sorghums. International Sorghum and Millets Newsletter 46:10–14.

Reynolds, T.L., M.A. Nemeth, K.C. Glenn, W.P. Ridley, and J.D. Astwood. 2005. Natural variability of metabolites in maize grain: Differences due to genetic background. Journal of Agricultural and Food Chemistry 53(26):10061–10067.

Rodriguez-Amaya, D.B., and M. Kimura. 2004. HarvestPlus Handbook for Carotenoid Analysis. HarvestPlus Technical Monograph Series 2.

Rogers, E.M. 1983. *Diffusion of Innovations,* 5th edition. New York: The Free Press.

Sautter, C., S. Poletti, P. Zhang, and W. Gruissem. 2006. Biofortification of essential nutritional compounds and trace elements in rice and cassava. Proceedings of the Nutrition Society 65(2):153–159.

Shewry, P.R., and H.D. Jones. 2005. Transgenic wheat: Where do we stand after the first 12 years? Annals of Applied Biology 147:1–14.

Shi, J.R., H. Wang, Y. Wu, J. Hazebroek, R.B. Meeley, and D.S. Ertl. 2003. The maize low-phytic acid mutant lpa2 is caused by mutation in an inositol phosphate. Plant Physiology 131(2):507–515.

Shukla, S., T.T. VanToai, and R.C. Pratt. 2004. Expression and nucleotide sequence of an INS (3) P_1 synthase gene associated with low-phytate kernels in maize (*Zea mays* L.). Journal of Agricultural and Food Chemistry 52(14):4565–4570.

Simonne, A.H., S.J. Kays, P.E. Koehler, and R.R. Eitenmiller. 1993. Assessment of β-carotene content in sweetpotato breeding lines in relation to dietary requirements. Journal of Food Composition and Analysis. 6(4):336–345.

Sison, M.E.G.Q., G.B. Gregorio, and M.S. Mendioro. 2006. The effect of different milling times on grain iron content and grain physical parameters associated with milling of eight genotypes of rice (*Oryza sativa* L.). Phillipines Journal of Science 135:9–17.

Stein, A., J.V. Meenakshi, M. Qaim, P. Nestel, H.P.S. Sachdev, and Z.A. Bhutta. 2005. Analyzing the health benefits of biofortified staple crops by means of the disability-adjusted life years approach: A handbook focusing on iron, zinc and vitamin A. International Food Policy Research Institute (IFPRI) and International Center for Tropical Agriculture (CIAT), Washington, D.C., and Cali (City), Columbia.

Strain, J.J., and K.D. Cashman. 2002. Minerals and trace elements. In M.J. Gibney, H.H. Voster, and F.J. Kok (eds.). *Human Nutrition.* Blackwell Science Ltd. Oxford, U.K. p. 177–224.

Taylor, M.A., W.L. Morris, L.J.M. Ducreux, H.V. Davies, and S. Millam. 2004. Metabolic engineering of carotenoid biosynthesis in potato tubers. Aspects of Applied Biology 72:155–162.

Tomlins, K., E. Rwiza, A. Nyango, R. Amour, T. Ngendello, R. Kapinga, D. Rees, and F. Jolliffe. 2004. The use of sensory evaluation and consumer preference for the selection of sweetpotato cultivars in East Africa. J Sci Food Agric. 84: 791–799.

Trethowan, R.M., M. Reynolds, K.D. Sayre, and I. Ortiz-Monasterio. 2005. Adapting wheat cultivars to resource conserving farming practices and human nutritional needs. Annals of Applied Biology 146(4):405–413.

Ullah, A.H.J., K. Sethumadhavan, E.J. Mullaney, T. Ziegelhoffer, and S. Austin-Phillips. 2003. Fungal phyA gene expressed in potato leaves produces active and stable phytase. Biochemical and Biophysical Research Communications 306(2):603–609.

UN SCN. 2004. 5th Report on the World Nutrition Situation: Nutrition for Improved Development Outcomes. United Nations System Standing Committee on Nutrition (SCN), Geneva.

Underwood, B.A. 2000. Overcoming micronutrient deficiencies in developing countries: Is there a role for agriculture? Food Nutr. Bull. 21:356–360.

Van Jaarsveld, P.J., M. Faber, S.A. Tanumihardjo, P. Nestel, C.J. Lombard, and A.J.S. Benade. 2005. β-Carotene-rich orange-fleshed sweetpotato improves the vitamin A status of primary school children assessed by the modified-relative-dose-response test. Am J Clin Nutr. 81:1080–1087.

Vasconcelos, M., K. Datta, N. Oliva, M. Khalekuzza-man, L. Torrizo, S. Krishnan, M. Oliveira, F. Goto, and S.K. Datta. 2003. Enhanced iron and zinc accumulation in transgenic rice with the ferritin gene. Plant Science 164(3):371–378.

Welch, R.M., W.A. House, S. Beebe, D. Senadhira, G.B. Gregorio, and Z. Cheng. 2000. Testing iron and zinc bioavailability in genetically enriched beans (*Phaseolus vulgaris* L.) and rice (*Oryza sativa* L.) in a rat model. Food Nutr. Bull. 21:428–433.

White, P.J., and M.R. Broadley. 2005, Biofortifying crops with essential mineral elements. Trends in Plant Science 10(12):586–583.

Wissuwa, M. 2005. Mapping nutritional traits in crop plants. In: M.R. Broadley and P.J. White (eds.), *Plant Nutritional Genomics*. Ames, Iowa: Blackwell Publishing. p. 220–241.

Wong, J.C., R.J. Lambert, E.T. Wurtzel, and T.R. Rocheford. 2004. QTL and candidate genes phytoene synthase and β-carotene desaturase associated with the accumulation of carotenoids in maize. Theoretical and Applied Genetics 108(2):349–359.

Zeng, Y.W., J.F. Liu, L.X. Wang, S.Q. Shen, Z.C. Li, X.K. Wang, G.S. Wen, and Z.Y. Yang. 2004. Analysis on mineral element contents is associated with varietal type in core collection of Yunnan rice. Rice Science 11(3):106–112.

Zhang, L.Y., and Y.Z. Xie. 1988. Inheritance of flesh colour and its correlation with other traits in sweet potato (Ipomoea batatas). Jiangsu Journal of Agricultural Sciences 4(2):30–34.

Chapter 4

Bioinformatics and Plant Genomics for Staple Crops Improvement

David Edwards

Plant bioinformatics

Agricultural productivity has increased dramatically during the past hundred years through the application of technology, and has continued to provide enough food for the expanding human population. This remarkable growth has been achieved through a combination of improved agronomic practices and advances in germplasm. The potential yield of modern germplasm is significantly greater now than at any time in history, and this is especially so for crops grown under less than optimal agro-climatic conditions. Even considering these advances, germplasm enhancement offers the greatest potential to further increase the yield and quality of crops and offers the potential to extend agricultural productivity to environments that are currently not viable due to agro-climatic constraints.

Technologies for germplasm enhancement have recently undergone a revolution with the dawning of the genomics era and the application of molecular tools. These advances have created new challenges for scientists who wish to apply the vast abundance of data being created by high throughput technologies for the enhancement of crop germplasm. The field of bioinformatics has evolved with the objective of integrating the increasingly diverse data types and applying this information for crop improvement. Bioinformatics is not yet mature and the definition of bioinformatics varies depending of the background of the user. Applied bioinformatics may be loosely defined as "the structuring of biological information to enable logical interrogation" (Edwards and Batley, 2004), and as such is inherent in virtually all crop research.

Objectives of plant bioinformatics

Applied plant bioinformatics is an integral part of crop research, and, as such, the objectives of plant bioinformatics reflect the objectives of crop research: the improvement of crop production, increased yield and quality, and a reduction in the environmental footprint. Bioinformatics plays an essential role, from upstream research in the sequencing and characterization of genomes to trait analysis and optimization of breeding strategies. The most promising areas for future crop improvement come from the application of the knowledge of plant genomes to crop improvement; many bioinformatics tools and systems are designed to exploit the information from gene and genome sequence for the identification of molecular genetic markers associated with agronomic traits or allelic forms of genes that are directly responsible for an improved agronomic phenotype (Fig. 4.1). More recently, the expansion of high throughput gene expression systems, such as gene expression microarrays (Brazma et al., 2003; Parkinson et al., 2005; Barrett et al., 2005; Craigon et al.,

93

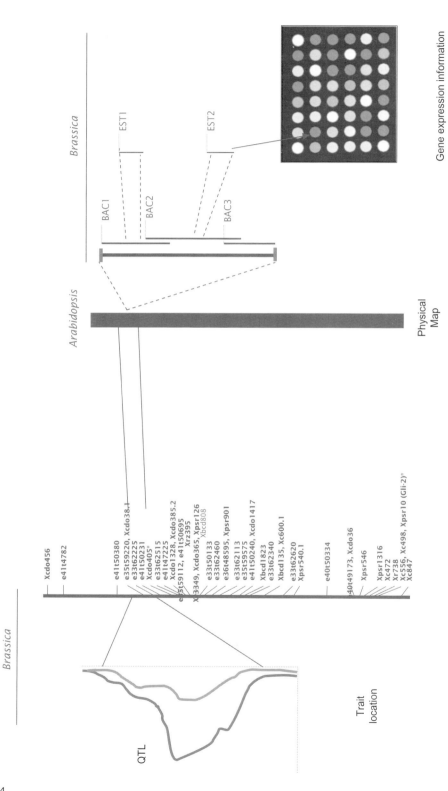

Fig 4.1. Identification of genes responsible for heritable traits. Heritable trait information may be mapped onto a genetic map using molecular genetic markers. Bioinformatics comparative genomics tools permit the identification of genome locations, possibly through completed model genome sequences. Candidate genes for the trait may be identified through expressed sequence annotation and gene expression data. Candidate genes may then be validated for association with the trait in the field.

2004), Serial Analysis of Gene Expression (SAGE) (Velculescu et al., 1995; Gibbings et al., 2003; Lee and Lee, 2003; Fizames et al., 2004; Robinson et al., 2004b; Poroyko et al., 2005; Coemans et al., 2005; Lorenz and Dean, 2002; Matsumura et al., 2003), and Massively Parallel Signature Sequencing (MPSS) (Brenner et al., 2000a; Brenner et al., 2000b, Meyers et al., 2004a), have added an extra dimension to genomic data. Developments in proteome and metabolome analyses (Fiehn et al., 2000; Fiehn, 2002; Sumner, Mendes, and Dixon, 2003; Trethewey, Krotzky, and Willmitzer, 1999) are providing further challenges in crop bioinformatics. The development of bioinformatics tools for the integration of current "-omic" data types with phenome information, represented by the variety of crop phenotypic data, provides the potential to bridge the gap from genome to phenome (Edwards and Batley, 2004). As bioinformatics matures, the standardization of systems and data formats enable greater integration and, hence, application of a broad range of biological data for practical applied crop improvement (The Plant Ontology Consortium, 2002; Bada et al., 2004; Camon et al., 2005; Clark, Brooksbank, and Lomax, 2005; Lee et al., 2005a; Lewis, 2005).

Genomic knowledge to improve germplasm

The scientific study of heritable traits and the field of genetics were established through the methodical analysis of plant species and important discoveries in genetics and heredity, such as dynamic genomes (McClintock, 1950). With the recent sequencing of the complete genome of the model plant *Arabidopsis thaliana* (The Arabidopsis Genome Initiative, 2000) and the model monocot rice (Li et al., 2002; Sasaki et al., 2002), and the ongoing sequencing projects for numerous crop and non-crop plant species, such as *Brassica, Medicago,* lotus, tomato, potato,

poplar, soybean, *Capsella,* papaya, *Eucalyptus,* grape, *Mimulus guttatus, Triphysaria versicolor,* banana, sorghum, maize and wheat (Jackson et al., 2006), plant genomics has truly come of age. In addition, the sequencing of numerous plant pathogens, including *Agrobacterium tumefaciens, Burkholderia cenocepacia, Clavibacter michiganensis, Erwinia carotovora, Erwinia chrysanthemi, Leifsonia xyli, Onion yellows phytoplasma, Pseudomonas syringae, Ralstonia solanacearum, Spiroplasma kunkelii, Xanthomonas axonopodis, Xanthomonas campestris* and *Xylella fastidiosa* provides a valuable resource to help combat these pathogens. The sequence data are crucial to our understanding of crop growth and development, as the sequence of the genes, or allelic variants of the sequenced genes, are responsible for almost all of the heritable differences between crop varieties and ecotypes. This information, often referred to as the genetic blueprint, is the foundation for all additional information from the genome to the phenome.

Genome and phenome

The genomics era started with the high throughput sequencing of expressed genes in the form of Expressed Sequence Tags (ESTs) (Adams et al., 1991). These ESTs are single-pass sequences of cloned cDNA fragments, each of which represents an expressed gene. While it is the objective of many crop researchers to access the complete sequence of their crop, EST sequencing remains the most cost-effective method for gene discovery for minor or orphan crops. Gene and genome sequence information in its raw state consists of a series of chromatograms with little direct biological significance. Only when these trace files are translated into the A, T, G, and C of the genetic code and subsequently to the amino acid code of expressed proteins do these sequences gain biological relevance.

Standard pipelines are now available for this translation of data (Ewing et al., 1998). However, the next stage, the assignment of potential biological function to these sequences, is still undergoing significant development. With some certainty, the genes that underlie the majority of heritable traits will have already been sequenced and are in gene and genomic databases. The challenge remains to associate traits with these genes. Several tools and pipelines are available for the prediction of biological function from gene and protein sequences. These are predominantly based on sequence identity or predicted structural identity with previously characterized genes. The implementation of standard vocabularies for gene annotation, such as Gene Ontology (Bada et al., 2004; Camon et al., 2005; Clark, Brooksbank, and Lomax, 2005; Lee et al., 2005; Lewis, 2005), has greatly enhanced the usability of these annotations. Further vocabularies for plant development, organs and phenotypes are being developed (The Plant Ontology Consortium, 2002), which will greatly enhance the computation of these complex data.

The challenge remains to link the information from the genome to the phenotype of inherited traits. Information that bridges this divide includes gene expression data, and proteome and metabolome data. Gene expression information is predominantly in the form of results from microarray experiments (Brazma et al., 2003; Craigon et al., 2004; Barrett et al., 2005; Parkinson et al., 2005; Shen et al., 2005; Tang, Shen, and Dickerson, 2005). However, large quantities of sequence-based gene expression data, such as SAGE (Velculescu et al., 1995; Matsumura, Nirasawa, and Terauchi, 1999; Gibbings et al., 2003; Lee and Lee, 2003; Fizames et al., 2004; Robinson et al., 2004b; Poroyko et al., 2005) and MPSS (Brenner et al., 2000a; Brenner et al., 2000b, Meyers et al., 2004a; Meyers et al., 2004b; Lu et al., 2005) are increasingly becoming available.

The application of proteomics and metabolomics to crop species is still relatively new; however, this information will become increasingly valuable, particularly where the metabolites analyzed are directly related to desired crop traits, for example, desirable flavor or aroma compounds. Lastly, the integration of phenotypic data is crucial for the exploitation of genomics for crop improvement. Methods have been developed for the high throughput capture and analysis of phenotypic information; however, the majority of phenotypic data will continue to require skilled crop scientists to evaluate the phenotypic assessment of germplasm under field conditions.

The implementation of genomic information for crop improvement requires an understanding of how the variation in the genetic material corresponds with observed heritable phenotypic differences. Gene and genome sequences, along with their associated information, provide an insight into the nature of the variation and how DNA polymorphisms are reflected in the phenotype (Powell, Machray, and Provan, 1996; Kashi, King, and Soller, 1997; Tautz, 1989; Gupta, Roy, and Prasad, 2001; Rafalski, 2002). Molecular genetic markers are key to understanding the association between the genotype and the phenotype; they can be applied to identify genes underlying traits and as tools for molecular breeding.

Genetic markers are one of the key tools for the genetic analysis of heritable traits. While the first generation of genetic markers was based on observed heritable traits, the majority of current markers are molecular, based on a direct measurement of genetic variation between individuals and populations. Several molecular genetic-marker systems have been developed and applied for crop research, including Amplified Fragment Length Polymorphisms (AFLP) (Vos et al., 1995), Simple Sequence Repeats (SSR) (Tautz, 1989; Powell, Machray, and

Provan, 1996; Kashi, King, and Soller, 1997; Tóth, Gáspári, and Jurka, 2000), and Single Nucleotide Polymorphisms (SNP) (Gupta, Roy, and Prasad, 2001; Rafalski, 2002). Each of these technologies has advantages and disadvantages relating to transferability between populations, license restrictions, cost of development, and high throughput implementation. Molecular genetic markers may be used to develop genetic maps, enabling the association of observed traits with specific regions of genomes. DNA sequence-based markers such as SSRs and SNPs, can provide a link from the genetic map to the physical sequence map. This enables the identification of the genes responsible for the observed trait. Variation within and around these genes in different individuals may then be associated with observed heritable phenotypic variation.

Knowledge of genetic variations associated with traits may then accelerate germplasm improvement through either molecular breeding using linked molecular genetic markers or through transgenic technologies if the underlying gene has been identified. Bioinformatics systems play a key role, from the discovery of molecular genetic markers through to their application and analysis of the resulting data, enabling the linking of the genome information with the observed phenome (Marth et al., 1999; Batley et al., 2003; Barker et al., 2003; Robinson et al., 2004a; Savage et al., 2005; Jewell et al., 2006).

Viewing and interrogating data

A wide variety of tools and databases have been developed to enable the viewing and integration of the multiple forms of biological data being generated in crop science, and in particular, through the high throughput technologies associated with the genomics revolution. These tools and databases vary from huge repositories, housing and integrating multiple data types for several species, to highly specific systems for the analysis of specific data types or individual species (Table 4.1). Early databases applied a variety of formats and ontologies that limited the transferability and integration of data from different sources. More recent systems have adopted broadly based standards for data, which permit the migration and integration of data from multiple sources, including archival data. There has also been a move toward web-based systems that are interrogated using a graphical interface. These systems are often more user-friendly than stand-alone systems that may require knowledge of specific commands to undertake interrogation or analysis. Examples of select databases and systems are listed in Table 4.1.

Current bioinformatics systems

Bioinformatics is still undergoing rapid evolution with new systems being developed and applied to a broad range of plant species and data types. At the same time, systems are coalescing as formats gradually become standardized, allowing the valuable comparison of data across species. DNA sequence data often forms the core of bioinformatics systems and the largest of the DNA sequence repositories is the International Nucleotide Sequence Database Collaboration (INSDC), made up of the DDBJ (DNA Data Bank of Japan) at The National Institute of Genetics in Mishima, Japan (Ohyanagi et al., 2006), GenBank at the National Center of Biotechnology Information in Bethesda, Maryland (Benson et al., 2006) and the EMBL Nucleotide Sequence Database, maintained at the European Bioinformatics Institute (EBI) in the United Kingdom (Cochrane et al., 2006). In addition, the databases at The Institute for Genomic Research (TIGR) based in Rockville, Maryland (Lee et al., 2005b) maintain various data types, including genomic sequence, annotation, and gene expression information. The UniProt

Table 4.1. A list of selected databases and systems, along with their Internet addresses and key references.

ArrayExpress	www.ebi.ac.uk/arrayexpress/	(Brazma et al. 2003; Parkinson et al. 2005)
AtEnsembl	www.ebi.ac.uk/arrayexpress/	(Love et al. 2006, Hubbard et al. 2005)
BarleyBase/PLEXdb	www.barleybase.org/	(Shen et al. 2005)
BASC	http://bioinformatics.pbcbasc.latrobe.edu.au/	(Love et al. 2006)
BGI-RIS Rice Information System	http://rice.genomics.org.cn	(Zhao et al. 2004)
EMBL	www.ebi.ac.uk/embl/	(Kanz et al. 2005)
Gene Ontology Annotation (GOA) database	www.ebi.ac.uk/GOA/	(Camon et al. 2005)
Graingenes	http://wheat.pw.usda.gov/GG2/index.shtml	(Matthews et al. 2003; Carollo et al. 2005)
Gramene	www.gramene.org	(Ware et al. 2002)
HarvEST	http://harvest.ucr.edu	
ICIS	w ww.icis.cgiar.org:8080/	(McLaren et al. 2005)
KEGG	www.genome.jp/kegg/	(Kanehisa et al. 2004)
Legume Information System (LIS)	www.comparative-legumes.org	(Gonzales et al. 2005)
Maize Genetics and Genomics Database (MaizeGDB)	www.maizegdb.org	(Lawrence et al. 2004)
MetaCyc, AraCyc	http://metacyc.org	(Zhang et al. 2005; Mueller, Zhang and Rhee 2003; Krieger et al. 2004a; Krieger et al. 2004b)
MIPS Plant Genome Information Resources (PlantsDB)	http://mips.gsf.de	(Mewes et al. 2004)
NASCArrays	http://affymetrix.arabidopsis.info	(Craigon et al. 2004)
NCBI Gene expression Omnibus (GEO)	www.ncbi.nlm.nih.gov/geo/	(Barrett et al. 2005)
National Center for Biotechnology Information (NCBI)	www.ncbi.nih.gov	
PlantGDB	www.plantgdb.org	(Dong et al. 2004; Dong et al. 2005)
SNPServer	http://hornbill.cspp.latrobe.edu.au/snpdiscovery.html	(Savage et al. 2005)
SSR Taxonomy Tree	http://bioinformatics.pbcbasc.latrobe.edu.au/cgi-bin/ ssr_taxonomy_browser.cgi	(Jewell et al. 2006)
SwissProt	www.ebi.ac.uk/swissprot	(Boeckmann et al. 2003)
The Arabidopsis Information Resource (TAIR)	www.arabidopsis.org	(Garcia-Hernandez et al. 2002; Huala et al. 2001; Weems et al. 2004; Rhee et al. 2003; Reiser and Rhee, 2005)
The Institute for Genomic Research (TIGR)	www.tigr.org	(Lee et al. 2005b)
UniProt Knowledgebase (UniProtKB)	www.uniprot.org	(Bairoch et al. 2005)

consortium maintains the largest protein sequence database (Bairoch et al., 2005), which developed in 2002 as a collaboration between the Swiss Institute of Bioinformatics (SIB), the EBI, and the Protein Information Resource (PIR). The UniProt Knowledgebase (UniProtKB) consists of two sections: UniProtKB/Swiss-Prot, hosting manually annotated entries; and UniProtKB/TrEMBL, hosting annotation of CoDing Sequences (CDS) extracted from the EMBL database (Boeckmann et al., 2003). The Munich Institute for Protein Sequences (MIPS), Germany, hosts another comprehensive database resource and maintains genome databases for *Arabidopsis thaliana* (MatDB), *Oryza sativa* (rice, MosDB), *Medicago truncatula* (UrMeLDB), *Lotus japonicus, Zea mays,* and *Solanum lycopersicum* (tomato) (Mewes et al., 2004).

Arabidopsis was the first plant to be fully sequenced and remains the model species for dicotyledonous crops. There has been a huge quantity of research on this plant and there have been many database systems developed to host and integrate the resulting data. The Arabidopsis Information Resource (TAIR) is a collaborative project between the Carnegie Institution, Washington; the Department of Plant Biology, Stanford University, California; and the National Center for Genome Resources (NCGR), Sante Fe, New Mexico, which provides an extensive web-based resource for *Arabidopsis thaliana* (Huala et al., 2001; Rhee et al., 2003; Weems et al., 2004; Reiser and Rhee, 2005). Data include genetic mapping, protein sequence, gene expression, and community data within a relational database. Several tools are available for viewing and analyzing these data, including SeqViewer (for the visualization of the genome sequence and associated annotations) and AraCyc, a database of *Arabidopsis* biochemical pathways with a graphical overview (Mueller, Zhang, and Rhee, 2003; Krieger et al., 2004a; Krieger et al., 2004b; Zhang et al., 2005).

The Nottingham Arabidopsis Stock Centre (NASC) has developed AtEnsembl (Love et al., 2006) using the Ensembl browser system (Hubbard et al., 2005) to provide a genome-based view on integrated data. The resource provides a broad range of Ensembl features, including gene and protein information, links to Affymetrix gene expression data, pointers to germplasm, and extensive data download capabilities.

While *Arabidopsis* is well supported by bioinformatics systems, bioinformatics support for other dicotyledonous crops is varied. Several systems have been developed for *Brassicas*, which share a significant amount of genomic similarity with *Arabidopsis* (Love et al., 2004; Love et al., 2005). These systems include comparative genetic and genomic viewers and tools for the transfer of the comprehensive information from *Arabidopsis* for *Brassica* crop improvement (Beckett et al., 2005; Love et al., 2006).

Leguminous crops are particularly well supported by the Legume Information System (LIS) (Gonzales et al., 2005). The LIS has been developed to enable utilization of genomic information from the model legume *Medicago truncatula* (barrel medic) for the improvement of major crops, such as *Glycine max* (soybean). The LIS includes genetic and physical maps as well as annotated expressed sequences for several legume species, enabling the comparison of quantitative trait loci (QTL), genetic, and sequenced genetic loci.

The cereal community is well supported by bioinformatics systems that aim to transfer the basic science of the "omics" technologies to delivering better varieties of the world's major food crops.

Rice is the only fully sequenced crop plant, with the sequence of two related varieties now publicly available. The Beijing Genome Institute (BGI) has established and updated a Rice Information System (BGI-RIS) that integrates information and resources for the comparative analysis of rice genomes (Zhao et al., 2004). The BGI-RIS combines the genomic data of *Oryza sativa* L. ssp. *indica* (by BGI) with *Oryza sativa* L. ssp. *japonica,* along with annotation, genetic markers, expressed genes, repetitive elements, and genomic polymorphisms, using graphical interfaces.

MaizeGDB (Lawrence et al., 2004) is a relatively new database system that combines information from the original MaizeDB and ZmDB repositories (Dong et al., 2003), with sequence data from PlantGDB (Dong et al., 2004; Dong et al., 2005). The system maintains information on maize genomic and gene sequences, genetic markers, literature references, as well as contact information for the maize research community. In contrast to many systems, MaizeGDB also hosts manually curated information from primary literature, bulletin boards, and many links related to maize research.

GrainGenes is a large collaborative project for scientists working in small-grain research (Matthews et al., 2003; Carollo et al., 2005) including pathologists, molecular biologists, geneticists, and small-grain breeders; its aim is to assist in the development of improved crop varieties. Information, which has been produced across several decades, is maintained for barley, wheat, rye, oat, and related wild species. Grain-Genes integrates genetic data for Triticeae and *Avena.* Information includes genetic markers, map locations, alleles, presence of alleles in different cultivars, key references, and disease symptoms.

The HarvEST system is an EST database and viewing software initially developed for cereals and now supporting several species, including barley, *Brachypodium* sp., citrus, *Coffea* sp., cowpea, rice, soybean, and wheat. HarvEST supports microarray content design, gene function annotation, and interfacing with physical and genetic maps. One feature of this system is that it can be run on a stand-alone laptop with a user-friendly interface and without the need for Internet

connectivity or a significant amount of computing power.

As well as species-based systems, several bioinformatics systems have been developed around specific data types. BarleyBase and the more recent PLEXdb (Plant Expression Database) (Shen et al., 2005) are systems for analyzing gene expression in plants and plant pathogens. The PLEXdb provides a web interface, integrating microarray data from several plant species, enabling the comparative analysis of gene expression and providing an insight into the agronomic importance of different genes.

Several tools have been developed for the discovery of molecular genetic markers from the abundance of DNA sequence data. These tools provide a rich source of markers, which can then be applied to genetic trait mapping and marker-assisted selection. The SSRPrimer is a web-based tool that enables the real-time discovery of SSRs within submitted DNA sequences, with the concomitant design of PCR (Polymerase Chain Reaction) primers for SSR amplification (Robinson et al., 2004a). Alternatively, users may browse an SSR Taxonomy Tree (Jewell et al., 2006) to identify predetermined SSR amplification primers for any species represented within the GenBank database. This system currently hosts almost 14 million SSR primer pairs for organisms from simple viruses and bacteria to plants and animals. The SNPServer is an online tool for the discovery of SNPs from DNA sequence data (Savage et al., 2005). Following submission of a sequence of interest, SNPServer uses BLAST (Altschul et al., 1990) to identify similar sequences, CAP3 (Huang and Madan, 1999) to cluster and assemble these sequences, and then the SNP discovery software autoSNP (Batley et al., 2003; Barker et al., 2003) to detect SNPs and insertion/deletion (indel) polymorphisms. Each of these tools increases the availability of these molecular markers for genetic studies and allows researchers to target molecular markers within specific candidate genes.

While bioinformatics has grown with the expansion of genomic and gene expression data, there still remain relatively few resources that manage the information that has the most relevance to crop improvement, in particular, phenotypic data relating to how crops perform in the field. This information is essential for the integration of agronomic data with the wealth of related genomic information that is being produced (Edwards and Batley, 2004). The International Crop Information System (ICIS) is arguably the most advanced database of this type and enables breeders to load and store information pertaining to genealogy and phenotype (McLaren et al., 2005). The ICIS was initially developed by the International Rice Research Institute's (IRRI) bioinformatics and biometrics unit and is widely used for rice and wheat and is being extended to include information on legumes. Predominantly, the International Centre for Agricultural Research in the Dry Areas (ICARDA), International Centre for Research in the Semi-Arid Tropics (ICRISAT), as well as the United States Department of Agriculture (USDA) provide data. Databases may be queried across multiple datasets to identify varieties that have agronomic qualities, such as disease resistance, yield, and response to stress.

With the development of metabolomic technologies and the increased interest in molecular pharming, there has been an expansion in the application of bioinformatics tools for the analysis of this information and its integration with genetic and genomic data. Kyoto Encyclopedia of Genes and Genomes (KEGG) is a key bioinformatics resource for the integration of biochemical pathways with genetic and genomic information (Kanehisa et al., 2004). The KEGG database enables the visualization of complex cellular pathways, such as metabolism, signal transduction and the cell cycle, fre-

quently with graphical pathway diagrams. Large-scale metabolite studies require their own specific systems for data analysis and integration. While this field of research is still in relative infancy, several tools have already been developed and the area is likely to expand greatly in the coming years (Fiehn, 2002; Sumner, Mendes, and Dixon, 2003).

One of the principle challenges in developing bioinformatics systems and integrating diverse biological data is the development of an agreed vocabulary of terms. This is essential for the computation of information and the transfer of information from one species or group to another. Several consortia are developing vocabularies or ontologies for different data types, with Gene Ontology (GO) probably being the most advanced (Bada et al., 2004; Camon et al., 2005; Clark, Brooksbank, and Lomax, 2005; Lee et al., 2005). The GO is a structured vocabulary for gene and protein annotation, which is designed to replace the various different nomenclatures used by specific databases, enabling integration of data between systems. The GO is now recognized as the standard vocabulary for the scientific community to describe the function and subcellular location of gene products. The GO Consortium (GOC) was founded in 1998 and now includes the TAIR (Huala et al., 2001; Garcia-Hernandez et al., 2002; Rhee et al., 2003), TIGR (Lee et al., 2005b; Quackenbush et al., 2000), Gramene (Ware et al., 2002), and UniProtKB (Boeckmann et al., 2003; Bairoch et al., 2005) groups, which undertake GO annotation for plant species. Other groups, including the Plant Ontology Consortium (The Plant Ontology Consortium, 2002), are developing vocabularies to describe attributes, such as phenotypes, structures, or developmental stages, and the maturation of these ontologies. Their application within bioinformatics analysis will enable the robust integration of a broader range of information than is currently possible.

Conclusions and future outlook

Although plant genomics is a relatively new field of study, it is already contributing toward crop improvement, assisted by developments in bioinformatics, which enable the translation of vast quantities of data into practical tools for crop improvement. The genomics revolution continues to grow at an ever-increasing rate and is now being complemented by post-genomic technologies. At the same time, the timeframe from discovery to practical delivery and germplasm enhancement continues to decrease. These factors suggest that genomics and post-genomics technologies, supported by advances in applied bioinformatics, will drive the future of crop improvement, facilitating the goals of increased, secure food production with a reduction in the environmental footprint of agriculture.

References

Adams, M.D., Kelley, J.M., Gocayne, J.D., Dubnick, M., Polymeropoulos, M.H., Xiao, H., Merril, C.R., Wu, A., Olde, B., Moreno, R.F., et al. 1991. Complementary DNA sequencing: Expressed sequence tags and human genome project. *Science* 21, 252(5013):1651–1656.

Altschul, S.F., Gish, W., Miller, W., Myers, E.W. and Lipman, D.J. 1990. Basic local alignment search tool. *J. Mol. Biol.* 215:403–410.

Bada, M., Stevens, R., Goble, C., Gil, Y., Ashburner, M., Blake, J.A., Cherry, J.M., Harris, M. and Lewis, S. 2004. A short study on the success of the Gene Ontology. *J. Web. Semantics.* 1:235–240.

Bairoch, A., Apweiler, R., Wu, C.H., Barker, W.C., Boeckmann, B., Ferro, S., Gasteiger, E., Huang, H., Lopez, R., Magrane, M., Martin, M.J., Natale, D.A., O'Donovan, C., Redaschi, N. and Yeh, L.S. 2005. The Universal Protein Resource (UniProt). *Nucleic Acids Res.* 33(Database issue):D154–D159.

Barker, G., Batley, J., O'Sullivan, H., Edwards, K.J. and Edwards, D. 2003. Redundancy-based detection of sequence polymorphisms in expressed sequence tag data using autoSNP. *Bioinformatics* 19:421–424.

Barrett, T., Suzek, T.O., Troup, D.B., Wilhite, S.E., Ngau, W.C., Ledoux, P., Rudnev, D., Lash, A.E., Fujibuchi, W. and Edgar, R. 2005. NCBI GEO:

Mining millions of expression profiles—database and tools. *Nucleic Acids Res.* 33(1):D562–D566.

Batley, J., Barker, G., O'Sullivan, H., Edwards, K.J. and Edwards, D. 2003. Mining for single nucleotide polymorphisms and insertions/deletions in maize expressed sequence tag data. *Plant Physiology* 132:84–91.

Beckett, P., Bancroft, I., and Trick, M. 2005. Computational tools for Brassica–Arabidopsis comparative genomics. Comparative and Functional Genomics 6(3):147–152.

Benson, D.A., Karsch-Mizrachi, I., Lipman, D.J., Ostell, J. and Wheeler, D.L. 2006. GenBank. *Nucleic Acids Res.* 34:D16–D20.

Boeckmann, B., Bairoch, A., Apweiler, R., Blatter, M.C., Estreicher, A., Gasteiger, E., Martin, M.J., Michoud, K., O'Donovan, C., Phan, I., Pilbout, S. and Schneider, M. 2003. The SWISS–PROT protein knowledgebase and its supplement TrEMBL in 2003. *Nucleic Acids Res.* 31:365–370.

Brazma, A., Parkinson, H., Sarkans, U., Shojatalab, M., Vilo, J., Abeygunawardena, N., Holloway, E., Kapushesky, M., Kemmeren, P., Lara, G.G., et al. 2003. ArrayExpress–a public repository for microarray gene expression data at the EBI. *Nucleic Acids Res.* 31(1):68–71.

Brenner, S., Johnson, M., Bridgham, J., Golda, G., Lloyd, D.H., Johnson, D., et al. 2000a. Gene expression analysis by massively parallel signature sequencing (MPSS) on microbead arrays. *Nat. Biotechnol.* 18:630–634.

Brenner, S., Williams, S.R., Vermaas, E.H., Storck, T., Moon, K., McCollum, C., et al. 2000b. In vitro cloning of complex mixtures of DNA on microbeads: Physical separation of differentially expressed cDNAs. *Proc. Natl. Acad. Sci. USA* 97:1665–1670.

Camon, E., Magrane, M., Barrell, D., Lee, V., Dimmer, E., Maslen, J., Binns, D., Harte, N., Lopez, R. and Apweiler, R. 2005. The Gene Ontology Annotation (GOA) Database: Sharing knowledge in UniProt with Gene Ontology. *Nucleic Acids Res.* 32:D262–D266.

Carollo, V., Matthews, D.E., Lazo, G.R., Blake, T.K., Hummel, D.D., Lui, N., Hane, D.L. and Anderson, O.D. 2005. GrainGenes 2.0. An Improved Resource for the Small–Grains Community. *Plant Physiologist* 139:643–651.

Clark, J.I., Brooksbank, C. and Lomax, J. 2005. It's All GO for Plant Scientists. *Plant Physiologist* 138:1268–1278.

Cochrane, G., Aldebert, P., Althorpe, N., Andersson, M., Baker, W., Baldwin, A., Bates, K., Bhattacharyya, S., Browne, P., van den Broek, A., Castro, M., Duggan, K., Eberhardt, R., Faruque, N., Gamble, J., Kanz, C., Kulikova, T., Lee, C., Leinonen, R., Lin, Q., Lombard, V., Lopez, R., McHale, M., McWilliam, H., Mukherjee, G., Nardone, F., Pastor, M.P.,

Sobhany, S., Stoehr, P., Tzouvara, K., Vaughan, R., Wu, D., Zhu, W. and Apweiler, R. 2006. EMBL nucleotide sequence database: Developments in 2005. *Nucleic Acids Res.* 34:D10–D15.

Coemans, B., Matsumura, H., Terauchi, R., Remy, S., Swennen, R. and Sági, L. 2005. SuperSAGE combined with PCR walking allows global gene expression profiling of banana (*Musa acuminata*), a non-model organism. *Theor. Appl. Genet.* 111: 1118–1126.

Craigon, D.J., James, N., Okyere, J., Higgins, J., Jotham, J. and May, S. 2004. NASCArrays: A repository for microarray data generated by NASC's transcriptomics service. *Nucleic Acids Res.* 32(1): D575–D577.

Dong, Q., Roy, L., Freeling, M., Walbot, V. and Brendel, V. 2003. ZmDB, an integrated database for maize genome research. *Nucleic Acids Res.* 31:244–247.

Dong, Q., Schlueter, S.D. and Brendel, V. 2004. Plant-GDB, plant genome database and analysis tools. *Nucleic Acids Res.* 32(1):D354–D359.

Dong, Q., Lawrence, C.J., Schlueter, S.D., Wilkerson, M.D., Kurtz, S., Lushbough, C. and Brendel, V. 2005. Comparative plant genomics resources at plantgdb. *Plant Physiologist* 139:610–618.

Edwards, D. and Batley, J. 2004. Plant bioinformatics: From genome to phenome. *Trends in Biotechnology* 22(5):232–237.

Ewing, B., Hillier, L., Wendl, M.C. and Green, P. 1998. Base-calling of automated sequencer traces using phred. I. accuracy assessment. *Genome Research* 8:175–185.

Fiehn, O. 2002. Metabolomics—the link between genotypes and phenotypes. *Plant Mol. Biol.* 48: 155–171.

Fiehn, O., Kopka, J., Dormann, P., Altmann, T., Trethewey, R. and Willmitzer, L. 2000. Metabolite profiling for plant functional genomics. *Nat. Biotechnol.* 18:1157–1161.

Fizames, C., Munos, S., Cazettes, C., Nacry, P., Boucherez, J., Gaymard, F., Piquemal, D., Delorme, V., Commes, T., Doumas, P., Cooke, R., Marti, J., Sentenac, H. and Gojon, A. 2004. The *Arabidopsis* root transcriptome by serial analysis of gene expression. Gene identification using the genome sequence. *Plant Physiologist* 134:67–80.

Garcia-Hernandez, M., Berardini, T.Z., Chen, G., Crist, D., Doyle, A., Huala, E., Knee, E., Lambrecht, M., Miller, N., Mueller, L.A., Mundodi, S., Reiser, L., Rhee, S.Y., Scholl, R., Tacklind, J., Weems, D.C., Wu, Y., Xu, I., Yoo, D., Yoon, J. and Zhang, P. 2002. TAIR: A resource for integrated *Arabidopsis* data. *Functional and Integrative Genomics* 2:239–253.

Gibbings, J.G., Cook, B.P., Dufault, M.R., Madden, S.L., Khuri, S., Turnbull, C.J. and Dunwell, J.M. 2003. Global transcript analysis of rice leaf and seed using SAGE technology. *Plant Biotechnol. J.* 1:271–285.

Gonzales, M.D., Archuleta, E., Farmer, A., Gajendran, K., Grant, D., Shoemaker, R., Beavis, W.D. and Waugh, M.E. 2005. The Legume Information System (LIS): An integrated information resource for comparative legume biology. *Nucleic Acids Res.* 33:D660–D665.

Gupta, P.K., Roy, J.K. and Prasad, M. 2001. Single nucleotide polymorphisms: A new paradigm for molecular marker technology and DNA polymorphism detection with emphasis on their use in plants. *Curr. Sci.* 80:524–535.

Huala, E., Dickerman, A., Garcia-Hernandez, M., Weems, D., Reiser, L., LaFond, F., Hanley, D., Kiphart, D., Zhuang, J., Huang, W., Mueller, L., Bhattacharyya, D., Bhaya, D., Sobral, B., Beavis, B., Somerville, C. and Rhee, S.Y. 2001. The *Arabidopsis* Information Resource (TAIR): A comprehensive database and web-based information retrieval, analysis, and visualization system for a model plant. *Nucleic Acids Res.* 29:102–105.

Huang, X. and Madan, A. 1999. CAP3: A DNA sequence assembly program. *Genome Res.* 9:868–877.

Hubbard, T., Andrews, D., Caccamo, M., Cameron, G., Chen, Y., Clamp, M., Clarke, L., Coates, G., Cox, T., Cunningham, F., Curwen, V., Cutts, T., Down, T., Durbin, R., Fernandez-Suarez, X.M., Gilbert, J., Hammond, M., Herrero, J., Hotz, H., Howe, K., Iyer, V., Jekosch, K., Kahari, A., Kasprzyk, A., Keefe, D., Keenan, S., Kokocinsci, F., London, D., Longden, I., McVicker, G., Melsopp, C., Meidl, P., Potter, S., Proctor, G., Rae, M., Rios, D., Schuster, M., Searle, S., Severin, J., Slater, G., Smedley, D., Smith, J., Spooner, W., Stabenau, A., Stalker, J., Storey, R., Trevanion, S., Ureta-Vidal, A., Vogel, J., White, S., Woodwark, C. and Birney, E. 2005. Ensembl 2005. *Nucleic Acids Res.* 33:D447–D453.

Jackson, S., Rounsley, S. and Purugganan, M. 2006. Comparative sequencing of plant genomes: Choices to make. *Plant Cell* 18(5):1100–1104.

Jewell, E., Robinson, A., Savage, D., Erwin, T., Love, C.G., Lim, G.A.C., Li, X., Batley, J., Spangenberg, G.C. and Edwards, D. 2006. SSR Primer and SSR Taxonomy Tree: Biome SSR discovery. *Nucleic Acids Res.* (in press).

Kanehisa, M., Goto, S., Kawashima, S., Okuno, Y. and Hattori, M. 2004. The KEGG resource for deciphering the genome. *Nucleic Acids Res.* 32:D277–D280.

Kanz, C., Aldebert, P., Althorpe, N., Baker, W., Baldwin, A., Bates, K., Browne, P., van den Broek, A., Castro, M., Cochrane, G., Duggan, K., Eberhardt, R., Faruque, N., Gamble, J., Garcia Diez, F., Harte, N., Kulikova, T., Lin, Q., Lombard, V., Lopez, R., Mancuso, R., McHale, M., Nardone, F., Silventoinen,

V., Sobhany, S., Stoehr, P., Tuli, M.A., Tzouvara, K., Vaughan, R., Wu, D., Zhu, W. and Apweiler, R. 2005. The EMBL Nucleotide Sequence Database. *Nucleic Acids Res.* 33:D29–D33.

Kashi, Y., King, D. and Soller, M. 1997. Simple sequence repeats as a source of quantitative genetic variation. *Trends Genet* 13:74–78.

Krieger, C.J., Zhang, P., Mueller, L.A., Wang, A., Paley, S., Arnaud, M., Pick, J., Rhee, S.Y. and Karp, P.D. 2004a. MetaCyc: A multiorgansim database of metabolic pathways and enzymes. *Nucleic Acids Res.* 23:D438–D442.

Krieger, C.J., Zhang, P., Mueller, L., Wang, A., Paley, S., Arnaud, M., Pick, J., Rhee, S.Y. and Karp, P. 2004b. MetaCyc: Recent enhancements to a database of metabolic pathways and enzymes in microorganisms and plants. *Nucleic Acids Res.* 32(database issue):D438–D442.

Lawrence, C.J., Dong, Q., Polacco, M.L., Seigfried, T.E. and Brendel, V. 2004. Maizegdb, the community database for maize genetics and genomics. *Nucleic Acids Res.* 32:D393–D397.

Lee, J.Y. and Lee, D.H. 2003. Use of serial analysis of gene expression technology to reveal changes in gene expression in *Arabidopsis* pollen undergoing cold stress. *Plant Physiologist* 132:517–529.

Lee, V., Camon, E., Dimmer, E., Barrell, D. and Apweiler, R. 2005a. Who tangos with GOA?—Use of Gene Ontology Annotation (GOA) for biological interpretation of "-omics" data and for validation of automatic annotation tools. *In Silico Biol.* 5:5–8.

Lee, Y., Tsai, J., Sunkara, S., Karamycheva, S., Pertea, G., Sultana, R., Antonescu, V., Chan, A., Cheung, F. and Quackenbush, J. 2005b. The TIGR Gene Indices: Clustering and assembling EST and known genes and integration with eukaryotic genomes. *Nucleic Acids Res.* 33:D71–D74.

Lewis, S.E. 2005. Gene Ontology: Looking backwards and forwards. *Genome Biol.* 6:103.

Li, W., Li, J., Liu, Z., Li, L., Liu, J., Qi, Q., Liu, J., Li, L., Li, T., Wang, X., Lu, H., Wu, T., Zhu, M., Ni, P., Han, H., Dong, W., Ren, X., Feng, X., Cui, P., Li, X., Wang, H., Xu, X., Zhai, W., Xu, Z., Zhang, J., He, S., Zhang, J., Xu, J., Zhang, K., Zheng, X., Dong, J., Zeng, W., Tao, L., Ye, J., Tan, J., Ren, X., Chen, X., He, J., Liu, D., Tian, W., Tian, C., Xia, H., Bao, Q., Li, G., Gao, H., Cao, T., Wang, J., Zhao, W., Li, P., Chen, W., Wang, X., Zhang, Y., Hu, J., Wang, J., Liu, S., Yang, J., Zhang, G., Xiong, Y., Li, Z., Mao, L., Zhou, C., Zhu, Z., Chen, R., Hao, B., Zheng, W., Chen, S., Guo, W., Li, G., Liu, S., Tao, M., Wang, J., Zhu, L., Yuan, L. and Yang, H. 2002. A draft sequence of the rice genome (*Oryza sativa* L. ssp. *indica*). *Science* 296:79–92.

Lomax, J. 2005. Get ready to GO! A biologist's guide to the Gene Ontology. *Brief Bioinform* 6:298–304.

Lorenz, W.W. and Dean, J.F. 2002. SAGE profiling and demonstration of differential gene expression along the axial developmental gradient of lignifying xylem in loblolly pine (*Pinus taeda*). *Tree Physiol.* 22:301–310.

Love, C.G., Batley, J., Lim, G., Robinson, A.J.R., Savage, D., Singh, D., Spangenberg, G.C. and Edwards, D. 2004. New computational tools for Brassica genome research. *Comparative and Functional Genomics* 5(3):276–280.

Love, C.G., Robinson, A.J., Lim, G.A.C., Hopkins, C.J., Bately, J., Barker, G., Spangenberg, G.C. and Edwards, D. 2005. *Brassica* ASTRA: An integrated database for *Brassica* genome research. *Nucleic Acids Res.* 33:D656–D659.

Love, C., Logan, E., Erwin, T., Kaur, J., Lim, G.A.C., Hopkins, C., Batley, J., James, N., May, S., Spangenberg, G. and Edwards, D. 2006. Integrating and Interrogating Diverse *Brassica* Data within an EnsEMBL Structured Database. *Acta Horticulturae* 706:77–82.

Lu, C., Tej, S.S., Luo, S., Haudenschild, C.D., Meyers, B.C. and Green, P.J. 2005. Elucidation of the Small RNA Component of the Transcriptome. *Science* 309:1567–1569.

Marth, G.T., Korf, I., Yandell, M.D., Yeh, R.T., Gu, Z.J., Zakeri, H., Stitziel, N.O., Hillier, L., Kwok, P.Y. and Gish, W.R. 1999. A general approach to single nucleotide polymorphism discovery. *Nat. Genet.* 23:452–456.

Matthews, D.E., Carollo, V.L., Lazo, G.R. and Anderson, O.D. 2003. GrainGenes, the genome database for small-grain crops. *Nucleic Acids Res.* 31:183–186.

Matsumura, H., Reich, S., Ito, A., Saitoh, H., Kamoun, S., Winter, P., Kahl, G., Reuter, M., Kruger, D.H. and Terauchi, R. 2003. Gene expression analysis of plant host–pathogen interactions by SuperSAGE. *Proc. Natl. Acad. Sci. USA* 100:15718–15723.

Matsumura, H., Nirasawa, S. and Terauchi, R. 1999. Technical advance, transcript profiling in rice (*Oryza sativa* L.) seedlings using serial analysis of gene expression (SAGE). *Plant J.* 20:719–726.

McClintock, B. 1950. The origin and behavior of mutable loci in maize. *Proceedings of the National Academy of Sciences* 36:344–355.

McLaren, C.G., Bruskiewich, R.M., Portugal, A.M. and Cosico, A.B. 2005. The international rice information system. A platform for meta-analysis of rice crop data. *Plant Physiology* 139(2):637–642.

Mewes, H.W., Amid, C., Arnold, R., Frishman, D., Güldener, U., Mannhaupt, G., Münsterkötter, M., Pagel, P., Strack, N., Stümpflen, V., Warfsmann, J. and Ruepp, A. 2004. MIPS: Analysis and annotation of proteins from whole genomes. *Nucleic Acids Res.* 32:D41–D44.

Meyers, B.C., Tej, S.S., Vu, T.H., Haudenschild, C.D., Agrawal, V., Edberg, S.B., et al. 2004a. The use of

MPSS for whole-genome transcriptional analysis in Arabidopsis. *Genome Res.* 14:1641–1653.

Meyers, B.C., Vu, T.H., Tej, S.S., Ghazal, H., Matvienko, M., Agrawal, V., et al. 2004b. Analysis of the transcriptional complexity of Arabidopsis thaliana by massively parallel signature sequencing. *Nat. Biotechnol.* 22:1006–1011.

Mueller, L.A., Zhang, P. and Rhee, S.Y. 2003. AraCyc. A Biochemical Pathway Database for *Arabidopsis. Plant Physiology* 132:453–460.

Ohyanagi, H., Tanaka, T., Sakai, H., Shigemoto, Y., Yamaguchi, K., Habara, T., Fujii, Y., Antonio, B. A., Nagamura, Y., Imanishi, T., Ikeo, K., Itoh, T., Gojobori, T. and Sasaki, T. 2006. The rice annotation project database (rap-db): Hub for *Oryza Sativa* ssp. japonica genome information. *Nucleic Acids Res.* 34:D741–D744.

Parkinson, H., Sarkans, U., Shojatalab, M., Abeygunawardena, N., Contrino, S., Coulson, R., Farne, A., Garcia Lara, G., Holloway, E., Kapushesky, M., et al. 2005. ArrayExpress—a public repository for microarray gene expression data at the EBI. *Nucleic Acids Res.* 33(1):D553–D555.

Poroyko, V., Hejlek, L.G., Spollen, W.G., Springer, G.K., Nguyen, H.T., Sharp, R.E. and Bohnert, H.J. 2005. The maize root transcriptome by serial analysis of gene expression. *Plant Physiologist* 138:1700–1710.

Powell, W., Machray, G.C. and Provan, J. 1996. Polymorphism revealed by simple sequence repeats. *Trends Plant Sci.* 1:215–222.

Quackenbush, J., Liang, F., Holt, I., Pertea, G. and Upton, J. 2000. The TIGR Gene Indices: Reconstruction and representation of expressed gene sequences. *Nucleic Acids Res.* 28:141–145.

Rafalski, A. 2002. Applications of single nucleotide polymorphisms in crop genetics. *Curr. Opin. Plant Biol.* 5:94–100.

Reiser, L. and Rhee, S.Y. 2005. Using the *Arabidopsis* Information Resource (TAIR) to find information about *Arabidopsis* Genes. In: *Current Protocols in Bioinformatics* (Baxevanis, A.D., et al., eds.). New York: John Wiley and Sons, 1.11.1–1.11.45.

Rhee, S.Y., Beavis, W., Berardini, T.Z., Chen, G., Dixon, D., Doyle, A., Garcia-Hernandez, M., Huala, E., Lander, G., Montoya, M., Miller, N., Mueller, L.A., Mundodi, S., Reiser, L., Tacklind, J., Weems, D.C., Wu, Y., Xu, I., Yoo, D., Yoon, J. and Zhang, P. 2003. The *Arabidopsis* Information Resource (TAIR): A model organism database providing a centralized, curated gateway to *Arabidopsis* biology, research materials and community. *Nucleic Acids Res.* 31:224–228.

Robinson, A.J., Love, C.G., Batley, J., Barker, G. and Edwards, D. 2004a. Simple sequence repeat marker loci discovery using SSRPrimer. *Bioinfomatics,* 20:1475–1476.

Robinson, S.J., Cram, D.J., Lewis, C.T. and Parkin,

I.A.P. 2004b. Maximizing the efficacy of SAGE analysis identifies novel transcripts in *Arabidopsis. Plant Physiologist* 136:3223–3233.

Sasaki, T., Matsumoto, T., Yamamoto, K., Sakata, K., Baba, T., Katayose, Y., Wu, J., Niimura, Y., Cheng, Z., Nagamura, Y., Antonio, B.A., Kanamori, H., Hosokawa, S., Masukawa, M., Arikawa, K., Chiden, Y., Hayashi, M., Okamoto, M., Ando, T., Aoki, H., Arita, K., Hamada, M., Harada, C., Hijishita, S., Honda, M., Ichikawa, Y., Idonuma, A., Iijima, M., Ikeda, M., Ikeno, M., Ito, S., Ito, T., Ito, Y., Ito, Y., Iwabuchi, A., Kamiya, K., Karasawa, W., Katagiri, S., Kikuta, A., Kobayashi, N., Kono, I., Machita, K., Maehara, T., Mizuno, H., Mizubayashi, T., Mukai, Y., Nagasaki, H., Nakashima, M., Nakama, Y., Nakamichi, Y., Nakamura, M., Namiki, N., Negishi, M., Ohta, I., Ono, N., Saji, S., Sakai, K., Shibata, M., Shimokawa, T., Shomura, A., Song, J., Takazaki, Y., Terasawa, K., Tsuji, K., Waki, K., Yamagata, H., Yamane, H., Yoshiki, S., Yoshihara, R., Yukawa, K., Zhong, H., Iwama, H., Endo, T., Ito, H., Hahn, J.H., Kim, H.I., Eun, M.Y., Yano, M., Jiang, J. and Gojobori, T. 2002. The genome sequence and structure of rice chromosome 1. *Nature* 420:312–316.

Savage, D., Batley, J., Erwin, T., Logan, E., Love, C. G., Lim, G.A.C., Mongin, E., Barker, G., Spangenberg, G.C. and Edwards, D. 2005. SNPServer: A real-time SNP discovery tool. *Nucleic Acids Res.* 33:W493–W495.

Shen, L., Gong, J., Caldo, R.A., Nettleton, D., Cook, D., Wise, R.P. and Dickerson, J.A. 2005. BarleyBase—an expression profiling database for plant genomics. *Nucleic Acids Res.* 33(1):D614–D618.

Sumner, L., Mendes, P. and Dixon, R. 2003. Plant metabolomics: Large-scale phytochemistry in the functional genomics era. *Phytochemistry* 62:817–836.

Tang, X., Shen, L. and Dickerson, J.A. 2005. BarleyExpress: A web-based submission tool for enriched microarray database annotations. *Bioinformatics* 21(3):399–401.

Tautz, D. 1989. Hypervariability of simple sequences as a general source for polymorphic DNA markers. *Nucleic Acids Res.* 17:463–6471.

Tóth, G., Gáspári, Z. and Jurka, J. 2000. Microastellites in different eukaryotic genomes: Survey and analysis. *Genome Res.* 10:967–981.

The Arabidopsis Genome Initiative. 2000. Analysis of the genome sequence of the flowering plant *Arabidopsis thaliana. Nature* 408:796–815.

The Plant Ontology Consortium. 2002. The Plant Ontology™ Consortium and Plant Ontologies. *Comparative and Functional Genomics* 3(2):137–142.

Trethewey, R.N., Krotzky, A.J. and Willmitzer, L. 1999. Metabolic profiling: A rosetta stone for genomics? *Current Opinion in Biotechnology* 2:83–85.

Velculescu, V.E., Zhang, L., Vogelstein, B. and Kinzler, K.W. 1995. Serial analysis of gene expression. *Science* 270:484–487.

Vos, P., Hogers, R., Bleeker, M., Reijans, M., van de Lee, T., Hornes, M., Freiters, A., Pot, J., Peleman, J., Kuiper, M. and Zabeau, M. 1995. *Nucleic Acid Res.* 23:4407–4414.

Ware, D., Jaiswal, P., Ni, J., Pan, X., Chang, K., Clark, K., Teytelman, L., Schmidt, S., Zhao, W., Cartinhour, S., et al. 2002. Gramene: A resource for comparative grass genomics. *Nucleic Acids. Res.* 30(1):103–105.

Weems, D., Miller, N., Garcia-Hernandez, M., Huala, E. and Rhee, S.Y. 2004. Design, implementation, and maintenance of a model organism database for *Arabidopsis thaliana. Comparative and Functional Genomics* 5:362–369.

Zhang, P., Foerster, H., Tissier, C., Mueller, L., Paley, S., Karp, P. and Rhee, S.Y. 2005. MetaCyc and AraCyc: Metabolic Pathway Databases for Plant Research. *Plant Physiology* 138:27–37.

Zhao, W., Wang, J., He, X., Huang, X., Jiao, Y., Dai, M., Wei, S., Fu, J., Chen, Y., Ren, X., et al. 2004. BGI-RIS: An integrated information resource and comparative analysis workbench for rice genomics. *Nucleic Acids Res.* 32:D377–D382.

Part 2

Cereals and Oil/Protein Crops

Chapter 5

Breeding Spring Bread Wheat for Irrigated and Rainfed Production Systems of the Developing World

Ravi P. Singh and Richard Trethowan

Introduction

Wheat is the primary source of calories for millions of people worldwide and is grown on some 225 million hectares, producing about 580 million tons annually (Aquino et al., 2002). Roughly half of this area and production can be attributed to developing countries. Nevertheless, developing countries also consume most of the wheat sold on the export market (Aquino et al., 2002), reflecting a huge number of consumers in these regions. In some countries, such as those in North Africa, per capita consumption of wheat is as high as 240 kg (FAO, 2001).

Wheat is one of the most drought-tolerant and water-use-efficient cereals. It can be grown from the equator to latitudes of 60°N and is found worldwide at altitudes ranging from sea level to more than 3,000 m. The only limitation to production is humid and high-temperature areas in the tropics and high-latitude environments where fewer than 90 frost-free days are available for crop growth.

The world's largest producers of wheat are China, India, and the United States, producing annually 100, 70, and 64 million tons with productivities of 3.8, 2.6, and 2.9 t/ha, respectively (Aquino et al., 2002). China has higher yields than India. Some of the underlying factors that contribute to these higher yields include a larger area in China devoted to growing winter wheat, facultative growth habit, cooler environments leading to a longer growing season for spring wheat, and better management practices. Only 10% of total wheat production is sold on the export market, the primary exporting countries being the U.S., Canada, Australia, and France.

Key adaptive mechanisms that allow wheat to adapt to diverse production environments are vernalization and photoperiod response. Fall-sown winter wheat survives low winter temperatures and snow cover through a vernalization response; it requires a period of low temperature before reproductive development is triggered. In some parts of the world, such as northern Kazakhstan and western Siberia, winter temperatures are too low for wheat to survive. In these environments, photoperiod-sensitive spring wheat that does not require vernalization is sown. The photoperiod response controls reproductive development via initiation of spike formation in response to increasing day length; this allows the crop to fully utilize the short summer growing season.

At the International Maize and Wheat Improvement Center (Centro Internacional de Majoramiento de Maiz y Trigo, or CIMMYT) bread and durum wheat germplasm are developed that target the diverse wheat-production environments found in the developing world. The CIMMYT spring wheat breeding program is based in Mexico and shuttles germplasm between two contrasting environments, thereby achieving two generations a year (Braun et al., 1996). The lines developed through this process are then tested widely around the world and selected materials, based on international performance, are identified for continued crossing. In order to better facilitate the crossing program in Mexico and the deployment of appropriate germplasm to these diverse production environments, the concept of mega-environments (ME) was introduced in the early 1990s (Rajaram et al., 1994). MEs are geographical areas where wheat adaptation can be expected to be similar in terms of climatic, disease, or crop-management constraints. This concept of ME has continued to be refined across the years as more sophisticated tools have become available and access to information

has improved. DeLacy et al. (1994), in an analysis of historical yield data from the International Spring Wheat Yield Nursery (ISWYN), concluded that ME definitions could be improved by integrating genotype × environment interaction (GEI) analyses. In recent years advances in GIS (global information systems) and the availability of spatial maps for factors as diverse as wheat production statistics, soil composition, rainfall, and annual temperature at critical growth stages have helped refocus ME-based breeding objectives (Trethowan et al., 2005a). Table 5.1 illustrates key spring wheat MEs and related traits that currently receive focus in wheat breeding at CIMMYT. Drawing on some specific examples from CIMMYT's wheat-breeding program, this

Table 5.1. Spring wheat mega-environments (ME) and main targeted traits for wheat improvement in current wheat improvement research at CIMMYT.

ME	Moisture regime	Temperature	% Area*	Main target traits
ME1	Irrigated	Temperate	36.1	High yield potential, lodging tolerance Water and nutrient use efficiency Resistance to three rusts Large white grain with leavened and flat bread quality
ME2	High rainfall (>500 mm)	Temperate	8.5	High yield potential and lodging tolerance Resistance to rusts, septoria tritici and fusarium head blight Large red grain and leavened bread quality
ME4	Low rainfall (<500 mm)	Temperate or hot	14.6	Drought tolerance with responsiveness to water availability Better root and emergence characteristics Adaptation to conservation agriculture Resistance to three rusts, septoria tritici, tan spot, root diseases Large white grain and leavened and flat bread quality
ME5	Irrigated or High rainfall	Warmer	7.1	High yield with early maturity, lodging tolerance Heat tolerance Resistance to rusts and spot blotch for low rainfall areas Resistance to rusts and fusarium head blight for high rainfall areas Large white or red grain depending on low or high rainfall, leavened and flat bread or noodle quality depending on country
ME6	Low rainfall	Cool	6.2	Drought tolerance, photosensitivity, tall stature Resistance to rusts Red grain with leavened bread quality

* Percent of developing country wheat area.

chapter describes options available to wheat breeders to improve wheat yield, adaptation to diverse environments, and processing and nutritional quality.

Cytogenetics of wheat

Bread wheat (*Triticum aestivum* L.) is an allohexaploid with 21 pairs of chromosomes and three genomes, A, B, and D. Based on chromosome pairing behavior, it was believed for a long time that einkorn wheat *Triticum monococcum* contributed the A genome to both tetraploid durum (*Triticum turgidum*) and hexaploid bread wheat. However, more recent molecular genetic evidences show that *T. urartu* Tumanian ex Gandilyan, *Aegilops speltoides* Tausch, and *Aegilops tauschii* Coss. are donors of A, B, and D genomes, respectively (Gill and Friebe, 2002). The hexaploid wheat shows disomic inheritance and behaves like a diploid. This has been possible mainly due to the presence of two major pairing regulator genes, *Ph1* (Riley and Chapman, 1958) and *Ph2* (Mello-Sampayo, 1971). Wild relatives of cultivated wheat have been used in transferring new genetic diversity not known to occur in cultivated wheat germplasm. Although mutants of *Ph* genes (Sears, 1977), which enhance pairing between wheat and chromosomes of related alien species, have been used successfully to induce translocations of chromosome segments onto wheat chromosomes, such translocations have also been achieved through radiation and tissue culture.

Over the years, numerous cytogenetic stocks have been developed. Sears (1954) and Sears and Sears (1978) identified numerous wheat aneuploids, such as monosomics, nullisomics, and telosomics, for different wheat chromosomes or chromosome arms. Such stocks were very useful initially in determining the location of genes governing a specific trait on wheat chromosomes and later in determining the locations of molecular markers. More recently, development of deletion stocks (Endo and Gill, 1996) has highlighted the relationship between the chromosomes' physical maps with that of genetic maps. Moreover, these deletion stocks have assisted in targeting a specific region of the wheat chromosome that is useful for identifying closely linked markers with genes of interest, positional cloning of genes, and determining synteny with other cereal species, such as rice, which has had its whole genome sequenced (Mateos-Hernandez et al., 2006). Although cytogenetic stocks are not used directly in wheat breeding, they have played a vital role in enhancing the understanding of wheat genetics, identification of molecular markers linked to genes of interest for marker-assisted selection, and perhaps more importantly in introducing new genetic diversity from alien species and genera.

Breeding objectives

Increasing grain yield, yield stability, resistance/tolerance to biotic and abiotic stresses, and end-use quality characteristics are among the most important breeding objectives at present and will remain so in the future, considering that most of the wheat produced in developing countries will be consumed locally by humans. In developing countries where population pressure continues to increase while land and water resources decline due to urbanization and unsustainable use, the only alternative to hunger is to enhance productivity either through genetic enhancement or through better crop-management practices. It is estimated that current global production of about 600 million tons must increase by about 2% annually and that by 2020 the global requirement will be at least 800 million tons. Increasing prosperity in highly populated countries like China and India has resulted in an increase in meat-based diets, which in turn creates an increased demand for grain for feed. The high cost of petroleum-based fuels in recent years has triggered the search for alternative energies, including bio-fuels. Whether or not wheat will compete with other high-biomass crops is not known yet.

The International Wheat Improvement Network led by CIMMYT in collaboration with other international centers and numerous national and advanced research institutions continuously adjusts and modifies breeding objectives and breeding schemes to maintain effectiveness and efficiency in tailoring germplasm products for current and future needs. For example, as water resources continue to decline, wheat will have to be produced with less water. This requires development of high-yielding cultivars that are water-efficient for irrigated areas or more drought-tolerant for rainfed areas. New cultivars must possess desirable end-use characteristics so that farmers can obtain fair price in local as well as global markets. At the same time, the cost of production must be reduced to increase profitability. Expansion of resource-conserving technologies, e.g., zero-tillage, in many countries would not only reduce production costs but would also increase long-term sustainability. However, it is evident that breeding objectives must be modified to develop a different kind of germplasm that has better emergence and growth characteristics and resistance to diseases and pests that survive on residues. Main breeding objectives, or targeted traits, for different spring wheat mega-environments are listed in Table 5.1.

Yield potential, yield stability and wide adaptation

Yielding capacity of a cultivar in a particular environment in the absence of biotic and abiotic stresses can be defined as the yield potential of the cultivar in that environment. Factors that lead to annual fluctuations in yield potential at a site are often associated with variations in temperature and radiation (Lobell et al., 2005). The yield potential of semidwarf wheat cultivars, regardless of their origin, has continued to increase at the rate of about 1% annually (Sayre et al., 1997). Comparison of 1990–1995 yield potential of eight semidwarf wheat cultivars released between 1962–1988 at CIMMYT's research station at Ciudad Obregón in northwestern Mexico, showed that yield potentials increased from 6.13 t/ha for 'Pitic 62' to 7.78 t/ha for 'Bacanora 88' (Sayre et al., 1997). According to Rajaram et al. (2002), yield potential of bread wheat in northwestern Mexico increased from 4.5 t/ha for the tall cultivar 'Yaqui 50' to 6.5 t/ha for the first semidwarf mega-cultivars 'Siete Cerros 66' and 'Sonalika,' which were associated with the widespread Green Revolution. This jump in yield potential is attributed mainly to the incorporation of dwarfing genes *Rht1* and *Rht2* that provided lodging tolerance, which allowed more fertilizers and irrigation to be applied to crops to increase production. In comparison with the first successful semidwarfs, yield potential of cultivar 'Baviacora 92,' released in 1992, was about 9 t/ha (Rajaram et al., 2002). Some of the new wheat genotypes developed in recent years have shown further increases in yield potential of 8–10% over 'Baviacora 92' (Singh et al., 2006).

Yield stability and wide adaptation are important traits that must be present together with yield potential to ensure that a genotype maintains its superiority in a range of environments, management practices, and biotic and abiotic stresses. Wide adaptation of CIMMYT-derived semidwarf wheat cultivars has been attributed to their photoinsensitivity, which resulted from shuttle breeding at two contrasting sites in Mexico (Rajaram et al., 2002). Breeding for specific adaptation has not been very successful because in most areas temperatures and rainfall patterns shift annually. Moreover, farmers are not able to maintain the same agronomic practices, and diseases and pest pressures vary from year to year. The wide adaptation and stable performance of CIMMYT-derived wheats are largely due to the shuttle breeding in Mexico where segregating populations were selected from two environments with differing disease and abiotic stresses; this was followed by international, multilocation testing of advanced lines. Using this approach, it is possible to identify the best stable performers in a single year of testing. Wide adaptation and stable

yield performance of a new breeding line 'Weebill 1' is shown in Figures 5.1 and 5.2. This genotype was yield-tested together with 49 other genotypes in each of the two international yield nurseries, 23rd ESWYT and 10th SAWYT; the first targeted for irrigated wheat environments, the second for semi-arid. Regression of yield performance of this line over the best local checks used by cooperators shows that 'Weebill 1' not only had superior yield potential but also highly stable performance in low, intermediate, as well as

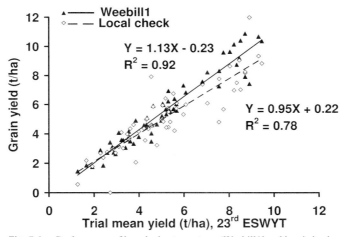

Fig. 5.1.   Performance of bread wheat genotype 'Weebill1' and local checks, regressed over trial mean at 62 international sites in 23rd ESWYT (Elite Spring Wheat Yield Trial).

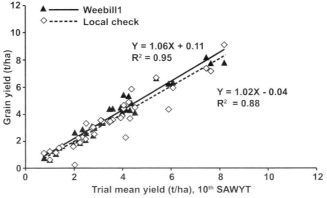

Fig. 5.2.   Performance of bread wheat genotype 'Weebill1' and local checks regressed over trial mean at 36 international sites in 10th SAWYT (Semi Arid Wheat Yield Trial).

Table 5.2. Grain yield potential of some of the 1st and 2nd generation derivatives of wheat cultivar Baviacora 92 developed through single-backcross approach at Ciudad Obregon, Sonora State, Mexico during 2004–2005 crop season.

Bread wheat genotype	Grain Yield	
	t/ha	% Baviacora 92
Baviacora 92	6.35	100.0
Kambara 1 (=Tacupeto 2001)	6.51	102.6
Fret 2 (Selection -12Y)	6.66	104.9
Weebill 1 (Selection -35Y)	6.71	105.7
Weebill 1*2/Kukuna (Selection -24Cel)	7.43*	117.2
Weebill 1*2/Kukuna (Selection -7Cel)	7.27*	114.6
Kambara 1*2/Kuukuna (Selection -9Cel)	6.80	107.2
l.s.d. ($p = 0.05$)	0.76	
C.V.	6.5%	

*Kambara 1 (released in Northwestern Mexico as Tacupeto 2001), Fret 2 and Weebill 1 are first generation derivatives of Baviacora 92 whereas other lines are 2nd generation derivatives.

high-yielding environments, regardless of irrigation or drought stress.

Widely adapted genotypes are the best parents for targeting future crosses for further genetic improvements. Strategy of using widely adapted parents was used in designing crosses involving 'Weebill 1,' which gave rise to new lines with significantly higher yield potential (Table 5.2).

Breeding strategies to increase yield potential

Various studies have shown that increases in yield potential are mainly associated with increased biomass, kernel number, and harvest index (Singh et al., 1998c; Sayre et al., 1997). Yield components, such as grain size and number, or harvest index in more recent germplasm do not correlate to increased yield potential. This means that selection for increased yield potential and higher kernel weight can proceed simultaneously. Large kernel size continues to be an important trait in some local markets of developing countries and appears to be associated with better emergence under poor

management. Some of the recent wheat germplasm developed at CIMMYT has not only shown increased grain yield potential but also kernel weight as high as 57 g/1,000 kernels in northwestern Mexico compared with about 40 g for most of the wheat germplasm developed during the 1980s and 1990s (Fig. 5.3).

Although the early increases in yield potential of semidwarf wheat cultivars came from the incorporation of dwarfing genes, subsequent progress can be attributed to quantitatively inherited additive genes. It is likely that intense breeding efforts during the last three decades in the post-Green Revolution era had already selected additive genes that greatly contributed to enhancing yield potential. If that is the case, then further progress is expected from selecting genes that have much smaller effects. Such progress would necessitate modification of commonly used traditional breeding schemes. Alternatively, introgression of new genetic diversity from unrelated wheat germplasm, including wide hybridization, can create a new genetic pool and bring in small-effect genes that may not be present in wheat

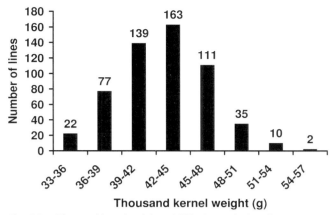

Fig. 5.3. Thousand kernel weights of 559 advanced wheat lines grown at Ciudad Obregon, Sonora State, Mexico during the 2004–2005 crop season.

germplasm commonly used in a breeding program.

Wheat improvement or improvement of other self-pollinated crops often utilizes simple, three-way (top), four-way (double), or repeated backcrossing approaches. Various wheat breeders also commonly practice pedigree or bulk methods of selection. Wheat breeding at CIMMYT in the 1960s and 1970s relied on simple, top, and double crosses, followed by the pedigree method of selection. It was realized later that best advanced breeding lines were rarely derived from double crosses, possibly because the genetic variation generated by such crosses was too large and the chances of recovering plants with desirable combinations of genes were reduced due to insufficient population sizes. During the early 1980s, CIMMYT breeders relied upon simple and three-way crosses and occasionally single backcrosses, followed by a modified bulk selection scheme where individual plants were harvested in the F_2 generation to grow the F_3 generation, however, bulk selection was practiced in the F_3-F_5 generations. Individual plants or spikes were once again harvested in the F_5 or F_6 generation (Rajaram et al., 2002).

Following the study by Singh et al. (1998c), which showed that selection schemes had little or no effect on the performance of progeny lines, but that it was the choice of parents that determined the progeny response, a selected bulk-breeding scheme was introduced in bread wheat improvement. Under this scheme, in all segregating generations until F_5 or F_6, one spike from each of the selected plants is harvested as bulk and a sample of seed is used to grow the next generation. Individual plants or spikes are harvested in the F_5 or F_6 generation. This scheme allows the retention of a larger sample of selected plants without increasing the cost and was found to be highly efficient in terms of operational costs. Moreover, retaining a large sample of plants in segregating populations increases the probability of identifying rare segregates that carry the most desired genes.

Based on the view "germplasm is paramount to increasing yield potential" (Rasmusson, 1996), at the CIMMYT, we began to utilize a single-backcross crossing approach that was initially aimed at incorporating resistance to rust diseases based on multiple additive, minor genes (Singh and Huerta-Espino, 2004). However, soon it became apparent that the single-backcross approach also favored selection of genotypes with higher yield potential as shown in Figure 5.4. The reason why single backcross shifts the progeny mean toward the higher side of the curve is that it favors the retention of most of the desired additive genes from the backcross or recurrent parent while simultaneously allowing the incorporation and selection of additional useful small-effect genes from the donor parent. As shown in Figure 5.4, the shift in mean results in a higher frequency of new breeding lines that have superior yield potential than for the check cultivar. Repeated backcrossing is not desirable as it was devised to incorporate a single—or only a few—major gene(s) with the least disturbance to the genetic make-up of the recurrent parent.

Crossing of spring wheat germplasm with winter wheat germplasm in the 1970s and 1980s not only increased the biomass of spring wheats, but it also increased yield potential, as observed in numerous cultivars released in many countries from 'Veery' and 'Attila' crosses of CIMMYT. Although thousands of crosses were made in this effort, the above two crosses with winter wheat parents 'Kavkaz' and 'Nord

Desperez,' respectively, can be considered the most successful because they led to the development of mega-cultivars that were subsequently grown on millions of hectares in the eastern African highlands, North Africa, the Middle East, Asia, and the Americas. These mega-cultivars showed 8–10% higher yield potential over previously developed cultivars (Sayre et al., 1997). This yield increase has often been associated with the presence of the alien translocation commonly known as 1B.1R, where the short arm of chromosome 1B is replaced by the short arm of chromosome 1R of rye (*Secale cereale*) (Rajaram et al., 2002). Alternatively, the presence of genes for resistance to major diseases, such as all three rusts and powdery mildew, on the rye chromosome arm could have provided superior disease resistance of Veery lines and some unknown small-effect genes from winter wheat germplasm could have provided increased yield potential (Singh et al., 1998b).

Alien translocations were often selected for incorporating genes that confer resistance to diseases and pests or tolerance to abiotic stresses (Gill and Friebe, 2002). However, often their effects on increasing

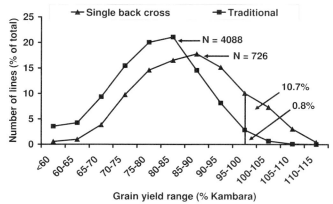

Fig. 5.4.   Grain yield potential, expressed as percentage of check cultivar 'Kambara,' of 4088 advanced lines derived from traditional (simple and three-way) crosses versus 728 lines derived from single-backcross breeding approach at Ciudad Obregon, Sonora State, Mexico during the 2004–2005 crop season.

grain yield potential remains undetermined. Alien translocation T7DS.7DL-7Ae#1L from *Thinopyrum elongatum* that carries leaf and stem rust resistance genes *Lr19* and *Sr25,* respectively, has been shown to increase yield potential ranging from almost non-significant levels to over 15% (depending on genetic background and under irrigated conditions) through increased biomass production caused by increased photosynthetic rate (Singh et al., 1998b; Reynolds et al., 2001). Thus its widespread incorporation could lead to a quantum jump in yield potential. The initial translocation carried a gene that caused higher endosperm pigmentation that is an undesirable quality trait for bread wheat. However, Knott (1980) developed a non-yellow mutant and this non-yellow variant translocation maintains the yield-enhancing effect (Huerta-Espino and Singh, 2005). Incorporation of rye translocation T4BS.4BL-2R#1L into three wheat genotypes showed that in a 'Weaver' background significant increases in yield potential, ranging between 5–15%, occurred in all three experimental environments: irrigated-normal sown, irrigated-late sown, and post-anthesis drought-stressed (Huerta-Espino and Singh, 2005). However, this translocation had negative effects when incorporated into two other genetic backgrounds. These studies exemplify the importance of genotype × translocation interactions, which are expected to occur when a portion of a wheat chromosome of a highly balanced genotype is replaced with an alien segment. Such imbalance could result in negative effects and needs to be rebalanced by compensating genes located in other homoeologous wheat chromosomes or the rest of the genome. A major effort needs to be made to study the effects of various other alien translocations that have been generated. Development of molecular markers for these translocations can aid their rapid identification and incorporation and could help enhance yield potential.

Breeding strategies to safeguard wheat crops from important diseases

Globally important fungal diseases of wheat caused by biotrophs (obligate parasites) include the three rusts, powdery mildew, and the bunts and smuts; those caused by hemi-biotrophs (facultative parasites) include Septoria tritici blotch, Septoria nodorum blotch, spot blotch, tan spot, and fusarium head blight (scab). The obligate parasites are highly specialized and significant variation exists in the pathogen population for virulence to specific resistance genes. Evolution of new virulence through migration, mutation, or recombination of existing virulence genes and their selection is more frequent in rust and powdery mildew fungi. Therefore, breeding for resistance to these diseases has always been more dynamic. Physiological races are known to occur for most bunts and smuts; however, evolution and selection of new races is less frequent. Because most bunts and smuts can easily be controlled by chemical seed treatment, little emphasis is currently placed on resistance breeding for these diseases. Changes in pathogen races are also less frequent for diseases caused by facultative parasites, possibly because there is no significant survival advantage for the new over the old races in crop residues during the off-season. The importance of some diseases is increasing in those developing countries where residue retention is becoming a common practice of conservation agriculture.

Strategies to control rust diseases

The three rust diseases—stem (or black), leaf (or brown), and stripe (or yellow)—are caused by fungi *Puccinia graminis* f.sp. *tritici, P. triticina,* and *P. striiformis* f.sp. *tritici.* They continue to cause losses—often major—in various parts of the world and hence receive more attention in breeding.

The wheat crop can be protected from rust, or at least the occurrence of epidemics can be reduced, by emphasizing the following: (1) regional cooperation in monitoring the evolution and migration of new races of rust fungi; (2) enhanced information on the genetic basis of resistance in important wheat cultivars for their deployment; and (3) a shift toward breeding and deploying wheat cultivars with durable resistance. The phenomenon of the erosion of race-specific resistance genes, or their combinations, has led scientists to look for alternative approaches to resistance management. Van der Plank (1963) was the first epidemiologist to clearly define the theoretical basis of the concepts of resistance. In the late 1960s and 1970s, there was a revival of the concept of general (race-nonspecific) resistance and its application in crop improvement (Caldwell, 1968). This concept was widely used for breeding leaf-rust resistance by Caldwell (1968), stem-rust resistance in wheat by Borlaug (1972), and yellow-rust resistance by Johnson (1988). The wide application of such a concept in breeding for leaf-rust resistance, commonly known as slow rusting, has dominated CIMMYT's bread wheat improvement program for almost 30 years, with major impact (Marasas et al., 2004). Today, we better understand the genetic

basis of durable resistance to rust diseases, and this knowledge is being applied in breeding. The genetic basis of durable resistance to three rusts is described below. We firmly believe that development and deployment of such resistance will provide a long-term genetic solution to rust control.

Lr34, Lr46 *and other minor genes for durable resistance to leaf rust*

The South American cultivar 'Frontana' is considered the best-known source of durable resistance to leaf rust (Roelfs, 1988). The Mexican-Rockefeller Program first used this variety in the 1950s. Later its derivatives, such as 'Penjamo 62,' 'Torim 73,' and 'Kalyan/Bluebird,' showed slow-rusting characteristics possibly derived from 'Frontana'. Genetic analysis of 'Frontana' and several CIMMYT wheats possessing excellent slow-rusting resistance to leaf rust worldwide has indicated that such adult-plant resistance is based on the additive interaction of *Lr34* and two or three additional slow-rusting genes (Singh and Rajaram, 1992). Leaf-rust severity observed in Mexico on most slow-rusting cultivars could be related to the number of minor genes they carry (Fig. 5.5). Under conditions in which susceptible cultivars display 100%

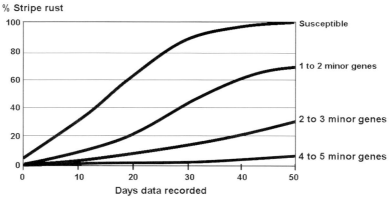

Fig. 5.5.   Relationship between the progress of stripe/leaf rust and the number of minor genes present in wheat cultivars.

leaf-rust severity, cultivars with only *Lr34* display approximately 40% severity; cultivars with *Lr34* and one or two additional minor genes display 10–15% severity; and cultivars with *Lr34* and two or three additional genes display 1–5% severity. Leaf rust may increase to unacceptable levels if disease continues to increase further on cultivars carrying only Lr34, or Lr34 and one or two additional genes. However, cultivars with Lr34 and two or three additional genes are referred to as "near immune" (Singh et al., 2000) and showed a stable response in all environments tested so far, with final leaf-rust ratings lower than 10% (Navabi et al., 2003). The presence of *Lr34* can be indicated by the presence of leaf tip necrosis in adult plants, which is closely linked with it (Singh, 1992a).

Slow-rusting resistance to leaf rust is common in spring wheat germplasm and at least 10–12 slow-rusting genes are involved in the adult-plant resistance of CIMMYT wheat germplasm (Singh and Rajaram, 2002). Lines in which *Lr34* is absent but that still possess high levels of slow-rusting resistance have been identified, indicating that durable resistance is feasible even in the absence of *Lr34*. A second slow-rusting resistance gene *Lr46* was recently identified in wheat cultivar 'Pavon 76' and located in the short arm of chromosome 1BL (Singh et al., 1998b; William et al., 2003b).

Yr18, Yr29 *and other minor genes for durable resistance to stripe rust*

Singh (1992b) and McIntosh (1992) have indicated that the moderate level of durable adult-plant resistance to stripe rust (caused by *Puccinia striiformis*) in the CIMMYT-derived U.S. wheat cultivar 'Anza' and winter wheats, such as 'Bezostaja,' is controlled in part by the *Yr18* gene. This gene is completely linked to the *Lr34* gene; the level of resistance it confers by itself is usually not adequate. However, combina-

tions of *Yr18* and three to four additional slow-rusting genes result in adequate resistance levels in most environments (Singh and Rajaram, 1994). Genes *Lr34* and *Yr18* occur frequently in germplasm developed at CIMMYT and elsewhere. Recently identified slow-rusting gene *Yr29* is completely linked to slow-leaf-rusting gene *Lr46* (William et al., 2003b).

Low disease severity to stripe rust is usually associated with at least some reduction in infection type because stripe rust grows systematically in leaf tissues. This phenomenon results in chlorotic or necrotic stripes and therefore creates difficulties in distinguishing slow-rusting resistance from race-specific resistance. Durability and acceptance of adult-plant resistance can be expected if the cultivar's low disease severity is due to the additive interaction of several (four to five) partially effective genes (Navabi et al., 2004).

New threat and challenge posed by the Ug99 race of stem rust pathogen

Stem, or black rust, caused by *Puccinia graminis tritici* (Pgt), is historically known to cause severe losses to wheat production. However, it has been controlled effectively through the use of genetic resistance in cultivars associated with the Green Revolution of the 1960s and 1970s. Over 80% of the spring wheat area in developing countries is currently sown with cultivars either derived directly from CIMMYT germplasm or from CIMMYT germplasm used as parents. For more than 30 years, a major proportion of the CIMMYT wheat germplasm and germplasm developed by other breeding programs has remained resistant to stem rust. Resistance gene *Sr31,* located on rye translocation 1B.1R, has contributed to high levels of resistance in several wheat cultivars developed worldwide in recent years. Consequently, stem rust disease is often not considered important and in many countries is ignored in wheat breeding; research in this area in the last two decades has also declined substantially.

Detection in Uganda in 1999 of Pgt race Ug99 that has a broad virulence, including virulence for *Sr31*, and its migration to Kenya and Ethiopia has been recognized as a highly significant event and led to the launch of the Global Rust Initiative in 2005. All major cultivars currently grown in North Africa, the Middle East, and Asia are moderately or highly susceptible to Ug99. Predominant wind patterns or human error are likely to introduce it to these regions and beyond. One of the major challenges in wheat breeding is identifying or developing, and then diffusing, adapted resistant culti-vars before this migration occurs. Although some race-specific resistance genes, mostly of alien origin, viz., *Sr22, 24, 25, 26, 27, 29, 32, 33, 35, 36, 39, 40, 44, R* (1A.1R translo-cation) and *Tmp,* can provide effective control, not all can be used in developing cultivars because some of these alien trans-locations are associated with negative effects on grain yield or quality. Shortening of these alien segments could make them more useful. Improved wheat germplasm that carry *Sr24, Sr25, Sr26, SrTmp*, and *SrR* genes have already been identified and can be used in breeding. Gene *Sr24* currently occurs with a relatively high frequency in wheat cultivars but it became ineffective in India and South Africa soon after the release of cultivars that carried this resistance gene. The best strategy therefore is to reconstitute durable adult-plant resistance that once pro-tected the Green Revolution and subsequent wheat cultivars. If race-specific genes need to be used, they must be deployed in com-bination to enhance their longevity.

Durable stem-rust resistance of some older U.S., Australian, and CIMMYT spring wheats is believed to be due to the deploy-ment of *Sr2* in conjunction with other unknown minor, additive genes. McFadden transferred gene *Sr2* to hexaploid wheat in the 1920s from tetraploid emmer wheat cultivar 'Yaroslav.' The slow-rusting gene *Sr2* confers by itself only moderate levels of resistance. Its presence can be detected through its complete linkage with the pseudo-black chaff phenotype. A large number of wheat lines were evaluated in 2005 in Kenya during the stem rust epidemic caused by Ug99. Genotypes with a pseudo-black chaff phenotype showed varying degrees of disease severity with a maximum severity reaching about 60% compared with 100% severity for highly susceptible materi-als. Moreover, the host reaction for these genotypes on the same internode varied from MR to S (moderately resistant to sus-ceptible). These observations clearly indi-cated that although slow-rusting resistance gene *Sr2* continued to confer at least some resistance, the level of resistance was not sufficient when this gene was present alone under high disease pressure in Kenya. *Sr2* was detected in several highly resistant old, tall Kenyan cultivars, including 'Kenya Plume' (Singh and McIntosh, 1986), and CIMMYT-derived semidwarf cultivar 'Pavon 76.' These cultivars have shown a maximum disease score of 15MR (15% disease severity with moderately resistant reaction). Because 'Pavon 76' is susceptible to race Ug99 at the seedling stage, its resis-tance, as speculated earlier (Rajaram et al., 1988), is based on multiple additive genes of which *Sr2* is an important component. Wide testing of improved wheat germplasm also has helped in identifying additional sources of adult-plant resistance. These sources are being used at CIMMYT to incor-porate durable stem-rust resistance into high yielding, widely adapted wheat cultivars using the methodology described below.

Breeding for durable resistance to rust diseases

Breeding for durable resistance based on minor additive genes has been challenging and often slow, for several reasons: (1) a sufficient number of minor genes may not be present in a single source genotype, (2) a source genotype may be poorly adapted, (3) there may be confounding effects from the segregation of both major and minor genes in the population, (4) crossing and selection schemes and population sizes are more suitable for selecting major genes, (5) reliable molecular markers for several minor genes are unavailable, and (6) the cost associated with identifying and utilizing multiple markers is high. One suggested approach is to use recurrent selection schemes to accumulate several minor genes in a single genetic background. But, such selection schemes have often been more of a scientific interest than actually being applicable to breeding. Selection for resistance alone will not generate important popular cultivars unless it is simultaneously combined with other traits such as high yield and quality. However, such germplasm carrying combinations of minor genes should be very useful in transferring these genes to modern cultivars.

A successful example of breeding for resistance based on minor genes is the development of resistance to leaf and stripe rusts in wheat, which took about 30 years of continuous effort at CIMMYT. In the early 1970s, S. Rajaram, influenced by the concept of slow-rusting resistance in wheat proposed by R. Caldwell (1968) and partial resistance to late blight of potato put forth by J. Niederhauser et al. (1954), made a strategic decision: to initiate selection for slow-rusting resistance to leaf rust in CIMMYT spring wheat germplasm. In the early phase of breeding he maintained plants and lines in segregating populations that showed 20–30% rust severity with compatible infection type. This strategy led to the release in Mexico and other countries of several successful wheat cultivars including 'Pavon 76' and 'Nacozari 76'. These slow-rusting lines were used heavily in the crossing program

and resulted in the wide distribution of minor genes within CIMMYT spring wheat germplasm.

The genetic basis of slow rusting resistance in CIMMYT wheats started to become clear in the early 1990s. High-yielding lines that combine four or five additives, minor genes for both leaf and stripe rusts, and show near-immune levels of resistance were developed in the 1990s (Singh et al., 2000). Three or four lines carrying different minor genes were crossed (3-way and 4-way crosses), and plants in large segregating populations were selected under artificially created rust epidemics. Races of pathogens that have virulence for race-specific resistance genes present in the parents were used to create the epidemics. The resulting highly resistant lines are now being used in a planned manner to transfer these minor resistance genes to well-adapted, "farmer's choice" cultivars that are currently grown across large areas but have become susceptible to rust races in Mexico. The crossing and selection scheme described below was developed and applied based on genetic information on the number of additive, minor genes that must be transferred to achieve the desired level of restistance. This strategy has allowed simultaneous transfer not only of resistance genes but also other minor genes with small effects that increase the yield potential or improve the grain quality of an adapted cultivar.

To transfer minor gene-based resistance into a susceptible adapted cultivar or any selected genotype, we use a "single back-cross-selected bulk" scheme, where the cultivar/genotype is crossed with a group of about 8–10 resistance donors; 20 spikes of the F_1 plants from each cross are then backcrossed to obtain 400–500 BC_1 seeds. Selection is practiced from the BC_1 generation onward for resistance and other agronomic features under high rust pressure. Because additive genes are partially dominant, BC_1 plants carrying most of the genes show intermediate resistance and can be selected visually. About 1,600 plants per cross are space-grown in the F_2, whereas about 1,000 plants are maintained in the F_3–F_5 populations. Plants with desirable agronomic features and low to moderate terminal disease severity in early generations (BC_1, F_2, and F_3) and plants with low terminal severity in later generations (F_4 and F_5) are retained. We use a selected-bulk scheme where one spike from each selected plant is harvested as bulk until the F_4 generation, and plants are harvested individually in the F_5. Bulking of selected plants poses no restriction on the number of plants that can be selected in each generation because harvesting and threshing are quick and inexpensive, and the next generation is derived from a sample of the bulked seed. Because high resistance levels require the presence of four to five additive genes, the level of homozygosity from the F_4 generation onward is usually sufficient to identify plants that combine adequate resistance with good agronomic features. Moreover, selecting plants with low terminal disease severity under high disease pressure means that more additive genes may be present in those plants. Selection for seed characteristics is carried out on seeds obtained from individually harvested F_5 plants. Small plots of the F_6 lines are then evaluated for agronomic features, homozygosity of resistance, etc., before conducting yield trials.

Resistant derivatives of several cultivars and genotypes were recently developed using the above methodology. In each case we could identify derived lines that not only carry high levels of resistance to leaf rust or yellow rust or both, but also show about 5–15% higher yield potential than the original cultivar. We believe this approach to wheat improvement allows us to maintain the characteristics of the original cultivar while improving its yield potential and rust resistance. It should be noted that having minor gene-based resistance in several backgrounds should ease future selection for these resistance genes.

Breeding for durable resistance to powdery mildew

Powdery mildew, caused by *Blumeria graminis* f.sp. *tritici,* is an important disease of wheat in highly productive areas of South America and Asia and North Africa. The evolutionary rate for this fungus is even higher than that for rusts because both asexual and sexual mechanisms are prevalent during the crop season. Since natural epidemics of powdery mildew do not occur in Mexico, where CIMMYT has its main breeding operations, most of the breeding materials sent from Mexico to other regions were moderately susceptible to susceptible when grown in powdery mildew-prone areas.

Because the use of race-specific genes, either singly or in combination, results in a rapid build-up of the new races, the best breeding strategy for a long-term control is the utilization of adult-plant resistance or slow mildewing. Several sources of such durable, adult-plant resistance have been identified (Wang et al., 2005). One such source is the North American cultivar 'Knox 62' and its derivatives, in which long-lasting resistance has been attributed to the presence of a few additive, minor genes (Griffey and Das, 1994). Lillemo et al. (2006) demonstrated that the high level of adult plant resistance at all tested international sites of CIMMYT spring wheat genotype 'Saar' was due to three or four additive genes. 'Saar' was initially developed for durable resistance to leaf and stripe rusts (Navabi et al., 2003, 2004) and evaluation of a common population with two rusts and powdery mildew showed high positive correlations in adult-plant responses to these three diseases (Lillemo et al., 2005). Recent studies using the near-isogenic lines of slow-rusting genes *Lr34/Yr18* and *Lr46/Yr29* have shown that these two genes reduce powdery mildew severities by about 50% and 30%, respectively, in Norway under high disease pressure (Lillemo et al., 2005). The effect of *Lr34/Yr18* pleiotropic locus in reducing powdery mildew severity was much higher in Australia (Spielmeyer et al., 2005). The pleiotropic effects of these loci, and perhaps others yet to be discovered, would ease simultaneous selection for resistance to wheat leaf and yellow rusts and powdery mildew. A similar breeding strategy, as discussed earlier, for rust diseases is currently being used at CIMMYT to improve the adult-plant resistance to powdery mildew where a systematic transfer is being made from adult plant sources to high-yielding wheats through a single-backcross approach.

Breeding for resistance to fusarium head blight (scab)

Several species of the genus *Fusarium* cause fusarium head blight (FHB), or scab, which is a main production constraint in environments where humid and semi-humid conditions coincide with flowering in wheat. The Yangtze River basin of China with about 7 million ha has traditionally been known to be highly prone to scab epidemics. Disease incidences leading to epidemics are now frequent in other developing countries, e.g., Argentina, Brazil, and Uruguay, where residue retention is a common practice for conservation agriculture. Introduction of maize–wheat crop rotation further increases the disease build-up. The most important species implicated in scab epidemics is *Fusarium graminearum* (telomorph: *Giberrella zeae*). In addition to crop losses, the fungus also produces mycotoxins, such as deoxynivalenol (DON) that accumulates in the grain, rendering it unsuitable for human and livestock consumption.

Sources of resistance to FHB have been divided into three groups: China and Japan, Argentina and Brazil, and Eastern Europe (Singh and Rajaram, 2002). More recently, additional sources, including some hexaploid synthetic-derived wheat lines, have been found to carry moderate resistance. Earlier genetic analysis indicated that a few additive genes confer resistance in Chinese and Brazilian wheats, but the genes present in Chinese sources are different from those in Brazilian sources (Singh et al., 1995; Van Ginkel et al., 1996). Although several genomic regions are now known to contribute quantitative resistance (Buerstmayr et al., 2002; Anderson et al., 2001), a gene from Chinese cultivar 'Sumai 3' in the short arm of chromosome 3B has shown the largest and most consistent effect in reducing disease severity and mycotoxin accumulation (Anderson et al., 2001). The Chinese sources are probably the best resistances currently available and must be combined with other sources of resistance. The Chinese cultivars that best combined with CIMMYT materials to transmit scab resistance are 'Sumai#3,' 'Ning 7,840,' 'Shanghai#5,' 'Yangmai#6,' 'Suzhoe#6,' 'Wuhan#3' and 'Chuanmai 18,'

Further progress in enhancing the level of resistance beyond the current level can come from a breeding strategy that favors the accumulation of multiple minor genes from various sources into a single genotype. This would involve intercrossing parents with resistance located in different genomic regions, followed by growing large segregating populations and using flanking markers linked to resistance loci for selection in early segregating generations. Conventional field screening must wait until the F_4 or F_5 generations when homozygosity has increased significantly. Progenies that show higher levels of resistance than the parental sources could be used for a targeted transfer to high-yielding cultivars that would already by then carry moderate levels of resistance from the ongoing breeding efforts. This strategy is currently practiced at CIMMYT to accumulate resistance genes from different sources.

Table 5.3.   Sources of high level of resistance to septoria tritici blotch evaluated at Toluca research station of CIMMYT during 2005.

Cross	Septoria Score	Source
OASIS/SKAUZ//4*BCN/3/PASTOR/4/KAUZ*2/YACO//KAUZ	42	Bobwhite
OASIS/SKAUZ//4*BCN*2/3/PASTOR	32	Bobwhite
TINAMOU	32	Bobwhite
ALD/CEP75630//CEP75234/PT7219/3/BUC/BJY/4/CBRD/5/TNMU/PF85487	31	Brazilian
BH1146*3/ALD//BUC/3/DUCULA/4/DUCULA	21	Brazilian
CATBIRD	52	Chinese
N894037	41	Chinese
CHUANMAI 107	32	Chinese
MILAN/S87230//BABAX	21	Milan
MINO	31	Milan
MURGA	21	Synthetic
REH/HARE//2*BCN/3/CROC_1/AE.SQUARROSA (213)//PGO/4/HUITES	21	Synthetic
FINSI/METSO	21	Synthetic
BL 1496/MILAN/3/CROC_1/AE.SQUARROSA (205)//KAUZ	21	Synthetic
ALTAR 84/AE.SQ//OPATA/3/2*WH 542	11	Synthetic

*Disease scores follow a 00–99 two-digit scale and evaluated when susceptible checks had scores between 96 and 98. The first digit represents the relative height of plants infected with the disease and the second digit represents the relative severity within that height.

Breeding for resistance to Septoria tritici blotch

CIMMYT initiated breeding for resistance to Septoria tritici blotch, caused by *Mycosphaerella graminicola* (anamorph *Septoria tritici*), in semidwarf wheat in early 1970. Steady progress has been made since then. Currently, several high-yielding semidwarf wheats with good resistance are available. Resistance in these wheat genotypes is derived from diverse sources, including synthetic wheats (Table 5.3). The main problem encountered in the early breeding work was breaking the association of resistance with lateness and tallness present in the above sources. Efforts are being made to combine these resistances. Two high-rainfall sites, Toluca (Mexico State) and Patzcuaro (Michoacan State), are used in Mexico for selection. Genetic studies conducted on CIMMYT wheats indicate that five to eight genes confer resistance to Septoria tritici blotch, depending on the source population (Briceno, 1992; Jlibene et al., 1992; Matus-Tejos, 1993). Two to three genes that have predominantly additive effects are generally

needed to confer an acceptable level of resistance. The selection methodology, therefore, is similar to that described for combining minor, additive genes for resistance to leaf and stripe rusts.

Some synthetic wheats developed at CIMMYT have shown excellent resistance that appears to be leading toward immunity to the disease. These sources offer new genetic diversity of resistance originating from durum wheat and/or *T. tauschii*. A high level of resistance from original synthetic parents was successfully transferred to derived, high-yielding lines (Table 5.3). It is not known whether this kind of resistance will remain durable or if a pathogen population can shift to overcome such major gene-based resistance.

Breeding for resistance to spot blotch and tan spot

The first crosses to incorporate spot blotch (caused by *Bipolaris sorokiniana*) resistance into CIMMYT wheats were made about 20 years ago. These crosses involved moderately resistant cultivars, such as BH1,146, from Brazil. However, the level of resistance in progenies was inadequate when tests were carried out at Poza Rica, Mexico, which is an ME5 testing site. In the mid-1980s, wheat genotypes carrying resistance to scab and obtained from the Yangtze River Valley of China showed varying levels of spot-blotch resistance when tested at Poza Rica. These Chinese lines included Suzhue 1 to 10, Wuhan 1 to 3, Shanghai 1 to 8, and certain Ningmai and Yangmai lines. About the same time, the wide-crossing program at CIMMYT produced resistant lines that contain *Thinopyrum curvifolium* in their pedigree (Villareal et al., 1995). Some of these lines and their derivatives possess good resistance and appear to be promising in Bangladesh, lowland Bolivia, and Nepal. Resistance in these wheats, such as 'Sabuf,' 'Chyria 1,' and 'Cugap,' appears to be controlled by two to three genes (Velazquez-Cruz, 1994), whereas 'Longmai 10' and 'Yangmai 6' may carry polygenic resistance with a relatively high narrow-sense heritability (Sharma et al., 1997). A few synthetic wheats developed at CIMMYT also carry resistance derived from the *T. tauschii* accessions. A key problem with selection for spot-blotch resistance is the negative correlation of disease severity with heading date and plant height (Duveiller and Gilchrist, 1994). Therefore, care must be taken if short types with early maturity are required. Current strategy followed at CIMMYT is to combine resistances from these diverse sources. Identification of some highly resistant lines from such crosses indicates that resistance is additive. However, a recent study has shown that durable leaf-rust resistance gene *Lr34* also confers a moderate level of resistance to spot blotch, which may be sufficient to prevent losses in areas where disease pressure is usually not too high (Joshi et al., 2004).

Tan spot (caused by *Drechslera tritici-repentis*) resistance is not widely dispersed in CIMMYT germplasm, but moderate resistance is known to occur (Rees and Platz, 1992). Some newer CIMMYT lines, such as 'Milan,' 'Attila,' 'Corydon,' and 'Tinamou,' and some Chinese wheats and their derivatives, such as 'Luan,' are also reported to carry moderate to high resistance (Diaz de Ackermann and Kohli, 1998). Tan spot is on the increase in areas where reduced tillage practices are being combined with residue retention, so adequate resistance to this disease has become necessary to reduce losses. Therefore, resistance to this disease must be incorporated, especially in germplasm targeted for conservation agriculture.

Breeding for tolerance to abiotic stresses

Wheat is increasingly being grown on marginal lands and in farming systems where inputs are limited. Depressed international and local prices have caused farmers to use their best land for higher-value crops, and in southern and central Asia, where wheat is grown under irrigation, water availability for irrigation is declining as demand for water from urban centers increases (Pingali and Rajaram, 1999). In addition, wheat production in low-latitude areas is likely to be most affected by climate change; some estimates predict wheat yield reductions of 12% if average global temperatures rise by 2°C (Rosenzweig and Hillel, 1995). Figures 5.6a and 5.6b generated by GIS modeling illustrate how currently favorable ME1 wheat growing areas of India would change to heat-stressed ME5 areas by 2050.

Wheat breeders can mitigate the effects of changing production environments to some extent. Farmers have been selecting wheat for drought tolerance for thousands of years, gradually accumulating those genes that confer adaptation to limited moisture, so variability already exists within the wheat gene pool for drought tolerance. Our understanding of the genetic basis of drought tolerance is poor. Nevertheless, considerable

(a)

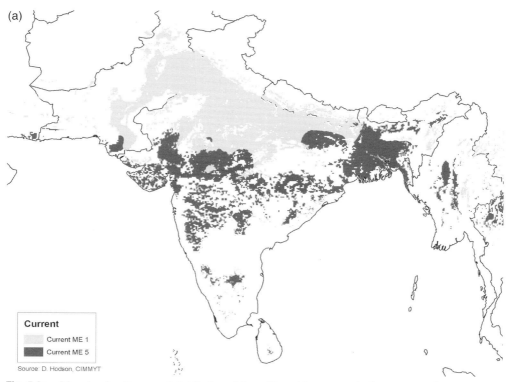

Current

Current ME 1

Current ME 5

Source: D. Hodson, CIMMYT

Fig. 5.6a. Map showing the current distribution of favorable and heat-stressed wheat mega-environments ME1 and ME5 in South Asia.

progress has been made in yield improvement under drought in recent decades using the wheat gene pool and selection under drought stress (Trethowan et al., 2002). The opportunity exists to improve tolerance further if new genetic variability can be combined with existing variability and if the underlying genetic control of tolerance can be better understood.

Synthetic hexaploid wheat, developed by crossing modern durum wheat (*Triticum durum*) with *Aegilops tauschii*, the probable donor of the D-genome in hexaploid wheat, has introduced new genetic variation into the wheat gene pool for many characters (Hatchett et al., 1981; Innes and Kerber, 1994; Villareal et al., 1994; Aguilar-Rincón et al., 2000). Not surprisingly, synthetic wheat has also been a source of variation for drought and heat tolerance (Gororo et al., 2002; Trethowan et al., 2003; Trethowan

et al., 2005b). The original crosses that gave rise to hexaploid wheat occurred some 8,000 years ago and probably involved only a few accessions of *A. tauschii*. Hundreds of accessions of *A. tauschii* have been collected from some of the harshest environments and are maintained by various gene banks, which represent considerable genetic variation. Once the synthetic hexaploid is made it can be easily crossed with cultivated wheat and target traits transferred. Although synthetic wheats can also introduce undesirable traits, some synthetic-derived advanced materials have improved wheat adaptation around the world, especially in drought-stressed environments (Ogbonnaya et al., 2006; Dreccer et al., 2006; Lage and Trethowan, 2006). Figure 5.7 represents the performance of a synthetic hexaploid derivative 'Vorobey' sown in 33 different environments, with mean yield ranging from less than 1 t/ha to

(b)

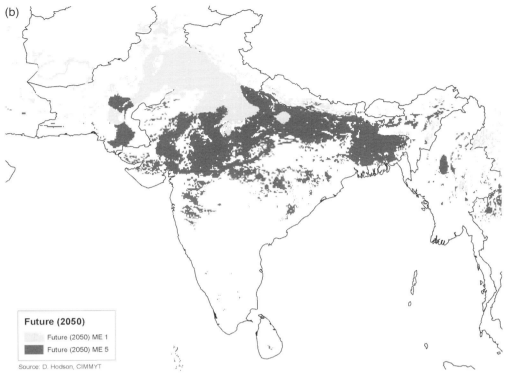

Future (2050)

Future (2050) ME 1

Future (2050) ME 5

Source: D. Hodson, CIMMYT

Fig. 5.6b. Map showing the predicted distribution of favorable and heat-stressed wheat mega-environments ME1 and ME5 in South Asia in 2050.

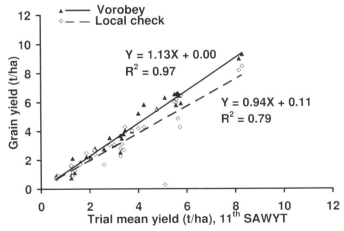

Fig. 5.7. Grian yield performance of a synthetic derived wheat genotype 'Vorobey' compared to local checks at 33 international locations grown in 11th SAWYT (Semi Arid wheat Yield Trial).

more than 9 t/ha compared to the best locally adapted cultivar at each location. Clearly the synthetic derivative out-yields each of 33 locally adapted check cultivars across the range of environments.

Recent research at CIMMYT has shown that considerable new genetic variation also exists among the wild tetraploid progenitors, *T. dicoccum*, of cultivated durum wheat; the resulting synthetic hexaploids when recombined with adapted wheat through limited backcrossing have greatly improved yield under drought conditions (Table 5.4) (Lage and Trethowan, 2006). Remember that the observed drought tolerance of the synthetic-derived materials is not just due to the genes introgressed from synthetic wheat. Many original primary synthetics do not have exceptional drought tolerance, but when recombined with existing genetic variation in adapted wheat, they can express high levels of tolerance; it is likely that new additive genes from synthetic wheat contribute to additive and additive × additive variation.

Drought tolerance is a complex character that is greatly influenced by GEI. However, water-use efficiency (WUE) can be increased by improving root resistance to soil-borne diseases as well as micronutrient imbalances (Trethowan et al., 2005b), because healthier roots will use available soil moisture more efficiently. Fortunately, many of these characters are simply inherited, and molecular markers are available to assist selection (Eagles et al., 2001; William et al., 2003a).

To improve breeding efficiency for drought tolerance, CIMMYT strategy is to ensure that drought-tolerant germplasm maintain responsiveness if more moisture becomes available in a season. As shown in Figures 5.1, 5.2, and 5.7, high yield potential and drought-stress tolerance are not mutually exclusive; they can be bred simultaneously by selecting segregating populations grown under favorable environments and drought-stressed environments in alternate generations—a practice used at CIMMYT. Moreover, to generate a more precise drought stress at different growth stages, drip-irrigation system has been installed to irrigate 17 ha of CIMMYT's experimental field at Ciudad Obregón in northwestern Mexico where rainfall during the crop season in most years is negligible. This system allows the application of exact amounts of water at chosen growth stages to generate different drought scenarios representing different parts of the world.

The genetic control of heat tolerance, like drought tolerance, is poorly understood. Nevertheless, significant variation for heat tolerance exists in the wheat gene pool (Pfeiffer et al., 2005). In many environments, late planting can expose crop and selection nurseries to high temperatures from flowering onward, giving wheat breeders the opportunity to select lines with high levels of heat tolerance. At CIMMYT, lines are selected during the segregating phase for adaptation to heat by late planting. A gravity table is used to separate bulk populations into those that can maintain seed weight under high temperature; the derived lines are then tested under heat stress in yield trials. Figure 5.8

Table 5.4. Grain yield of *Triticum dicoccum* based synthetic backcross derivatives compared to their drought tolerant recurrent parents under drought stress at Ciudad Obregon, Sonora State, Mexico during 2005–2006 crop season.

Pedigree	Yield (t/ha)	Yield (% of the recurrent parent)
*T.dicoccum*PI225332// *Ae. sq.*(895)// 3*Weebill 1	2.623	146 a*
*T.dicoccum*PI94625// *Ae. sq.*(373)// 3*Pastor	2.830	161 a
Weebill 1	1.790	100 b
Pastor	1.756	100 b

*Means in columns followed by the same letter are not significantly different at P < 0.05.

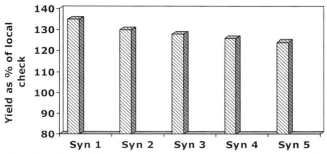

Fig. 5.8. Grain yield performance, expressed as percentage over the check cultivar 'Kambara,' of five synthetic derived bread wheat genotypes grown under heat stress created by late planting at Ciudad Obregon, Sonora State, Mexico; synthetic derivatives were significantly better than the check at P < 0.05.

compares the yield of synthetic-derived lines tested under late heat stress with the highest yielding local check cultivar in the Sonoran desert in northwestern Mexico; the derived lines are clearly higher yielding than the locally adapted cultivar. Physiological tools, such as the infrared thermometer that measures canopy temperature depression (CTD), are also available to assist the plant breeder in discriminating among progenies (Reynolds et al., 1998). The CTD is easy to measure, improves the heritability of selection, and can be used to select in segregating generations (Trethowan, 2006b).

Heat avoidance or early maturity is an extremely important trait for wheat to have in order to circumvent the effects of high temperature at grain filling. As shown in Figure 5.6a, the northeastern Gangetic Plains of India are classified as a heat-stressed environment. All currently grown popular cultivars are earlier maturing than cultivars popular in the northwestern Gangetic Plains. A simultaneous improvement in heat tolerance and yield potential for earlier maturing germplasm is the only option to increase production in heat-stressed environments. Figure 5.9 illustrates the performance of two new CIMMYT-derived high-yielding, early-maturing lines and the best check, popular cultivar 'HUW234,' at the research station

of Banaras Hindu University, Varanasi and in farmers' fields under zero-tillage in six villages in a heat-stressed ME5 environment. The two lines performed better than the check at all sites and showed an average advantage of about 18% over 'HUW234.' Selecting segregating populations for improved heat tolerance in the northeastern Gangetic Plains may provide future wheat germplasm for the northwestern Gangetic Plains (see Fig. 5.6b) and other parts of the world where heat is expected to cause major stress in coming years due to global warming.

There is also significant variation in the wheat gene pool for tolerance to soil toxicities and deficiencies, such as salinity (Salam et al., 1999), aluminum (de Sousa, 1998), boron (Moody et al., 1990), and zinc (Cakmak and Braun, 2001). Nevertheless, new genetic variation will need to be found or created and introgressed if adaptation to some extreme environments is to be improved.

Breeding for adaptation to conservation agriculture

Farmers are increasingly adopting conservation agriculture (CA) in many parts of the world. CA reduces soil loss from erosion, conserves water, improves soil organic

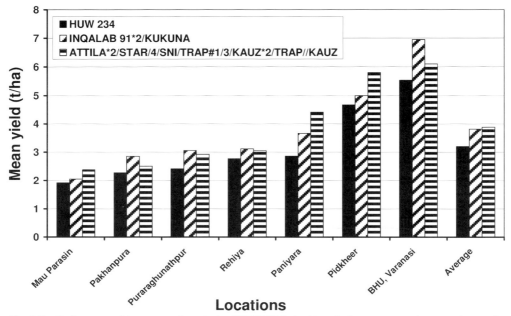

Fig. 5.9. Performance of two new early maturing CIMMYT derived bread wheat genotypes in comparison to the best check cultivar 'HUW 234' at research station of BHU, Varanasi and six farmers' fields in different villages in the northeastern Gangetic Plains of India.

matter, reduces inputs, and limits CO_2 emissions (Reeves et al., 2001). This largely involves zero-tillage and the retention of stubble from the previous crop in an appropriate crop rotation. One of the primary impediments to the adoption of CA by farmers has been access to proper machinery; however, recent developments in seeder design and improved access have seen more and more farmers from both developing and developed countries adopting the practice. This movement to CA has implications for wheat breeding because significant genotype × tillage interactions (GTI) have been found (Cox, 1991; Trethowan and Sayre, unpublished results). Much of the earlier literature was inconclusive regarding GTI, largely because few genotypes were tested and all tested materials were developed under conventional tillage (Dao and Nguyen, 1989; Ditsch and Grove, 1991). Nonetheless, adoption and performance of current and

new wheat materials has been very successful in many areas of the world; selection under zero-tillage is expected to further improve adaptation to CA.

The retention of crop residues has led to a change in the disease spectrum in some areas (Duveiller and Dubin, 2002). Diseases such as tan spot and crown rot (caused by *Fusarium pseudograminearum*) which survive on crop residues can become limiting in some environments. Adequate genetic resistance to most of these diseases exists in the wheat gene pool and among the newly developed synthetic wheats. Genotypes with better emergence and establishment characteristics will also cope better with standing residues (Trethowan et al., 2005b). In many developing countries farmer access to herbicides, so necessary in zero-tillage systems, can be limited. Therefore genotypes that not only emerge well, but that also have vigorous early growth, will help suppress weeds.

At CIMMYT, segregating populations are grown under zero-tillage to skew gene frequency in favor of adaptation to this farming system. Preliminary evidence from a study in which populations derived from the same crosses were either always subjected to zero-tillage or always to conventional tillage shows that lines developed under zero-tillage perform better when yield tested under zero-tillage than when tested under regular tillage (Trethowan and Sayre, unpublished data).

Breeding for input-use efficiency

Wheat with better N- and P-use efficiency will reduce the amount of these essential elements that need to be applied by farmers, thereby increasing the profitability of the farming system and reducing leaching to ground water, river systems, and the sea. There is significant variation in the wheat gene pool for both N- (Ortiz-Monasterio et al., 1997) and P-use efficiency (Jones et al., 1992).

To illustrate how these traits can be improved genetically, the case of N-use efficiency (NUE) will be discussed citing a study conducted by Van Ginkel et al. (2001). The NUE is comprised of two components, N-uptake efficiency (UPE) and N-utilization efficiency (UTE). In crosses between lines with high UPE and high UTE they found that the selection environment was critical in identifying genotypes with improved NUE. They tested four environments for the selection of segregating populations: (1) always sown under low nitrogen, (2) always sown under medium nitrogen, (3) always sown under high nitrogen and, (4) alternating low nitrogen with high nitrogen. The alternating treatment identified lines that performed well at intermediate and high nitrogen levels. No differences were found among selection regimes when the resulting lines were evaluated at low nitrogen. Generally if nitrogen is limiting, the contribution of UPE to overall NUE is greater than UTE (Ortiz-Monasterio et al., 1997).

Breeding for industrial and nutritional quality

Wheat provides essential nutrients to human beings and a range of different food is prepared from it. Bread wheat is generally milled into flour (both refined and whole meal) and made into leavened breads, flat breads, biscuits, and noodles, whereas durum wheat is milled into semolina and consumed as pasta, cooked grits, or couscous, and in some Middle Eastern countries it is baked into bread (Qarooni, 1994). The genetics of wheat industrial quality is well understood and our understanding continually increases as more alleles and their effects are discovered. Some attributes, such as protein content and alpha-amylase activity, are influenced by environmental factors, whereas others are not. Protein content tends to be higher when the plant is under stress and lower under well-watered or N-limiting conditions. Protein content affects qualities such as dough strength, dough mixing time, and loaf volume. The largest fraction of total protein is gluten, made up, in turn, of glutenins and gliadins (MacRitchie, 1994). Gluten influences the viscoelastic properties of wheat flour and largely determines how a particular variety is used.

While a relatively small portion of total variation in protein content across years and locations is genetic (Peña et al., 2002), the quality of protein is controlled by known high and low molecular weight glutenins and gliadins (Shewry et al., 1992; Weegels et al., 1996). As shown in Figure 5.10, loci that control both high and low molecular weight glutenins can be determined using ID SDS-PAGE (Peña et al., 2002) and the information can be used to make better crosses and ultimately, higher quality cultivars.

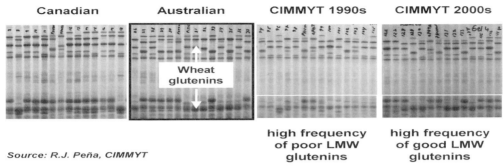

Fig. 5.10. Comparison of high and low molecular weight glutenin profiles associated with bread making quality of Canadian, Australian, and bread wheat germplasm of CIMMYT developed in the 1990s and 2000s.

A number of rapid, indirect quality tests are available to the plant breeders that can be applied in the early generations to increase the probability of identifying progeny with the desired quality profiles. These include near-infrared reflectance (NIR) to measure protein content, grain hardness, and even milling yield (Burridge et al., 1994; Zeleny, 1947) and SDS-sedimentation (Axford et al., 1979) tests can be used to approximate gluten content and—indirectly—gluten strength. Dough rheological properties can be measured in different ways; some methods are time-consuming but accurate, e.g., the Alveograph, and others are faster and less expensive, but slightly less accurate, e.g., the Mixograph. Peña et al. (2002) provided an overview of the application of these and other methods in wheat breeding. The decision about which tests to apply and at what stage in the breeding process they should be applied depends on cost, time, heritability, and program objectives. If the primary objective of a breeding program is yield and industrial quality is secondary, then quality tests should be introduced after yield testing. Trethowan et al. (2001) showed that selection for quality, particularly protein content, before yield testing significantly reduced the number of high-yielding lines in a breeding program. The CIMMYT currently places a very high emphasis on improvement of leavened and flat bread quality characteristics. Improved wheat materials developed in

recent years are more similar to Canadian and Australian wheats, which are known to possess good bread quality characteristics and carry very similar high and low molecular weight glutenin subunits (Fig. 5.10).

Millions of people worldwide suffer from micronutrient deficiencies, particularly vitamin A, Fe, and Zn (Graham et al., 2001). In recent years there has been a realization that not only should processing and product quality be high, but so should the nutritional value of wheat products. To date there has been no vitamin A, conditioned by beta-carotene, found in wheat. All yellow-pigmented materials have contained only lutein. There is significant variation for Fe and Zn grain concentration (Monasterio and Graham, 2000), however this variation within the range that most nutritionists agree is required to impact upon human health is known to occur only in unadapted wheat relatives, such as spelt wheat, diploid *Aegilops tauschii* and some wild tetraploids (Cakmak et al., 2002). To complicate wheat breeding further, the inheritance of Fe and Zn is quantitative in nature, affected by environment and influenced by seed shriveling (Trethowan, 2006a). As much of the Zn and Fe are associated with the aleurone layer of the seed (Graham et al., 2001) there is a dilution effect of these nutrients as yield potential increases. Nevertheless, Trethowan (2006a) showed that it is possible to simultaneously increase yield and Fe content

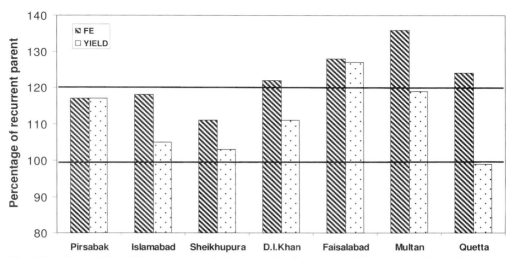

Fig. 5.11. Grain yield and Fe concentration in the grain of backcross derived lines compared to their recurrent parent at seven sites in Pakistan during the 2004–2005 crop season.

across a broad geographic area; as shown in Figure 5.11, wheat lines developed in Mexico with higher yield and Fe content maintained their performance when grown at seven sites in Pakistan.

Application of molecular markers in wheat breeding

Recent improvements in molecular-marker assay techniques, reduced cost, and projected improvements expected from single nucleotide polymorphism (SNP) technology, coupled with increasing numbers of markers for traits of importance, have resulted in increased use of marker-assisted selection (MAS) in applied wheat breeding (Eagles et al., 2001). The SNPs and large-scale sequencing will enable the development and deployment of more gene-based markers (Brookes, 1999), thereby improving the detection and application of markers in breeding. Molecular markers offer wheat breeders considerable scope to improve low-heritability traits (bioassay or field test), traits that cannot be evaluated at the breeding site, or traits that require gene pyramiding, as is the case with some disease

resistances. Most of the currently available wheat markers identify simply inherited Mendelian traits, although quantitative trait loci (QTL) analyses have identified and improved our understanding of some complex traits. William and Trethowan (2006) provide a description of many of the markers currently used in applied wheat breeding (Table 5.5).

An immediate application of molecular markers is the characterization of parents. At CIMMYT, markers are used to create gene profiles of all parental material. All known markers for traits are applied at this stage and parents are selected for crossing on the basis of predicted segregation of different characters (Trethowan, 2006b). If a trait has a high value and the alternative assays have low heritability, as is the case with many root-disease characters, then markers can be applied through the segregating phase. However, cost and time constraints will limit the number of assays that can be applied, and the wheat breeder must identify the most effective and efficient intervention points for MAS in the breeding process. If multiple markers for multiple traits are to be applied it is important that they be used as

Table 5.5.   Molecular markers currently in use for breeding applications at CIMMYT.

Trait	Gene	Type of marker	Chromosome	Source	Reference
Root stress					
Heterodera avenae	*Cre1*	STS	2BL	CSIRO*	Ogbonnaya et al. 2001
Heterodera avenae	*Cre3*	STS/SSR	2DL	CSIRO*	Ogbonnaya et al. 2001
Crown rot	Qtl-2.49	SSR	1DL	Public	Collard et al. 2005
Flour color/*P.neglectus*	*Rlnn-1*	STS	7BL	Public	Sharp et al. 2001
Boron tolerance	*Bo-1*	SSR	7BL	Public	Jefferies et al. 2000
Rust					
Stem rust	*Sr2*	STM/SSR	3BS	Public	Hayden et al. 2004
Stem rust	*Sr24*	STS		Public	Mago et al. 2005
Stem rust	*Sr25*	STS		Public	Prins et al. 2001
Stem rust	*Sr26*	STS/SSR	6AL	Public	Mago et al. 2005
Stem rust	*Sr38*	SSR	2AS	Public	http://maswheat.ucdavis.edu
Stem rust	*Sr39*	SSR	2B	Public	http://maswheat.ucdavis.edu
Durable leaf & brown rust	*Lr34/Yr18*	STS	7DS	CSIRO*	
Quality					
Quality	GBSS-null	SSR	4B	Public	Mclauchlan et al. 2001
Quality	Hardness	SSR	5ABD	Public	Giroux and Morris 1998
Quality	*Glu1BX*	STS	1BS	Public	Juhasz et al. 2003
Insect					
Russian wheat aphid	*Dn2*	SSR	7D	Public	http://maswheat.ucdavis.edu
Russian wheat aphid	*Dn4*	SSR	1D	Public	http://maswheat.ucdavis.edu
Hessian fly	*H25*	SSR	4A	Public	http://maswheat.ucdavis.edu
Other					
Barley yellow dwarf	*BDV2*	STS/SSR	7DS	Public	Stoutjesdijk et al. 2001
Fusarium head blight	QTL-Sumai3	SSR	3BS	Public	Anderson et al. 2001
Agronomy	*Rht-B1b*	STS	4B	Public	Ellis et al. 2002
Agronomy	*Rht-D1b*	STS	4D	Public	Ellis et al. 2002
Agronomy	*Rht8*	SSR	2D	Public	Korzun et al. 1998
Pairing homolog	*ph1b*	STS	5B	Public	Qu et al. 1998.

*Markers from sources identified have been obtained with material transfer agreements.

early as possible in the breeding process to improve the probability of identifying germplasm carrying not only the MAS target traits but also the suite of yield and quality attributes necessary for cultivar release.

If markers are expected to segregate in a topcross, the F_1 topcross progeny can be screened for markers segregating in the original bi-parental cross; those in the topcross parent are not screened because they will all be positive and heterozygous. Following this period of allele enrichment, markers can be applied at an intermediate stage of development, such as F_4, or once the fixed-line progeny have been identified; in this instance all target markers are applied. In simple crosses in which several markers are segregating, MAS can be applied in the F_2 or F_3 generation. The number of plants selected for screening from each population will depend on the number of markers segregating, the cost of the assay, and the availability/effectiveness of alternative assays. The majority of the currently available wheat markers are dominant in nature, which makes it impossible to identify heterozygotes. Once fixed-line progeny are identified that are positive for all target alleles it will be necessary to screen individual plants within each line or conduct progeny tests to ensure the target gene is homozygous before crossing.

Conclusion

Demand for wheat in developing countries continues to rise in the 21st century as population and prosperity expand. Increasing productivity by developing and deploying cultivars that have higher yield potential, resistance, and tolerance to biotic and abiotic stresses, combined with better crop management practices, can fill the gap between production and demand, thus reducing the need for imports. A properly focused, impact-oriented breeding effort by the International Wheat Improvement Network, which involves partnerships of international centers (CIMMYT and ICARDA), national programs, and advanced research institutions and is financed by various governments and agencies, is well positioned to continue developing wheat cultivars that have superior yield potential and can resist pressure from emerging biotic and abiotic stresses. Shuttle breeding between Mexico and key hot-spot locations worldwide, combined with international multilocation testing, is a fast way to develop widely adapted germplasm. Infusion of new technologies, such as MAS, should further enhance the selection efficiency for traits that are otherwise difficult to phenotype. Wide incorporation of durable resistance to rusts and other important diseases will not only continue to protect yield potential gains but will also benefit the environment and profitability by reducing dependency on agrochemicals. Wide utilization of synthetic wheats has introduced new genetic diversity for drought and heat tolerance and resistance to certain diseases and pests previously limited in wheat germplasm. Heat-stressed wheat areas in India and various other countries are projected to increase significantly by 2050 due to global warming. All possible approaches must be explored to improve heat tolerance for wheat to remain viable in such heat-stressed environments. Improved end-use quality characteristics of new high-yielding wheat germplasm should help farmers receive fair price in local as well as international markets. International germplasm and information exchange must be maintained—or even enhanced—to sustain wheat production.

References

Aguilar-Rincón, V.H., Singh, R.P. and Huerta-Espino, J. 2000. Inheritance of resistance to leaf rust in four synthetic hexaploid wheats. Agrociencia 34:235–246.

Anderson, J.A., Stack, R.W., Liu, S., Waldron, B.L., Fjeld, A.D., Coyne, C., Moreno-Sevilla, B., Mitchell, F.J., Song, Q.J., Cregan, P.B. and Frohberg, R. C. 2001. DNA markers for *Fusarium* head blight resistance QTLs in two wheat populations. Theor. Appl. Genet. 102:1164–1168.

Aquino, P., Carrion, F. and Calvo, R. 2002. Selected wheat statistics. In *CIMMYT 2000–2001 World Wheat Overview and Outlook: Developing No-till Packages for Small Scale Farmers*, edited by J. Ekboir, pp. 52–62. Mexico, D.F.:CIMMYT.

Axford, D.W.E., McDermott, E.E. and Redman, D.G. 1979. Note on the sodium dodecyl sulfate test of bread-making quality: Comparison with Pelshenke and Zeleny tests. Cereal Chemistry 56:582–584.

Borlaug, N.E. 1972. A cereal breeder and ex-forester's evaluation of the progress and problems involved in breeding rust resistant forest trees. In *Proc. of a NATO-IUFRO Advanced Study Institute: "Moderator's Summary," Biology of Rust Resistance in Forest Trees*, Aug. 17–24, 1969, pp. 615–642. USDA Forest Service Misc. Publication 1221.

Braun, H.J., Rajaram, S. and Van Ginkel, M. 1996. CIMMYT's approach to breeding for wide adaptation. Euphytica 92:175–183.

Briceno, F.G. 1992. Inheritance of resistance to Septoria leaf blotch in selected spring bread wheat genotypes. M.S. thesis. Oregon State Univ., Corvallis, USA, 86 pp.

Brooks, A.J. 1999. The essence of SNPs. Genetics 8:177–186.

Buerstmayr, H., Lemmens, M., Hartl, L., Doldi, L., Steiner, B., Stierschneider, M. and Ruckenbauer, P. 2002. Molecular mapping of QTLs for Fusarium head blight resistance in spring wheat. I. Resistance to fungal spread (type II resistance). Theor. Appl. Genet. 104:84–91.

Burridge, P.M., Palmer, G.A. and Hollamby, G.J. 1994. Developing a strategy for rapid wheat quality screening. In *Proceedings of the 7th Assembly of the Wheat Breeding Society of Australia*, edited by J. Paull, I.S. Dundas, K.J. Shepherd and G.J. Hollamby, pp. 57–60. Adelaide, Australia.

Cakmak, I. and Braun, H.J. 2001. Genotypic variation for Zn efficiency. In *Application of Physiology in Wheat Breeding*, edited by M.P Reynolds, J.I. Ortiz-Monasterio and A. McNab, pp. 200–207. Mexico, D.F.: CIMMYT.

Cakmak, I., Graham, R.D. and Welch, R.M. 2002. Agricultural and molecular genetic approaches to improving nutrition and preventing micronutrient malnutrition globally. In *Encyclopedia of Life*

Support Systems Vol. 2, pp. 1075–1099. UNESCO publishing.

Caldwell, R.M. 1968. Breeding for general and/or specific plant disease resistance. In *Proc. 3rd Int. Wheat Genetics Symp.*, pp. 263–272. Canberra, Australia.

Collard, B.C.Y., Grams, R.A., Bovill, W.D., Percy, C.D., Jolley, R., Lehmensiek, A., Wildermuth, G. and Sutherland, M.W. 2005. Development of molecular markers for crown rot resistance in wheat: Mapping of QTLs for seedling resistance in a '2-49' × 'Janz' population. Plant Breeding 124:532–537.

Cox, D.J. 1991. Performance of hard red winter wheat cultivars under conventional-till and no-till systems. North Dakota Farm Research 48:17–20.

Dao, T.H. and Nguyen, H.T. 1989. Growth response of cultivars to conservation tillage in a continuous wheat cropping system. Agronomy Journal 81: 923–929.

de Sousa, C.N.A. 1998. Classification of Brazilian wheat cultivars for aluminum toxicity in acid soils. Plant Breeding 117:217–221.

DeLacy, I.H., Fox, P.N., Corbett, J.D., Crossa, J., Rajaram, S., Fischer, R.A. and Van Ginkel, M. 1994. Long-term association of locations for testing spring bread wheat. Euphytica 72:95–106.

Diaz de Ackermann, M. and Kohli, M.M. 1998. Research on *Pyrenophora tritici-repentis* tan spot of wheat in Uruguay. In *Helminthosporium Blights of Wheat: Spot Blotch and Tan Spot*, edited by E. Duveiller, H.J. Dubin, J. Reeves and A. McNab, pp. 134–141. CIMMYT, Mexico.

Ditsch, D.C. and Grove, J.H. 1991. Influence of tillage on plant populations, disease incidence, and grain yield of two soft red winter wheat cultivars. Journal of Production Agriculture 4:360–365.

Dreccer, F.M., Borgognone, M.G., Ogbonnaya, F.C., Trethowan, R.M. and Winter, B. 2006. CIMMYT-selected synthetic bread wheats for rainfed environments: Yield evaluation in Mexico and Australia. Australian Journal of Agricultural Research (in review).

Duveiller, E. and Dubin, H.J. 2002. Helminthosporium leaf blights: Spot blotch and tan spot. In *Bread Wheat Improvement and Production*, edited by B. C. Curtis, S. Rajaram, H. Gomez-Macpherson, pp. 285–299. Plant Production and Protection Series No. 30. Food and Agriculture Organization of the United Nations, Rome.

Duveiller, E. and Gilchrist, L. 1994. Production constraints due to *Bipolaris sorokiniana* in wheat: Current situation and future prospects. In *Wheat in Heat Stressed Environments: Irrigated, Dry Areas and Rice-Wheat Farming Systems*, edited by D.A. Saunders and G.P. Hettel, pp. 343–352. CIMMYT, Mexico.

Eagles, H.A., Hollamby, G.L., Eastwood, R.F. and Podlich, D.W. 2001. Use of markers for monogenically and polygenically inherited traits. In *Proc. 10th Australian Wheat Breeding Assembly*, pp. 35–37. Mildura, Australia.

Ellis, M.H., Spielmeyer, W., Gale, K.R., Rebetzke, G. J. and Richards, R.A. 2002. "Perfect" markers for the *Rht-B1b* and *Rht-D1b* dwarfing genes. Theor. Appl. Genet. 105:1038–1042.

Endo, T.R. and Gill, B.S. 1996. The deletion stocks of common wheat. J. Hered. 87:295–307.

FAO. 2001. Food Outlook. Rome, Italy: FAO.

Gill, B.S. and Friebe, B. 2002. Cytogenetics, phylogeny and evolution of cultivated wheats. In *Bread Wheat Improvement and Production*, edited by B.C. Curtis, S. Rajaram and H. Gomez Macpherson, pp. 71–88. FAO, Rome.

Giroux, M.J. and Morris, C.F. 1998. Wheat grain hardness results from highly conserved mutations in the puroindoline a and b. Proc. Natl. Acad. Sci. 95: 6262–6266.

Gororo, N.N., Eagles, H.A., Eastwood, R.F., Nicolas, M.E. and Flood, R.G. 2002. Use of *Triticum tauschii* to improve yield of wheat in low-yielding environments. Euphytica 123:241–254.

Graham, R.D., Welch, R.M. and Bouis, H.E. 2001. Addressing micronutrient malnutrition through enhancing the nutritional quality of staple foods: Principles, perspectives and knowledge gaps. Advances in Agronomy 70:77–142.

Griffey, C.A. and Das, M.K. 1994. Inheritance of adult-plant resistance to powdery mildew in Knox 62 and Massey winter wheats. Crop Sci. 33:641–646.

Hatchett, J.H., Martin, T.J. and Livers, R.W. 1981. Expression and inheritance of resistance to Hessian fly in synthetic hexaploid wheats derived from *Triticum tauschii* (Coss.) Schmal. Crop Science 21:731–734.

Hayden, M.J., Kuchel, H. and Chalmers, K.J. 2004. Sequence tagged microsatellites for the *Xgwm533* locus provide new diagnostic markers to select for the presence of stem rust resistance gene *Sr2* in bread wheat (*Triticum aestivum* L). Theor. Appl. Genet. 109:1641–1647.

Huerta-Espino, J. and Singh, R.P. 2005. Effects of 7DL.7Ag (mutated for reduced flour yellowness) and 4BL.2R translocations on grain yield of wheat under irrigated and stressed environments. In *Abstr. of 7th International Wheat Conference: Wheat Production in Stressed Environments,* Nov 27–Dec. 2, p. 339. Mar de Plata, Argentina.

Innes, R.L. and Kerber, E.R. 1994. Resistance to wheat leaf rust and stem rust in *Triticum tauschii* and inheritance in hexaploid wheat of resistance transferred from *T. tauschii*. Genome 37:813–822.

Jefferies, S.P., Pallotta, M.A., Paull, J.G., Karakousis, A., Kretschmer, J.M., Manning, S., Islam, A.K.M.R., Langridge, P. and Chalmers, K.J. 2000. Mapping and validation of chromosome regions conferring boron toxicity tolerance in wheat (*Triticum aestivum*). Theor. Appl. Genet. 101:767–777.

Jlibene, M., Gustafson, J.P. and Rajaram, S. 1992. A field disease evaluation method for selecting wheat resistant to *Mycosphaerella graminicola*. Plant Breeding 108:26–32.

Johnson, R. 1988. Durable resistance to yellow (stripe) rust in wheat and its implications in plant breeding. In *Breeding Strategies for Resistance to the Rusts of Wheat*, edited by N.W. Simmonds and S. Rajaram, pp. 63–75. CIMMYT, Mexico.

Jones, G.D.P., Jessop, R.S. and Blair, G.J. 1992. Alternative methods for the selection of phosphorous efficiency in wheat. Field Crops Res. 30: 29–40.

Joshi, A.K., Chand, R., Kumar, S. and Singh, R.P. 2004. Leaf tip necrosis: A phenotypic marker associated with resistance to spot blotch disease in wheat. Crop Sci. 44:792–796.

Juhasz, A., Gardonyi, M., Tamias, L. and Bedo, Z. 2003. Characterization of the promoter region of *Glu-1Bx7* gene from overexpressing lines of an old Hungarian wheat variety. *Proc. 10th Int. Wheat Genetics. Symp.*, edited by N.E. Pogna, M. Romano, E.A. Pogna and G. Galterio, pp. 1448–1350. Pasteum, Italy.

Knott, D.R. 1980. Mutation of a gene for yellow pigment linked to *Lr19* in wheat. Can. J. Genet. Cytol. 22:651–654.

Korzun, V., Roder, M.S., Ganal, M.W., Worland, A.J. and Law, C.N. 1998. Genetic analysis of the dwarfing gene (*Rht8*) in wheat. Part 1. Molecular mapping of *Rht8* on the short arm of chromosome 2D of bread wheat (*Triticum aestivum* L.). Theor. Appl. Genet. 96:1104–1109.

Lage, J. and Trethowan, R.M. 2006. Synthetic hexaploid wheat improves bread wheat adaptation to rainfed environments globally. Field Crops Res. (under review).

Lillemo, M., Chen, X.M., He, Z.H. and Singh, R.P. 2005. Leaf rust resistance gene *Lr34* is involved in powdery mildew resistance of CIMMYT bread wheat line Saar. In *Abstr. of 7th Int. Wheat Conf.: Wheat Production in Stressed Environments*, Nov 27–Dec. 2, p. 17. Mar de Plata, Argentina.

Lillemo, M., Skinnes, H., Singh, R.P. and Van Ginkel, M. 2006. Genetic analysis of partial resistance to powdery mildew in bread wheat line Saar. Plant Dis. 90:225–228.

Lobell, D.B., Ortiz-Monasterio, I., Asnar, G.P., Matson, P.A., Naylor, R.L. and Falcon, W.P. 2005. Analysis of wheat yield and climatic trends in Mexico. Field Crops Res. 94:250–256.

MacRitchie, F. 1994. Role of polymeric proteins in flour functionality. In *Wheat Kernel Proteins: Molecular and Functional Aspects*, pp. 145–150. Bitervo, Italy: Universita degli studi della Tuscia.

Mago, R., Bariana, H.S., Dundas, I.S., Spielmeyer, W., Lawrence, G.J., Prior, A.J. and Ellis, J.G. 2005. Development of PCR markers for the selection of wheat stem rust resistance genes *Sr24* and *Sr26* in diverse wheat germplasm. Theor. Appl. Genet. 111:496–504.

Marasas, C.N., Smale, M. and Singh, R.P. 2004. The economic impact in developing countries of leaf rust resistance breeding in CIMMYT-related spring bread wheat. Economics Program Paper 04-01. Mexico, DF: CIMMYT. p. 36.

Mateos-Hernandez, M., Singh, R.P., Hulbert, S.H., Bowden, R.L., Huerta-Espino, J., Gill, B.S. and Brown-Guedira, G. 2006. Targeted mapping of ESTs linked to the adult plant resistance gene *Lr46* in wheat using synteny with rice. Functional and Integrative Genomics 6:122–131.

Matus-Tejos, I.A. 1993. Genetica de la resistencia a *Septoria tritici* en trigos harineros. M.S. thesis. Colegio Postgrad., Montecillos, Mexico. 82 pp.

McIntosh, R.A. 1992. Close genetic linkage of genes conferring adult-plant resistance to leaf rust and stripe rust in wheat. Plant Pathol. 41:523–527.

Mclauchlan, A., Ogbonnaya, F.C., Hollingsworth, B., Carter, M., Gale, K.R., Henry, R.J., Holten, T.A., Morell, M.K., Rampling, L.R., Sharp, P.J., Shariflou, M.R., Jones, M.G.K. and Appels, R. 2001. Development of PCR-based DNA markers for each homoeo-allele of granule-bound starch synthase and their application in wheat breeding programs. Aust. J. Agric. Res. 52:1409–1416.

Mello-Sampayo, T. 1971. Genetic regulation of meiotic chromosome pairing by chromosome 3D of *Triticum aestivum*. Nature New Biol. 230:22–23.

Monasterio, I. and Graham, R.D. 2000. Breeding for trace minerals in wheat. Food and Nutrition Bulletin 21:393–396.

Moody, D.B., Rathjen, A.J. and Cartwright, B. 1990. Yield evaluation of a gene for boron tolerance using backcross-derived lines. In *Proc. 6th Assembly of Wheat Breeding Society of Australia*, edited by L. O'Brien, pp. 117–121.

Navabi, A., Singh, R.P., Tewari, J.P. and Briggs, K.G. 2003. Genetic analysis of adult-plant resistance to leaf rust in five spring wheat genotypes. Plant Dis. 1522–1529.

Navabi, A., Singh, R.P., Tewari, J.P. and Briggs, K.G. 2004. Inheritance of high levels of adult-plant resistance to stripe rust in five spring wheat genotypes. Crop Sci. 44:1156–1162.

Niederhauser, J.S., Cervantes, J. and Servin, L. 1954. Late blight in Mexico and its implications. Phytopathology 44:406–408.

Ogbonnaya, F., Dreccer, F., Ye, G., Trethowan, R., Lush, D., Shepperd, J. and Van Ginkel, M. 2006. Yield of synthetic backcross-derived lines. In *Extended Abstracts of the International Symposium on Wheat Yield Potential: Challenges to Interna-*

tional Wheat Breeding, March 20–24th, 2006, edited by M.P. Reynolds and D. Godinez, p. 12. Ciudad Obregón, Mexico,

Ogbonnaya, F.C., Subrahmanyam, N.C., Moullet, O., Majnik, J. de, Eagles, H.A., Brown, J.S., Eastwood, R.F., Kollmorgen, J., Appels, R. and Lagudah, E.S. 2001. Diagnostic DNA markers for cereal cyst nematode resistance in bread wheat. Aust. J. Agric. Res. 52: 1367–1374.

Ortiz-Monasterio, R.J.I., Sayre, K.D., Rajaram, S. and McMahon, M. 1997. Genetic progress in wheat yield and nitrogen use efficiency under four nitrogen rates. Crop Sci. 37:898–904.

Peña, R.J., Trethowan, R.M., Pfeiffer, W.H. and Van Ginkel, M. 2002. Quality (End-Use) Improvement in Wheat. Compositional, Genetic, and Environmental Factors. In *Quality Improvement in Field Crops*, edited by A.S. Basra and L.S. Randhawa, pp.1–137. Haworth Press, Inc. N.Y.

Pfeiffer, W.H., Trethowan, R.M., van Ginkel, M., Ortiz-Monasterio, I. and Rajaram, S. 2005. Breeding for abiotic stress tolerance in wheat. In *Abiotic Stresses: Plant Resistance through Breeding and Molecular Approaches*, edited by M. Ashraf and P.J.C. Harris, pp. 401–489. The Haworth Press, Inc. N.Y.

Pingali, P.L. and Rajaram, S. 1999. Global wheat research in a changing world: Options for sustaining growth in wheat productivity. In *CIMMYT 1998–99 World Wheat Facts and Trends. Global Wheat Research in a Changing World: Challenges and Achievements*, edited by Pingali, P.L.Mexico, D.F.: CIMMYT.

Prins, R., Groenewald, J.Z., Marais, G.F., Snape, J.W. and Koebner, R.M.D. 2001. AFLP and STS tagging of *Lr19*, a gene conferring resistance to leaf rust in wheat. Theor. Appl. Genet. 103:618–624.

Qarooni, J. 1994. Historic and present production, milling and baking industries in the countries of the Middle East and North Afrcia. Manhattan, K.S.: Department of Grain Science and Industry, Kansas State University.

Qu, L.J., Foote, T.N., Roberts, M.A., Money, T.A., Aragon-Alcaide, L., Snape, J.W. and Moore, G. 1998. A simple PCR-based method for scoring the *ph1b* deletion in wheat. Theor. Appl. Genet. 96:371–375.

Rajaram, S., Borlaug, N.E. and Van Ginkel, M. 2002. CIMMYT international wheat breeding. In *Bread Wheat Improvement and Production*, edited by B. C. Curtis, S. Rajaram and H. Gomez-Macpherson, pp. 103–117. FAO, Rome.

Rajaram, S., Singh, R.P. and Torres, E. 1988. Current CIMMYT approaches in breeding wheat for rust resistance. In *Breeding Strategies for Resistance to the Rust of Wheat*, edited by N.W. Simmonds and S. Rajaram, pp. 101–118CIMMYT. D.F.: Mexico.

Rajaram, S., Van Ginkel, M. and Fischer, R.A. 1994. CIMMYT's wheat breeding mega-environments (ME). In *Proc. 8th Intl. Wheat Genet. Symp.*, edited by Z.S. Li and Z.Y. Xin, pp.1101–1106. Chinese Academy of Agricultural Sciences, Beijing, China.

Rasmusson, D.C. 1996. Germplasm is paramount. Increasing Yield Potential in Wheat: Breaking the Barriers, edited by M. Reynolds, S. Rajaram and A. McNab, pp. 28–35. CIMMYT, Mexico.

Rees, R.G. and Platz, G.J. 1992. Tan spot and its control—some Australian experiments. In *Advances in Tan Spot Research. Proc. of the 2nd Int. Tan Spot Workshop*, edited by L.J. Francl, J.M. Krupinsky and M.P. Mc Mullen, p. 1. Fargo, ND, USA, North Dakota Agric. Expt. Stn.

Reeves, T.G., Pingali, P.L., Rajaram, S. and Cassaday, K. 2001. Crop and natural resource management strategies to foster sustainable wheat production in developing countries. In *Wheat in a Global Environment,* edited by Z. Bedo and L. Lang, pp. 23–36. Kluwer Academic Publishers, the Netherlands.

Reynolds, M.P., Calderini, D.F., Condon, A.G. and Rajaram, S. 2001. Physiological basis of yield gains in wheat associated with the *Lr19* translocation from *Agropyron elongatum*. In *Wheat in a Global Environment*, edited by Z. Bedo and L. Lang, pp. 345–351. Kluwer Academic Publishers, the Netherlands.

Reynolds, M.P., Singh, R.P., Ibrahim, A., Ageeb, O.A., Larqué-Saavedra, A. and Quick, J.S. 1998. Evaluating physiological traits to compliment empirical selection for wheat in warm environments. Euphytica 100:85–94.

Riley, R. and Chapman, V. 1958. Genetic control of the cytologically diploid behavior of hexaploid wheat. Nature 182:713–715.

Roelfs, A.P. 1988. Resistance to leaf rust and stem rust in wheat. In *Breeding Strategies for Resistance to the Rusts of Wheat*, edited by N.W. Simmonds and S. Rajaram, pp. 10–22. CIMMYT, Mexico.

Rosenzweig, C. and Hillel, D. 1995. Potential impacts of climate change on agriculture and food supply. Consequences 1:2. U.S. Global Change Research Information Office, Washington, D.C.

Salam, A., Hollington, P.A., Gorham, J., Jones, R.G.W. and Gliddon, C. 1999. Physiological genetics of salt tolerance in wheat (*Triticum aestivum* L.): Performance of wheat varieties, inbred lines and reciprocal F-1 hybrids under saline conditions. Journal of Agronomy and Crop Science 183:145–156.

Sayre, K.D., Rajaram, S. and Fischer, R.A. 1997. Yield potential progress in short bread wheats in Northwest Mexico. 37:36–42.

Sears, E.R. 1954. The aneuploids of common wheat. Mo. Agric. Exp. Sta. Res. Bull. 572:1–58.

Sears, E.R. 1977. Induced mutant with homoeologous pairing in common wheat. Can. J. Genet. Cytol. 19:585–593.

Sears, E.R. and Sears, L.M.S. 1978. The telocentric chromosomes of common wheat. In *Proc. 5th Int. Wheat Genet. Symp.*, edited by S. Ramanujam, pp. 389–407. New Delhi, Indian Agricultural Research Institute.

Sharma, R.C., Dubin, H.J., Devokota, R.N. and Bhatta, M.R. 1997 Heritability estimates of field resistance to spot blotch in four spring wheat crosses. Plant Breeding 116:64–68.

Sharp, P.J., Johnston, S., Brown, G., McIntosh, R.A., Pallotta, M., Carter, M., Bariana, H.S., Khatkar, S., Lagudah, E.S., Singh, R.P., Khairallah, M., Potter, R. and Jones, M.G.K. 2001. Validation of molecular markers for wheat breeding. Aust. J. Agric. Res. 52:1357–1366.

Shewry, P.R., Halford, N.G. and Tatham, A.S. 1992. High molecular weight subunits of wheat glutenin. Journal of Cereal Science 15:105–120.

Singh, R.P. 1992a. Association between gene *Lr34* for leaf rust resistance and leaf tip necrosis in wheat. Crop Sci. 32:874–878.

Singh, R.P. 1992b. Genetic association of leaf rust resistance gene *Lr34* with adult plant resistance to stripe rust in bread wheat. Phytopathology 82:835–838.

Singh, R.P. and Huerta-Espino, J. 2004. The use of "single-backcross, selected-bulk" breeding approach for transferring minor genes based rust resistance into adapted cultivars. In *Proc. of 54th Australian Cereal Chemistry Conference and 11th Wheat Breeders Assembly*, 21–24 September, 2004, Canberra, Australia, edited by C.K. Black, J.F. Panozzo and G.J. Rebetzke, pp. 48–51. Cereal Chemistry Division, Royal Australian Chemical Institute, North Melbourne, Australia.

Singh, R.P., Huerta-Espino, J. and Rajaram, S. 2000. Achieving near-immunity to leaf and stripe rusts in wheat by combining slow rusting resistance genes. Acta Phytopathlogica Hungarica 35:133–139.

Singh, R.P., Huerta-Espino, J., Rajaram, S. and Crossa, J. 1998a. Agronomic effects from chromosome translocations 7DL.7Ag and 1BL.1RS in spring wheat. Crop Sci. 38:27–33.

Singh, R.P., Huerta-Espino, J., Sharma, R. and Joshi, A.K. 2006. High yielding spring bread wheat germplasm for irrigated agro-ecosystems. In *Extended Abstracts of the International Symposium on Wheat Yield Potential:Challenges to International Wheat Breeding*, edited by M.P. Reynolds and D. Godinez, p. 5. CIMMYT, Ciudad Obregón, Mexico. Mexico, D.F.

Singh, R.P., Kazi-Mujeeb, A. and Huerta-Espino, J. 1998b. *Lr46*: A gene conferring slow rusting resistance to leaf rust in wheat. Phytopathology. 88:890–894.

Singh, R.P., Ma, H. and Rajaram, S. 1995. Genetic analysis of resistance to scab in spring wheat cultivar Frontana. Plant Dis. 79:238–240.

Singh, R.P. and McIntosh, R.A. 1986. Genetics of resistance to *Puccinia graminis tritici* and *Puccinia recondita tritici* in Kenya plume wheat. Euphytica 35:245:256.

Singh, R.P. and Rajaram, S. 1992. Genetics of adult-plant resistance to leaf rust in 'Frontana' and three CIMMYT wheats. Genome 35:24–31.

Singh, R.P. and Rajaram, S. 1994. Genetics of adult plant resistance to stripe rust in ten spring bread wheats. Euphytica 72:1–7.

Singh, R.P. and Rajaram, S. 2002. Breeding for disease resistance in wheat. In *Bread Wheat Improvement and Production*, edited by B.C. Curtis, S. Rajaram and H. Gomez-Macpherson, pp. 141–156. FAO, Rome.

Singh, R.P., Rajaram, S., Miranda, A., Huerta-Espino, J. and Autrique, E. 1998c. Comparison of two crossing and four selection schemes for yield, yield traits, and slow rusting resistance to leaf rust in wheat. Euphytica 100:35–43.

Spielmeyer, W., McIntosh, R.A., Kolmer, J. and Lagudah, E.S. 2005. Powdery mildew resistance and *Lr34/Yr18* genes for durable resistance to leaf and stripe rust cosegregate at a locus on the short arm of chromosome 7D of wheat. Theor. Appl. Genet. 111:731–735.

Stoutjesdijk, P., Kammholz, S.J., Kleven, S., Matssy, S., Banks, P.M. and Larkin, P.J. 2001. PCR-based molecular marker for the *Bdv2 Thinopyrum intermedium* source of barley yellow dwarf virus resistance in wheat. Aust. J. Agric. Res. 52:1383–1388.

Trethowan, R.M. 2006a. Breeding wheat for high iron and zinc at CIMMYT: State of the art, challenges and future prospects. In *Proc. 7th International Wheat Conference, Mar del Plata, Argentina*, November 27–December 2, 2005 (in press).

Trethowan, R.M. 2006b. Mejoramiento de cultivos para ambientes con factores desfavorables abióticos: Qué herramientas podemos aplicar para mejorar la productividad de las plantas y ejemplos de su aplicación. In *Proc. "A todo trigo: Conocimiento y produccion," Federación de centros y Entidades Germiales de Acopiadores de Cereales, Mar del Plata, May 17–18, Argentina*, pp. 59–68.

Trethowan, R., Borja, J. and Kazi-Mujeeb, A. 2003. The impact of synthetic wheat on breeding for stress tolerance at CIMMYT. Annual Wheat Newsletter 49:67–69.

Trethowan, R.M. and Hodson, D., Braun, H.J. and Pfeiffer, W.H. 2005a. Wheat breeding environments. In *Impacts of International Wheat Breeding Research in the Developing World: 1988–2002*, edited by J. Dubin, M.A. Lantican and M.L. Morris, pp. 4–11. Mexico, D.F.: CIMMYT.

Trethowan, R.M., Peña, R.J. and Van Ginkel, M. 2001. The effect of indirect tests for quality on grain yield and industrial quality of bread wheat. Plant Breeding 120:509–512.

Trethowan, R.M., Reynolds, M.P., Sayre, K.D. and Ortiz-Monasterio, I. 2005b. Adapting wheat cultivars to resource conserving farming practices and human nutritional needs. Annals of Applied Biology 146:404–413.

Trethowan, R.M., Van Ginkel, M. and Rajaram, S. 2002. Progress in breeding for yield and adaptation in global drought affected environments. Crop Science 42:1441–1446.

Van der Plank, J.E. 1963. Plant Diseases: Epidemics and Control. Academic Press, New York and London.

Van Ginkel, M., Ortiz-Monasterio, I., Trethowan, R.M., Hernandez, E. 2001. Methodology for selecting segregating populations for improved N-use efficiency in bread wheat. Euphytica 119:223–230.

Van Ginkel, M., van der Schaar, W. and Zhuping, Y. 1996. Inheritance of resistance of scab in two wheat cultivars from Brazil and China. Plant Dis. 80:863–867.

Velázquez-Cruz, C. 1994. Genetica de la resistencia a *Bipolaris sorokiniana* en trigos harineros. M.S. thesis. Colegio Postgrad., Montecillos, Mexico. 84 pp.

Villareal, R.L., Mujeeb-Kazi, A., Fuentes-Davila, G., Rajaram, S. and Toro, E.D. 1994. Resistance to Karnal bunt (*Tilletia indica* Mitra) in synthetic hexaploid wheats derived from *Triticum turgidum* × *T. tauschii*. Plant Breeding 112:63–69.

Villareal, R.L., Mujeeb-Kazi, A., Gilchrist, L. and Del Toro, E. 1995. Yield loss to spot blotch in spring bread wheat in warm non-traditional wheat production areas. Plant Dis. 79:893–897.

Wang, Z.L., Li, L.H., He, Z.H., Duan, X.Y., Zhou, Y.L., Chen, X.M., Wang, H., Lillemo, M., Singh, R.P. and Xia, X.C. 2005. Seedling and adult-plant resistance to powdery mildew in Chinese bread wheat cultivars and lines. Plant Dis. 89:457–463.

Weegels, P.L., Hamer, R.J. and Schofield, J.D. 1996. Critical Review: Functional properties of wheat glutenin. Journal of Cereal Science 23:1–18.

William, H.M., Crosby, M., Trethowan, R., Van Ginkel, M., Mujeeb-Kazi, A., Pfeiffer, W., Khairallah, M. and Hoisington, D. 2003a. Molecular marker service laboratory—an interface between the laboratory and the field. In *Proc. 10ᵗʰ Int. Wheat Genetics Symp.*, edited by N.E. Pogna, M. Romano, E.A. Pogna and G. Galterio, pp. 852–854.

William, M., Singh, R.P., Huerta-Espino, J., Ortiz Islas, S. and Hoisington, D. 2003b. Molecular marker mapping of leaf rust resistance gene *Lr46* and its association with stripe rust resistance gene *Yr29* in wheat. Phytopathology 93:153–159.

William, H.M. and Trethowan, R.M. 2006. Wheat molecular breeding: A case for optimism. In *Proc. of an International Symp. on Wheat Yield Potential: Challenges to International Wheat Breeding*, 20–24 March 2006, edited by M.P. Reynolds and D. Godinez. Ciudad, Obregón, Mexico. CIMMYT, Mexico, D.F. (in press).

Zeleny, L. 1947. A simple sedimentation test for estimating the bread-making and gluten qualities of wheat flour. Cereal Chemistry 24:465–475.

Chapter 6

Rice Breeding for Sustainable Production

Sant S. Virmani and M. Ilyas-Ahmed

Introduction

Rice (*Oryza sativa* L.) is staple food for half of humanity. More than three billion people, most of them in Asia, depend on rice as their major source of calories. Rice is the only major cereal crop that is consumed almost exclusively by human beings. It is cultivated worldwide in 114 of the world's 193 countries: 30 in Asia, 28 in America, 41 in Africa, 11 in Europe, and four in Oceana. It occupies about 11 percent of the world's arable land. Rice is grown annually on more than 150 million hectares with an annual production of around 600 million tons.

Rice is probably the most diverse of all food crops. It is grown as far north as Manchuria (50° N) in China and as far south as Uruguay and New South Wales (35° S) in Australia (Khush, 2005). Rice grows at more than 3,000 meters elevation in Nepal and Bhutan and three meters below sea level in Kerala, India. Rice is cultivated in five major ecosystems; irrigated, rainfed lowland, upland, deep water, and tidal wetlands. Global area, production, and productivity under different ecosystems are given in Table 6.1.

Rice is cultivated in six continents across the world, viz., Asia, Africa, South America (Latin America), North America (United States), Europe, and Australia. Area, production, and productivity of rice in these continents are given in Table 6.2.

About 90% of the global rice area is in the Asian continent, where more than 90% of the world's rice is produced and consumed. It is home for 60% of the world's population and two-thirds of the world's poor reside there. Major rice-producing countries in Asia are China, India, Indonesia, Bangladesh, Vietnam, Thailand, Myanmar, the Philippines, and Japan. Rice is cultivated in Asia in irrigated (57.4%), shallow lowland (33.5%), upland (6.4%), and deep-water (2.7%) ecosystems. In Africa, rice area has been increasing during the last couple of decades. The major rice-producing countries in Africa are Egypt, Nigeria, Madagascar, Sierra Leone, Liberia, and the Ivory Coast. Egypt has the second highest productivity in the world (9.1 t/ha) after Australia (9.66 t/ha).

In Latin America, major rice-producing countries are Brazil, Colombia, Peru, Ecuador, and Uruguay. In North America, rice is cultivated in five southern and southwestern states of the U.S., with a relatively high productivity of 7.0 t/ha. In Europe, rice is cultivated in a very limited area in Italy, Spain, and France, with an average productivity of about 5.0 t/ha. In Australia, though rice is cultivated only on 0.15 million ha, the productivity of the rice crop (9.66 t/ha) is the highest in the world.

The top 10 countries in the world, according to rice-growing area, are given in Table 6.3.

Table 6.1.   Global rice area ecosystems.

Ecosystem	Area (m.ha)	Area (%)	Total production (m. tons)	Average productivity (t/ha)
Irrigated	78	50	390	5.0
Rainfed	53	34	185	3.5
Deep-water	10	07	11	1.1
Upland	13	09	14	1.1
Total	154	100	600	3.9

Adapted from Rice Almanac, Third Edition (2002), IRRI.

Table 6.2.   Area, production, and productivity of rice in different continents (2000).

Continent	Area (m.ha)	% of global area	Production (m. tons)	% of global production	Average productivity (t/ha)
Asia	137.60	89.49	545.5	91.1	3.96
Africa	7.78	5.06	17.2	2.9	2.21
South America	6.39	4.16	23.0	3.8	3.61
North America (USA)	1.23	0.80	8.7	1.5	7.04
Europe	0.61	0.40	3.1	0.5	5.05
Australia	0.15	0.09	1.4	0.2	9.66
World	153.76	100	598.9	100	3.89

Source: Rice Almanac, Third Edition (2002), IRRI.

Table 6.3.   Countries with largest rice area (2000).

S. No.	Country	Rice area (m. ha)	Production (m. tons)	Productivity (t/ha)
1.	India	44.60	134.15	3.01
2.	China	30.50	190.16	6.23
3.	Indonesia	11.52	51.00	4.43
4.	Bangladesh	10.70	35.82	3.35
5.	Thailand	10.05	23.40	2.33
6.	Vietnam	7.65	32.55	4.25
7.	Myanmar	6.21	20.12	3.24
8.	Philippines	4.04	12.41	3.08
9.	Brazil	3.67	11.17	3.04
10.	Pakistan	2.31	7.00	3.03

Source: Rice Almanac, Third Edition (2002), IRRI.

The 10 countries account for more than 85% of the global rice area; the first five countries account for more than 70% of the global rice-growing area (Table 6.3). Nine of these 10 countries are in Asia. The tenth is in South America (Brazil), with 3.67 million hectares devoted to rice crops.

With respect to production, China ranks first and India second. This is partly due to large-scale adoption of hybrid rice in China. China also has higher productivity of inbred rices due to rich soils, 100% irrigated area, and favorable climatic conditions. Thailand's productivity is below that for Vietnam, mainly due to cultivation of aromatic, exportable rice varieties in Thailand, which are comparatively lower yielding. Japan (11.86 m tons) makes this list of top 10 rice-producing countries, replacing Pakistan (Table 6.4). A large area in Pakistan is planted with aromatic, long slender exportable Basmati rices, which again are comparatively lower yielding. When the highest productivity of rice is considered worldwide, many countries with very limited rice area make the list (Table 6.4).

China devotes by far the most area to rice—30 million hectares. The other nine countries each have less than or around one million hectares. Australia with 0.145, Egypt with 0.660, and the U.S., with 1.232 million hectares of rice, top the list in rice productivity.

Between 1948–2000, there was a considerable increase in global rice area, production, and productivity worldwide (Table 6.5).

Table 6.4. Top 10 countries with highest rice productivity (t/ha) (2000).

S. No.	Country	Productivity (t/ha)	Production (m. tons)	Area (m. ha)
1.	Australia	9.66	1.40	0.145
2.	Egypt	9.09	5.99	0.660
3.	U.S.A	7.04	8.37	1.232
4.	Japan	6.70	11.86	1.770
5.	Republic of Korea	6.59	7.07	1.072
6.	Uruguay	6.35	1.17	0.185
7.	China	6.23	190.17	30.50
8.	Italy	5.89	1.30	0.221
9.	Argentina	5.74	1.66	0.289
10.	Peru	5.55	1.67	0.300

Adapted from Rice Almanac, Third Edition (2002), IRRI.

Table 6.5. Global rice increase (%) during 1948–2000.

Global rice	1948	2000	% increase
Area (m. hectares)	86.70	153.76	177
Production (m. tons)	145.40	598.85	411
Productivity (t/ha)	1.68	3.89	231

Adapted from FAOSTAT (2006).

It is gratifying to note that while globally the area under rice cultivation has increased 1.77 times during the 52-year period from 1948 to 2000, as a result of a remarkable 2.31 increase in worldwide productivity, total global rice production has increased more than four times.

Despite the remarkable progress achieved in rice production and productivity during the last 50-plus years, there cannot be any complacency on this front. About 75 million people are added to the global population annually. Thus, the biggest challenge facing humanity in the 21st century is ensuring food security for the burgeoning billions. The present world population of 6.55 billion is expected to reach 7.52 billion by 2020 and 9.08 billion by 2050 (U.S. Census Bureau, online information). Ninety-five percent of this population increase will take place in developing countries of Asia and Africa, where rice is a staple food. According to various estimates, 40% more rice must be produced by 2030 to satisfy the growing demand (Khush, 2005). This increase in production is to be achieved with an ever-shrinking resource base, i.e., with less land, less water, less fertilizers, and with reduced use of pesticides to prevent further deterioration of the environment. The enormity of the challenge of ensuring food security in the decades ahead becomes clear from the fact that during the next 50 years, more food will have to be produced than what has been cumulatively produced during the last 10,000 years, i.e., ever since organized agriculture began. Rice-breeding efforts will continue to play a major role in meeting this enormous challenge of food production.

In this chapter, after briefly considering the origin, taxonomic and species relationships, and evolution of rice, the rice-breeding achievements from the past 50-plus years are highlighted. Future breeding strategies to be adopted, including the use of biotechnological tools, are considered for their role in increasing rice production to meet the growing demand for rice.

Rice crops

The basic aspects of rice crops, such as origin, taxonomy and species relationships, evolution of cultivated rices, and floral biology, which are directly related to rice breeding, are summarized in this section.

Origin of rice

Rice was first domesticated in southern China and northeastern India—probably independently—about 8,000 years ago (Khush, 1997). The origin of cultivated rice has been a subject of debate ever since De Candolle (1882). However, as data and evidence have accumulated over the years, our understanding of this issue has become clearer. The major recent contributions in this field have come from Oka (1964; 1974; 1988), Shastry and Sharma (1973), Chang

(1976), Morishima (1984) and Sharma et al. (2000).

The genus *Oryza* probably originated 130 million years ago and spread as a wild grass in Gondwanaland—the supercontinent that broke up into present-day Asia, Africa, the Americas, Australia, and Antarctica during the early Cretaceous period (Fig. 6.1). The genus *Oryza* is found on all these continents except Antarctica (Khush and Virk, 2000).

Oryza sativa and *Oryza glaberrima* are the only two cultivated species in O*ryza*, which comprises 22 species. The cultivated species originated from a common ancestor with AA genome. The common Asian rice, *O. sativa*, grown around the world, and the African rice, *O. glaberrima*, grown to a

limited extent in a few African countries, are considered to be examples of parallel evolution.

The wild progenitor of *O. sativa* is the common Asian wild rice *O. rufipogon*, which varies from perennial to annual types. Annual types, called *O. nivara*, were domesticated, resulting in *O. sativa*. Based on archeological and historical evidence, the period of domestication of *O. sativa* began 7,000 to 10,000 years ago. This conclusion appears to be reasonable, given the fact that the oldest remains of Asian rice found in eastern China and northern India date back to 7000 B.C. This domestication of *O. sativa* must have occurred independently and concurrently at several locations in a broad belt

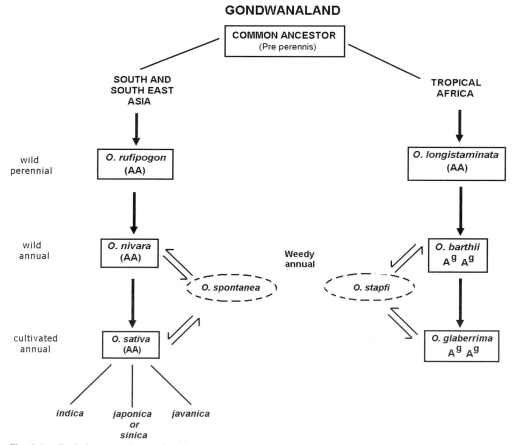

Fig. 6.1. Evolutionary pathway of cultivated rices.

that extends from the plains below the eastern foothills of Himalayas in India to upper Myanmar, northern Thailand, Laos, Vietnam, and beyond to southwest or southern China (Chang, 1976; Khush, 1997).

Domestication of *O. glaberrima* is believed to have occurred much later than that of *O. sativa* (~3,500 years later), probably in the Niger-river delta (Porteres, 1956), from annual species *O. breviligulata*, which in turn evolved from perennial *O. longistaminata*. Centers of diversity for *O. glaberrima* are swampy upper Niger-river basins and some areas around the Guinean coastline. In all probability, cultivars were first domesticated in their centers of diversity, where favorable conditions for the cultivars prevailed.

Taxonomy and species relationship

Rice belongs to genus *Oryza* of the tribe Oryzae of the family Graminea. While Roschevicz (1931) recognized only 20 species in the genus, Chatterjee (1948) recognized 23. Based on re-examination of the specimens in major herbaria of the world, Tateoka (1963, 1964) recognized 22 valid species, which is now a more or less universally accepted view. These 22 species are broadly grouped into four complexes, viz., Sativa, Officinalis, Ridleyi, and Meyeriana (Table 6.6).

The relationships among these 22 species, as revealed through extensive cytological investigations, indicate that the genus has diploid species (*O. sativa*, *O. glaberrima* and their wild relatives, *O. officinalis*,

Table 6.6. Species complexes in the genus *Oryza* and their genomic groups.

Complex/Species	Ploidy	Genome	Distribution
Sativa complex			
O. sativa	Diploid	AA	Worldwide
O. glaberrima	Diploid	$A^g A^g$	Africa
O. barthii	Diploid	$A^g A^g$	Africa
O. longistaminata	Diploid	$A^g A^g$	Africa
O. nivara	Diploid	AA	Tropical Asia
O. rufipogon	Diploid	AA	Tropical Asia
O. mesidionalis	Diploid	AA	Tropical Australia
O. glumacpetula	Diploid	AA	South America
Officinalis complex			
O. officinalis	Diploid	CC	Tropical Asia to New Guinea
O. eichingeri	Diploid	CC	East and West Africa
O. rhizomatis	Diploid	CC	Sri Lanka
O. minuta	Tetraploid	BBCC	Philippines, Papua New Guinea
O. punctata	Diploid	BB	
	Tetraploid	BBCC	Africa
O. latifolia	Tetraploid	CCDD	Latin America
O. alta	Tetraploid	CCDD	Latin America
O. grandighumis	Tetraploid	CCDD	South America
O. australiensis	Diploid	EE	Australia
Meyeriana complex			
O. granulata	Diploid	F	South & Southeast Asia
O. meyeriana	Diploid	F	Southeast Asia
O. brachyantha	Tetraploid	FF	Africa
O. schlechteri	Tetraploid	—	New Guinea
Ridleyi complex			
O. longiglumis	Tetraploid	—	New Guinea
O. ridleyi	Tetraploid	—	Southeast Asia

Source: Siddiq and Viraktamath (2000).

Table 6.7.   Genomic classification of rice species.

Genome	Genomic Constitution	Species	Distribution
A	AA	O. sativa, O. nivara, O. rufipogon (Formerly O. balunga), O. sativa f. spontanea	Asia
	A^lA^l	O. longistaminata, O. satpfii	Africa
	A^g A^g	O. glaberrima, O. barthii (formerly O. breviligulata)	
	A^{gd} A^{gd}	O. glumaepetula (Formerly O. cubensis)	America
C	CC	O. officinalis, O. cichengeri	Asia
B&C	BBCC	O. minuta, O. punctata	Asia
C&D	CCDD	O. latifolia, O. alta, O. grandiglunis	America
E	EE	O. australiensis	Australia
F	FF	O. brachyantha	Africa

Source: Siddiq and Viraktamath (2000).

O. australiensis, and *O. punctata*) with 2n = 24 chromosomes and tetraploid species (*O. minuta, O. latifolia, O. alta, O. malampuzhaensis, O. grandiglumis, O. longiglumis, O. punctata*, and *O. ridleyi*) with 2n = 48 chromosomes. Based on chromosome pairing and fertility of inter-specific crosses, these 22 species have been grouped into six distinct genomes, viz., A, B, C, D, E, and F (Table 6.7).

Cultivated rice is a secondarily balanced polyploid derived from an ancestral graminaceous species with a basic chromosome number of 5. During the evolutionary process, two of the chromosomes were duplicated resulting in two types of plants, viz., 2n = 14 and 2n = 10. The amphidiploid of crosses (2n = 12) between these two variants resulted in cultivated rice and its wild ancestors with 2n = 24 chromosomes (Siddiq and Viraktamath, 2000).

Evolution of cultivated rices

Based on geographical distribution of related species, cytomorphological features and sexual affinity, *O. perennis* (possibly *pre-perennis*, now extinct) appears to be the common ancestor of both the cultivated species. The Asian (*O. rufipogon*) and African (*O. longistaminate*) *perennis* are the result of parallel evolution aided by minor chromosomal re-arrangements and a series of macro and micro mutations, followed by natural selection. A single macro mutation might have led to the loss of its rhizomatous nature, giving rise to the annual weedy *O. nivara* and *O. barthii*. Intervention by man at this stage in the form of selection for early maturity and higher yield led, gradually over a period of time, to the emergence of the first forms of cultivated *O. sativa* in Asia and *O. glaberrima* in Africa. The evolutionary path of both the cultivated species suggests that they are of monophyletic origin.

Asian cultivated rice, subsequent to its origin and domestication, further differentiated into three eco-geographical races or sub-species, viz., *indica, japonica (sinica)*, and *javanica*. *Indica* rices are widely adapted and occur throughout humid regions of tropical and subtropical Asia (India, southern China, Vietnam, Thailand, Bangladesh, Myanmar, the Philippines, and Sri Lanka), whereas *japonicas* are adapted to sub-tropical and temperate regions of northern China, Korea, and Japan. The *javanicas* are adapted to the equatorial region of Indonesia. Evolution of sub-species in rice was due to natural and human selections in different geographical regions with different climatic conditions and consumer preferences. Characteristic features of these three sub-species,

Table 6.8.   Characteristic features of the subspecies of *O. sativa*.

Character	japonica	javanica	indica
Grain shape	Short	Long	Narrow
Length of 2nd leaf blade	Short	Long	Long
Angle between 2nd leaf and stem	Small	Small	Large
Texture of plant parts	Hard	Hard	Soft
Angle between flag leaf and stem	Medium	Large	Small
Flag leaf	Short, narrow	Long, wide	Long, narrow
Tiller number	Large	Small	Large
Tiller habit	Erect	Erect	Spreading
Leaf pubescence	None	Little	More
Glume pubescence	Dense	Dense	Sparse
Awns	Usually absent	More prominent	Less prominent
Threshability	Difficult	Difficult	Easy
Panicle length	Short	Long	Medium
Panicle branching	Few	Many	Medium
Panicle density	High	Moderate	Moderate
Panicle weight	Heavy	Heavy	Light
Plant height	Short	Taller	Tall
Grain (spikelet)	Short	Long	Long

Source: Siddiq and Viraktamath (2000).

indica, *japonica*, and *javanica*, are given in Table 6.8.

As far as the genetic affinity among these three subspecies is concerned, *indica* and *japonica* show the least genetic compatibility as compared with *indica-javanica* or *javanica-japonica* sub-species, as revealed by chromosome pairing behavior, F_1 sterility, and F_2 segregation pattern.

Floral biology

Rice is a self-pollinated crop, though outcrossing of 6–8% has been reported in some varieties under specific conditions. However, higher outcrossing is seen in wild relatives of rice, like *O. longistaminata* (100%) and *O. sativa f. spontanea* (50%). Male-sterile plants also show significant outcrossing (40–60%).

The reproductive phase starts with panicle initiation, which occurs 30–35 days before panicle emergence. Within a single day, the panicle emerges completely out of a flag leaf sheath. Anthesis starts immediately after panicle emergence in a definite sequence.

Spikelets on primary branches at the upper portion of the panicle open first, followed by the spikelets on secondary and tertiary branches. Spikelets on the lower-most secondary/tertiary branches open last. Complete anthesis in a panicle, from top to bottom, is completed in about 7–10 days. After completion of anthesis, it takes about 30 days for grain maturity in tropical rices, and 35–45 days in temperate rices.

Weather conditions greatly influence anthesis. On sunny days, anthesis starts around 9:00 am and continues till 2:00 pm. Maximum anthesis occurs between 10:00 and 11:00 am. On cloudy days, anthesis occurs in the afternoon. At the time of anthesis, lemma and palea separate, filaments of stamens elongate and protrude out, and anthers dehisce, releasing pollen. Most of the pollen is shed on protruded stigma of the same spikelet or neighboring spikelets, thus ensuring self-pollination. After 20–30 minutes of anthesis, anthers wither and spikelets close, leaving stamens sticking out from the seam of the lemma and palea. The pollen grains remain viable for only 2–5 minutes, depending upon the temperature, whereas stigmas continue to be receptive for 2–3 days after spikelet opens.

Rice breeding

Though the process of selecting desirable plants useful to human beings started with the domestication of the rice plant some 7,000–10,000 years ago, systematic, scientific rice breeding started only in the beginning of the 20th century. In this section, starting with a brief history of rice breeding, breeding methodologies, and salient achievements of breeding for enhanced yield potential, resistance to biotic and abiotic stresses, grain and nutritional quality, reduction in the maturity period, etc., are described. Major sources of information on the history of rice breeding are Pathasarathy (1972) and Siddiq and Viraktamath (2000), whereas for other aspects of rice breeding, such as breeding methodologies and achievements made in rice breeding, notable sources of information include Khush and Virk (2000, 2002, 2005), Khush and Brar (2002), Siddiq and Viraktamath (2000), Khush (2004), Coffman et al. (2004), Nanda (2000), and Nanda and Agrawal (2006).

History of rice breeding

Early breeding efforts were aimed at improvement of locally adapted popular varieties, which resulted in a large number of narrowly adapted varieties. Increased production came mainly through increased area. Early selection work was limited to purification of the varieties by removal of off-types. The next step was mass selection. Most of the varietal collections were limited to long-duration varieties grown in lowlands during monsoons. Rediscovery of Mendel's laws focused the attention of some of the rice breeders, particularly in India and Indonesia, on studying Mendelian ratios for qualitative characters, such as inheritance of anthocyanin pigmentation, awning, and other mutant traits. One unintended advantage of such studies was that it involved study of anthesis, time of flower opening, and emasculation and hybridization procedures, which was useful when subsequent work on recombination breeding was initiated. Field-plot techniques were improved for critical evaluation of genotypes. In the beginning, breeders grew individual plants in uniform rows to select better ones, unlike the prevalent practice of planting bunches of seedlings. Use of appropriate statistical tools enhanced breeding efforts with time.

The idea of the use of regional stations for breeding more widely adapted varieties was first conceived and implemented in Indonesia. Selections were made from the common hybrid material at six regional stations as well as at the central station at Bogor. Selections made at all the stations were evaluated at each of the experimental stations and subsequently in farmers' fields across Java. This procedure led to the development of varieties adapted to the whole of Java, covering different soil types and climates. In rice-exporting countries, such as Thailand and Burma (now Myanmar), much emphasis was placed on grain type and milling qualities. This restricted the number of varieties grown by farmers.

Breeders held the view that long-duration monsoon varieties grown in lowlands had higher yield potential; therefore, a large number of such varieties was developed. Limited work on breeding for resistance to blast disease was done in India during the late 1920s. One of the earliest releases from pure-line selection, CO-4, from *anaikomban*, was resistant to blast. It was crossed with the land race *Korangu samba*, a highly blast-susceptible but popular variety of the Tanjore delta during the then Madras presidency. Selection from a segregating population resulted in the identification of varieties CO-25 and CO-26. The earliest reference to breeding for insect pests is from Uttar Pradesh (India) for developing resistance against the rice bug.

The first three to four decades of modern rice breeding, up until 1950, had insufficient impact because the breeders produced too many narrowly adapted varieties, adaptation of varieties to wider regions was never tested, and there was negligible use of fertilizers. Rice breeders in different countries had practically no communication with one another, except through the very limited literature published on rice. Exchanges of seed material were very rare. World War II disrupted breeding activities in some major tropical Asian countries.

Immediately after World War II, food supplies fell short of demand and the threat of widespread famine loomed large. In response, the Food and Agriculture Organization (FAO) of the United Nations founded, within its framework, the International Rice Commission (IRC) in 1948 to find ways and means of increasing rice production. This was the first international cooperative effort to enhance rice production globally. The IRC recognized non-lodging habit, fertilizer responsiveness, early maturity, and wider adaptability as the important varietal characteristics for achieving higher and more stable yields. Accordingly, the FAO launched several regional network projects, including cataloging and maintenance of rice genetic stocks, *indica-japonica* hybridization, cooperative varietal trials, wide adaptability trials, variety-fertilizer interaction studies for *indica* varieties, uniform blast nursery development, and an international training program in rice breeding.

The ambitious *indica-japonica* hybridization program, initiated with an aim to recombine the high yield and fertilizer responsiveness of *japonicas* with the grain quality and disease and pest resistance of *indica*s, proved to be a disappointment, except for three varieties (ADT-27, 'Mashuri,' and 'Malinja') that were found to be promising for Malaysia. Mashuri was later used extensively as a parental line in developing several popular varieties.

In 1960, the advent of short-statured *indica* variety 'Taichung (Native)-1,' a derivative from a cross between spontaneous dwarf mutant 'Dee-gee-wu-gen' and the Taiwanese tall variety 'Tsai-Yuan-Chung,' provided the much-needed base for developing many high yielding varieties. TN-1, a short-duration, high-yielding variety that can be grown year-round, heralded the era of dwarf, short-statured, high-yielding varieties in tropical Asia.

Another significant development at this juncture was the establishment of the International Rice Research Institute (IRRI) at Los Baños, near Manila, Philippines. Breeding efforts were intensified at IRRI during 1961–62 to develop short-statured, fertilizer-responsive, high-yielding varieties. Within five years, the miracle rice IR-8 was developed and released, which subsequently gave impetus to the Green Revolution, along with dwarf wheat varieties in the Indian sub-continent. IR-5, a semi-tall variety, was released at the same time for less favorable ecosystems. IR-8 provided the momentum for the development of a series of dwarf, high-yielding varieties at IRRI, including IR-20, IR-26, IR-36, IR-42, IR-64, and IR-72. Several National Agricultural Research Systems (NARS) in Asia, utilizing these semi-dwarf, high-yielding IRRI varieties and the materials available through the International Rice Testing Program (IRTP, now the International Network for Genetic Enhancement of Rice or INGER), developed many semi-dwarf, high-yielding, locally adapted varieties that are now grown on large hectarages in different countries.

The next major advancement was the release of rice hybrids in China in 1976, though the world outside China came to know about this development only during the 1980s. Hybrids yielding 1.0 to 1.5 tons more than high-yielding varieties (HYV), and occupying more than 50% of the rice area in China, have contributed substantially to high production in China during the past

30 years. Encouraged by the example of China, IRRI revived research to develop hybrids for the tropics during the 1980s in collaboration with selected NARS. During the 1990s, tropical hybrids were developed and commercialized on a small scale in India, Vietnam, the Philippines, Indonesia, and Bangladesh. Presently, a large-scale adoption of hybrids in these countries appears to be a practically feasible and immediately adoptable option for enhancement of rice production and productivity.

Another attempt at breaking the yield barrier experienced in semi-dwarf HYV was initiated at IRRI and is now being followed in some of the selected NARS. Using *indica* × tropical *japonica/javanica* crosses, IRRI developed a New Plant Type (NPT) that has sturdy culms, erect broad leaves, a few—but all productive—tillers, and panicles with high grain numbers. It is expected that NPTs will give 10–15% higher yield than HYVs. The progress has been slow, however, though three NPT-developed lines have recently been released as varieties in China.

Since the 1990s, biotechnological tools have been increasingly applied in rice breeding. With the unraveling of the rice genome and the application of functional genomics and bioinformatics, much progress in rice breeding is expected in the decades ahead.

Genetics of some important economic traits

Precise knowledge of genetics of some economic traits of interest to rice breeders is as important as the variability available in breeding programs. Experimental designs and selection strategies are now planned on the basis of mode of inheritance of the characters.

Studies conducted over the years reveal that broadly complex characters, such as yield, yield components, tolerance and resistance to abiotic stress, and grain quality characters, are controlled by polygenes, whereas simple traits, such as morphological characters and resistance to biotic stresses, e.g., some of the diseases and pests, are controlled by one or a few genes.

Collection, conservation, and use of germplasm

Rice is one of the few crops in which the genetics of various agro-botanical and morphological characters and their linkage relationships have been studied in detail and well documented in the last 100 years (refer to *Rice Genetics Newsletter*, 1989 onward, for comprehensive information on genetics of various traits). Rice germplasm, one of the richest among crop species, consists of more than 100,000 entries, including a large number of accessions from 20 wild species. These are contributed by the Asian cultivated species *O. sativa* through its further differentiation into the subspecies of *indicas*, *japonicas*, and *javanicas*, or tropical *japonicas*, which have been dispersed across a wide geographical region due to human migration. Humans have introduced further diversity via natural selection under different agro-climatic conditions in various regions and through selection under different socio-economic backgrounds and cultivation methods.

A large portion of this diversity was sampled and conserved by different national systems well before the advent of semi-dwarf HYVs in the 1960s. There was apprehension in the 1970s and later that a small number of HYVs with a narrow genetic base, introduced in major rice-growing areas, would soon replace thousands of landraces nurtured by nature through millions of years and cultivated and protected by humans during the past 8,000–10,000 years. This led to massive germplasm collection efforts, both nationally and internationally. The gene bank at IRRI now maintains a collection of over 100,000 entries (Table 6.9) and the major rice-growing countries, including

Table 6.9. Origin of the accessions in the
International Rice Genebank collection at IRRI.

Country	No. of accessions
India	16,013
Lao PDR	15,280
Indonesia	8,993
China	8,507
Thailand	5,985
Bangladesh	5,923
Philippines	5,515
Cambodia	4,908
Malaysia	4,028
Myanmar	3,335
Vietnam	3,039
Nepal	2,545
Sri Lanka	2,123
7 countries with >1,000 and <2,000 accessions	10,241
105 countries <1,000 accessions	11,821
Total	108,256

Source: IRRI online information.

China, India, Indonesia, Bangaladesh, Myanmar, Japan, Korea, and Thailand, each maintain a national collection comprising 10,000 to 40,000 entries. The germplasm collected at IRRI and in many NARS is conserved under medium-term and long-term storage. The free distribution under the Material Transfer Agreement (MTA) has greatly helped rice breeding around the world.

Valuable genetic resources for breeding purposes include genes for resistance to various biotic and abiotic stresses, genes for yield and yield components, and genes for grain quality traits; in addition, genes for various special purposes are required in specific breeding programs. Extensive germplasm collection and its systematic evaluation and cataloging have also greatly enhanced rice-breeding efforts aound the world.

Breeding methodologies

Initial breeding methodology was simply line selection, which involved removal of various off-types found in popularly grown cultivars for uniformity in height, flowering time, plant type, and grain type. The initial phase of pure-line selection was followed by introductions as an important breeding methodology for meeting varietal requirements for various regions; this active phase of recombination breeding was initiated prior to the Green Revolution era, and it continues to be a major breeding method in many NARS. There was a brief phase during the 1960s and 1970s when mutation breeding was resorted to, but because of its limited success, it was discontinued.

Convinced by the potential of hybrids, as evidenced by China's success, breeders began employing heterosis breeding in selected NARS during the 1980s and 1990s. Ideotype breeding to develop NPTs was initiated at IRRI in the 1980s and was subsequently followed in a few NARS. Genetic male sterility-facilitated recurrent selection was practiced to develop diverse breeding populations for extraction of parental lines in recombination breeding, particularly in South American countries. Since the 1990s, the application of biotechnological tools for various breeding objectives has become very popular. Though different breeding methodologies were initiated and extensively adopted in various phases, most of them are being used to-date in one part of the world or another. Information about some of the major plant-breeding methodologies used in rice breeding is briefly given below.

Varietal introduction

"Introduction" is the oldest and simplest method of plant breeding. Besides fulfilling the varietal needs of a given situation, region, or country, this method enlarges and enriches genetic diversity available for further breeding efforts. A variety popular in a country, region, or state can be introduced to another country, region, or state having similar growing conditions. Maximum possible care is exercised to ensure that no new diseases, pests, or weeds are introduced via plant or seed material. Strict phytosanitary/quarantine regulations are in effect in many countries and only after issuance of the mandated requisite phytosanitary certificates may plant material or seed be introduced into another country. Before releasing the introduced variety to farmers for general cultivation, a thorough evaluation is undertaken for various agronomic characters, including yield and resistance/susceptibility to regional pests and diseases.

The breeding methodology of introduction continues to play a significant role in meeting the varietal requirements of many regions of the world. During the 1960s, introduction of 'Mashuri' from Malaysia, Taichung Native-1, 'Taichung-65,' and 'Tainan-3' from Taiwan, and IR-8 and other varieties, from IRRI in the Philippines, to several tropical Asian countries laid the foundation for the Green Revolution during the 1970s and subsequent years. This resulted in enormous increases in rice production, which averted the large-scale famines forecast for these countries.

The Green Revolution has been the major contribution of the introduction methodology. Using this method, the INGER program has played a significant role in the introduction of elite breeding material and has released varieties in several rice-growing countries during the past 30 years. The INGER trials became a major and regular source of exotic varieties/germplasm for many of the participating countries. Many promising genotypes, from INGER trials as well as some released directly as varieties, have been extensively used as parental lines in crossing programs. This utilization of several introduced genotypes as parental material has greatly strengthened rice-breeding programs of several NARS in Asia, Africa, and Latin America.

The 11 varieties first developed and released by IRRI directly (IR-5 to IR-72)

and subsequently the 23 varieties developed by IRRI and released by the Philippine Seed Board have been introduced into many rice-growing countries, and some of them have been released as varieties for general cultivation in more than 50 countries.

Similarly, some of the elite breeding lines developed at IRRI, numbering more than 100, have been released as varieties in 75 countries. Thirty-three varieties developed in India have been released in 46 other countries for general cultivation. In addition to the introduction of elite varieties across countries, within a country itself introduction breeding methodology plays a major role in the exchange of varieties developed in one region, state, or province, with others.

Pure-line selection

Pure-line selection basically involves development of homogeneous populations with desired characteristics from highly heterogeneous local varieties grown by farmers. It involves selection of desirable plants from local varieties grown by the farmers either in the farmers' fields or at research stations, multiplication of selected plants by panicle, plant-to-progeny rows, critical evaluation and assessment of progeny rows for 2–3 seasons, selection of progeny rows possessing a set of desirable characters and, finally, their multiplication and release as a pure-line variety. This was the most popular method in the initial phase of rice breeding (1910–1940) throughout the world.

By the end of World War II, several Asian tropical countries developed hundreds of pure-line varieties for a narrow range of growing conditions. Some of these turned out to be outstanding varieties. They continue to be grown in small pockets even today and have been extensively used as parental lines in recombination breeding. Examples of such varieties are: GEB-24; Basmati-370 from India; 'Khow Dak Mali' from Thailand;

C288-16 from Myanmar; salt-tolerant SR 26B and 'Nonabhokra' from India; pest-resistant Ptb-18, Ptb-33, Co-43, Co-44, and 'Latisail' from South India; deep-water-tolerant FR-13A from eastern India; and the popular 'Manoharsali T 141,' and 'Neang Mas,' Pure-line breeding, though one of the earliest breeding methodologies, is still relevant today in specific situations, particularly in unfavorable ecosystems. Some of the comparatively recently developed and released varieties through pure-line selection in India are: 'Taroari' basmati, grown around Karnal in Haryana state, from the traditional basmati cultures; 'Jalapriya,' 'Jitendra,' 'Jogon-17,' and 'Nalini,' bred for semi-deep and deep-water situations; and 'Lunisree' for saline-water conditions. The pure-line varieties developed formed a rich and diverse genetic base for recombination breeding.

Recombination breeding

Desirable traits found in different genotypes, as happens often in nature, are combined and brought together in one genotype through recombination breeding. Single-, double-, or three-way crosses are made between parental lines possessing desirable traits, and selection for a combination of desirable traits is made in segregating generations from the F_2 generation onward. This has been the most common rice-breeding methodology since World War II, particularly after the advent of semi-dwarf, high-yielding varieties in the Green Revolution era. Unlike in pure-line selection, where plant breeder can select the best genotype from the existing variability only, in recombination breeding, new and additional variability is deliberately created via crossing the selected desirable parental lines.

Depending on facilities, manpower, and screening techniques, segregating generations are handled either through pedigree or bulk method. Pedigree method is commonly followed for both qualitative and quantita-

tive characters, where facilities and manpower are adequate. Bulk method for handling segregating populations is adopted when the selection environment is not appropriate for identifying desired traits. In such situations, segregating generations are bulked and advanced for 4–5 generations from F_2 onward, ultimately followed by pedigree selection. For incorporation of resistance to an insect pest of variable occurrence when no appropriate method or facility is available for screening under artificially created conditions, bulking of segregating populations is continued until an appropriate degree of pest incidence is achieved under field conditions or advanced generations are screened at a place where required facilities are available. Such bulking of segregating generations is also practiced when breeding for drought resistance. In situations where the F_2 population size is relatively small or there is a likelihood of losing valuable segregants under harsh environments, viz., salinity/alkalinity or severe drought conditions, the F_2 generation is raised under optimal conditions and bulked; F_3 or later generations are screened under harsh conditions.

When a simply inherited trait is to be transferred to an otherwise elite genotype, backcross breeding is adopted. A good example of this is the transfer of grassy stunt virus resistance gene (*GSV*) from *O. nivara* to IRRI's advanced breeding elite lines (Khush, 1984). When the transfer of a desired single gene is from an unadapted and agronomically poor source, as was the case in this example, several backcrosses to the recurrent parent are needed to recover much of the genetic complement of the original adapted variety. On the other hand, if the source of a desired gene in backcrossing is a reasonably good adapted genotype, fewer backcrosses may be needed and additional desirable genetic recombination may also be expected. Backcross breeding is also a very effective tool to break undesirable linkages between desirable and undesirable characters.

One of the major limitations of recombination breeding is the long period of time required for fixing the desirable combination of traits. This breeding procedure, though very useful, becomes time-consuming, particularly in locations where only one crop of rice can be grown per year or in the case of long-duration rice varieties.

Wide hybridization

In addition to two cultivated species in the genus *Oryza*, viz., *O. sativa* and *O. glaberrima*, there are 20 wild species. The wild species belonging to the *O. sativa* complex (AA genome) can easily be crossed with the cultivated species. For the other distantly related wild species, embryo-rescue techniques, in addition to other precautions, need to be adopted to obtain hybrids. A number of useful genes have been introgressed into cultivated rice from wild germplasm. Transfer of resistance to grassy stunt virus from *O. nivara*, resistance to brown plant hopper from *O. officinalis*, and resistance to bacterial leaf blight from *O. longistaminata* are chief examples of the usefulness of wide hybridization in rice breeding. Diversification of sources of cytoplasmic male sterility is another area of contribution from wide hybridization. Male sterility-inducing cytoplasm has been used from *O. nivara* and *O. rufipogon* for developing new cytosterile lines as an alternative to widely used wild abortive (WA) source (Dalmacio et al., 1995; Hoan et al., 1997). A list of potentially useful traits found in wild species is given in Table 6.10.

Mutation breeding

Mutation breeding is a tool for creating additional variability and is particularly useful where natural genetic variability in a crop plant is limited. This methodology has been used for the improvement of several crops, including rice. Though the first attempt to use induced mutagenesis for rice improvement was made in 1936 by Ramaiah and Pathasarathy (Siddiq and Viraktamath, 2000), such efforts were intensified during the 1960s and 1970s worldwide due to massive support for mutagenesis projects from the International Atomic Energy Agency (IAEA), which enabled the generation of basic knowledge and material of theoretical and practical value.

For rice crops, X-rays, gamma rays and fast neutrons, among physical mutagens, and ethyl methane sulfonate (EMS) and nitrosomethyl urea (NMU), among chemical mutagens, were found to be effective (mutations/unit dose) and efficient (mutation rate relative to undesirable biological effects) for inducing point mutations. *Japonicas* were found to be more sensitive than *indicas* or *javanicas* for both physical and chemical mutagenesis. EMS was found to be more efficient than gamma rays (Siddiq and Swaminathan, 1966). A large number of point mutations affecting plant height, leaf, panicle, and grain have been isolated. Some of them have been either released directly as mutant varieties or used as donor parents in crossing programs for improvement of specific characters.

Among the notable mutants used in rice breeding are the early-maturing 'Remei' in Japan, RD-15 of Thailand, the dwarf-statured and fertilizer-responsive mutant 'Jagannath' of India, calrose-76 of the U.S., and the blast-resistant and glutinous RD-6 of Thailand. A cytoplasmic male-sterile mutant was induced at IRRI, resulting in CMS line IR 68885A (Virmani and Voc, 1991), which has been used to diversify the CMS base of IRRI-bred CMS lines for hybrid rice breeding programs in the tropics. However, because less progress was made than expected, mutation breeding is not currently being used. Although no significant progress has been made for quantitatively inherited traits

Table 6.10. Potential useful traits of *Oryza* species.

Species	Number of accessions*	Distribution	Useful or potentially useful traits
O. sativa complex			
O. sativa L.	84,186	Worldwide	Cultigen
O. glaberrima Steud.	1,299	West Africa	Cultigen, tolerance to drought, acidity and iron toxicity, resistance to RYMV, African gall midge, nematodes, and weed competitiveness
O. nivara Sharma et Shastry	1,130	Tropical and subtropical Asia	Resistance to grassy stunt virus
O. rufipogon Griff	858	Tropical and subtropical Asia, tropical Australia	Resistance to BB, tungro virus, tolerance to aluminum and soil acidity, source of CMS
O. breviligulata A. Chev. Et Roehr. (O. barthii)	214	Africa	Resistance to GLH, BB, drought avoidance
O. longistaminata A. Chev. Et Roehr.	203	Africa	Resistance to BB, nematodes, drought avoidance
O. meridionalis Ng	46	Tropical Australia	Elongation ability, drought avoidance
O. glumaepatula Steud.	54	South and Central America	Elongation ability, source of CMS
O. officinalis complex			
O. punctata Kotschy ex Steud	59	Africa	Resistance to BPH, zigzag leafhopper
O. minuta J.S. Presi. Ex C.B. Presl.	63	Philippines and Papua New Guinea	Resistance to BB, blast, BPH, GLH, tolerant to Shb
O. officinalis Wall ex Watt	265	Tropical and subtropical Asia, tropical Australia	Resistance to thrips, BPH, GLH, WBPH, BB, stem rot
O. rhizomatis Vaughan	19	Sri Lanka	Drought avoidance, rhizomatous
O. eichingeri A. Peter	29	South Asia and East Africa	Resistance to BPH, WBPH, GLH
O. latifolia Desv.	40	South and Central America	Resistance to BPH, high biomass production
O. alta Swallen	6	South and Central America	Resistance to striped stem borer, high biomass production
O. grandiglumis (Doell) Prod.	10	South and Central America	High biomass production
O. australiensis Domin	36	Tropical Australia	Resistance to BPH, BB, drought avoidance
O. Meyeriana complex			
O. granulate Nees et Arn. Ex Watt	24	South and South-East Asia	Shade tolerance, adaptation to aerobic soil
O. meyeriana (Zoil. Et Mor. Ex Steud.)	11	South-East Asia	Shade tolerance, adaptation to aerobic soil
Baill.			
O. ridleyi complex			
O. longiglumis Jansen	6	Irian Jaya, Indonesia, and Papua New Guinea	Resistance to blast, BB
O. ridleyi Hook. F.	15	South Asia	Resistance to BB, blast, stem borer, whorl maggot
Unclassified			
O. brachyantha A. Chev. Et Roehr.	19	Africa	Resistance to BB, yellow stem borer, leaf-folder, whorl maggot, tolerance to lateritic soil
O. schlechteri Pilger	1	Papua New Guinea	Stoloniferous
Related genera	15	—	—

Source: Brar and Khush (2002).

*Accessions maintained at IRRI, The Philippines (BPH = brown plant hopper, GLH = green leafhopper; WBPH = white = backed plant hopper; BB = bacterial blight; Shb = sheath blight; CMS = cytoplasmic male sterility; RYMV = rice yellow motle virus).

155

through mutation breeding in rice, basic information on various aspects of mutation breeding was developed and a few point mutations of various morphological characters of some value in rice breeding were isolated.

Heterosis breeding

Heterosis was exploited commercially in rice in the 1970s—50 years after its discovery in 1926 by Donald F. Jones. During the intervening 50 years, heterosis to the extent of 20–150% over the better parent was confirmed (Virmani and Edwards, 1983). But the skepticism regarding commercial feasibility of large-scale hybrid seed production in the predominately self-pollinated rice crop might be responsible for the delayed advent of rice hybrids as a field reality till 1976.

Meanwhile, the role of cytoplasm in causing male sterility was reported during the 1950s (Sampath and Mohanty, 1954; Katsuo and Mizushima, 1958), and the first usable cytoplasmic-genetic male sterility-fertility restoration system was developed by Shinjyo and Omura (1966) by incorporating nuclear genes of *japonica* variety 'Taichung-65' into the cytoplasm of the *indica* variety 'Chinsura Boro-II.' The first commercially usable CMS line was developed in China in 1973 from a spontaneous male-sterile plant isolated in a population of wild rice, *O. sativa f. spontaneae* on the southern Chinese island of Hainan (Yuan, 1977). Discovery of this source of cytoplasmic male sterility designated as "WA type" is considered to be a landmark in the history of hybrid rice breeding.

Chinese scientists pioneered the heterosis-breeding approach in rice. Systematic work on heterosis breeding in China was initiated in 1964 under the leadership of Professor Yuan Long Ping, who is recognized as the Father of Hybrid Rice. The first rice hybrid was released for commercial cultivation in 1976. Since then, China has been leading the world in the development and use of this technology. Presently, rice hybrids are cultivated on about 15 million hectares (50% of the total rice area in China) with a yield advantage of about 1.5 t/ha over the HYVs. During the past 30 years, hybrids have contributed more than 300 million tons of additional paddy production, besides providing large-scale employment in rural areas through the hybrid seed industry. Because of higher rice production due to adoption of hybrids, 4–6 million hectares of rice lands could be diverted in China to other high-value crops.

Inspired by China's remarkable success with hybrid rice, IRRI revived its hybrid rice program in the late 1980s under the leadership of the senior author to develop hybrids for the tropics. Intensive efforts for more than two decades resulted in the development and adoption, since the 1990s, of tropical hybrids in India, Vietnam, the Philippines, Bangladesh, Indonesia, and Myanmar. Though, presently, around 2.5 million hectares are under hybrid rice in these countries, the potential area for production of hybrids is more than 25 million hectares. Hybrids are also being cultivated on a smaller scale in the U.S., Brazil, and Egypt. Many more countries are expected to develop and adopt this technology in the years ahead.

Rice hybrids are being developed either by using the cytoplasmic-genetic male-sterility (CMS) system (or three-line system, because three lines, viz., A, B, and R, are used in breeding hybrids) or the environment-sensitive genic male-sterility (EGMS) system (or two-line system, because only two lines, either a temperature-sensitive genetic male sterile line or phototperiod-sensitive genetic male-sterile line and a male parent, are involved in development of hybrids). Currently, the CMS system is most commonly used in China and other countries. Recently, two-line hybrids have become popular in China.

The WA source of sterility-inducing cytoplasm is most commonly used in China and elsewhere, although some other sterility-inducing cytoplasms have been discovered (Virmani, 1994; 1996). Frequency of restorers in *indica* rice germplasm is around 25%, but maintainer frequency is 1–15%. In *japonica* and basmati rices, frequency of restorers is almost absent.

Hybrid rice breeding involves: (1) development of appropriate parental lines, (2) seed production of parental lines, (3) development of experimental hybrid combinations, and (4) critical multilocational evaluation of experimental hybrids (Fig. 6.2). For efficient development of parental lines, breeding material is grouped into different nurseries, viz., source nursery, testcross nursery, retest-cross nursery, backcross nursery, CMS-evaluation nursery, etc. Source nursery includes elite breeding lines, released varieties, introduced exotic material, and CMS and EGMS lines. Crosses are made between different CMS lines and selected lines from source nursery, and these are called "testcrosses", whether the male parents used are maintainers, restorers, partial maintainers, or partial restorers. If the F_1s are completely sterile, the male parents are classified as maintainers. If the F_1s are highly fertile (>80% spikelet fertility), the respective male parents are classified as restorers. Those in-between these two classes are called either "partial restorers" or "partial maintainers." Promising genotypes among maintainers are converted into CMS lines by repeated backcrossing (5–6 backcrosses) to sterile F_1s in backcross nursery and promising genotypes among restorers are evaluated once again in retest-cross nursery by producing new experimental hybrids. Hybrid combinations found to be promising are subsequently evaluated in observational yield trials (OYT), preliminary yield trials, initial hybrid rice trials (IHRT), and national trials. Evaluation of hybrids is done at the national level by the participating countries and at the international level by IRRI. Details of hybrid rice breeding procedures are published in a manual by Virmani et al. (1997). A general outline of the hybrid rice breeding program adopted at IRRI and in China is given in a flow chart (Fig. 6.2).

In the two-line system of heterosis breeding, the EGMS system is used. Male sterility/fertility is governed by gene(s), the expression of which is influenced by environmental factors, such as temperature, photoperiod, or temperature × photoperiod interaction.

The male-sterile genotypes, in which alteration of fertility/sterility occurs due to change in temperature at the sensitive, early reproductive phase of crop growth (around 5–15 days after panicle initiation), are called temperature-sensitive genic male-sterile (TGMS) lines. The photoperiod-sensitive genotypes, which respond in a similar manner to change in photoperiod at the sensitive stage, are called photoperiod-sensitive genic male sterile (PGMS) lines. In some genotypes, change in fertility/sterility is also due to interaction of temperature and photoperiod. Monogenic recessive genes control temperature/photoperiod-sensitive male sterility. So far, five TGMS and three PGMS genes have been identified; some of these genes have been mapped and molecular markers associated with them have been identified. These markers can be useful when using MAS (Marker-Aided Selection) in the development of new TGMS/PGMS lines. During the sterile phase, these lines are used as females in hybrid seed production plots, whereas in the fertile phase, they are used for self-multiplication.

The two-line system of heterosis breeding has the following advantages over the three-line system:

• Choice of parental lines is much wider. Any genotype can be converted into a TGMS/PGMS line, and any other genotype

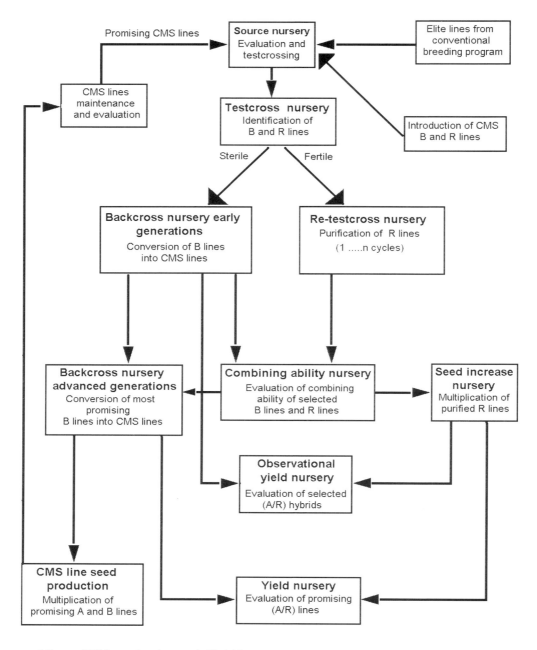

Source nursery
Evaluation and testcrossing

Promising CMS lines

Elite lines from conventional breeding program

CMS lines maintenance and evaluation

Testcross nursery
Identification of B and R lines

Introduction of CMS B and R lines

Sterile   Fertile

Backcross nursery early generations
Conversion of B lines into CMS lines

Re-testcross nursery
Purification of R lines
(1 .....n cycles)

Backcross nursery advanced generations
Conversion of most promising B lines into CMS lines

Combining ability nursery
Evaluation of combining ability of selected B lines and R lines

Seed increase nursery
Multiplication of purified R lines

Observational yield nursery
Evaluation of selected (A/R) hybrids

CMS line seed production
Multiplication of promising A and B lines

Yield nursery
Evaluation of promising (A/R) lines

A lines = CMS line or female parent of hybrid
B line = Maintainer lines
R line= Restorer lines or male parent of hybrid
A/R = Experimental hybrid

Fig. 6.2. Organization of hybrid rice breeding program in the two-line system of heterosis breeding, EGMS system.

with good combining ability can be used as a male parent. Because in the three-line system only maintainers and restorers (30–35% of the germplasm) can be used, the choice of germplasm in the two-line system is three times that in the three-line system.

- Seed production system is simpler and more efficient with involvement of only two lines.
- Negative effects of sterility-inducing cytoplasm are not encountered.
- Hybrids can be developed in *japonica* and basmati rices—the two groups in which frequency of restorers is negligible.
- The magnitude of heterosis in two-line hybrids is reported to be 5–10% higher than in three-line hybrids.

Despite the above-mentioned advantages, the three-line system continues to be the approach most used for developing hybrids in rice. The work on the two-line system was initiated in China during the 1980s. By 1995, a few hybrids had been developed and commercialized. Presently, out of 15 million hectares of area under hybrid rice in China, around three million hectares are under two-line hybrids. Some of the super hybrids developed in China recently, capable of yield potential of 100 kg/day/ha of crop duration, are reported to be two-line hybrids. At IRRI and in some NARS, though research work on two-line hybrids has been initiated, such hybrids have not yet been commercialized.

Interspecific and intersubspecific hybridization

Successful efforts in interspecific crosses between the two cultivated species of rice, viz., *O. sativa* and *O. glaberrima*, have resulted in the development of New Rice for Africa (NERICA). The NERICA rices are likely to usher in a Green Revolution in the uplands of sub-Saharan Africa. This is the first instance of a successful progeny being raised, and desirable selections made, from crosses between two cultivated species that evolved separately for millions of years and developed reproductive barriers. Deployment of biotechnological tools and embryo-rescue techniques enabled plant breeders to overcome the reproductive hurdles to achieve this remarkable success (Jones, et al., 1997).

The NERICA rices refer to a group of genotypes developed from the interspecific *O. sativa* × *O. glaberrima* crosses. These rices combine the yield potential and grain quality characteristics of *O. sativa* with the weed competitiveness, drought resistance, and tolerance to adverse soil conditions of *O. glaberrima*. The NERICA rices were first developed in 1999 and the first set of seven NERICA varieties were released from the African Rice Centre, WARDA, in 2000. Eleven more varieties were released in 2005. These 18 NERICA varieties have been extensively evaluated in more than 30 sub-Saharan African countries; local farmers were actively involved in the evaluation through the Farmers' Participatory Varietal Selection (PVS) and a community-based seed production system.

Popular NERICA varieties are early maturing (75–100 days), high yielding (2.5 t/ha against the check yield of 1.0 t/ha), weed competitive, resistant or tolerant to Africa's major pests and diseases, drought tolerant, and tolerant to soil acidity and iron toxicity. Grains of NERICA varieties have relatively high protein content. Presently, 150,000 hectares are under NERICA varieties, mostly in Guinea, the Ivory Coast, and Uganda. Seed shortage is a current constraint to the more widespread use of this miracle rice in Africa. NERICA varieties are also being developed for shallow lowland areas of Africa.

To combine the desirable traits from the subspecies of *O. sativa*, crosses were made among the subspecies, viz., *indicas*, *japonicas*, and *javanicas*. As described earlier, due to sterility of F_1 hybrids, progress was limited in earlier efforts. From an FAO-sponsored international project on *indica-japonica* crosses during the 1950s, only three varieties, ADT-27, 'Mahsuri,' and 'Malinga,' could be released. There was a better appreciation and understanding of the problem of hybrid sterility after the publication of the work of Ikehashi and Araki (1986).

A notable development in this area was the development of NPTs by introgressing useful genes from tropical *japonicas* and *javanicas* into *indica* varieties. To break the yield ceiling of semi-dwarf, high-yielding varieties, efforts to develop NPTs were initiated in 1990 at IRRI. Subsequently, efforts to develop NPTs were also initiated in China, India, and in some other NARS. Progress made in the development of NPTs is described in a following section.

Tissue culture

Tissue culture is an all-embracing term used to describe in vitro culture of gamete cells, tissues, organs, and isolated protoplasts. Relevant tissue culture techniques in rice breeding are: (1) anther culture for greater speed and efficiency in developing a variety, (2) somatic cell culture for efficient screening of a large number of cell populations for identifying variants with resistance to biotic and abiotic stresses, (3) embryo culture for rescuing hybrid embryos in interspecific and intergeneric crosses for transfer of useful traits from alien sources, and (4) recovery of novel genetic variants by somaclonal variation in regenerated plants. The various applications of tissue culture in rice breeding are summarized in Table 6.11.

Anther culture

Production of doubled haploids (DH) through anther culture is a rapid approach to attaining homozygosity; it shortens the time needed for the development of inbred cultivars and increases selection efficiency through early expression of recessive genes. Application of anther culture has contributed to the development of more than 100 varieties in China (Meifang, 1992) and over 42 cultivars in Korea. Compared with F_2 populations, fewer DH plants are needed for the selection of desired recombinants. Anther culture also has potential applications in wide crosses and in developing substitution and addition lines (Chu, 1982). Though

Table 6.11. Areas of possible application of tissue culture in rice breeding.

Culture technique	Types	Areas of application
Anther culture	a	Varietal improvement
	a.1	Shortening the breeding cycle
	a.2	Increasing selection efficiency
	a.3	Gametoclonal variation
	a.4	Production of chromosomal variants
	a.5	Genetic mutations
	b	Interspecific hybridization, development of addition and substitution lines
Somatic cell and tissue culture	a	*In vitro* selection and somaclonal variation for tolerance to salt stress and aluminum toxicity, high lysine and protein, disease resistance
	b	Micropropagation of cytosterile stocks
Embryo culture		Interspecific and intergeneric hybridization
Protoplast culture	a	Genetic transformation
	b	Somatic hybridization for production of cybrids

Source: Siddiq and Viraktamath (2000).

anther culture techniques for obtaining regenerants in *japonica* rices have progressed considerably, culturability in *indica* rices has not yet been successful due to genotype-culture media specificities. This has restricted the progress of rice breeding in tropical countries. Nevertheless, several DH lines from *indica* × *indica* and *indica* × *japonica* hybrids have been developed and evaluated in the field. Efforts to develop hybrids through chromosome doubling have not been successful. However, production of doubled haploids is a rapid way of developing mapping populations for construction of genetic linkage maps.

Somaclonal variation

Somaclonal variation refers to the variation arising through tissue culture in regenerated plants and their progeny. It occurs for various agronomic traits, such as plant height, tiller number, maturity, and disease resistance, along with various biochemical traits. The technique consists of growing callus or cell suspension cultures for several cycles and regenerating plants from these long-term cultures. The regenerated plants and their progenies are evaluated to identify individuals with altered phenotype. By this approach, some useful somaclonal variants, including those for disease resistance and male sterility, have been identified in rice.

The variation induced via in vitro culture may be useful in a few cases in rice breeding. In somatic cell culture, the advantage is that a large number of cells can be screened in a very limited space, with considerably less expenditure, giving an increased probability of isolating a rare variant possessing resistance to a particular biotic or abiotic stress. The effectiveness of cell selection in exploiting the genetic variation induced in tissue culture depends on the ability to regenerate plants from selected cell variants and the stability of expression of the desired trait at the whole-plant level.

Salt-tolerant rice plants have been obtained through in vitro selection in tissue culture (Zapata and Abrigo, 1986), and lines with increased lysine and protein content as well as tolerance to aluminum toxicity have been identified (Abrigo et al., 1985). Phytotoxin-resistant cell lines have been obtained for bacterial leaf blight, brown spot, and sheath blight diseases. One somaclonal variant from anther culture, possessing resistance to blast and good grain quality, has been released as 'Dama' (Heszky and Simon-Kiss, 1992).

Use of biotechnological tools

Among the many applications of molecular and biotechnological tools in rice breeding, the use of molecular markers and the development of transgenics are the most direct and immediate uses. These methodologies are becoming increasingly popular in rice breeding and are likely to play a major role in the decades ahead because of the complete unraveling of rice genomic sequences and the availability of information from the intensive efforts underway on rice functional genomics. In the section "Rice Breeding Strategies for the Future," later in this chapter, we discuss further the possible application of molecular/biotechnological tools in rice breeding and a few achievements already made.

Major Breeding Objectives in Rice

The major objectives and goals in rice breeding during the first 100 years of rice breeding have been:

- Enhancement of yield potential
- Incorporation of resistance/tolerance to biotic stresses
- Incorporation of resistance/tolerance to abiotic stresses
- Improvement of grain and nutritional quality
- Reduction of growth duration

Enhancement of yield potential

Various strategies adopted for increasing rice yield potential are:

- Conventional hybridization and selection
- Heterosis breeding
- Ideotype breeding (New Plant Type)
- Wide hybridization
- Genetic engineering

Conventional hybridization and selection

On average, during the past 35 years, about a one percent per year increase in yield potential of rice has been achieved since the development and release of the first improved semi-dwarf variety of rice, IR-8 (Peng et al., 2000).

The development of short-statured varieties involving spontaneous dwarf mutant 'Dee-gee-wu-gen' (DGWG) through recombination breeding is a landmark in the history of rice breeding. The semi-dwarf varieties, characterized by a non-lodging habit (making them responsive to higher doses of fertilizers), photoperiod insensitivity coupled with earliness enabling year-round cultivation, and upright foliage facilitating efficient utilization of solar energy and a high-leaf area index, enabled all of tropical Asia to achieve a major yield breakthrough in the mid-1960s. Released in 1949–50, 'Taichung (Native)-1,' a derivative of a cross between the spontaneous dwarf mutant 'DGWG' and a traditional tall variety 'Tsai-Yuan-Chung,' was the first dwarf variety with a high yield potential. It is not known exactly when or where DGWG was spotted first, but this spontaneous dwarf mutant had been grown in Taiwan during the 1940s. The semi-dwarf, high-yielding miracle variety IR-8, a derivative of a cross between DGWG and 'Peta,' an Indonesian variety, with its large-scale adoption throughout tropical Asia within a very short time, heralded the Green Revolution. Subsequently, rice breeders all across

Asia have developed other semi-dwarf, high-yielding varieties, using T(N)-1 and IR-8 as donors for short stature and high yield potential. Presently, more than 90% of the semi-dwarf HYVs being cultivated in tropical Asia are based on the *DGWG* gene.

Use of *DGWG* source for yield improvement has unique advantages for rice breeders. Inheritance of dwarf stature is simple, monogenic recessive, despite the fact that height is a polygenically controlled trait, thereby making it easier to combine desired plant type, along with other traits, such as photoperiod-insensitivity, earliness, etc., which are strongly associated with dwarf stature. In semi-dwarf varieties, some of the physiological features favoring high yield potential are a harvest index of 0.45–0.5 (as against 0.3–0.35 in tall traditional varieties) with little change in total biomass, erect upright leaves utilizing sunlight more efficiently, and delayed senescence allowing more time for grain filling. Photo-insensitivity in these varieties conferred wide adaptability, whereas earliness enabled them to fit into multiple cropping systems. These semi-dwarf, high-yielding varieties became very popular not only in tropical Asia, but also in Africa and Latin America. They have a higher panicle number, rather than the higher panicle weight of traditional tall varieties. Hence, selection criterion for yield potential in these varieties is population performance and not individual plant performance. Panicle number per unit area, optimum grain number per panicle, and grain test weight are used as further selection criteria. A variety possessing 400 panicles per square meter with an average grain number of 125–150 and a test weight of 25 g is estimated to yield around 8.0 t/ha.

A large number of varieties have been developed using the conventional method of hybridization, followed by selection in almost all the rice-growing NARS. In India alone, since the introduction of IR-8 in the late 1960s, more than 600 released varieties were developed using this method.

Heterosis breeding

China pioneered in developing and using rice hybrids to increase productivity. Hybrids have given a yield advantage of 15–20% over the highest yielding inbred varieties in China. Now hybrids are cultivated on 15 million hectares in China, with an average yield of 7.0 t/ha as compared with average yield of inbred lines of 5.6 t/ha (Yuan, 2004). Thus, average yield advantage by cultivation of hybrids in China has been 1.4 t/ha.

Two-line hybrids were commercialized in China in 1995. The area under two-line hybrids in 2002 was 2.6 million hectares (Yuan, 2004). Their yield advantage is 5–10% higher than that of three-line hybrids.

A project to develop super hybrids, capable of yielding 100 kg/day/ha, was initiated in 1996. Several super hybrids have been developed with a yield advantage of 20% over existing three-line hybrids on a commercial scale. The area planted with super hybrids in 2002 was 1.4 million hectares with an average yield of 9.1 t/ha (Yuan, 2004). A two-line super hybrid, P64S/E-32, and a three-line super hybrid, II 32-A/Ming-86, registered record yields of 17.1 t/ha and 17.9 t/ha, respectively, in 1999 and 2001 (Yuan, 2004). Additionally, super hybrids possess better grain quality. Second generation super hybrids, capable of yielding 12 t/ha, on a large scale, are under development.

Due to the successful efforts of IRRI, heterosis breeding has become a very effective strategy for enhancing yield potential in several tropical Asian countries. Hybrids are now commercially cultivated in Vietnam, India, the Philippines, Bangladesh, Indonesia, and Myanmar. Hybrids have shown a yield advantage of 20–30% in these countries (Table 6.12) at research stations and in on-farm trials (Virmani, 2003).

Table 6.12. Yield advantages of hybrids over inbred lines released for cultivation in four Asian countries.

Country	Yield (t/ha)		% yield advantage
	Hybrid	Inbred	
Bangladesh	7.48	5.79	29.2
India	6.33	5.22	21.3
Philippines	7.80	6.50	20.0
Vietnam	6.30	4.84	30.2

Source: Virmani (2003).

Estimated total area under hybrid rice outside China during 2005 was around 2.0 million hectares. In addition to the nine countries where hybrids have been commercialized, 11 countries are actively engaged in hybrid rice development. By 2020, the area under hybrid rice outside China is expected to exceed 20 million hectares. Thus, heterosis breeding has been an effective breeding methodology in rice for enhancement of yield potential. This technology is expected to play a much bigger role in enhancement of yield potential globally in the decades ahead.

Ideotype breeding (new plant type)

Another strategy used for enhancing yield potential in rice has been "ideotype breeding," which modifies plant architecture. Yield potential is determined by total biomass and harvest index (HI). Traditional tall rice varieties had HI of around 0.3 and total biomass of around 12 t/ha, with a maximum yield of 3.6 t/ha. Biomass in traditional varieties could not be increased by application of nitrogenous fertilizers because they grew excessively tall and lodged, thereby causing a decrease rather than an increase in yield. To increase the yield potential of tropical rices, it was necessary to increase HI and nitrogen responsiveness to increase biomass. This was achieved by reducing plant stature through the incorporation of the monogenic recessive gene sd_1 for short stature.

The first short-statured popular variety IR-8, developed at IRRI, also had a combination of other desirable traits, such as profuse tillering, dark green and erect leaves for good canopy architecture, and sturdy stems. It responded well to nitrogenous fertilizers and had higher biomass (18 t/ha) and an HI of 0.45. Its yield potential was between 8.0 and 9.0 t/ha (Chandler, 1969).

Ever since the development of short-statured, high-yielding rice varieties, which breached the centuries-old genetic yield barrier, rice scientists have been trying to find an even better morpho-physiological frame to provide the next quantum jump in yield potential. Early efforts to locate exploitable variability for physiological traits that directly or indirectly contributed to higher yield potential proved futile. No donor source was found that could alter sink-source equilibrium.

An NPT was conceptualized at IRRI in 1988. Modern semi-dwarf varieties produce a large number of unproductive tillers and an excessive leaf area, which cause mutual shading, thereby reducing canopy photosynthesis and sink size. To increase the yield potential of the semi-dwarf rices, further modification of plant architecture with the following characteristics was proposed (Khush et al., 1998):

- Low tillering (9–10 tillers)
- No unproductive tillers
- 200–250 grains per panicle
- Dark green, thick, erect leaves
- Vigorous and deep root system

Breeding efforts to develop NPTs were initiated in 1990. The objective was to develop improved germplasm with 15–20% higher yield than the existing HYVs. Donors for the target traits were identified in the *bulu* (or *javanica*) germplasm mainly from Indonesia. This germplasm is now referred to as tropical *japonicas*. By the mid-1990s, more than 500 NPTs had been evaluated in observational yield trials (Khush, 1995).

As envisioned, the NPTs possessed large panicles, few unproductive tillers, and lodging resistance. However, the first prototype lines had lower grain yield because of lower biomass production and poor grain filling. Hence, the next logical step was to increase biomass and improve grain filling. Poor grain filling in the NPTs was inherited. Therefore, in subsequent crosses, only parents with good grain filling were used (Khush, 1996). Many improved NPTs have normal grain filling and have yield potential better than that of the high-yielding *indica* varieties. One of the improved NPT lines, IR 64446-7-10-5 (MD 2 STR-1/Pakisan), has been released in China and another improved NPT line has been released in Indonesia for production.

Despite these successful efforts, further refinement of the original NPT design was necessary. An increase in tillering capacity was needed to increase biomass production and to compensate for any yield loss when loss of tillers resulted from insect attack, particularly stem borers. Moreover, in tropical *japonica* germplasm that was used for developing NPT lines, there were no donors for resistance to brown plant hopper and tungro virus. Improved NPT lines were, therefore, crossed with high-yielding *indica* varieties to slightly increase tillering ability and to introduce resistance to insect pests and diseases. Lines developed now have 10–12 tillers, resistances to some of the pests and diseases, and better yielding ability and grain quality than elite *indica* varieties. These improved NPT lines are presently being evaluated at IRRI and in several NARS.

Ideotype breeding was also carried out in some NARS, viz., China and India. Chinese rice breeders have developed a new plant type capable of giving 10–15% higher yield than the semi-dwarf HYVs. Similarly in India, the breeding work initiated at the Indian Agricultural Research Institute, New Delhi, during the late 1970s, has paid rich dividends. The breeding strategy adopted was to initially exploit the rich variability available in *indica* itself and, depending on the need, to use the variability available in other sub-species of *O. sativa*. New plant types developed have 8–10 synchronous tillers, plant height of 100 cm, extremely thick culm, 150–175 grains per panicle, a test weight of 25 g, and a slightly longer grain filling period (Siddiq and Viraktamath, 2000). Among the lines at advanced stages of testing, Pusa 1021, Pusa 1286, and Pusa 1077 series are promising. Though the success achieved so far is limited, the ideotype breeding approach appears to be promising for enhancing yield potential in the not-too-distant future.

Breeding for major biotic stresses

The FAO estimates that diseases and pests cause yield losses of up to 25% a year. Rice crops harbor about 100 pests, of which at least 20 are considered economically important. Pest problems assumed serious proportions only after the introduction and widespread adoption of semi-dwarf HYVs and the associated management practices. In the post-IR-8 era, a small number of improved varieties replaced thousands of traditional cultivars in large areas, thereby reducing the genetic variability of the crop. The application of higher doses of fertilizers, closer planting, development of irrigation facilities, and the availability of short-duration photoperiod-insensitive varieties enabled farmers in tropical Asia to grow successive rice crops throughout the year. But these same factors allowed many of the minor pests to become major ones, and these major pests produced new strains of virulence to cope with the changing severity of selection pressure. In addition to the age-old problems of blast and brown spot, the following diseases began to appear in rice crops: sheath blight, sheath rot, bacterial leaf blight, bacterial leaf streak, rice tungro virus, grassy stunt virus, and ragged stunt virus. Occurrence of new races in bacterial leaf blight and grassy stunt virus has increased the complexity of management of diseases in rice. Among insect pests, in addition to stem borers and gall midge, the more destructive ones are plant hoppers, leaf hoppers, leaf folder, and hispa. Biotype variations are found in gall midge and brown plant hoppers. Breeding efforts made to incorporate resistance/tolerance to some of the major pests and diseases are briefly given in the following section.

Developing host-plant resistance has been given top priority in tropical Asian countries, as chemical control is expensive and detrimental to the environment. In most tropical and sub-tropical Asian rice-growing countries, five diseases (blast, bacterial leaf blight, sheath blight, tungro, and grassy stunt virus) and four insect pests (brown plant hopper, green leaf hopper, stem borers, and gall midge) are of importance. At IRRI and in many national programs, breeding efforts to develop accessions possessing multiple resistances have been initiated. Screening of germplasm accessions resulted in the discovery of resistance sources for biotic stresses. At IRRI, with the use of identified donors, improved varieties with multiple resistances to as many as four diseases and insect pests have been developed. IR-26, the first variety with multiple resistances, was released in 1973. Previously released varieties, IR-5, IR-8, IR-20, IR-22, and IR-24 were susceptible to most of the diseases and insect pests. Since 1973, IRRI has developed many varieties with resistances to multiple diseases and insect pests (Table 6.13). Large-scale cultivation of varieties with multiple resistances has helped stabilize the world rice production. As and when the resistance is breached by the emergence of new races, pathotypes, or biotypes, new resistance genes are incorporated (Table 6.14).

In several national programs, breeding for resistance to major pests and diseases has

Table 6.13. Disease and insect reaction of IR varieties of rice.

				Reaction[a]						
						BPH biotype			Stem	Gall
IR variety	Blast	Bacterial Blight	Grassy Stunt	Tungro	GLH	1	2	3	Borer	Midge
IR5	MR	S	S	S	R	S	S	S	MS	S
IR8	S	S	S	S	R	S	S	S	S	S
IR20	MR	R	S	MR	R	S	S	S	MR	S
IR22	S	R	S	S	S	S	S	S	S	S
IR24	S	S	S	S	W	S	S	S	S	S
IR26	MR	R	MR	MR	R	R	S	R	MR	S
IR28	R	R	R	R	R	R	S	R	MR	S
IR29	R	R	R	R	R	R	S	R	MR	S
IR30	MS	R	R	MR	R	R	S	R	MR	S
IR32	MR	R	R	MR	R	R	R	S	MR	R
IR34	R	R	R	R	R	R	S	R	MR	S
IR36	R	R	R	R	R	R	R	S	MR	R
IR38	R	R	R	R	R	R	R	S	MR	R
IR40	R	R	R	R	R	R	R	S	MR	R
IR42	R	R	R	R	R	R	R	S	MR	R
IR44	R	R	S	R	R	R	R	S	MR	S
IR46	R	R	S	MR	MR	R	S	R	MR	S
IR48	R	R	R	R	R	R	R	S	MR	—
IR50	MS	R	R	R	R	R	R	S	MR	—
IR52	MR	R	R	R	R	R	R	S	MR	—
IR54	MR	R	R	R	R	R	R	S	MR	—
IR56	R	R	R	R	R	R	R	R	MR	—
IR58	R	R	R	R	R	R	R	S	MR	—
IR60	R	R	R	R	R	R	R	R	MR	—
IR62	MR	R	R	R	R	R	R	R	MR	—
IR64	MR	R	R	R	R	R	MR	R	MR	—
IR65	R	R	R	R	R	R	R	S	MR	—
IR66	MR	R	R	R	R	R	R	R	MR	—
IR68	MR	R	R	R	R	R	R	R	MR	—
IR70	R	S	R	R	R	R	R	R	MR	—
IR72	MR	R	R	R	R	R	R	R	MR	—
IR74	R	S	R	R	R	R	R	R	MR	—

Source: Khush and Virk (2000).
R = Resistant, MR = Moderately Resistant, S = Susceptible.

been a major activity since the introduction of semi-dwarf HYVs. Several donors for resistance to major pests and diseases have been identified (Table 6.15), and a large number of varieties possessing resistance/tolerance have been developed in India and other rice-growing countries (Table 6.16).

In the oldest and most widespread disease—blast—resistance is needed at the seedling and late vegetative stages and, for neck blast, in the reproductive stage. Correlation for resistance between leaf blast and neck blast suggests that genes for resistance may be common (Siddiq and Viraktamath, 2000). So far, 12–15 resistance genes at seven different loci have been discovered. The resistance conferred by a single dominant Mendelian gene is highly race-specific. The spectrum of blast races varies from region to region and the range of resistance imparted by different resistance genes also varies. Though without much success, varietal deployment strategies have been attempted, including sequential release of resistant varieties and planting of a mixture of varieties carrying resistance genes. The

Table 6.14.   Genes for resistance in IR varieties.

Variety	Bacterial blight[a]	Blast	Grassy stunt	BPH[b]	GLH[c]
IR5	0	Pita	0	0	Glh3
IR8	0	Pi20(Pi b, Pi k-s)	0	0	Glh3
IR20	Xa4	(Pi b, Pi k-s)	0	0	Glh3
IR22	Xa4	Pi20(Pi b, Pi k-s)	0	0	0
IR24	—	Pi20(Pi b, Pi k-s)	0	0	—
IR26	Xa4	Pi20(Pi b, Pi k-s)	0	Bph1	—
IR28	Xa4	(Pi b, Pi k-s)	Gs	Bph1	Glh9
IR29	Xa4	(Pi b, Pi k-s, Pi z-t)	Gs	Bph1	Glh9
IR30	Xa4	(Pi b, Pi k-s)	Gs	Bph1	Glh3
IR32	Xa4	Pita (Pi b)	Gs	bph2	—
IR34	Xa4	(Pi b, Pi k-s, Pi z-t)	Gs	Bph 1	Glh9
IR36	Xa4	Pita (Pi b)	Gs	bph2	glh 10
IR38	Xa4	Pita (Pi b)	Gs	bph2	—
IR40	Xa4	Pita (Pi b)	Gs	bph2	—
IR42	Xa4	Pita	Gs	bph2	glh4
IR43	Xa4	Pi 20 (Pi b)	0	0	—
IR44	Xa4	Pita	0	Bph 1	—
IR45	Xa4	(Pi b, Pi k-s)	0	Bph 1	Glh3
IR46	Xa4	Pita, Pi 20	Gs	Bph 1	—
IR48	Xa4	Pita, Pi 20	Gs	bph2	—
IR50	Xa4	Pita (Pi b)	Gs	bph2	Glh9
IR52	Xa4	Pita (Pi b)	Gs	bph2	Glh9
IR54	Xa4	Pita (Pi b)	Gs	bph2	Glh9
IR56	0	Pik*, Pita	Gs	Bph 3	Glh9
IR58	Xa4	Pita (pi b)	Gs	Bph 3	Glh9
IR60	Xa4	Pita (Pi b)	Gs	Bph 3	Glh9
IR62	Xa4	Pita (pi b)	Gs	Bph 3	—
IR64	Xa4	Pita, Pi 20	Gs	Bph 1	—
IR65	Xa4	Pita (Pi b)	Gs	bph 2	Glh9
IR66	Xa4	(Pi b, Pi k-s)	—	Bph 3	—
IR68	Xa4	Pita (Pi b)	—	Bph 3	—
IR70	Xa4	Pi K, Pita	—	Bph 3	—
IR72	Xa4	Pita (Pi b)	—	Bph 3	—
IR74	0	Pi K, Pi 20	—	Bph 3	—

Source: Khush and Virk (2005).
[a]0 = no gene, [b]BPH = brown planthopper, [c]GLH = green leafhopper, — = not determined.

multi-line concept, though theoretically a sound strategy, has not been a success. Pyramiding of resistance genes for a region is more practical. This approach could be practiced successfully after the advent of molecular-marker-assisted selection (MAS) during the 1990s (Sridhar et al., 1999a; b).

Of the dozen major insect pests, brown plant hopper (BPH) is extremely destructive. Serious epidemics in India, the Philippines, Indonesia, and Vietnam in 1973–1976 caused crop losses estimated at more than US$400 million. Gall midge, confined to

small pockets until recently, has now spread to several non-endemic areas. A serious outbreak in 1986 in Andhra Pradesh, India caused a crop loss estimated at over US$10 million. Leaf roller is yet another pest that has assumed serious proportions lately.

Identification of resistance gene (*bph-1*) against BPH in TKM-6, and its incorporation into IR-26, provided a great relief to the rice-farming community everywhere. But this relief was short lived, because in a couple of years, IR-26, IR-28, IR-30, IR-34, with the same resistance gene, succumbed in

Table 6.15. Donors for resistance to insect pests and diseases. A: Insect pests.

Resistance gene	Donor source
Brown Planthopper	
Bph.1	Mudgo, Balamavee, Co 10, Co 20, Mtu 15, Tum.6
bph.2	ASD 7, Ptb 18, Ptb21, Ptb 33, Madayal, Dikwee 328, Malkora
Bph.3	Ptb 21, Ptb 33, Rathu Heenati, Gangal, Muthumanikm. Horna mawee, Ptb.19
bph.4	Babawee, Senawee, Hotel Samba, Sulai, Thirissa, Lexham Samba.
bph.5	ARC '10550
Bph.6	Swarnalata
bph.7	T12
bph.8	China Saba, Col.5. Thailand
Bph.9	Pokkadi, Kaharamana, Balanawee.
Green Leaf hopper	
Glh.1	Pankhari203, ARC 10313, ARC 7320
Glh.2	ASD 7, Tendeng, Sefa, Badal -2
Glh.3	IR 8, Chiknal, Laki 659
glh.4	Ptb 8, ARC 7012
Glh.5	ASD 8, Bhawalia, Bazal
Glh.6	TAPL 796
Glh.7	Maddai Karuppan
glh.8	DV 85
White Backed Planthopper	
Wbph.1	N 22, Senawee, Siam Garden, Oha
Wbph.2	ARC 10239, Chempan
Wbph.3	ADR 52, Chai Anaser
Wbph.4	Podiwi −A8
Wbph.5	N. Diang Marie
Wbph.6	Bian -Gu, Da-Hua-Gu, Da-qi-Gu.
Zigzag Leaf hopper	
Zlh.1	Rathu Heenati
Zlh.2	Ptb 21
Zlh.3	Ptb 33
Gall midge	
Gm.1	W1263, Eswarakora, Usha, Samridhi, NHTA-8
Gm.2	Surekha, IET 6286, RP435-107
gm.3	RP.2068-18-3.5, Bhumansan
Gm.4	Abhaya, Bhumansan
Gm.5	ARC 5984, Banglei
Gm.6	Dukong. 1
Gm.7 (k) and Gm.8 (k)	Bhumansan
Gm.9 (k) and Gm.10 (k)	NHTA-8
Gm.7 (k), Gm. 9 Gm.11 (k)	T 1432
Bacterial leaf blight	
Xa-1	Kogyoku, IR22, IR29, IR30, Java14
Xa-2	Te-tep, Rantai Emas-II
Xa-3	Zenith, Sateng, Java.14
Xa-4	Tkm 6, IR.20, IR.22
Xa-5	DZ.192, IR.1545-339
Xa-7	DV 85
Xa-8	P1 231129
Xa-10	Cas 209
Xa-11	RP 9-3, IR8, Elwe
Xa-12	Kogyoku, Java 14

Table 6.15.   Donors for resistance to insect pests and diseases. B: Diseases (*Continued*)

Resistance gene	Donor source
xa-13	BJ1, Chinsurah Boro.II
Xa-14	T(N)1
xa-15	M 41
Xa-16	Te-tep
Xa-17	Asominori
Xa-18	Toyonishiki
xa-19	Xm.5
xa-20	Xm.6
Xa-21	*O.longistaminata*
Blast (*Pyricularia oryzae*)	
Pi.1	LAC 23
Pi.Z^s	5173
Pi.3, Pi.ta	Pai-kan-tao
Pi.5 (t), Pi.7 (t)	Moroberekan
Pi.6 (t)	Apura
Pi.8 (t)	IRAT 13
Pi.9 (t)	*O. minuta*
Pi.10 (t)	Tongil
Pi.11(t)	Zhaiyeing
Rice Tungro virus	Latisail, Vikramarya, Utrimerah, Katari Bhog, Gundil Kuning, Jimbrug. Ptb 18, Ptb 21, ARC 10599, ARC 14320

Source: Siddiq and Viraktamath (2000).

Table 6.16.   List of important disease/insect pest-resistant varieties of India.

Disease/insect	Resistant varieties
Blast	IR 64, Vani Sasyasree, Improved Sona, Rasi, Rasmi, Pinakini, VLK 39, Tikkana, VL8, Rajarajan, IR 36, Swarnadhan, NLR 9672, Pant Dhan.10 VL Dhan.221, K332, K 333.
Bacterial leaf blight	Ajaya PR 110, PR 113
Rice tungro virus	Srinivas, Vikramarya, IR 20, Pankhari 203. Kataribhog.
Brown planthopper	Sonasali, Manasarovar, Vajram, Krishnaveni, Chaitanya, Nagarjuna, Pavizam, Chandana, Pratibha, Co-42, Nagarjuna, HKR 120, Karthika, Aruna, Rasmi, Jyoti, Makom, Remya, Annanga, Daya, Kanaka,
Green leafhopper	Vikramarya, Lalat, Khaira
Gall midge	Karna, Lalat, Sarasa, Abhaya, Kavya, Ruchi, Kshira, Sneha, Divya, Bhuban, Phalguna, Surekha, Shakti, Vibhava, Pothana, Mahaveer, IR 36, Dhanyalakshmi.
WBPH	HKR 120
Stem borer	Ratna, Sasyasree, Vikas
Sheath blight	Manasarovar, Pankaj, Swarnadhan

Source: Siddiq and Viraktamath (2000).

Indonesia and India. Utilizing a new resistance gene (*bph-2* from Ptb-18), IR-32, IR-36, IR-40, and IR-42 were developed. During the next 3–4 years, the resistance of *bph-2* also broke down in parts of Indonesia and the Philippines due to the emergence of biotype-3. Still stronger sources of resistance, like *bph-3* and *bph-4* from Rathu Heenathi and Balawee, were deployed in IR-56, IR-60, and IR-72. In India, the damage caused by BPH has been contained by the release of 'Krishnaveni,' 'Chaitanya,' 'Jyothi,' and 'Sonasali,' incorporating diverse resistance sources from Ptb-10, Ptb-18, Ptb-21, and ARC 5984.

A breeding strategy in the case of pests like stem borer and leaf folder, against which there is no strong source of resistance in germplasm, has been the use of moderate sources of resistance found in varieties like TKM-6 for stem borer and in Ptb-12, 'Derukasail,' and ARC 11128 for leaf folder. Breeding strategies adopted to pool the minor genes for resistance are selective diallel mating and recurrent selection. The much slower progress achieved so far necessitated adoption of the transgenic approach in such cases.

Wild species of rice are a rich source of genes for resistance breeding. Resistance has been introduced from related wild species; for example, resistance to tungro virus was incorporated into IR-64 from *O. rufipogon* (Khush, 1999). Similarly, resistance to grassy stunt virus was transferred from *O. nivara* through backcrossing. This dominant gene for resistance has now been incorporated into many improved varieties. Embryo rescue is also employed to produce interspecific hybrids. Genes for resistance to three biotypes of BPH were transferred from *O. officinalis* to an elite breeding line (Jena and Khush, 1990). Multani et al. (1994) transferred genes for resistance to brown plant hopper from *O. australiensis* to cultivated rice. Similarly, genes for resistance to blast and bacterial leaf blight have been transferred from *O. minuta* to cultivated, improved germplasm (Brar and Khush, 1997).

Molecular marker-assisted breeding is useful in some specific cases of breeding for resistance. If resistance genes can be tagged with tightly linked molecular markers, time, money, and effort can be saved in transferring such genes from one varietal background to another. Molecular markers have been used for pyramiding the resistance genes for blast as well as for bacterial leaf blight at IRRI (Huang et al., 1997) and in India (Singh et al., 2001). Marker-assisted selection is also being used for transferring resistance genes from pyramided lines to new plant types (Sanchez et al., 2000). Protocols for rice transformation have been developed to allow transfer of foreign genes from diverse biological systems into rice (Khush, 2004). A major target for genes encoding toxins from *Bacillus thuringiensis* (*Bt*) has been introduced for resistance to yellow stem borer with excellent levels of resistance in the laboratory and greenhouse (Datta et al., 1997; Tu et al., 2000). Xu et al. (1996) reported transgenic rice carrying a cowpea trypsin inhibitor (*cpti*) gene with enhanced resistance against striped stem borer and pink stem borer. IRRI scientists have also developed transgenic IR 58025 B and IR 68899 B, possessing *Xa21* and *Bt* gene, respectively. These lines should be useful in developing transgenic CMS lines and transgenic hybrids.

Breeding for abiotic stresses

Bio-physical constraints affect normal growth and development of rice crops in almost 70% of the rice-production area. The entire rainfed ecosystem consists of unfavorable environments: uplands, lowlands and deep water and flood-prone areas, toxic soils due to acidity, alkalinity, salinity, and sodicity across ecosystems, and extremely low and/or high temperatures during crop growth. Each of these environments is complex and is characterized by more than one constraint.

While in upland ecologies, major problems are moisture stress, weeds, impoverished soils, and phosphorous deficiency, the lowland ecologies pose anoxic conditions due to stagnant flood-waters, submergence, intermittent drought, and low light. In coastal and irrigation command areas, salinity and sodicity are the limitations. Low temperatures are the major problem in high-altitude hilly areas, whereas high temperatures during the reproductive phase pose problems in certain plains. Yield losses suffered due to various abiotic stresses are quite high, particularly in parts of Asia and in Africa and Latin American countries.

Of the two approaches to reduce crop losses due to abiotic stresses, viz., (1) breeding of varieties to suit the stress environment and (2) improvement of stress environment, the former is the preferred strategy due to its cost-effectiveness and reliability. Breeding for tolerance to hostile and harsh environments is not easy, not only because of the complex nature of the stressed environment, but also due to different kinds of mechanisms conferring tolerance to components of stress. The genetics of traits conferring tolerance is also complex.

Component traits of an adaptive value to stressed environment are believed to be inherited as a co-adapted complex, evolved and sustained by natural and human selection across time. Hence, recombining them with high yield and better quality is a challenging task. Pure-line selection in native, well-adapted varieties and recombination breeding are the short-term and long-term strategies, respectively. Factors contributing to breeding varieties for unfavorable ecosystems are: site characterization; identification and use of appropriate donors; breeding selective component(s) (for example,

drought tolerance in upland rice, submergence and drought tolerance in lowland rice, salinity and sodicity tolerance for coastal areas, and irrigated command areas); and development of appropriate screening techniques.

Drought tolerance

Three mechanisms by which crops can cope with drought are escape, avoidance, and tolerance. Upland rice varieties in general survive drought with an escape mechanism facilitated by early maturity or avoidance through drought-induced elongation of roots to reach comparatively deeper moisture zones. A tolerance mechanism operates through reducing transpiration losses by leaf rolling, early closure of stomatal openings, and cuticular resistance. Another mechanism is the ability to recover quickly when moisture is replenished after a prolonged drought spell. Because most of the characteristics of stress tolerance are independently inherited, there is a possibility of recombining them through convergent improvement. Though some sources for breeding for drought resistance are available (Table 6.17), progress has been slow due to the lack of rapid and reliable screening techniques. Screening for absolute grain yield, relative grain yield, and spikelet fertility percent appears to be the key (Maurya and O'Toole, 1986). Recently developed NERICA varieties from *O. sativa* × *O. glaberrima* have

Table 6.17. Promising sources of desired traits for breeding drought tolerant varieties.

Trait	Sources
High drought tolerance with good root system	Moroberekan, OS4, OS6, E425, 20A, Lac23, 63–83, Seratus Malam, Arias Halus, Khao Lo, Khao Youth, Padi Tatakir, KU86, KU-70-1, Kinandang Patong, Salummpikit, Azucena
Moderate drought tolerance and early maturity	Black Gora, Brown Gora, Kalakeri, Bala, Lalankanda, N 22, IAC 25, Dular
Improved drought tolerance and good recovering ability	IET 1444, IR43, IR4575 selection (semi-dwarf), C22, UPLR-5 IR6023-10-1-1 (Intermediate tall)

Source: Chang and Loresto (1986).

shown good tolerance to drought in West African countries. The genes for tolerance to drought have been derived from *O. glaberrima*.

Genetic engineering techniques hold great promise for developing rice varieties with drought tolerance. Garg et al. (2002) introduced *Ots* A and *Ots* B genes for trehalose biosynthesis from *Escherichia coli* into rice. Transgenic rices accumulated disaccharide trehalose at 3–10 times the rate of non-transgenic controls. Trehalose is a non-reducing disaccharide of glucose that functions as a compatible solute in the stabilization of biological structures under abiotic stress. The transgenic rice lines displayed increased tolerance for abiotic stresses, such as drought and salinity.

Submergence tolerance

In the lowland ecosystem, submergence is as serious a constraint as drought. More than 10–15 million hectares of rice worldwide are subjected to this biophysical constraint. Submergence tolerance is defined as the ability of rice crop to survive under complete submergence for as many as 10 days. Submergence due to flash floods or excessive rain-caused inundation, coupled with poor drainage in low-lying areas, can occur at any stage of crop growth. Mechanisms for tolerance to submergence are: (1) ability to survive under water without any growth till the constraint is removed and (2) escape from submergence by growing along with rising water levels.

Screening techniques used for assessing submergence tolerance in Asia include: evaluation at seedling stage under simulated conditions in the greenhouse for germplasm and selective seed-bed method for F_2 and bulk populations before transplanting, screening in artificial tanks at different stages of growth for segregating populations and breeding lines and, finally, testing in a typical field environment. Scoring for rela-

tive tolerance is done on the basis of two important parameters, viz., the percentage survival of plants immediately after draining the water and the recovering ability of stressed plants 10 days after draining water. However, these screening techniques are laborious and cumbersome. Though the genetics of submergence tolerance is complex, landraces like FR 13A, FR 43B, 'Kurkaruppan,' and 'Goda Heinati,' have been reported to possess at least one major tolerance gene (Mohanty et al., 1982). In spite of years of breeding efforts, the majority of popular varieties in the flood-prone lowlands are still either traditional varieties or slightly improved versions produced through pure-line selection. The donors for tolerance to submergence have been FR13A, FR43B, and 'Kurkaruppan.' Varieties derived for flood-prone areas (e.g., Madhukar) are not as tolerant or adapted as donors.

Utilizing FR13A through marker-assisted selection, some improved lines with submergence tolerance have been developed at IRRI (Xu et al., 2006). Hybrids perform better than the inbred varieties (Virmani, 2003). The Regional Rainfed Rice Lowland Consortium coordinated by IRRI is currently addressing these issues.

Salinity tolerance

The growing area under coastal salinity is static, while that under inland salinity has been increasing due to faulty water management. Breeding varieties for salt tolerance has been much slower due to poorly understood mechanisms of tolerance and inadequate screening techniques. Traditional landraces or improved varieties from pure-line selection are predominant in such areas. Recombination breeding has been attempted with limited success so far.

Coastal areas are saline due to accumulation of sodium chloride and sodium sulfate; inland areas are seldom sodic (alkaline). The

Table 6.18.   Genetics of salt tolerance in rice.

Index	Mode of Inheritance	Reference
Panicle sterility	Dominant, Trigenic	Akbar & Yabano (1977)
Shoot length, dry weight of shoot & root; Na & Cl content in shoot	Polygenic, Predominantly additive	Akbar et al. (1986)
Days to flowering; plant height; tiller number, spikelets/panicle, test grain weight, dry matter accumulation	Polygenic, additive & dominance effects	Narayanan & Srirangasamy (1991)
Low Na to K ratio	Polygenic, additive & dominance effects	Gregorio & Senadhira (1993)
IRRI's tolerance score	Polygenic, major influence of minor genes	Mishra et al. (1998)
Seedling shoot growth	Six basic genes interacting in varied combinations	Garg and Siddiq (in press)
Shoot proline accumulation	Predominantly non-additive	
Shoot ionic ratio	Dominance, recessiveness not clear	

Source: Siddiq and Viraktamath (2000).

composition of salts and their severity vary with region, and tolerance is growth-stage specific. These complexities deserve attention while identifying donor sources and evolving breeding strategies. The genetics of salt tolerance depends on the chosen index (Table 6.18).

The mechanism of tolerance to sodium chloride salinity is the retention of sodium and chlorides in the root rather than in the shoot. In most of the tolerant and well-adapted donors, tolerance is either through exclusion of salts or compartmentalization at the cellular or tissue level, in association with rapid pace of growth. By and large, the ability of a plant to regulate either sodium uptake or selective accumulation of potassium, translocation of salts from root to shoot, and ionic balance (ratio of sodium to macro/micronutrient) rather than absorbed sodium content determine the level of tolerance (Siddiq et al., 1999). Decades of research demonstrated that the tolerance is polygenically controlled. Hybrids have shown better performance than inbred varieties (Virmani, 2003).

Screening breeding populations either in hot spots or in lysimeter/microplots under artificially created stress conditions have produced some appreciable results. Marker-associated QTL (quantitative trait loci) for tolerance are expected to greatly facilitate breeding efforts. Similarly, transgenics with trehalose genes (Garg et al., 2002) also offer opportunities.

Breeding for grain and nutritional quality

Grain quality is consumer specific. Most consumers in the tropics and subtropics prefer slender grains, long or medium long, and in temperate countires, preference is for short, bold and round grain. In the U.S., long grains are preferred. Preference is based on end-product use, inlcuding the manner of preparation and consumption. Cooking quality of rice is largely determined by amylose content and the gelatinization temperature of rice starch. In the tropics and subtropics, rice grain with intermediate amylose (~23%) is preferred; it gives a non-sticky, flaky cooked product. In Japan, Korea, and China, preference is for low amylose content and gelatinization temperatures, which results in a moist and sticky cooked product. Glutinous or waxy types, used mostly for preparation of rice cakes, have no amylose. The entire starch fraction is amylopectin. Another important cooking quality index is gel consistency, which determines how long the cooked rice remains soft. Varieties with medium or low gel consistency with medium or low gelatinization temperature are most preferred, as they remain soft longer after cooking.

Nutritional quality of rice is judged by protein content and quality, vitamins, and minerals. In-depth research carried out at IRRI and a few NARS revealed the genetics and breeding of some of the major quality traits. (For details, see the monograph by Juliano, *Rice Chemistry and Technology* [1985].)

For traders, percentage milling outturn (ratio of rice/paddy), head rice recovery (percentage of unbroken, whole grains), and color of rice grain are important. The indices for grain quality are many and, except for aroma, most of them are polygenically controlled, making breeding for specific traits very complex. The genetics of various indices for quality traits in rice is given in Table 6.19.

Improved varieties like IR-5 and IR-8 had poor grain quality. They cooked dry because of high amylose content and so had very poor consumer acceptance. Moreover, they had bold and chalky grains of very poor appearance, which frequently broke during milling. All varieties released from IRRI have slender and translucent grains with good milling recovery. However, improvements in cooking quality were achieved slowly, because all the donors for disease and pest resistance used in the hybridization

Table 6.19. Genetics of various indices of quality in rice.

Characteristics	Mode of inheritance (predominant gene action)
Physical features	
Grain length	Polygenic (Somrith, 1974)
Grain shape (L/B ratio)	Polygenic (Somrith, 1974)
Waxy rice	Monogenic recessive (*wx*)
Chalkiness (white belly/white center)	Monogenic recessive (*wb/we*)
	Polygenic (Chang & Li, 1980)
Cooking behavior	
Water absorption	Polygenic/non-additive (Zaman et al., 1987)
Volume expansion	Polygenic/non-additive (Zaman et al., 1987)
Kernel elongation	Polygenic/non-additive (Sood et al., 1982)
Amylose content	Oligogenic + modifiers
	Polygenic/non-additive gene action (Puri & Siddiq, 1982)
Gelatinization temperature	Oligogenic + modifiers (McKenzie & Rutger, 1983);
	Polygenic-additive gene action (Kablon, 1965; Stensil, 1986; Somrith et al., 1979)
	Polygenic/additive gene action (Puri & Siddiq, 1982);
	Oligogenic + modifiers (Chang & Li, 1980).
Gel consistency	Polygenic/additive gene action (Zaman et al., 1985)
Aroma	Monogenic recessive (Sood & Siddiq, 1979) complimentary 2-3 genic interaction (Singh & Mani, 1987; Tripathi & Rao, 1979).
Nutritive quality	
Protein content	Polygenic/non-additive & additive gene action (Sood & Siddiq, 1986); dominance of additive effects (Chang & Lin, 1974)

Source: Siddiq and Viraktamath (2000).
High amylose partially dominant; High GT partially dominant;
High water absorption & volume expansion dominant;
Low kernel elongation over-dominant; Low protein partially dominant.

program had high amylose content and low gelatinization temperature. IR-48 was the first variety with intermediate amylose content, but it had low gelatinization temperature. Similarly, IR-20, IR-32, IR-36, and IR-46 have an intermediate gelatinization temperature but high amylose content. The first IR variety with intermediate amylase and gelatinization temperatures was IR-64. It has been widely accepted as a high-grain-quality rice in India, Indonesia, Vietnam, the Philippines, and several other countries. In fact, IR-64 is the most widely planted variety of rice in the world today. It is planted on about 8 million hectares worldwide (Khush, 1995).

Nutritive quality of rice grain is as important as cooking quality for those who consider rice as their staple food. Rice accounts for 40% of the protein consumed in Asia. Rice has an inherently low protein percentage, and it loses 15–20% of its protein during the milling process. Rice lacks vitamin A and is deficient in iron, zinc, and lipids. Fortification of rice with such elements is essential to ensure nutritional security. IRRI researches ways to enhance iron and zinc content through the exploitation of natural variability and genetic engineering. A recombination breeding program to develop iron- and zinc-rich, high-yielding elite breeding lines resulted in IR 68144-2B-2-2-5, which is now under extensive field-testing.

Transgenics are being tested to develop lines with higher levels of β-carotene (precursor of vitamin A). These lines have golden yellow grains popularly known as "Golden Rice" (Al-Babili and Beyer, 2005). Because no reliable gene source for β-carotene is found in rice, by taking advantage of the existence of an incomplete pathway leading to biosynthesis of β-carotene, rice has been successfully engineered by splicing the missing genes from the bacterium *Erwinia* and daffodil (*Narcissus sp.*), thus producing β-carotene in the endosperm (Ye et al., 2000). The transgenic Golden Rice has been developed in the background of *japonica* variety 'Taipei 309.'

In India, breeding efforts during the last three decades yielded improved basmati rices (grown in north and northwestern parts of the Indian subcontinent) that are highly valued the world-over because of their pleasant aroma and super fine, extra-long grains. These high quality tall rices are low yielding. It took more than two decades to combine high quality traits of traditional basmati with the semi-dwarf, high-yielding attribute that resulted in the world's first high-yielding dwarf basmati, known as 'Pusa basmati-1,' with 50–100% yield increment. It is very widely grown in the Indian states of Haryana, Punjab, Uttaranchal, and Himachal Pradesh. Recently, a super fine-grained basmati-like hybrid, 'Pusa Rice Hybrid-10,' has been released that has 40% yield advantage over Pusa basmati-1 and is early maturing by 15–20 days. Other such varieties from India are 'sugandha-2,' 'sugandha-3,' and 'sugandha-5'.

Breeding for short growth duration

Pre-Green Revolution varieties in tropical and sub-tropical Asia matured in 160–170 days and many were photoperiod-sensitive. Because of this long duration, multiple cropping was not possible. IR-8 and subsequent varieties, such as IR-20, IR-26 (and many HYVs developed in NARS during the 1970s and 1980s), mature in about 130 days. Most of the current varieties belong to the medium, early-to-medium maturity group (115–130 days). But even these varieties could be grown only once a year. Therefore, development of varieties with shorter growth cycles became desirable. IR-28, IR-30, and IR-36 mature in 110 days, whereas IR-50 and IR-58 mature in 105 days. While short- and medium-duration varieties produced 53 kg/day, the later-maturing IR-74 produced only 35 kg/day. The availability of short-duration varieties has led to major changes in cropping patterns in Asia (Khush, 1995). In the Philippines, many rice farmers grow an upland crop either before or after their rice crop under rainfed conditions. In some cases, two crops of rice are regularly grown in a rainy season. In some irrigated areas, farmers grow three crops of rice in one year. In Indonesia and Vietnam, short-duration varieties are very popular.

Short-duration varieties grow very rapidly during the vegetative phase; thus, they are more competitive with weeds. Consequently, weed control costs are reduced. These varieties also use less irrigation water, further reducing costs. In addition, they lead to greater on-farm employment due to increased cropping intensity.

Impact of Rice Breeding

The impact of rice breeding and germplasm enhancement by several NARS and IRRI, popularly described as the "Green Revolution" has led not only to major increases in food production, but has also improved socio-economic conditions and environmental sustainability (Pingali and Hossain, 1998; Khush and Virk, 2005), especially in tropical Asian countries (Table 6.20).

Farmers now harvest 5–7 tons/ha from the improved HYVs in irrigated ecosystems

against 1–3 tons/ha earlier. Since 1966, when the first semi-dwarf HYV was released, the global area under rice has increased only marginally, from 126 m ha to 152 m ha (20%), but productivity increased from 2.1 t/ha to 3.9 t/ha (86%), and the total production increased from 257 m tons in 1966 to 600 m tons in 2006 (133%).

In many countries, the growth in production outstripped increases in population, leading to a substantial increase in cereal consumption and calorie intake. During 1965–1990, the daily calorie supply (in relation to the requirement) improved from 81% to 120% in Indonesia, from 86% to 110% in China, and from 82% to 94% in India (UNDP, 1994). At the global level, the increase in per capita availability and decrease in the cost of production contributed to declines in price. The unit cost of production is 20–30% lower for the HYVs, and the price adjusted to inflation is 40% lower than during the mid-1960s. The diffusion of HYVs has also contributed to growth in income for rural, landless workers (Hossain, 1998) because labor requirements increased due to the high intensity of cropping. Marketing a larger volume of farm produce generated additional employment in rural trade, transport, and construction activities. If 1961 yields prevailed today, three

Table 6.20. Total area, coverage of high-yielding varieties, and increase in rice production in selected countries of Asia.

Country	Total area planted (M. hectares)		Area planted to HYV %	Production (M. tons)		Increase in production %
	1966	1996		1966	1996	
Bangladesh	9.1	10.3	46	14.3	28.0	96
China	31.3	31.4	100	98.5	190.1	93
India	35.2	42.7	75	45.6	120.0	163
Indonesia	7.7	11.3	77	13.6	51.2	276
Myanmar	4.5	6.5	58	6.6	20.9	217
Pakistan	1.4	2.3	41	2.0	5.6	180
Philippines	3.1	4.0	94	4.1	11.3	176
Sri Lanka	0.5	0.8	94	1.0	2.2	120
Thailand	7.3	9.2	13	13.5	21.8	61
Vietnam	4.7	7.3	80	8.5	26.3	209

Source: Khush and Virk (2002)

times more land in China and two times more land in India would have been necessary to achieve the 2000 rice harvest. As an example, to produce the 2000 world rice production of 600 m tons, at the yield level of 1965, an additional 135 m ha would have been required.

Future Outlook on Rice Breeding

Since the release 40 years ago of the first semi-dwarf HYV, IR-8, annual global rice production has been enhanced by 350 m tons, from 257 to 600 m tons. An important question is, "Can we accomplish a similar remarkable feat, during the next 40 years, wherein the demand is estimated to be an additional 400 m tons?"

Challenges and Limitations

The current global population of 6.5 billion is estimated to reach 9.0 billion by 2050. More than 90 percent of this increase will be in developing countries of Asia and Africa. To meet consequent rising needs, rice production needs to be enhanced by at least 1.5%/year, and global rice production needs to be enhanced by approximately 10 m tons/ year. Enhancement of 600 m tons to almost 1,000 m tons by 2050 is going to be a phenomenal challenge. Production steadily declined during the 1980s and 1990s. The growth rate in rice production, which was above 3%/year during the 1970s, declined to 2% during the 1980s and to 1.4% during the 1990s. This is a matter of great concern. Other challenges will also need to be met: declining water and land resources, global climate change, emerging pest scenarios, and restricted exchange of germplasm. Productivity of rice crops is highest in the irrigated ecosystem, and around 75% of global production comes from 55% of the irrigated areas. Adequate availability of water for rice cultivation is going to be a major concern in

the decades ahead. As the population increases alarmingly, in Asia and Africa per capita rice land is decreasing because of its transformation for housing, urbanization, roads, etc. The quality and health of soil is deteriorating; sustaining productive rice land is emerging as a major challenge for the forthcoming years.

Global climate change is another major concern. Enormous quantities of greenhouse gases are released into the atmosphere through mining and combustion of fossil fuels, deforestation, maintenance of livestock herds and, to a much smaller extent, through rice production. The emission of methane and nitrous oxide gases from lowland rice cultivation and deforestation entailed in upland rice production under slash-and-burn shifting cultivation are considered contributors to global climate change.

The accumulation of greenhouse gases in the atmosphere has warmed the planet and caused changes in the global climate. In 2001, the UN-sponsored Intergovernmental Panel on Climate Change (IPCC) reported that worldwide temperatures have increased by more than 0.6°C in the past century. The IPCC estimates that, by 2100, average temperatures will increase by between 1.4° and 5.8°C. The IPCC also reported that sea levels have risen by between 10 and 20 cm worldwide, and precipitation patterns have also progressively changed.

Temperate regimes greatly influence growth duration, growth pattern, and productivity. Studies on rice productivity under global warming suggest that productivity of rice and other tropical crops will decrease as global temperatures increase. Peng et al. (2004) reported that the yield of the dry-season rice crop in the Philippines decreased as much as 15% for each 1°C increase in mean temperature during the growing season. Mohandrass et al. (1995) predicted a yield decrease of 14.5% for summer rice crops across nine experimental stations in

India during 2005. Temperature increase in sub-tropical and temperate areas may have positive or negative effects, depending on location. For example, temperature increase would improve crop establishment in Mediterranean areas, where cool weather usually causes poor crop establishment, while a similar temperature increase combined with low night temperatures in northern Japan would reduce the beneficial effects on grain production.

Rising sea levels are likely to affect rice production, especially in the low-lying deltas of the Ganges, the Mekong, the Nile, the Yangtze, the Yellow, and other major river systems that are regularly affected by tidal waves. Nearly 650,000 ha of the Mekong and 350,000 ha of Red River (Vietnam) are affected by salinity, which has reduced the potential for wetland rice production. In the Mekong, the effect of ocean tides could sometimes be observed as far as 200 km from the sea. Changes in precipitation patterns are also likely to affect rice production. Worldwide, around 45% of rice area is classified as rainfed. Variability in the amount and distribution of rainfall is the most important yield-limiting factor of rainfed rice. Variability in the onset of the rainy season leads to variation in the start of sowing/transplanting operations of rainfed rice. In freely drained uplands, heavy precipitation of 200 mm on one day, followed by a rain gap for the next 20 days, severely damages or even kills rice plants. Complete crop failure usually occurs when severe drought stress is experienced during the reproductive phase.

Flood is the most damaging constraint in low-lying areas. Most of the lowland and deep-water varieties can tolerate complete submergence for a week, but submergence for longer periods results in 50–100% crop destruction. Changes in patterns of rainfall distribution may lead to more frequent occurrences of intense floods and droughts in different parts of the world (Depledge, 2002). This is going to be, again, a major constraint for enhancing rice production on a sustainable basis. How can these and other emerging challenges be met?

Rice Breeding Strategies for the Future

In the immediate future, extending the coverage of HYVs in many rice-growing countries where, at present, the coverage of HYVs is less than 50%, can enhance production. In some countries, where the percentage of HYV coverage is already high, hybrid rice technology and NPT can be extended to rice farmers. Bridging the yield gap between "potential" and "achieved" is another strategy to enhance rice production.

The time-tested traditional plant breeding methodologies will continue to be useful in developing new varieties for various ecosystems. Varieties that are specifically adapted to aerobic cultivation and System of Rice Intensification (SRI) cultivation are being developed to meet water shortages. Similarly, genotypes that can withstand higher temperatures and grow well under increased ozone and methane concentrations are being developed to meet global climate change. But all these approaches may give only marginal increases in production. During the past 2–3 decades, the yield increase achieved through traditional breeding methods was 1.0% per year. Given the necessary funding, it seems likely that conventional crop breeding, genetic engineering, and natural resource management can improve productivity (Huang et al., 2002). A major paradigm shift is needed, however, to enhance rice production to meet the anticipated demand. Hopefully, the recent advances in rice biotechnology could provide the needed breakthrough (Khush and Peng, 1996; Khush 2001a,b).

Biotechnological tools will be used increasingly in rice breeding and are expected to play a major role in ushering in the second and everlasting Green Revolution. In fact, rice is recognized as a model system for the study of the cereal genome (Shimamoto and Kyozuka, 2002). Rice has many advantages: its small genome size of 430 Mb (Armuganathan and Earle, 1991); the availability of its whole genome sequence (Sasaki et al., 2002); a large public germplasm collection (over 100,000 diverse accessions); and the development of several key genomic mapping resources (McCouch et al., 2002; Gowda et al., 2003). Intensive studies on rice functional genomics are underway now in several laboratories around the world (Leung and An, 2004). Because the basic information about genes and pathways learned from a species can be applicable to other species, rice genomics promises fundamental insights into the genes of many other crop species.

Some of the remarkable developments in rice biotechnology that can assist rice breeding are: production of sporophytic plants with gametophytic chromosome number from microspore cultures; plant regeneration from protoplast culture of *japonica* and *indica* rices; transgenic rices; construction of comprehensive molecular maps of the rice genome; mapping and tagging of genes of economic importance (particularly for resistance to major diseases and pests); molecular characterization of biodiversity in rice germplasm; and the ability to combat rice pests and pathogens (Coffman et al., 2004). Biotechnological applications that have been of direct relevance in rice breeding are in the areas of wide hybridization, molecular markers, and in development of transgenics. Possible uses of biotechnological tools in rice breeding are briefly considered below.

Wide hybridization

The wild gene pool contains useful genes for resistance to stresses. The barriers most commonly encountered are lack of crossability and abortion of hybrid embryos resulting from genic and chromosomal differences (Brar and Khush, 1997). Embryo rescue and protoplast fusion have been employed to produce interspecific hybrids. More recently, molecular techniques have been employed for precise monitoring of alien gene introgression (Brar and Khush, 2002).

As indicated previously, there are 20 wild species in the genus *Oryza* in addition to the two cultivated species *sativa* and *glaberrima*. There are diploid and tetraploid species belonging to six different genomes. While wild species with AA genome can be easily crossed with the cultivated species, embryo rescue is necessary for crosses between cultivated species and wild species. A number of useful genes have been introgressed into cultivated rice from the wild germplasm (Table 6.21). Transfer of resistance to grassy stunt virus from *O. nivara* (Khush, 1977) and resistance to brown plant hopper from *O. officinalis* and resistance to bacterial blight from *O. longistaminata* are some of the important examples.

Wide hybridization is also very useful in the diversification of sources of male sterility-inducing cytoplasm. New sources of male sterility in rice—that are alternatives to the widely used WA cytoplasm—have been developed using cytoplasm from *O. nivara*, *O. rufipogon*, *O. perennis* and *O. glumaepatula* (Dalmacio et al., 1995; Hoan et al., 1997; Khush and Brar, 2002). Some backcross derivatives from a cross between *O. rufipogon* (Malaysian accessions) and cultivated rice out-yielded the recurrent parent by 18% (Xiao et al., 1996). Xiao et al. (1996) identified two QTL contributing to yield from wild species. These QTL are now being transferred to several semi-dwarf HYVs and parental lines of hybrids.

Use of molecular markers

The discovery during the 1980s that the differences in restriction endonuclease sites could be used as markers to construct genetic

Table 6.21. Introgression of genes from wild *Oryza* species into cultivated rice.

| Trait | Wild species | Donor Oryza species | |
		Genome	Accession number
Transferred to *Oryza sativa*			
Grassy stunt resistance	*O. nivara*	AA	101508
Bacterial blight resistance	*O. longistaminata*	AA	—
	O. officinalis	CC	100896
	O. minuta	BBCC	101141
	O. latifolia	CCDD	100914
	O. australiensis	EE	100882
	O. brachyantha	FF	101232
Blast resistance	*O. minuta*	BBCC	101141
Brown planthopper resistance	*O. officinalis*	CC	100896
	O. minuta	BBCC	101141
	O. latifolia	CCDD	100914
	O. australiensis	EE	100882
Whitebacked planthopper resistance	*O. officinalis*	CC	100896
Cytoplasmic male sterility	*O. sativa* f. *spontanea*	AA	—
	O. perennis	AA	104823
	O. glumaepatula	AA	100969
Tungro tolerance	*O. rufipogon*	AA	105908
	O. rufipogon	AA	105909
Progenies under evaluation for introgression			
Yellow stem borer	*O. longistaminata*	AA	—
Sheath blight resistance	*O. minuta*	BBCC	101141
Increased stem elongation ability	*O. rufipogon*	AA	CB751
Tolerance to acidity, iron and aluminum toxicity	*O. glaberrima*	AA	many
Weed competitiveness	*O. glaberrima*	AA	many

Source: Khush and Brar (2002).

maps brought a paradigm shift in rice breeding, causing attention to shift from morphological markers (phenotypes) to DNA markers. Cornell University developed the first Restriction Fragment Length Polymorphism (RFLP) map in 1988 (McCouch et al., 1988), and high-density linkage maps were developed subsequently (Kurata et al., 1997). Polymorphism in the nucleotide sequence is the basis for these DNA markers to serve as selection tools. Such polymorphisms are revealed by different markers, e.g., RFLPs, Random Amplified Polymorphic DNA (RAPD), Amplified Fragment Length Polymorphism (AFLP), Micro-satellites or Simple Sequence Repeats (SSR). The DNA markers are linked to gene loci controlling desired trait(s) and these markers co-segregate with the desired trait across generations. Ten traits in rice have been tagged and mapped (Table 6.22).

The majority of genes tagged and mapped have been those associated with resistance to biotic stresses, and these have been assigned to chromosomes 11 and 12, which share a duplicate segment (Kurata et al., 1997). The molecular markers are useful in rice breeding in: (1) marker-assisted selection (MAS), (2) map-based cloning of genes, (3) determining allelism of gene(s) conferring identical phenotypes, (4) study of genetic similarity or dissimilarity, and (5) tracking the introgression of transgene(s). Here, selection is based on genotype rather than phenotype, with the ability to ascertain the genotype at seedling or even at the seed stage and to simultaneously screen for multiple genes/traits in a single plant.

Table 6.22.   Number of genes tagged and mapped for various traits in rice.

Trait	Number of genes tagged	Number of genes mapped
Blast resistance	15	15
Bacterial leaf blight resistance	09	08
Rice Tungro virus resistance	01	01
Rice yellow mosaic virus resistance	01	01
Gall midge resistance	08	08
Brown plant hopper resistance	03	02
Whiteback plant hopper resistance	02	01
Thermo-sensitive genic male sterility	04	03
Photoperiod-sensitive genic male sterility	03	03
Wild-abortive cytoplasmic male sterility	03	03

Source: Sarma and Sundaram (2003).

It is worthwhile to discuss some of the successful examples of the use of molecular markers in rice breeding. Two of the most serious and widespread diseases in rice production are blast (*Pyricularia oryzae*) and bacterial leaf blight (*Xanthomonas oryzae*). Several races, or pathotypes, of these pathogens occur and genes imparting resistance to these pathotypes have been identified, mapped, and tagged. By using molecular markers, these genes imparting resistance have been pyramided into elite breeding lines. For bacterial leaf blight, *Xa4*, *Xa13*, and *Xa21* have been pyramided in popular varieties at IRRI (Huang et al. 1997) and at Punjab Agricultural University, Ludhiana, India (Singh et al., 2001). At the Directorate of Rice Research, Hyderabad, India, also, these three genes have been pyramided, by using molecular markers, into the most popular and widely grown varieties in India, viz., 'Swarna' and BPT-5204. Pyramided lines are presently under multilocational evaluation in the national coordinated testing program. Similarly, three genes imparting resistance to blast have been pyramided into an elite line (Hittalamani et al., 2000). The pyramided lines showed a wider spectrum/ higher level of resistance. Another example is the transfer of a thermosensitive male-sterility gene to aromatic Thai varieties (Lopez et al., 2003). Molecular markers are routinely being used in rice breeding now.

Genetic transformation

The first transgenic cereal to be developed was rice, using a protoplast-based transformation system (Toriyama et al., 1988). Subsequently, biolistic techniques have increasingly been used for rice transformation, as they circumvent genotype specificity (Christou et al. 1991). Using the biolistic technique, co-transformation of a number of genes is possible and this opens up the possibility of transforming rice with genes for a whole biosynthetic pathway. Demonstration of successful transformation in rice using *Agrobacterium* has been a major breakthrough in genetic engineering (Hei et al., 1994). For successful rice transformation, regeneration through tissue culture is a prerequisite.

Transgenic rices carrying insecticidal protein genes from *Bacillus thuringiensis* (*Bt* rice), chitinase gene from bacterial and other sources, and bacterial leaf blight resistance gene *Xa21* from *O. longistaminata* have been developed. Engineered rice with better nutritional quality (vitamin A, Golden Rice), submergence tolerance, and male sterility are currently undergoing evaluation before their expected release. A list of useful genes transferred is given in Table 6.23.

By using the desirable genes from other species or genera of plants—and even animals—several useful transgenics have

Table 6.23. List of useful genes transferred into rice.

Sl. No.	Useful trait	Target Pest	Novel Gene	Rice sub-species transformed	Reference(s)
1	Insect pest Resistance				
	Bt-σ-endotoin	Stem borer	*Cry 1A(b)*	*indica, japonica* Aromatic	Fujimoto et al. (1993) Datta et al. (1998) Alam et al. (1998) Ghareyazie et al. (1997)
	Bt-σ-endotoin	Stem borer	*Cry 1A(b) and Cry 1A(c)*	*japonica*	Cheng et al. (1998)
	Bt-σ-endotoin	Stem borer	*Cry 1A(c)*	*indica*	Nayak et al. (1997)
	Bt-σ-endotoin	Stem borer	*Cry 2 A*	*indica, basmati*	Maqbool et al. (1998)
	Potato protease inhibitor II	Leaf folder	*Pin II*	*japonica*	Duan et al. (1996)
	Cowpea trypsin inhibitor	Leaf folder	*CP Ti*	*japonica*	Xu et al. (1996)
	Corn cystatin		*CC*	*japonica*	Irie et al. (1996)
	Oryza cystatin		*Oc*	*japonica*	Hosoyama (1995)
	Snowdrop lectin	BPH	*GNA*	*japonica*	Rao et al. (1998)
	Cysteine proteinase inhibitor	Nematode	*OC-IΔD86*	African *indica*	Vain et al. (1998)
2	Disease resistance				
	Xa 21	BLB	*Xa 21*	*japonica*	Wang et al. (1996)
	Xa 21	BLB	*Xa 21*	*indica*	Tu et al. (1995)
	Chitinase gene		*Chi 11*	*indica*	Lin et al. (1995)
	grapevine stilbene synthase gene	Blast		*japonica*	Stark et al. (1997)
		Capsid Brome mosaic virus disease	*RNA-2,3 genes from BMV*	*japonica*	Huntley et al. (1996)
	Coat protein gene	Rice dwarf virus	Outer coat protein (S8) gene of RDV	*japonica*	Zheng et al. (1997)
	Coat protein gene	Rice tungro spherical virus	RTSV coat protein genes	*indica, japonica*	Sivamani et al. (1999)
	RNA polymerase gene	Rice yellow mottle virus	RNA-dep. RNA polymerase of RYMV	African *indica*	Pinto et al. (1999)
3	Abiotic stress resistance				
	LEA protein gene etc.,	Salt and drought	*P5CS*	*japonica*	Zhu et al. (1997)
			HVA1 (LEA3)	*japonica*	Xu et al. (1996)
				Basmati	Jain et al. (1996)
			CodA	*japonica*	Sakamoto et al. (1998)
			Adc	*japonica*	Capell et al. (1998)
			GPAT	*japonica*	Yokoi et al. (1998)
4	Nutritional quality	Vit. A- (β carotene)		*japonica*	Burkhardt et al. (1997)

Source: Siddiq and Viraktamath (2000).
Adc, arginine decarboxylase, Coda, Choline Oxidase gene from a soil bacterium Arthobacter globiformis; HVA1, barley late embroyogenesis-abundant (LEA-3) protein gene; OC IΔD86, an engineered cysteine proteinase inhibitor (Oryza cystatin-IΔD 86) gene; PSCS, Δ'-Pyrroline-5-carboxylase synthetase; RTSV, rice tungro spherical virus; RYMV, rice yellow mottle virus.

been developed in rice that are likely to enhance rice productivity across ecosystems. They have not been commercialized in any country as yet. While specific information about the status of transgenic technologies in rice is difficult to locate, some information can be gained through the patents filed and the field trials being conducted. In 2002 307 rice biotechnology patents were filed by 404 organizations (Brookes and Barfoot, 2003). Pioneer Hi-Bred International has the largest number of patents (68), followed by Monsanto (33), Syngenta (32), and Bayer (19).

Transgenic technologies under development in rice can be classified into five classes: yield enhancement, resistance to biotic stresses, resistance to abiotic stresses, nutritional enhancement, and herbicide tolerance.

Transgenics for yield enhancement

ADP-glucose pyrophosphorylase (ADPGPP) is a critical enzyme for regulating starch biosynthesis in plant tissues. Following the breakthrough attained in potato tubers of the plants transformed with *glgC16* gene from *E. coli* encoding ADPGPP (Stark et al., 1992), the same gene was introduced into rice recently.

With the aim of improving photosynthetic efficiency, efforts are being made to transfer C_4 traits into C_3 rice. Ku et al. (1999) used *Agrobacterium*-mediated transformation to introduce the gene for phosphoenolpyruvate carboxylase (PEPC) from maize, which catalyses initial fixation of atmospheric CO_2 in C_4 plants. Many transgenic rice plants showed high levels of expression of the maize gene. The activities of PEPC in the leaves of some of the transgenic plants were 2–3 times higher than those in maize, and the enzyme accounted for up to 12% of the total leaf soluble protein. This demonstrated a successful strategy for introducing a key biochemical component of the C_4

pathway of photosynthesis into a C_3 pathway of rice. These findings are only preliminary, and the transgenics are still far from being commercialized.

Transgenics for resistance to biotic stresses

A major target for *Bt* deployment in transgenic rice is the yellow stem borer, *Scirpophaga incertulas*. The pest is widespread in Asia and causes substantial crop losses. Improved rice cultivars are either susceptible or have only moderate resistance. Transgenic rice with *Bt* is, therefore, promising for reducing the damage caused by stem borer.

Fujimoto et al. (1993) introduced truncated d-endotoxin gene *Cry 1A(b)* into rice. Transgenic plants in R_2 generation expressing the cry 1A(b) protein showed increased resistance to striped stem borer and leaf folder. Wunn et al. (1996) introduced *Cry 1A(b)* gene into IR-58 through particle bombardment. The transgenic plants in R_0, R_1, and R_2 generations showed resistance to several lepidopteran insect pests. Nayak et al. (1997) transformed IR-64 through particle bombardment using the *Cry 1A(c)* gene and identified six independent transgenic lines with high expression of crystal protein. Transgenic rice plants proved highly toxic for stem borer larvae.

Cheng et al. (1998) produced more than 2,600 transgenic rice plants in nine strains through *Agrobacterium*-mediated transformation using two synthetic genes—*Cry1A(6)* and *Cry1A(c)*. Transgenic plants were highly toxic to striped stem borer and yellow stem borer. Maqbool et al. (1998) transformed basmati-370 and M-7 cultivars using novel endotoxin *Cry 2A* gene.

The *Bt* rices have recently been released in Iran and are likely to be released in China shortly. Efforts are also underway for developing transgenics with resistance to brown plant hopper, white-backed plant hopper, green leaf-hopper, etc.

Among diseases, sources of resistance to sheath blight have not been found in rice germplasm, and there are only a few known donors for rice tungro virus resistance. A highly successful strategy known as Coat Protein (CP)-mediated protection has been deployed against certain virus diseases, such as tobacco mosaic virus (TMV) in tobacco and tomato. A CP gene for rice stripe virus was introduced into two *japonica* varieties by electroporation of protoplasts (Hayakawa et al., 1992). The resultant transgenic plants had high levels of CP (up to 0.5% of the total soluble protein) and exhibited significant levels of resistance to the virus, which was heritable.

For resistance to sheath blight of rice (*Rhizoctonia solani*), about six chitinase genes have been identified in rice and are being manipulated to increase the level of resistance (Zhu and Lamb, 1991). A 1.1 kilobase rice genomic DNA fragment containing a chitinase gene was introduced by Lin et al. (1995) through PEG-mediated transformation. Zhang et al. (1998) introduced the *Xa21* gene, from *O. longistaminata* accession from Mali, using projectile bombardment of cell suspensions of elite *indica* varieties IR-64, IR-72, MH-63, and BG 90–2. Six of the 55 R_0 lines carrying the *Xa21* gene showed high levels of resistance to bacterial leaf blight in subsequent generations.

Transgenic rice for abiotic stress tolerance

Transgenic rice plants producing 3–10 times higher amounts of trehalose have been developed; the plants also show tolerance to drought and salinity (Garg et al., 2002). These have not yet been field-tested for commercialization.

Sukamoto and Murata (1998) introduced the *Cod A* gene for choline oxidase from *Arthrobacter globiformis*. The *Cod A* gene was inherited into the second generation of transgenic rice and its expression was stably maintained at levels of mRNA, the protein, and enzyme activity. Levels of glycinebetaine were estimated to be as high as 1 and 5 mmol/g of fresh leaves in two types of transgenic plants (Chl COD and Cyt COD plants) in which cholin oxidase was targeted to the chloroplast and cytosol, respectively. Preliminary results showed that transgenic plants grew better than normal plants during recovery from treatment with 0.15 m NaCl for seven days. Further analysis of transgenic plants demonstrated their ability to synthesize betaine and confer enhanced tolerance to salt and cold.

Transgenics for nutritional quality

Rice contains neither β-carotene (provitamin A) nor C40 carotenoid precursors in its endosperm. Rice in its milled form (as it is usually consumed) is therefore entirely without vitamin A and its carotenoid precursors. Millions of consumers who depend on rice for a large proportion of their calories suffer from vitamin A deficiency. Ye et al. (2000) produced transgenic rice (Golden Rice) with provitamin A biosynthetic pathway engineered into its endosperm. Three genes, viz., phytoene synthase (*pry*), phytoene destaurase (*Crt1*) and lycopene cyclase (*lcy*) were introduced via *Agrobacterium*-mediated transformation. High-performance Liquid Chromatography (HPLC) analysis revealed the presence of β-carotene in transgenic endosperm. The transformed *japonica* rice variety 'Taipie 309' is now under cultivation on a small scale. Efforts are underway at IRRI, in the Philippines, at DRR in Hyderabad, India, and at several other laboratories to transfer the genes for β-carotene into widely grown popular varieties through conventional backcrossing and transformation. The progress has been very slow and, due to various complications and complexities involved, Golden Rice is not likely to be available for mass consumption for at least another five years.

Table 6.24. Some examples of transgenic rice plants with agronomically important genes.

Transgene	Transfer method	Useful trait
Bar	Microprojectile bombardment	Tolerance to herbicide
Bar	PEG- mediated	Tolerance to herbicide
Coat protein	Electroporation	Tolerance to stripe virus
Coat protein	Particle bombardment	Tolerance to rice tungro spherical virus
Chitinase	PEG-mediated	Sheath blight resistance
cryIA(b)	Electroporation	Tolerance to striped stemborer
cryIA(b)	Particle bombardment	Tolerance to yellow stemborer and striped stemborer
cryIA(c)	Particle bombardment	Tolerance to yellow stemborer
cry1A(b), cryIA(c)	*Agrobacterium* mediated	Tolerance to striped stemborer and yellow stemborer
cry1A(b), cryIA(c)	Particle bombardment	Tolerance to yellow stemborer
cry1A(c), cry2A, gna	Particle bombardment	Tolerance to stemborer, leaf folder and brown planthopper
CpTi	PEG-mediated	Tolerance to striped stemborer and pink stemborer
Gna	Particle bombardment	Insecticidal activity for brown planthopper
Corn cystatin (CC)	Electroporation	Insecticidal activity for *Sitophilus zeamais*
Xa21	Particle bombardment	Resistance to bacterial blight
coda	Electroporation	Increased tolerance to salt
Soybean ferritin	*Agrobacterium* mediated	Increased iron content in seed
Psy	Particle bombardment	Phytoene accumulation in rice endosperm
psy, crt1, lcy	*Agrobacterium* mediated	Provitamin A

Modified from Khush and Brar (2002).

Goto et al. (1999) introduced the entire gene sequence for the soybean ferritin gene into the rice variety 'Kita-ake,' via *Agrobacterium*-mediated transformation. The introduced ferritin gene was regulated by the rice seed storage protein glutelin promoter (*Glub-1*) and was terminated by the *Nos* Polyadenylation signal. Synthesis of soybean ferritin protein was confirmed in each of the transformed rice seeds by Western blot analysis, and specific accumulation in endosperm was determined by immunological tissue printing. The iron content of T_1 seeds was up to three times higher than in untransformed seeds.

Transgenics for herbicide tolerance

Herbicide tolerance has been the priority trait for the major multinational private companies, as evidenced by the number of field trials conducted in the U.S., Europe, and other developed countries. Brookes and Barfoot (2003) reported that in the U.S., Monsanto and Bayer accounted for more than 80% of herbicide-tolerant rice field trials. They also reported testing of herbicide resistant transgenics in Europe (Italy), South America (Brazil and Argentina), and Japan. Though no specific information is available, it is possible that this technology is also being tested in China. A list of transgenic rice plants developed with agronomically important genes is given in Table 6.24.

Potential Impact of Biotechnological Tools

Recent advances in cell and molecular biology of rice offer new opportunities for enhancing the efficiency of rice breeding. Biotechnology is now becoming an important component of rice breeding (Table 6.25 and 6.26).

Anther culture has become an important technique to shorten the breeding cycle, to

Table 6.25.   Overcoming some of the constraints of conventional rice breeding using biotechnology tools.

Constraint	Biotechnology applications
Limited genetic variability for resistance to stemborers	Transgenic rice carrying *Bt* genes shows enhanced resistance to stemborers
Limited genetic variability for resistance to sheath blight	Transgenic rice carrying chitinase genes shows tolerance to sheath blight
Lack of genetic variability for β-carotene	Transgenic 'golden' rice carrying phytoene synthase (*psy*), phytoene desaturase (*crt1*) and lycopene cyclase (*lcy*) shows pro-vitamin activity in rice seeds
Low selection efficiency to pyramid genes for durable resistance to pests	Marker assisted selection practiced to pyramid genes for bacterial blight, blast and gall midge resistance
Longer breeding cycle of rice varieties	Varieties developed through anther culture in shorter period
Narrow gene pool for resistance to pests	Genes from wild species with broad spectrum of resistance to bacterial blight, brown planthopper and tungro incorporated into elite breeding lines of rice
Characterization of pathogen population-difficult and laborious	DNA fingerprinting practiced to characterize pathogen populations and for gene deployment for durable resistance

Source: Khush and Brar (2002).

Table 6.26.   Some examples of the biotechnology products in rice.

Biotechnology tools	Product(s)
Anther culture	Several improved cultivars and elite breeding lines developed and released for commercial cultivation in China, Korea, and Philippines.
Molecular marker assisted selection	Marker-assisted selection practiced. Pyramided lines with durable resistance carrying genes for resistance to bacterial blight, blast and gall midge developed in India, China, Philippines, and Indonesia. Some of the pyramided lines have been field-tested and are in the process of release for commercial cultivation.
Alien introgression	Elite breeding lines of rice carrying genes from wild species for resistance to bacterial blight, blast, brown planthopper, tungro and tolerance to acidic conditions have been developed. Varieties resistant to brown planthopper and tolerant to acidity have been released in Vietnam.
Somaclonal variation	Early maturing, high yielding and blast resistant varieties released in Hungary and Japan.
Transformation	Novel genes inserted into rice cultivars for tolerance to herbicide and resistance to stemborer, sheath blight and bacterial blight and pro-vitamin A for improved nutritional quality. Field tests of transgenic rice carrying *Bt* gene for stemborer resistance have been made in China. However, no commercial release of transgenic rice has been made so far.

Source: Khush and Brar (2002).

fix recombinants, and to overcome sterility in distant crosses. Doubled-haploid populations are useful in mapping genes governing agronomical traits, including QTL.

Molecular markers have been helpful in the tagging of numerous genes for tolerance to major biotic and abiotic stresses. Marker-assisted selection has become an integral tool in rice breeding for moving desirable genes from one varietal background to another and for pyramiding of genes for development of desirable pest-cultivars. Fine mapping of QTL should provide means to pyramid QTL for tolerance to major abiotic stresses. Map-based cloning has made it possible to isolate useful genes governing important agronomic traits and incorporation of these genes into elite rice cultivars through transformation. Advances in tissue culture and molecular-marker technology have made it possible to broaden the gene pool of rice and to enhance the efficiency of introgression of useful genes from wild species across crossability barriers. Advances in genetic engineering have facilitated introduction of cloned, novel genes into rice through transformation. Transgenic rice has been developed that enhanced resistance to diseases and insect pests, as well as nutritional qualities.

Public and private efforts in sequencing the rice genome have added new dimensions to research in functional genomics to precisely reveal the functions of rice genes. Identification of genes and their manipulation present another major breakthrough in rice genetics and breeding. So far, a few rice cultivars have been developed through the application of biotechnological tools; many more are expected to be developed in the future. There are, however, several challenges to the application of biotechnological tools in rice breeding, such as public opinion, intellectual property issues, sharing of benefits, biosafety and environmental protection issues, and protection of genetic diversity. It is important that rice biotechnologists and breeders work hand in hand to use the available biotechnological tools judiciously to solve the breeding problems that remain unsolved by conventional rice breeding or to significantly increase the efficiency of conventional breeding. Future food and nutritional security will depend upon the availability of rice cultivars with higher yield potential, durable resistance to major pests and diseases, tolerance to abiotic stresses, and higher level of desirable nutrients in the grain. Conventional breeding approaches judiciously supplemented with innovative biotechnology tools should help meet future production demands.

Summary and Conclusion

Rice—a staple food for half of humanity—is cultivated on about 150 m ha annually with a production of 600 m tons. More than 90% of rice is produced and consumed in Asia. It is grown in ecologies ranging from hills to deep-water. Despite the remarkable progress made during the past 50 years in keeping rice production abreast of population growth, the task of ensuring food security, in view of the shrinking resource base, during the next 50 years is challenging. Rice breeding should continue to play a major role in meeting this challenge.

There has been a significant impact of rice breeding during the last 40 years. The tall traditional rice varieties have been replaced by the semi-dwarf, high-yielding varieties in more than 70% of the rice-growing areas worldwide. Now, rice farmers harvest 5–7 tons per hectare of paddy in irrigated areas as compared with 1–3 tons per hectare earlier. From 1966, when the first semi-dwarf, high-yielding rice variety was released, rice area has increased only marginally from 126 to 152 million hectares, whereas the global rice production has increased from 257 million tons to 600 million tons. This increased production has led to food security in many rice-growing countries. Development of high-yielding, short-duration varieties has increased cropping intensity, thereby increasing labor requirement and providing more employment opportunities in rural areas. The availability of rice varieties with multiple pest and disease resistance has reduced the need for application of pesticides and facilitated the adoption of Integrated Pest Management, thereby sustaining environmental quality.

In the immediate future, adoption of hybrid rice, NERICA rices, and NPT may contribute to enhancing productivity. In the long run, new tools of biotechnology will contribute to increased effectiveness and efficiency in rice breeding. With the unraveling of structural genomics and the intensive efforts underway now on functional genomics, much is expected from the application of biotechnology. Application of molecular markers will make rice breeding more precise and efficient. Development of transgenics will ensure incorporation of desired gene(s) from any source. It is hoped that a paradigm shift brought about by the new knowledge and skills in rice-breeding methodologies and techniques will lead to a second Green Revolution, thereby ensuring food and livelihood security for the billions who depend on rice as their staple food.

References

Abrigo, W.M., Novero, A.U., Coronal, V.P., Cabuslay, G.S., Blanco, L.C., and Parao, F.T. 1985. Somatic cell culture at IRRI. *In: Biotechnology in International Agricultural Research*, 149–158. IRRI: Los Banos, Philippines.

Al-Babili, S., and Beyer, P. 2005. Golden Rice—five years on the road—five years to go? *Trends in Plant Sciences* 10:565–573.

Armuganathan, K., and Earle, E.D. 1991. Nuclear DNA content of some important plant species. *Plant Mol. Biol. Reporter* 9:208–218.

Brar, D.S., and Khush, G.S. 1997. Wide hybridization for rice improvement: Alien gene transfer and molecular characterization of introgression. *In:* Jones, M.P., Dingkhun, M., Johnson, D.E., and Fagade, S.O., eds. *Interspecific Hybridization: Progress and Prospects*, 21–29, WARDA.

Brar, D.S., and Khush, G.S. 2002. Transferring genes from wild species into rice. *In:* Kang, M.S., ed. *Quantitative Genetics, Genomics and Plant Breeding*, 197–217, CABI Publishing: Wallingsford, U.K.

Brookes G., and Barfoot, P. 2003. GM rice: Will this lead the way for global acceptance of GM crop technology? *Brief No. 28, ISAAA:* Los Banos, Philippines.

Chandler, R.F. Jr. 1969. Plant morphology and stand geometry in relation to nitrogen. *In:* Eastern, J.D., Haskin, F.A., Sullivan, C.Y., and Van Bul, C.H.M., eds. *Physiological Aspects of Crop Yield*, 265–285, ASA Publication: Madison, WI, USA.

Chang, T.T. 1976. The origin, evolution, cultivation, dissemination and diversification of Asian and African rices. *Euphytica* 25:425–441.

Chatterjee, D., 1948. A modified key and numeration of the species of genus *Oryza* Linn. *Indian J. Agric. Sci.* 8:185–192.

Cheng, X., Sardana, R., Kaplan, H., and Altosaar, I. 1998. *Agrobacterium*-transformed rice plants expressing synthetic *Cry 1A(b)* and *Cry 1A(c)* genes are highly toxic to striped stem borer and yellow stem borer. *Proc. Natl. Acad. Sci. USA* 95:2767–2772.

Christou, P., Ford, T.L., and Kofron, M. 1991. Production of transgenic rice plants from agronomically important *indica* and *japonica* varieties via electric discharge particle acceleration of exogenous DNA into immature zygotic embryos. *Bio/Technology* 9:957–962.

Chu, C.C. 1982. Anther culture of rice and its significance in distant hybridization. *In:* Polland, C.E.R., ed. *Proc. of Rice Tissue Conference*, 47–53, IRRI: Los Banos, Philippines.

Coffman, R., McCouch, S.R., and Herdt, R.W. 2004. Potentials and limitations of biotechnology in rice. *Intl. Rice Comm. Newsletter.* 53(sp. ed.):26–40.

Dalmacio, R., Brar, D.S., Ishii, T., Stich, L.A., Virmani, S.S., and Khush, G.S. 1995. Identification and transfer of a new cytoplasmic male sterility source from *O. perennis* into *indica* rice (*O. sativa*). *Euphytica* 82:221–225.

Datta, S.K., Torrizo, L., Tu, J., Olivia, N., and Datta, K. 1997. Production and molecular evaluation of transgenic rice plants. *IRRI Discussion Paper Series* No. 21, IRRI: Manila, Philippines.

De Candolle, 1882. Origin of cultivated plants (fase. Ed.) Hafner: New York, 1967.

Depledge, J. 2002. Climate change in focus: The IPCC third assessment report. *Briefing Paper News Series* No.29, Royal Institute of International Affairs.

Fujimoto, H., Itoh, K., Yamamoto, M., Kyozuka, J., and Shimamoto, K. 1993. Insect-resistant rice generated by introduction of a modified d-endotoxin gene of *Bacillus thuringiensis*. *Bio/Technology* 11:1151–1155.

Garg, A.K. Ju-Kon Kim., Owens, T.G., Ranwala, A.P., Yang Do Choi, Kochian, L.V., and Wu, R.J. 2002. Trehalose accumulation in rice plants confers high tolerance levels to different abiotic stresses. *Proc. Natl. Acad. Sci. USA* 89:9865–9869.

Goto, F., Yoshihara, T., Shigemoto, N., Toki, S., and Takaiwa, F. 1999. Iron fortification of rice seed by soybean ferratin gene. *Nature Biotechnology* 17:282–286.

Gowda, M., Venu, R.C., Roopalakshmi, K., Sreerekha, M.V., and Kulkarni, R.S. 2003. Advances in rice breeding, genetics and genomics. *Molecular Breeding* 11:337–352.

Hayakawa, T., Zhu, Y., Itoh, K., and Kumura, Y. 1992. Genetically engineered rice resistant to rice stripe virus, an insect transmitted virus. *Proc. Natl. Acad. Sci. USA* 89:9865–9869.

Hei, Y., Ohta, S., Komari, T., and Kumashiro, T. 1994. Efficient transformation of rice mediated by *Agrobacterium* and sequence analysis of the boundaries of T-DNA. *Plant J.* 6:271–282.

Heszky, L.E., and Simon-Kiss, I. 1992. "DAMA" the first plant variety of biotechnology origin in Hungary, registered in 1992. *Hung. Agric. Res.* 1:30–32.

Hittalamani, S., Parco, A., Mew, T.W., Zeigler, R.S., and Huang, N. 2000. Fine mapping and DNA marker-assisted pyramiding of the three major genes for blast resistance in rice. *Theor. Appl. Genet.* 100:1121–1128.

Hoan, N.T., Sarma, N.P., and Siddiq, E.A. 1997. Identification and characterization of new sources of cytoplasmic male sterility in rice. *Plant Breeding* 116:547–551.

Hossain, M. 1998. Nature and impact of green revolution in Bangladesh. Intl. Food Policy Research Institute, Report No. 67, Washington, D.C., USA.

Huang, J., Pray, C., and Rozelle, S. 2002. Enhancing crops to feed the poor. *Nature* 41878–41684.

Huang, N., Angeles, E.R., Domingo, J., Magpantay, G., Singh, G., Zhang, G., Kumaravadivelu, N., Bennet, J., and Khush, G.S. 1997. Pyramiding of bacterial blight resistance genes in rice: Marker assisted selection using RFLP and PCR. *Theor. Appl. Genet.* 95:313–320.

Ikehashi, H., and Araki, H. 1986. Genetics of sterility in remote crosses of rice. *In:* Rice Genetics, 119–130, IRRI: Manila, Philippines.

IRRI. 2002. *Rice Almanac*, 3rd Edition. Maclean, J.L., Dawe, D.C., Hardy, B., and Hettel, G.P., eds. IRRI and CABI publishing.

Jena, K.K., and Khush, G.S. 1990. Introgression of genes from *Oryza officianalis* to cultivated rice. *Theor. Appl. Genet.* 80:737–745.

Jones, M.P., Dingkuhn, M., Aluko, G.K., and Semen, M. 1997. Interspecific *O. sativa* L. × *O. glaberrima* Steud. progenies in upland rice improvement. *Euphytica* 92:237–246.

Juliano, B.O., ed. 1985. Rice Chemistry and Technology. Revised Edition. American Association of Cereal Chemists: St. Paul, Minnesota, USA.

Katsuo, K., and Mizushima, U. 1958. Studies on the cytoplasmic difference among rice varieties, *Oryza sativa L.* I. On the fertility of hybrids obtained reciprocally between cultivated and wild varieties. *Japanese J. Breed.* 8:1–5.(in Japanese).

Khush, G.S. 1977. Disease and insect resistance in rice. *Adv. Agron.* 29:265–341.

Khush, G.S. 1984. Terminology for rice growing environments. *In: Terminology for Rice Growing Environments*, 5–10, International Rice Research Institute: Manila, Philippines.

Khush, G.S. 1995. Breaking the yield barrier of rice. *Geo. Journal* 35:329–332.

Khush, G.S. 1996. Prospects of and approaches to increasing the genetic yield potential. *In:* R.E. Evenson, R.W. Herdt, and M. Hossain, eds. *Rice Research in Asia: Progress and Priorities*, 59–71, IRRI: Manila, Philippines.

Khush, G.S. 1997. Origin, dispersal, cultivation and variation of rice. *Plant Molecular Biology* 35(3)2:5–34.

Khush, G.S., 1999. Green Revolution: Preparing for 21st century. *Genome* 42:646–655.

Khush, G.S. 2001a. Challenges for meeting the global food and nutrient needs in the new millennium. *Proc. of Nutrition Society* (2001), 60:15–26.

Khush, G.S. 2001b. Green Revolution: The way forward. *Nature Reviews (Genetics)* 2:815–822.

Khush, G.S. 2004. Harnessing science and technology for sustainable rice-based production systems. *Int. Rice Comm. Newsletter* 53(sp. ed.):17–23.

Khush, G.S. 2005. What it will take to feed 5.0 billion rice consumers in 2030. *Plant Molecular Biology* 59(1):1–6.

Khush, G.S., and Peng, S. 1996. Breaking the yield frontier in rice. *In:* Reynold, M.P., Rajaram, S., and McNab A., eds. *Increasing the Yield Potential in Wheat: Breaking the Barriers,* 36–51, CIMMYT: Mexico.

Khush, G.S., Peng, S.S., and Virmani, S.S. 1998. Improving yield potential by modifying plant type and exploiting heterosis. *In:* Waterlow, J.C., Armstrong, D.G., Fowden, L., and Riley R., 1998. *Feeding the World Population of more than Eight Billion People: A Challenge To Science*, 150–170, Oxford University Press: New York.

Khush, G.S., and Virk, P.S. 2000. Rice breeding: Achievements and future strategies. *Crop Improvement* 27(2)1:15–144.

Khush, G.S., and Virk, P.S. 2002. Rice improvement: Past, present and future. *In:* Kang M.S., ed. *Crop Improvement: Challenges in the Twenty-First Century*, 17–42, Food Product Press: New York. London.

Khush, G.S., and Brar, D.S. 2002. Biotechnology for rice breeding: Progress and impact. *In: Progress in Rice Genetic Improvement. Proceedings of the 20th session of International Rice Commission.* Bangkok, July 23–26, 2002. FAO corporate document repository.

Khush, G.S., and Virk, P.S. 2005. *IR Varieties and their Impact.* International Rice Research Institute: Los Banos, Philippines.

Ku, M.B.S., Agarie, S., Nomura, M., Fukayama, H., Tsuchida, H., Ono, K., Horose, S., Toki, S., Miyao, M., and Matsouka, M. 1999. High-level expression of maize phosphoenolpyruvate in transgenic rice plants. *Nature Biotech* 17:76–80.

Kurata, N., Umehara, Y., Tanoue, H., and Sasaki, T. 1997. Physical mapping of rice genome with YAC clones. *Plant Mol. Biol.* 35:101–113.

Leung, H., and An, G. 2004. Rice functional genomics: Large-scale gene discovery and applications to crop improvement. *Adv. Agron.* 82:56–102.

Lin, W., Anuratha, C.S., Datta, K., Portrykus, I., Muthukrishnan, S., and Datta, S.K. 1995. Genetic engineering of rice for resistance to sheath blight. *Bio/Technology* 13:686–691.

Lopez, M.T., Tooyinda, T., Vanavichit, A., and Tragoonrung, S. 2003. Microsatellite markers flanking the *tms*2 gene faciliated tropical TGMS rice line development. *Crop Sci.* 43:2267–2271.

Maqbool, S.B., Riazuddin, S., Loc, N.T., Gatehouse, A.M.R., Gatehouse, J.A., and Christou, P. 1998. Expression of multiple insecticidal genes confers broad resistance against a range of different rice pests. *Mol. Breeding* 7:85–93.

Maurya, D.M., and O'Toole, J.C. 1986. Screening upland rice for drought tolerance. 245–261 *In: Progress in Upland Rice Research.* IRRI: Manila, Philippines.

McCouch, S.R., Kochert, G., Yu, Z.H., Wang, Z.Y., Khush G.S., Coffman, D.R., and Tanksley, S.D. 1988. Molecular mapping of rice chromosomes. *Theor. Appl. Genet.* 76:815–829.

McCouch, S., Teytelman, L., Xu, Y., Lobos, K., Clare, K., Walton, M., Fu, B., Maghirang, R., Li, Z., Xing, Y., Zhang, Q., Kono, I., Yono, M. Fjellstrom, R., DeClarck, G., Schneider, D., Cartinhour, S., Ware, D., and Stein, L. 2002. Development of 2240 new SSR markers for rice. *DNA Res.* 91:99–207.

Meifang, L.T., 1992. Anther culture breeding of rice at CAAS. *In:* Zheng, K., and Murashige, T., eds. *Anther Culture for Rice Breeders,* 75–85, Hangzhou, China.

Mohandrass, S., Kareem, A.A., Ranganathan, T.B., and Jayaraman, S. 1995. Rice production in India under current and future climate. *In:* Mathews, R.B., Kropf, M.J., Bachelet, D., and Vaan Laar, H.H., eds. *Modeling the Impact of Climate Change on Rice Production In Asia,* 165–181, CAB International: UK.

Mohanty, H.K., Suprihatno, B., Khush, G.S., Coffman, W.R., and Vergara, B.S. 1982. Inheritance of submergence tolerance in deep water rice. 121–131. *In: Proc. of International Deep Water Workshop 1981, Bangkok, Thailand.* IRRI: Los Banos, Philippines.

Morishima, H. 1984. Tsunudo, S., and Takahashi, N., eds. Wild plants and domestication. *In: Biology of Rice,* 3–30, Elsevier Science Publishers: Amsterdam.

Multani, D.S., Jena, K.K., Brar, D.S., Delos Reyes, B.G., Angeles, E.R., and Khush, G.S. 1994. Development of monosomic alien addition lines and introgression of genes from *Oryza australiensis* to cultivated rice. *Theor. Appl. Genet.* 88:102–109.

Nanda, J.S. 2000. Rice breeding and genetics: Research perspective. *In:* Nanda, J.S. (ed.) *Rice Breeding and Genetics.* Oxford & IBH Publishing Company Pvt. Ltd.: New Delhi.

Nanda, J.S., and Agrawal, P.K. 2006. *Rice.* Kalyani Publishers: New Delhi, India.

Nayak, P., Basu, D., Das, S., Basu, A., Ghosh, D., Ramakishnan, N.A., Ghosh, M., and Sen, S.K. 1997. Transgenic elite indica rice plants expressing *Cry 1Ac* d-endotoxin of *B. thuringiensis* are resistant against yellow stem borer. *Proc. Natl Acad. Sci. USA* 94:2111–2116.

Oka, H.I. 1964. Pattern of interspecific relationships and evolutionary dynamism in *Oryza sativa* L. and five wild diploid forms of *Oryza. Crop Sci.* 1:445–450.

Oka, H.I. 1974. Experimental studies on the origin of cultivated rice. *Genetics* 78:475–486.

Oka, H.I. 1988. *Origin of Cultivated Rice.* Japanese Scientific Societies Press: Tokyo and Elsevier Science Publishers, Amsterdam.

Pathasarathy, N. 1972. Rice breeding in tropical Asia up to 1960. *In: Rice Breeding,* 5–29, International Rice Research Institute: Los Banos, Philippines.

Peng, S., Laza, R.C., Visperas, R.M., Sanico, A.L., Cassman, K.G., and Khush, G.S. 2000. Grain yield of rice cultivars and lines developed in the Philippines since 1966. *Crop. Sci.* 40:307–314.

Pingali, P.L., and Hossain, M., eds. 1998. *Impact of Rice Research.* IRRI/Thailand Development Research Institute.

Porteres, R. 1956. Taxonomic agrobotanique des riz cultivés *O. sativa* L. et *O.glaberrima* Steudel. *J. Agr. Trop. Bot. App.* 3:341–384.

Roschevicz, R.J. 1931. A., Contribution to knowledge of rice. *Appl. Bot. Genet. Pl. Breed. Bull* 27(4):3–133.

Sampath, S., and Mohanty, H.K. 1954. Cytology of semi-sterile rice hybrid. *Curr. Sci.* 231:82–183.

Sanchez, A.C., Brar, D.S., Huang, N., Li, Z., and Khush, G.S. 2000. STS marker-assisted selection for three bacterial blight resistance genes in rice. *Crop Sci.* 40:792–797.

Sarma, N.P., and Sundaram, R.M. 2003. Molecular markers in rice breeding. 64–74. *In:* Pakki Reddy, G.and Janaki Krishna, P.S., eds. *Biotechnology Interventions for Dryland Agriculture: Opportunities and Constraints.* B.S. Publications: Hyderabad, India.

Sasaki et al. 2002. The genome sequence and structure of rice chromosome 1. *Nature* 420:312–316.

Sharma, S.D., Tripathy, S., and Biswal, J. 2000. Origin of *O. sativa* and its ecotypes. 349–369. *In:* Nanda J.S., ed. *Rice Breeding and Genetics.* Oxford and IBH Publishing Company Pvt. Ltd.: New Delhi.

Shastry, S.V.S., and Sharma, S.D. 1973. Origin of Rice. *In:* Hutchinson, J., ed. *Evolutionary Studies in World Crops: Diversity and Change in the Indian Subcontinent,* 55–63, Cambridge University Press: London.

Shimamoto, K., and Kyozuka, J. 2002. Rice as a model for comparative genomics of Plants. *Ann. Rev. of Plant Biol.* 53:399–419.

Shinjyo, C., and Omura, T. 1966. Cytoplasmic male sterility in cultivated rice *O. sativa* L. I-Fertilities of F_1, F_2 and offspring obtained from their mutual reciprocal backcrosses and segregation of completely male sterile plants. *Japanese J. Breed.* 16(Suppl.):179–180.

Siddiq, E.A., and Swaminathan, M.S. 1966. Induced mutations in relation to breeding and phylogenetic differentiation of *Oryza sativa. In: Mutations in*

Rice Breeding. International Atomic Agency Technical Report, series No. 86, 25–52.

Siddiq, E.A., and Viraktamath, B.C. 2000. Rice. *In:* Chopra, V.L., ed. *Plant Breeding: Theory and Practice*, 1–85, Kalyani Publishers: New Delhi, India.

Singh, S., Sidhu, J.S., Huang, N., Vikal, Y., Li, Z., Brar, D.S., Dhaliwal, H.S., and Khush, G.S. 2001. Pyramiding three bacterial blight resistance genes (xa5, xa13 and Xa21) using marker-assisted selection into *indica* rice cultivar PR106. *Theoretical and Applied Genetics* 102:1011–1015.

Sridhar, R., Reddy, J.N., Singh, U.D., and Agrawal, P.K. 1999a. Usefulness of combination of bacterial blight resistant genes at Cuttack. *IRRN* 24(2): 24–25.

Sridhar, R., Singh, U.D., Agarwal, P.K., Reddy, J.N., Chandrawanshi, S.S., Sanger, R.B.S., Bhat, J.C., Ratnaiah, Y., and Row, K.V.S.R.K. 1999b. Usefulness of blast-resistant genes and their combinations in different blast-endemic locations in India. *IRRN* 24(2):22–24.

Stark, D.M. Timmerman, K.P., Bary, G.F., Priess, J., and Kishore, G.M. 1992. Regulation of amount of starch in plant tissue by ADP glucose pyrophosphorylase. *Science* 285:287–292.

Sukamoto, A., and Murata, N. 1998. Metabolic engineering of rice leading to biosynthesis of glycine betaine and tolerance to environmental stress. *In: Intl. Workshop on Breeding and Biotechnology for Environmental Stress in Rice*, 164–165, Sopporo, Japan.

Tateoka, T. 1963. Taxonomic studies of Oryza. III. Key to species and their enumeration. *Bot. Mag.* 76:165–173.

Tateoka, T. 1964. *Report of exploration in East Africa and Madagascar*. National Science Museum: Tokyo.

Toriyama, K., Arimoto, Y., Uchimiya, Y., and Hinata, K. 1988. Transgenic rice plants after direct gene transfer into protoplasts. *Bio/Technology* 6:1072–1074.

Tu, J., Zhang, G., Datta, K., Xu, C., He, Y., Zhang, Q., Khush, G.S., and Datta, S.K. 2000. Field performance of transgenic elite commercial hybrid rice expressing *Bacillus thuringiensis* endoprotein. *Nature Biotech.* 18:1101–1104.

UNDP. 1994. Human Development Report. Oxford University Press: U.K.

U.S. Census Bureau. Available at: www. uscensusbureau.org.

Virmani, S.S. 1994. *Heterosis and Hybrid Rice Breeding* (Monographs on Theoretical and Applied Genetics 22). IRRI and Springer Verlag: Berlin.

Virmani, S.S. 1996. Hybrid rice. *Adv. Agron.* 57:377–462.

Virmani, S.S. 2003. Advances in hybrid rice research and development in the tropics. *In:* Virmani, S.S.,

Mao, C.X., and Hardy, B., eds. *Hybrid Rice for Food Security, Poverty Alleviation and Environmental Protection. Proceedings of the 4th International Symposium on Hybrid Rice*, Hanoi, Vietnam, May 14–17. 2002, 7–20, IRRI: Los Banos, Philippines.

Virmani, S.S., and Edwards, I.B. 1983. Current status and future prospects for breeding hybrid rice and wheat. *Adv. Agron.* 361:45–214.

Virmani, S.S., and Voc, P.C. 1991. Induction of photo and thermo-sensitive male sterility in *indica* rice. *Agronomy Abstracts* 119.

Wunn, J., Kloti, A., Burkhardt, P.K., Ghosh Biswas, G.C., Launis, K., Iglesias, V.A., and Portrykus, I. 1996. Transgenic *indica* rice breeding line IR-58 expressing a synthetic *Cry 1A(b)* gene from *B. thuringiensis* provides effective insect pest control. *Bio/Technology* 14:171–176.

Ye, X., Al-babili, S., Kloti, A., Zhang, J., Lucca, P., Beyer, P., and Portrykus, I. 2000. Engineering the pro-vitamin A (β-carotene) biosynthesis pathway into (carotenoid free) rice endosperm. *Science* 287:303–305.

Yuan, L.P. 1977. The execution and theory of developing hybrid rice. *Chinese Agric. Sciences* 12:7–31 (in Chinese).

Yuan, L.P. 2004. Hybrid rice technology for food security in the world. *Int. Rice Comm. Newsletter*. 53(sp. ed.):24–25.

Yuan, L.P., and Virmani, S.S. 1988. Organisation of hybrid rice breeding program. 33–37 *In: Hybrid Rice,Proceedings of the International Symposium on Hybrid Rice Held during October*, 6–10, 1986 at Changsha, Hunan, China. IRRI: Manila, Philippines.

Xiao, J., Grandillo, S., Ahn, S.N., McCouch, S.R., and Tanksley, S.D. 1996. Genes from wild rice improve yield. *Nature* 384:1223–1224.

Xu, D., Xue, Q., McElroy, D., Mawal, Y., Hilder, V.A., and Wu, R. 1996. Constitutive expression of cowpea trypsin inhibitor gene "*cpti*" in transgenic rice plants confers resistance to two major rice insect pests. *Mol. Breeding* 2:167–173.

Xu, K., Xu, X., Takeshi, F., Patrick, C., Reycel, M., Heuer, S., Ismail, A., Julia, B., Ronald, P.C., and Mackill, D.J. 2006. Sub-1A is an ethylene-response-factor-like gene that confers submergence tolerance to rice. *Nature* 442:705–708.

Zapata, F.J., and Abrigo, E.M. 1986. Plant regeneration and screening for long term NaCl-stressed rice callus. *IRRN* 11:24–25.

Zhang, S., Song, W.Y., Chen, L., Raun, D., Taylor, N., Ronald, P., Beachy, R., and Fauquet, C. 1998. Transgenic elite *indica* rice varieties resistant to *Xanthomonas oryzae. Mol. Breed.* 4:551–558.

Zhu, Q., and Lamb, C.J., 1991. Isolation and characterization of a rice gene encoding a basic chitinase. *Mol. Gen. Genetics.* 226:289–296.

Chapter 7

Barley Breeding for Sustainable Production

Salvatore Ceccarelli, Stefania Grando, Flavio Capettini and Michael Baum

Introduction

Barley (*Hordeum vulgare* L.) is grown on more than 56 million hectares, about 15 million of which are in developing countries. It is grown for animal feed, human food, and malt, and in many different types of environments. However, in developing countries, most barley is grown in marginal environments, often on the fringes of deserts and steppes, or at high elevations in the tropics, receiving modest or no inputs. This partly explains why yields there are nearly half of those in developed countries (Table 7.1).

Botany

Vernacular names

Barley (English), Orge (French), Gerste (German), Orzo (Italian), Cebada (Spanish), Cevada (Portuguese), Yachmen' (Russian), Shai'r (Arabic), Kritari (Greek), Jau (India, Iran, Pakistan, Afghanistan), Läua Mach (Vietnam), Kao (Thailand), Gebse (Amaric), and Segem (Tigrigna).

Barley belongs to the tribe Triticeae of the grass family Poaceae together with the other important cereals, wheat and rye. The main distinction from other members of the tribe is that each spike node bears three one-flowered spikelets ("triplets") of which one or all three are fertile. The genus *Hordeum* includes about 30 species (Bothmer, 1992a). According to Bothmer, 45 taxa of the genus are mostly diploid (2n = 2x = 14 chromosomes, 28 taxa), but also tetraploid (2n = 4x = 28 chromosomes, 16 taxa) and hexaploid (2n = 6x = 42 chromosomes, 8 taxa), with a basic chromosome number x = 7.

Plants are usually about 1 m tall; some landraces can be taller, up to 1.5 m. Culms are mostly erect, up to 3 mm thick, with 3 to 4 nodes. Leaves are glabrous, rarely hairy, with long, prominent auricles surrounding the culm; however, liguleless forms also exist. Leaf sheaths are split almost to the base. Spikes with awns are 5 to 25 cm long; their color ranges from white yellowish to purplish to blackish. There are 10 to 30 nodes, each bearing three one-flowered spikelets. All three spikelets are fertile in six-rowed types, but only the central spikelet is fertile in two-rowed types. In the latter type, the lateral spikelets are non-fertile with anthers (type *nutans*) or fully reduced (type *deficiens*). The central spikelet is sessile, with more or less flat, narrow glumes; the lemma is glabrous or scabrid with five nerves, usually tightly covering the kernel together with the palea (covered or husked or hulled barley). In "naked" or "hull-less" barley (due to the major recessive gene *n* or *nud* on chromosome 7H), the kernel threshes free. The awn of the lemma is rough or glabrous, 3–18 cm long. Sometimes awns are absent or transformed into "hoods." Lateral spikelets are usually sessile, but may be pedicellate (3 mm long) when fertile. Glumes are 1–2 cm long, more or less flattened.

Table 7.1. Area, yield, and production of barley (2001–2005 means).

Regions	Area ('000 ha)	Yield (t/ha)	Production ('000 mt)
Europe	29,001	3.772	68,206.9
North America	5,963	2.537	19,594.2
Latin America	1,175.2	1.516	1,638.4
Asia	12,088.4	1.437	18,834.0
Oceania	4,311.4	1.562	4,331.4
Africa	4,491.6	0.892	4,197.4
Developing	15,377.2	1.311	119,609.3
Developed	41,318.6	2.324	22,765.2
World	56,696.8	2.063	142,374.6

Source: FAOSTAT Database, http://www.fao.org.

Kernels may be yellow, gray, green, violet or black and have a furrow on the dorsal side, which is more evident in the naked types. Contrary to other cereals, the aleurone layer has more than one row of cells (two to four). Plants are essentially autogamous; diploids, 2n = 14; tetraploids, 2n = 28, have been produced artificially.

Origin of the species

Archeological evidence from the Near East indicates that barley was first domesticated from its wild progenitor, *Hordeum vulgare* L. subsp. *spontaneum* (C. Koch) Thell., in the Fertile Crescent more than 10,000 years ago. The earliest cultivated forms with non-brittle rachis were two-rowed, hulled or naked barley, which were found in Pre-pottery Neolithic A layers at two sites, dated 8000 BC. The six-rowed forms appeared some 1,000–2,000 years later. Barley and emmer wheat appear to be the first cultivated crops and, together with two domesticated animals, sheep and goats, they formed the basis of the Neolithic farming system, which spread to other parts of the Old World. With the Neolithic "package," barley spread throughout Europe, around the Mediterranean, eastward to the Indus and southward to Ethiopia, but it did not reach China before the second millennium BC (Harlan, 1991;

Zohary and Hopf, 1988). The archeological evidence is supported by the contemporary geographical distribution of subsp. *spontaneum*. The area of its massive stands in primary habitats includes or is located close to the Neolithic archeological sites of the Near East. Other regions that have been proposed as centers of origin for barley are Morocco (Molina-Cano et al., 1987, 2005), Ethiopia (Bekele, 1983; Negassa, 1985) and Tibet (Murphy and Witcombe, 1981; Xu, 1982).

Bothmer et al. (1991) provided a logical explanation of barley origin, arguing that the one-seeded diaspora in subsp. *spontaneum* was more functional for seed dispersal in the natural habitat than three seeds linked together. Therefore, the wild progenitor of all cultivated barley forms would be a two-rowed *spontaneum* type with a brittle rachis from which a diversity of tough-rachis forms developed through mutations and domestication. Traits, such as brittle/tough rachis, row number, seed cover, or awn development, are controlled by relatively few genes. Additional genetic diversity has been generated through repeated cycles of differentiation and hybridization (Harlan, 1992). As there are no crossing barriers between cultivated barley and wild *H. spontaneum*, gene introgression in both directions is a common process at the field borders and/or within fields in the Fertile Crescent. Therefore, mutation, hybridization, and gene recombination generated new genetic diversity, including new brittle types. This type of barley evolution was first suggested by Alphonse de Candolle in 1884 and later advocated by a number of other authors.

The monophyletic two-rowed progenitor hypothesis has been questioned by some authors, who propose an alternative hypothesis that assumes as progenitor six-rowed types that originated from an extinct six-rowed progenitor with a brittle rachis, similar to the present *agriocrithon* forms. These are, however, secondary products of hybridization of subsp. *spontaneum* with six-rowed culti-vated forms (Zohary, 1959), not adapted to survival in undisturbed habitats.

Four other hypotheses on the origin of cultivated barley have been postulated by Åberg, (1938, 1940), Takahashi (1955), Bakhteev (1960), and Shao and Li (1987). Åberg suggested that *H. agriocrithon* was a common ancestor of both cultivated barley, *H. vulgare*, and two-rowed wild barley, *H. spontaneum*. Takahashi (1955) favored the diphyletic theory of barley origin. Accord-ing to this, the two-rowed cultivated barley originated from *H. spontaneum* and the six-rowed cultivated form was derived from *H. agriocrithon*. Bakhteev (1960) identified the bottle-shaped wild barley form, *H. laguncu-liforme* Bacht., to be the ancestor of all cul-tivated *H. vulgare* forms. However, Shao and Li (1987) questioned the latter hypoth-esis, claiming that *H. lagunculiforme*, was an unstable intermediary form in evolution from two-rowed to six-rowed wild barley. In their opinion, two-rowed wild barley was the oldest ancestor, which was the first stage of evolution of cultivated barley. The six-rowed wild barley, *H. agriocrithon*, was the second stage, whereas the cultivated form was the third and final stage of evolution.

The chromosomes of *H. vulgare* and of *H. spontaneum* are symbolized by the letter H followed by the letters S and L for short and long chromosome arms, respectively. The seven barley chromosomes are currently designated as follows: 1 H, 2 H, 3 H, 4 H, 5 H, 6 H, and 7 H. This replaces the Burnham and Hagberg (1956) designation: 5, 2, 3, 4, 7, 6, and 1.

Genetic resources

Global holdings

The barley gene pool, held in *ex situ* gene bank collections worldwide, is the second largest after wheat with nearly 400,000 accessions (FAO, 1997). As with other

cereals, the major part of the barley gene bank holdings is duplications; there are fewer distinct accessions. An earlier esti-mate (Plucknett et al., 1987) indicated that only 20% of barley global collections might be distinct (unique) accessions. Chapman (1987) made an inventory of unique acces-sions of barley and concluded that both land-races and cultivars were represented in the global collections by the same number of unique accessions, i.e., 25,000. Genetic diversity of cultivated barley is relatively well represented in gene banks; it is believed that about 85% of landraces have been col-lected, in contrast with only 20% of wild *Hordeum* spp. (Plucknett et al., 1987).

Barley holdings in major gene banks are presented in Table 7.1. The highest propor-tion of landraces/obsolete cultivars is in the collections of the Plant Genetic Resources Centre in Ethiopia (PGRCE) (100%) and the International Center for Agricultural Research in the Dry Areas (ICARDA) (87%), which is one of 15 centers of the Consultative Group on International Agri-cultural Research (CGIAR). Among these international centers, ICARDA (www. icarda.org) has the global mandate for barley improvement in developing countries. In spite of the impressive number of barley accessions in germplasm collections, only 10% are stored under long-term conditions, i.e., at temperatures below −10°C, and less than a half of the total holdings are kept in medium-term type cold stores (FAO, 1997).

Barley gene pool

Harlan and deWet (1971) classified the crop gene pool into primary, secondary, and ter-tiary pools, according to the relationships with the cultivated species. The primary barley gene pool has three major compo-nents: (1) cultivars and breeding lines; (2) landraces; and (3) the barley wild progeni-tor, *H. vulgare* ssp. *spontaneum* (Bothmer et al., 1991). As there are no crossability barriers within the primary gene pool, gene transfer to adapted cultivars is feasible.

The barley wild progenitor, *H. vulgare* subsp. *spontaneum*, belongs to the primary gene pool. There are no crossing barriers with cultivated barley and their chromosomes are fully homologous. As the species has evolved in the harsh and diverse environments in West Asia, it is a valuable source of genes for stress tolerance and adaptation to marginal environments and low-input farming systems (Grando et al., 2001), and a source of novel genes for disease and insect resistance (Williams, 2003). Natural populations of wild barley in many parts of the Near East—a center of origin and primary genetic diversity—are continuously eroding by heavy overgrazing by small ruminants, but wild barley can tolerate overgrazing better than wild wheat species.

Although the total number of wild barley, *H. vulgare* subsp. *spontaneum* accessions held in gene banks, i.e., 16,100, is quite impressive, most of them (86%) are single-seed progenies from a very limited number of accessions from Palestine held at Plant Breeding International (PBI), Cambridge, UK. Other collections represent diversity of specific areas, and the area of distribution of the species is insufficiently represented (Bothmer, 1992b).

A single species, *H. bulbosum*, belongs to the secondary gene pool. Crosses with cultivated barley mostly result in *bulbosum* chromosome elimination and production of *vulgare* haploids. This phenomenon has been widely exploited in the production of doubled haploids used as mapping populations for genetic studies and for shortening the breeding cycle. Sometimes the *bulbosum* chromosomes are not eliminated and the embryo develops into a true hybrid. Chromosome meiotic pairing in such hybrids may be high, but the fertility is extremely low, making gene transfer possible, but difficult.

Utilization of other wild *Hordeum* spp. in cultivated barley breeding has been very limited because of crossability problems, sterility of hybrids, resistance to chromosome doubling, and linkage of desirable with undesirable traits. *Hordeum bulbosum* L. is the single species in the secondary gene pool, and its global holdings are approximately 1,000 accessions. It has been used as a source of resistance to a number of diseases, such as powdery mildew, leaf rust, and scald, and as source of resistance to barley mosaic virus (Pickering et al., 2004).

All other Hordeum species, rather distantly related to the cultigen, belong to the tertiary gene pool. Strong crossability barriers and no or very low chromosome homology with *vulgare* chromosomes have prevented their contribution to barley breeding so far. However, this situation may change with new developments in biotechnology and as new, interesting methodologies become available to barley breeders.

It is estimated that 3,200 accessions are held in 11 gene banks, but a large majority (95%) is concentrated in the Canadian-Scandinavian Collection. The attractiveness of this previously neglected part of the barley gene pool will certainly increase with new advances in biotechnology.

Ecology

Barley is a cool-season crop in countries with a Mediterranean climate and is well adapted to stressful and extreme environments. Barley fields can be seen as high as 4800 m asl in the Himalayas, in latitudes >60°N in Iceland and Scandinavia, and in the rainfed semi-arid regions of West Asia and North Africa with less than 250 mm annual rainfall. Barley is a main food crop in highlands and marginal areas where other cereals cannot grow, as well as being animal feed and forage around the world. It is also an important industrial crop providing raw material for malt and beer production. Its straw is of better quality than that of wheat and is, therefore, a valuable complement of the cattle and small ruminant diet. Barley is grown in a wide range of environments, but nearly two-thirds of the world's production occurs in sub-humid or semi-arid regions. Barley is regarded as a drought-tolerant crop and its water-use efficiency is higher than that of other cereals; however, it is responsive to supplementary irrigation. Barley is more tolerant than other cereals to alkaline soils and less tolerant to acid ones (Poehlman, 1986).

Barley is cultivated in a wider range of environments than any other cereal; therefore, timely heading and appropriate crop duration are essential attributes of adaptation. There are three main factors that control the development of the crop: photoperiod, temperature, and vernalization by cold or, more rarely, by short days.

Photo-thermal response in barley has been studied in detail at the University of Reading, UK (Roberts et al., 1988; Ellis et al., 1988, 1989), and the genetic basis of the seasonality was determined in many barley accessions held in the University of Okayama collection (Takahashi and Yasuda, 1970; Yasuda et al., 1993).

Uses of barley

Barley grain is used for different purposes: as feed for animals, malt, and food for human consumption. About 70% of world barley is used for animal feed and 20% for malting, and the remaining 5% for direct food use. World malting barley trade accounts for 30% of the total world barley trade, which is estimated at 15.1 t annually (FAOSTAT, 2005). This accounts for about 11% of the total barley production. A large body of research has been directed toward the understanding of barley carbohydrate chemistry and the function of individual components in terms of feed, malt, and beer quality (Fox et al., 2003).

Feed barley

Barley is an important feed source for livestock, including beef and dairy cattle, pigs, poultry, and sheep. Most of the barley used as animal feed is not necessarily bred to be used as feed; often it is simply malting barley that in a particular location and year did not meet the malting industry standards. The majority of feed barley is hulled, even though hull-less barley is becoming more popular for monogastrics.

Traits related to feed quality are high starch level, acid detergent fiber (ADF), neutral detergent fiber (NDF) and dry matter digestibility (DMD). All these traits are influenced by both the genotype and the environment.

One factor that has limited the use of barley in poultry and some swine rations is the high proportion of insoluble hull fiber attached to the seed. Hull-less barley cultivars have the potential to markedly increase energy content if used as a feed ingredient. Another recent advance that has the potential to increase the use of barley in feed applications is the development of low-phytate (LP) barley, which may reduce phosphorus levels in animal feces. The majority of phosphorus in plants is found in a form that animals are not able to digest. As a result, diets are supplemented with available inorganic phosphorus and the undigested organic fraction is passed into the feces, which can result in environmental issues where intensive livestock production occurs. Although low-phytate cultivars are not yet commercially available, they will likely be an attractive option for farmers, given the pressure to reduce the impact of livestock production on the environment.

One type of animal feed that is very common in several developing countries in West Asia and North and sub-Saharan Africa is barley straw. In West Asia and North Africa, barley is mostly grown for animal feed, particularly for small ruminants, such as sheep and goats, which in turn play an important role in the economy of farmers in the drier areas of the region. In these countries, barley is normally combine-harvested, but when it is too short because of drought, it may either be grazed at maturity or hand-harvested by uprooting it. In this case, the whole crop is collected and the straw is chopped up for feed. In the case of combine-harvesting, the barley stubble is grazed during summer and early autumn and, together with the straw, provides between one-quarter and one-half of the metabolizable energy fed to sheep (Goodchild et al., 1997).

Farmers consider characteristics associated with straw quality (mostly softness and thin stems) very important. These qualities play a major role in the acceptance and adoption of new varieties (Traxler and Byerlee, 1993).

Thomson and Ceccarelli (1990) found large effects of environmental factors, such as rainfall, temperature, soil fertility, and crop management, on straw quality. As many of these factors are unpredictable, pragmatic breeding based on phenotypic assessment of straw quality is not expected to be very effective.

Recently, a number of quantitative trait loci (QTL) for traits associated with barley straw quality have been identified in a cross between *Hordeum vulgare* 'Arta' and an *H. spontaneum* line (Grando et al., 2005).

Malting barley

In most industrialized nations, the production of beer is decreasing, whereas in countries such as China, Vietnam, India, Mexico, Columbia, and Russia it is rapidly increasing. Globally, beer production is growing at a steady pace; in 2003 it was 1.48 billion hl, 26% more than in 1991.

Barley selected for use in the malting industry must meet special quality standards, the most important of which are protein content, hot water extract, viscosity, Kolbach index, β-glucan content, fermentability, and diastatic power. Other important traits are grain size, grain uniformity and dormancy, thermostability of enzymes, such as α- and β-amylase (both highly correlated with diastatic power), milling energy, and thin husk.

Malting quality is a complex character, depending on the interactions of a number of genes expressed during grain development and during the malting process. Many of the traits affecting malting quality, particularly protein content, are known to show strong interaction with the environment. For example, a good malting variety may become unacceptable to the breweries in a dry year, when both protein content and grain size are adversely affected.

In most of the European countries and Australia, malting barley varieties are predominantly two-rowed, whereas in Canada and the United States, both six-rowed and two-rowed genotypes are used, with a predominance of the former.

Because of the economic importance of malting barley, there has been a major effort made to identify genes controlling malting traits. Information on these has been published in the Barley Genetics Newsletters and in the Proceeding of the International Barley Genetics Symposia. A recent list of genes controlling malting quality traits with their chromosomal location has been published by Fox et al. (2003). In addition, eight regions of the barley genome have been found to be associated with one of the most important malting quality traits, namely, malt extract (Collins et al., 2003), particularly on chromosomes 2 H and 5 H. Diastatic power is considered to be a quantitative trait and is affected by the loci controlling β-amylase activity such as *Bym2* on chromosome 2H (Coventry et al., 2003).

A major effort has been made to map QTL for malting quality mostly by three groups around the world, the North

American Barley Genome Mapping Project, the Australian National Barley Molecular Marker Program, and the European group. There is a major interest in using marker-assisted selection for malting traits and one of the examples is the conversion of elite feed varieties into malting quality in South Australia. The strategy is based on back-crossing, with extensive use of molecular markers for malt extract on chromosomes 1 H, 2 HS, 2 HL and 5 H, and the β-amylase gene on 4 H, which influences diastatic power and fermentability.

Food barley

Barley has historically been the energy food of the masses, with a reputation for building strength. It was awarded to the champions of the Eleusian games in Greece; and Roman gladiators were called *hordearii*, or barley-men, because barley was the main component of their training diet.

Early barley remnants from Mesopotamia and Egypt are much more abundant than those of wheat, and the early literature suggests that barley was more important than wheat for human consumption. The Sumerians had a god for barley but none for wheat. In the Near East and Mediterranean, the shift to wheat as human food came in classical times, and by the first century A.D., barley was already mostly fed to animals. In northern Europe, barley remained the main food cereal until the 16th century.

Barley is still a major staple food in several regions characterized by harsh living conditions; these regions are home to some of the poorest farmers in the world, who depend on low productivity systems.

On the contrary, very little barley is used at present as human food in developed countries. However, in the last two decades, we have seen a rediscovery of food preparations with barley. In several countries, there has been a renewed interest in food barley as a health food in general and as a source of

soluble fiber, which can help people in lowering both blood cholesterol level and postprandial blood glucose content in non-insulin-dependent diabetes.

In addition, barley is being recognized as an excellent source of antioxidants, vitamins, minerals, and phytonutrients such as phenolics and lignans. These components have biological activities that can reduce the risk of coronary heart disease, diabetes, and certain cancers.

In Morocco, Algeria, Libya, and Tunisia, average annual national consumption of food barley in 2002 was 35.6, 15.4, 12.7, and 6.5 kg/person, respectively (FAOSTAT, 2005). Although specific statistics for regional consumption are lacking, it is widely recognized that consumption is much higher in certain areas within a country, e.g., the Gamal region in Algeria, Tensift in Morocco, and Sfax in Tunisia. Barley is also consumed as food in the highlands of Ethiopia, Eritrea, Peru, and Ecuador (12.9, 3.9, 4.1, and 1.7 kg/person/y, respectively); in East Europe (Estonia, Moldova, Latvia, and Lithuania: 13.4, 19.5, 19.5, and 17.8 kg/person/y, respectively); and in the highlands of Central Asia (3.4 kg/person/y).

Food-barley consumption has decreased considerably in the last 40 years along with the increase of urban populations and, often, with the introduction of national policies supporting wheat consumption. This is the case in Morocco where food-barley consumption has decreased from 87 kg/person/y in 1961 to 36 kg/person/y in 2002. In the case of Algeria, Libya, and Tunisia, food-barley consumption in 1961 was 27, 35, and 15 kg/person/y, respectively.

In Europe the average consumption of food barley decreased from 1.6 kg/person/y in 1961 to a minimum of 0.9 in 1991, when it started to resurge and reached 1.6 kg/person/y in 2002.

Over the years, there has been accumulated local knowledge on preparation and the health and nutritious attributes of food barley

(Grando and Gomez Macpherson, 2005). Food-barley cultivars have particular characteristics appreciated by consumers that make them irreplaceable by feed or malting barley. This local knowledge and the unique genetic material need to be preserved for future generations.

Barley products utilized for traditional preparations can be classified as:

- Whole grain
- Cracked grain
- Raw-grain flour (fine and coarse)
- Whole roasted grain
- Roasted-grain flour (fine and coarse)

Morocco is the largest consumer of barley as food. There, the crop has played a significant role in food security of households throughout history. Since the beginning of the second millennium, succeeding dynasties have relied on large barley grain storage facilities as bulwarks against hunger. About 20% of barley grain in Morocco is used as food, mainly in the mountainous and southern parts of the country, with an average annual consumption of 54 kg/person, compared with 5.5 kg/person in the cities (Belhadafa et al., 1992). Barley is used as flour, as semolina, and as whole, dehulled grain (Amri et al., 2005). A large variety of dishes, including soups, breads, and couscous are made from barley products. Preparations include both product from fully mature grains and grains harvested at physiological maturity (Azenbou). Barley grain stored underground for over three years is called Aballagh and is used to produce both flour and semolina (coarse flour) (Amri et al., 2005).

Barley accounts for more than 60% of the food of the people in the highlands of Ethiopia (Yirga et al., 1998); it is used in diverse recipes that have deep roots in culture and tradition. Some recipes, such as Besso (fine flour of well-roasted barley grain moistened with water, butter, or oil), Zurbegonie

(same type of flour as used for besso dissolved in cold water with sugar), and Chiko (besso soaked with butter alone), which have long shelf life, can only be prepared from barley grain. Other recipes, such as Genfo (thick porridge), Kolo (de-hulled and roasted barley grain served as a snack), and Kinche (thick porridge), are most popular when made from barley grain, but can be prepared from other cereals also. Barley is the preferred grain, after tef, for making the traditional bread called Injera, which can be used either alone or in combination with tef flour or other cereal flours. Other recipes, such as Dabbo (bread), Kitta (thin, unleavened, dry bread), and Atmit (soup) can be prepared only with barley or barley blended with other cereal flours. Among local beverages, Tella and Borde are prominent, and best made from barley grain. Barley spikes, both unripe at milk or dough stage and ripe and dry, are also roasted over flame, and the grain is eaten as a snack called Eshete or Wotelo, if the spikes are unripe, or Enkuto, if the roasted barley spikes are dry (Bekele et al., 2005). Barley is also traditionally used in the preparation of a gruel used as weaning food.

In Yemen, barley is grown at 1,800–3,000 m asl, and the grain is used in various dishes and local drinks. Maloog and Matany are two types of bread, the first made from a blend of barley flour and bread wheat flour and the second from a blend of barley flour and lentil flour. Nakia is a local drink made from boiled barley grain (Lutf, 2005).

In the Andes of Colombia, Ecuador, Peru, and Bolivia, barley is the staple food of farmers at altitudes of 2,200–4,000 m asl. It is the crop best adapted to high altitudes, drought, salinity, and aluminum toxicity. Its earliness and cold tolerance make it suited to the short frost-free growing seasons in high altitudes (Capettini, 2005). In this area, barley is roasted and finely ground into Machica or Pito; barley rice, which is coarsely cracked barley, is used for soups;

and barley flakes, which are a relatively recent product, are eaten for breakfast. Hull-less barley is preferred and earns a higher price than regular barley. For example, in Ecuador, variety 'Atahualpa,' with its larger and lighter hull-less kernels, fetches 10% more than other varieties (Capettini, 2005).

Hull-less barley is the main staple food crop in Tibet, where it accounts for 56% of the total food production, and about 2.1 million people consume barley (Tashi, 2005). The main product of hull-less barley is a roasted barley flour known as "Tsangpa." Chang, brewed from hull-less barley is the major alcoholic beverage. In addition barley is used for cakes, soups, porridge, and snack foods. To prepare Tsangpa, the grain is carefully cleaned, washed, and roasted with fine sand. The sand is needed to distribute the heat evenly and prevents the barley kernels from burning. After roasting the sand is sieved off and the remaining roasted barley grain (called Yue) is ground into Tsangpa using a water mill. Tsangpa can be eaten as such (mixed with sugar) or can be the basic ingredient for the preparation of soups, cakes, and beverages.

In most developed countries, the use of barley for human food is very limited—less than 5% of the total production (Jadhav et al., 1998). However, in the past two decades there has been a renewed interest in food uses of barley, even in countries where barley never had a prominent position as a food crop. The whole barley grain can be processed to produce blocked, pot, and pearled barley, barley flakes, and flour (Newman, 1985). Barley products are suitable for use in many food preparations, including different types of bread, pasta, rice extender, and for baby foods, although most of the use has been largely confined to pot or pearled barley in soup and to flakes in breakfast cereals.

Roasted barley can be used as coffee-substitute. "Barley coffee" is very popular in Europe. In Italy, known as Caffè d'orzo, it is commonly used as a breakfast drink for children, often mixed with milk.

The potential benefit of soluble dietary fiber, such as the effect β-glucan can have in lowering cholesterol level and postprandial blood glucose and insulin response, has been reported by several authors (Brennan and Cleary, 2005). Tocols (tocopherols and tocotrienols) can also reportedly lower total cholesterol and low density lipoprotein cholesterol (Wang et al., 1993). Barley grain is a good source of both β-glucan and tocols.

Most efforts in barley breeding have focused on feed and malting cultivars. International and national agricultural research has almost completely neglected the improvement of food barley, particularly the quality aspects. Attributes, such as kernel weight, size, color, and protein content, are often recorded routinely in breeding program, but many more characters are associated with the use of barley as human food. In response to the request from several national programs, the Barley Project at ICARDA began a program to improve the adaptation and quality of food barley, based mainly, but not exclusively, on hull-less germplasm. β-glucan content, kernel hardness, husk percentage, and cooking time have been introduced as screening elements. Information on milling qualities of barley and the use, functionality, and interaction of barley preparations is still fragmented or lacking.

One active area of research is the production of biofortified crops—varieties bred for increased mineral and vitamin content—to mitigate the problem of micronutrient deficiency that affects hundreds of millions of people in developing countries, especially women and children. The HarvestPlus Challenge Program is an international, interdisciplinary, research program that seeks to reduce micronutrient malnutrition by harnessing the powers of agriculture and nutrition research to breed nutrient-dense staple foods. The ICARDA receives support from HarvestPlus to identify barley germplasm with high concentrations of iron and zinc in the grain. A collection of barley landraces and improved lines was screened recently at ICARDA. The content of zinc in the grain varied from 23 to 50 ppm, the content of iron from 26 to 67 ppm. The hull-less barley variety 'Atahualpa' with an average of 47 ppm of zinc and 55 ppm of iron, is particularly promising.

Eventually, the development of micronutrient-dense barley varieties will supply more micronutrients in those areas where barley is already in the food system.

Diseases and pests

Barley is affected by many diseases. The ones that are economically most important are caused by fungi, viruses and bacteria:

- Leaf blights: powdery mildew (*Erysiphe graminis f.sp. Hordei*), scald (*Rhyncosporium secalis*), spot blotch (*Bipolaris sorokiniana*), net blotch (*Dreschlera teres*), bacteria blight (Xanthomonas campestris, *Pseudomonas syringae*)
- Rusts: stripe or yellow rust (*Puccinia striiformis*), leaf rust (*Puccinia hordei*), stem rust (*Puccinia graminis* f.sp. *hordei*)
- Viruses: Barley Yellow Dwarf Virus (BYDV) and Barley Stripe Mosaic Virus (BSM)
- Smuts: covered smut (*Ustilago hordei*), loose smut (*Ustilago nuda*)
- Head blight: Fusarium head blight (FHB) (*Fusarium graminearum*)
- Nematodes, root rots
- Russian Wheat Aphid (RWA) (*Diuraphis noxia*)

For most of these diseases, the genetics of resistance is relatively well known. Barley yellow dwarf virus (family *Luteoviridae*) is the most prevalent and one of the most economically important viruses in barley (Niks et al., 2004). This virus causes a symptom-complex consisting of stunted growth, late heading, and discoloration of leaves. Several yield components can be affected. Control of BYDV infection can best be achieved by growing resistant or tolerant cultivars. Only a few genes have been reported to protect barley cultivars sufficiently. One of these, *Yd2*, may be considered a major gene, and it has been mapped to chromosome 3H (Collins et al., 1996). The gene has been commonly used by barley breeders and occurs in many present-day cultivars.

Powdery mildew is usually destructive in northern Europe, Japan, North America, and in areas with a cool and humid climate during the growing period; the most sustainable control is the use of resistant varieties. At least two genetically separable pathways control resistance to powdery mildew in barley (Jørgensen, 1994). In the first pathway, resistance is mediated by recessive alleles at the *Mlo* locus (*mlo* resistance alleles). This resistance is effective against all tested powdery mildew isolates and requires for its function at least two further host genes, designated *Ror1* and *Ror2* (Freialdenhoven et al., 1996; Büschges et al., 1997). The second resistance pathway can be triggered by a number of race-specific resistance genes (R genes; e.g., *Mla, Mlg, Mlk*) (Jørgensen, 1994). The *Mla* resistance locus on chromosome 1H has as many as 32 different alleles (Jørgensen, 1994) and can be selected with closely linked markers (Wei et al., 1999). Net blotch, which occurs as spot (spot blotch) and net forms, is distributed wherever barley is grown and is favored by warm, wet conditions; its severity can be reduced by chemical seed treatment, crop rotation, and use of resistant varieties. Resistance to the net form of net blotch is controlled by *Rpt.CI9819* gene located on chromosome 6H, and *Pt.a* located on chromosome 3HL, whereas resistance to the spot form of net blotch is controlled by *Rpt4* on chromosome 7HL. Scald is a serious foliar disease of barley in cool, moist conditions, where yield losses can be considerable.

Means of control are crop rotation, resistant varieties, and elimination of crop residues (the fungus survives on infected straw, seed, and dead leaves). Chromosome 3 H has a major gene complex, *Rh*, controlling resistance to scald. Fusarium head blight, also known as scab, is common in warm, humid areas. It can make a potentially profitable barley crop unusable for malting, and infected grain may be harmful when fed to animals and humans. Leaf and stripe rust are common in developing countries in Latin America and sub-Saharan Africa (Capettini et al., 2003). Leaf rust is controlled by several genes, such as *Rph.Hb* on chromosome 2 H, *Rph4* on chromosome 1HS, *Rph7* on chromosome 3 H, and *Rph19* on chromosome 7 H. At least 40 major genes for resistance have been mapped to date, and QTL have also been found for resistance to all major diseases (Williams, 2003).

Several nematodes are capable of parasitizing barley: root gall nematode, cyst nematode, root knot nematode, and root lesion nematode. Rotation with non-host crops reduces nematode incidence. Resistance to cereal cyst nematodes is controlled by *Ha2* and *Ha4* genes located on chromosomes 2 H and 5 H, respectively.

The availability of several molecular markers sufficiently close to the resistance gene makes conducting marker-assisted selection possible. The transfer of the gene *Ha2* for resistance to cereal cyst nematode (CCN) from the feed variety 'Chebec' to the malting quality variety 'Sloop' in Australia is one successful example of marker-assisted selection; the result was the release of Sloop SA. Other examples are given in the section "Marker-assisted selection" later in this chapter.

The Russian wheat aphid (*Diuraphis noxia*) has been described as one of the most destructive pests of small grains; it is a serious pest of wheat and barley in North America, South America, South Africa, and Australia, where it causes major economic losses (Liu et al., 2002). It is also widespread in North Africa and Ethiopia. The six-row variety STARS-9577B (PI 591617) is one of the best known sources of resistance, together with some landraces from Afghanistan. The resistance of STARS-9577B is controlled by dominant alleles at two loci, with alleles at one locus conferring a high level of resistance and alleles at the other locus conferring an intermediate level of resistance only when recessive alleles are present at the first locus (Mornhinweg et al., 2002).

Breeding

Modern barley breeding started at the end of the last century and on several occasions barley research and breeding "pioneered" the way for other crops. Within this relatively short breeding history, breeders have employed mutation breeding, hybrid breeding, and population improvement programs involving composite crosses and evolutionary breeding. Male-sterile-facilitated recurrent selection was used in breeding for disease resistance and interspecific and intergeneric crosses were used for the same purpose. Haploid breeding based on the *bulbosum* system or anther culture became routine programs focused on accelerated development of homozygous lines from segregating populations. Innovative biotechnological methods, including transformation, provide new possibilities, unthinkable some 30 years ago.

The development of superior barley cultivars is part of a continuous dynamic process that culminates with their adoption by farmers in different production areas of the world.

Breeding methods

Barley is a self-pollinating crop and breeding programs are based on methods that are typical for self-pollinators. Before scientific breeding began, farmers grew populations of barley and usually they or their neighbors maintained them. These populations, which are still cultivated today in many developing countries, are composed of many homozygotes that are different from each other and are known as landraces.

Occasional cross-pollination produces a few heterozygotes, which are rapidly brought to homozygosity by self-pollination, generating new genotypes. Until shortly before 1950, the predominant method was either mass selection of landraces or pure line selection within landraces. In the first method, the best plants or the best spikes of landraces are bulked together to produce a new, improved landrace. In the second, the selected spikes are kept separate and each gives a line that eventually becomes a new, uniform cultivar. Many cultivars between 1930 and 1940 were produced, mainly with the second method. Both methods exploit the genetic variability present in nature. New variability can be created by crossing or by mutation and both methods have been used in barley. Crossing requires the emasculation of the plant to be used as female parent by clipping the top of the lemma and palea with scissors and then removing with tweezers the anthers from each floret of the spike before the pollen reaches maturity and pollinates the stigma of the same floret. The spike is bagged and after 1–3 days the stigma is pollinated with the pollen of 1–3 spikes of the male parent. The F_1 seed is often planted off-season to produce an F_2 population which is the starting point for different selection methods. The method most widely used is the pedigree method, which consists of selecting individual plants within the F_2, growing F_3 families of the selected plants, selecting again the best F_3 plants within the best F_3 families, and so on. The lines are tested for yield and other characteristics and the best become cultivars.

The pedigree method works well for characters that can be easily identified in the early segregating generations. However, for characters with low heritability, the pedigree system can be a form of pseudo-random selection, a situation which must be carefully guarded against (Andersen and Reinbergs, 1985).

A modification of the pedigree method is the F_2 progeny method, where a similar number of F_3 families are selected within each cross and yield tested as F_4 and F_5, and final plant selection is made in F_6. In the bulk-pedigree method, the F_2 population derived from each cross is kept as a bulk for a number of generations (usually five or six) without any selection. During this period homozygosity increases and in F_6 or F_7, selection of nearly homozygous plants is done within the best bulks. The efficiency of the method can be improved by discarding bulks at each generation or by an early evaluation of the merit of each bulk by testing a similar number of randomly extracted lines from each bulk.

Sometimes bulking spikes from selected plants helps increase the efficiency of the method. This is especially used when selection for disease resistance is possible within the bulk. Bulk handling of crosses has been used by many programs as an inexpensive way to grow early generation material with or without selection. At the endpoint of bulk breeding systems, selections are made and tested as individual lines. Bulk breeding systems are well adapted to mass selection. For example, screening for kernel plumpness and weight can be done by mass-selecting bulk materials; this also indirectly selects for disease tolerance and resistance when disease pressure is applied on the bulk. This methodology has been successfully applied by the Alberta Agriculture, Food and Rural Development barley-breeding program based at Lacombe, Province of Alberta, Canada.

Composite crosses are a modification of the original bulk method of breeding to some degree. Composite crosses result from combining a number of single crosses into one large mixture or composite (Harlan and Martini, 1929; Harlan et al., 1940; Suneson and Stevens, 1953). Composite cross breeding utilizes large germplasm pools with minimum time and labor demands. Composite crosses are an efficient way to do long-term breeding, especially when genetic male sterility is to facilitate the mating system. When a composite cross is grown in the intended-use area long enough to allow natural selection to influence gene frequencies, most of the selected plants are well adapted to the area (Wiebe, 1979). Generally, objectives in composite cross breeding should be long-term to encourage potential recombination of many factors, and thus germplasm should be as broad-based as possible.

Recurrent selection is used to increase recombination and avoid rapid fixation of gene combinations. This can be done by crossing among F_1s and by repeating the crosses among a random number of F_2 plants or by using male-sterility genes to facilitate continuous intercrossing in a scheme known as male sterility-facilitated recurrent selection (MSFRS). The single seed descent (SSD) method consists of advancing a large number of plants (up to 2,000) from each single F_2 under conditions that accelerate growth (low nutrients, small pots) in a greenhouse. Under these conditions, plants complete their cycle in about three months, produce only 3–4 seeds/plant and three generations/year can be obtained. One seed is used for the following generation until the desired level of homozygosity is reached. No selection is practiced during this process because it only aims at reaching homozygosity as quickly as possible. With this method, each cross produces a large number of lines and selection is practiced only among homozygotes. This method is very efficient in exploiting the variability in genetically narrow crosses, or in identifying rare recombinants. Also since most SSD populations are advanced beyond the F_3 to fairly high

levels of homozygosity before selection is practiced, heritabilities are expected to be relatively high for most traits.

Backcross breeding is practiced when one parent (A) possesses most desirable attributes except one, which is present in another parent (B). After an initial cross, the F_1 is backcrossed to parent A (recurrent parent) and selection is practiced for the character to be transferred from parent B (donor parent). The backcrossing is repeated until full recovery of the genetic background of parent A. A recurrent cyclic breeding method is sometimes used as a choice of the backcross method when selecting for quantitatively inherited traits. In this method, the recurrent parent is substituted for by the best available germplasm in a productive breeding program and a new cycle of selection is practiced. This method capitalizes on the enhancements obtained in the program during the breeding process, instead of coming back to a parent that might have been surpassed by a newly-produced, superior genotype. (Gebhardt et al., 1992).

Mutation breeding using both radiation and chemicals has been widely used in barley (Hockett and Nilan, 1985). Mutation breeding is the use of mutants either directly as new cultivars or indirectly via crosses with other genotypes to develop improved germplasm sources and cultivars. The use of mutants as parents in crossbreeding enhances variability in a crop species. Mutants can easily be incorporated into conventional pedigree, bulk, backcross, and recurrent-selection breeding schemes. Genetic variability available in germplasm collections can be supplemented with induced mutations. For some traits, germplasm sources may be inadequate or only present in poor genetic backgrounds. Mutant variation can provide new genes at previously unknown loci, as well as new series of alleles with similar phenotypic expression, but in improved adapted genetic backgrounds. Thus, in most situations, mutation breeding is used to supplement and complement other breeding methods to achieve desired goals. Although a number of cultivars have been produced, the method has to be considered as a complement rather than as an alternative to other breeding methods.

The use of hybrids, like in most self-pollinated crops, has never been highly developed commercially in barley, and no information is available about any commercial variety release.

Hybrid production depends on the degree of effective heterosis, the availability of a practical system for making crosses for production of parental lines (e.g., through inducing male sterility and restoring fertility), and real gains in yield and/or quality that exceed the additional costs related to seed production and establishment of a hybrid system. Probably the most popular male-sterility method for barley was the "balanced tertiary trisomic system," based on a recessive gene (Ramage, 1975). Although successfully used experimentally, the advantages of the hybrids produced never were high enough to justify the additional costs of the system when compared with available self-pollinated cultivars.

The choice of the breeding method to be used in a breeding program depends on the objectives, the target area, the resources available and all the particular conditions that may be present in a program (administrative, political, etc.). The pedigree method and all its modifications used to be the most popular method used by barley breeders, but it is usually also the most resource-demanding. Decreases in resources in programs worldwide forced scientists to find less expensive methods with similar selection efficiencies, e.g., obtaining the highest genetic progress per unit of time with similar or less resources. Independent of the method of breeding adopted, a program must be able to efficiently select for the traits that are in the objectives and obtain measurable progress through the generations. This may be obtained by carefully choosing the environments where selection for the desired traits can be carried out (e.g., water-stressed environments when selecting for drought resistance or hot spots for diseases when selecting for biotic stresses) or creating the necessary environment by artificially controlled conditions (e.g., irrigated and inoculated nurseries when selecting for particular diseases).

Breeding for disease resistance

Developing genetic resistance to plant diseases is a major objective of most plant breeding programs worldwide. Genetic resistance is the most environmentally friendly and durable method of control of crop diseases, as well as the only affordable method for low-income farmers. Carrying out effective selection depends on the available genes or sources of resistance, as well as on the screening environments where it is possible to maximize the response to selection. When disease is not regularly present every season, intervention to decrease the environmental variation may be needed.

Providing optimal conditions for disease development (misting, inoculation) in screening nurseries assures maximum efficiency when selecting for disease resistance (Capettini et al., 1999). When available, quantitative resistance, determined by multiple minor genes, should be chosen. Quantitative resistance is believed to be more durable since the chance of breaking the resistance genes due to changes in the pathogen across time is lower. The introgression of different major and/or minor genes for resistance into the same cultivar is also a useful strategy to assure long-term prevalence of the resistance and is described as "gene pyramiding."

Although quantitative resistance is preferable due to its higher probability of persistence across time, "durable resistance," usually inherited in a quantitative fashion, can only be confirmed a posteriori. There are examples in barley where qualitative resistance conferred by one major gene has been durable. Probably one of the most well known cases has been the *Rpg1* gene (also known as *T* gene), which has given resistance to stem rust for more than 50 years and is still functional. Every breeder should carefully select breeding strategy according to the resources and resistance sources available, the number of loci involved and their mode of inheritance, and whether the objectives are short-term or long-term protection. Long-term resistance should be the ultimate objective, but an urgent need to develop resistant cultivars and the ease of the breeding method may take precedence (Sharp, 1985).

Resistance sources have been identified, are available for all of the commercially important diseases mentioned above (Sharp, 1985; Lundqvist et al., 1997), and can be requested from gene banks. Some of these sources may be in agronomically non-adapted backgrounds. From the breeding point of view, it is best to use resistance genes already incorporated into adapted varieties that have undergone several cycles of breeding, to avoid the genetic gap with local cultivars. The global barley improvement program of ICARDA can provide, upon request, diverse germplasm with resistance.

Two different approaches have been used to induce haploidy in barley: interspecific crosses with *H. bulbosum* L. and anther or microspore in vitro culture. Crosses between *H. bulbosum* (2n = 2x = 14) as the male and *H. vulgare* ssp. *vulgare* as the female parent lead to preferential elimination of *H. bulbosum* chromosomes in the first eight days after fertilization, thus leaving haploid *H. vulgare* plants. This method is very efficient for haploid-barley production (Chen and Hayes, 1989; Inagaki et al., 1991; Pickering and Devaux, 1992).

Anther culture techniques have been developed as an alternative to the *bulbosum* method. Anthers are cultured on solid or liquid induction medium to develop calli with induced morphogenesis (embryoids), which are transferred to a regeneration medium to regenerate green haploid plants. Improvement of media and culture conditions, transfer of genes for regeneration capacity, and improvement of donor plant conditions have made it possible to use this method for a wide range of barleys. With barley genotypes showing an excellent response to androgenesis (such as Igri), production of up to 10 green plants per anther has been reported (Ziauddin et al., 1990; Olsen, 1991; Jähne et al., 1991).

The culture of isolated microspores is a recent development in anther culture technique (Jähne and Lörz, 1995; Picard et al., 1995). The main advantages of this technique are the unicellular status and the haploid nature of the targets, which allows clean selection and regenerability into homozygous plant after chromosome doubling, without loss or decrease of fertility.

The DH lines are very successfully produced by private companies. Most of these companies serve barley breeding programs that are located in the western world.

Breeding for stress tolerance

Formal, or institutional, breeding has been highly efficient in improving barley yield levels, malting quality, and disease resistance. However, its efficiency has remained largely confined to favorable environments, or to environments that could be made favorable by providing fertilizer and irrigation, and by chemically controlling weeds, pests, and diseases. In many developing countries, the use of modern varieties is limited, and farmers still rely on landraces.

The ICARDA has given considerable emphasis to barley improvement for stress-affected, low-input farming systems in developing countries, recognizing that resource-poor farmers, who practice approximately 60% of global agriculture, and produce 15–20% of the world food (Francis, 1995), have not known the benefit of the Green Revolution and that some 1.4 billion people are dependent on agriculture practiced in stressful environments (Pimbert, 1994).

To better serve those farmers who were neglected in the first Green Revolution, ICARDA breeders and researchers developed a new breeding strategy based on selection of segregating populations in the target environments. The new strategy incorporates farmers' participation (Ceccarelli and Grando, 2005). Whenever possible, selection is done on farmers' fields, using their level of inputs, their technology, their experience, and their preferences.

Many formal breeding programs for crops have common features: (1) they generally produce genetically uniform cultivars (pure lines, clones, hybrids); (2) they are largely conducted either in favorable environments or in well-managed experiment stations where growing conditions are optimum or near-optimum; (3) selection in crops such as barley is almost exclusively for grain yield, malting quality, and disease resistance, either using individual or combined selection criteria; (4) they promote cultivars that can be grown across large areas (widely adapted in a geographical sense); and (5) they do not involve the clients (the farmers) in any of the steps that will eventually lead to new cultivars, except perhaps in the final field testing of a few promising lines.

These types of breeding programs are based on the following assumptions: (1) selection must be conducted under good growing conditions where heritability is high, and therefore response to selection is also high; (2) yield increases can only be obtained through replacement of locally adapted landraces (Brush, 1991) that are low yielding and disease susceptible; (3) breeders know better than farmers the characteristics of a successful cultivar; and (4) when

farmers do not adopt improved cultivars, it is usually attributed to ineffective extension and/or inefficient or insufficient seed production capabilities—that the breeder might have bred the wrong varieties is rarely considered.

Because of the success that breeding has had in favorable environments, these features and assumptions are not questioned, even when the objective is to improve yield and yield stability for poor farmers in stressful environments. The implicit assumption is that what has worked well in favorable conditions must also be appropriate under unfavorable conditions. Very little attention has been given to the need for developing new breeding strategies specifically for dry areas. (Ceccarelli and Grando, 2002). In the last few years, there has been mounting evidence that these assumptions are not valid, particularly for crops such as barley, which in developing countries are predominantly grown in areas that are too marginal for other crops (Ceccarelli, 1994, 1996).

Unfavorable environments are defined as those in which crop yields are commonly low due to the effects of abiotic and biotic stresses. The semiarid areas of Syria, where barley/livestock is the predominant farming system, are a good example of such environments, which are also found in other developing countries. The main characteristics of these dry areas are:

1. Complex environmental stresses, including low annual rainfall, poor rainfall distribution, low winter temperatures, and high temperatures and hot winds from anthesis to grain filling. The frequency, timing, intensity, and duration of each of these stresses, as well as their specific combinations, vary from year to year. However, low yields of barley are common, crop failures occur one out of ten years, and yields of 3.0 t/ha or more are expected less than 15% of the time.

2. The use of inputs, such as fertilizers, pesticides and weed control, is uneconomical and risky for resource-poor farmers because of the probability of low yields and crop failures in unfavorable environments.

Many resource-poor farmers practicing agriculture in dry areas have adopted a strategy based on both intraspecific and interspecific diversity (Martin and Adams, 1987). They grow different crops in the same field at the same time (interspecific diversity); they also grow genetically heterogeneous and/or different cultivars of the same crops (intraspecific diversity) (Haugerud and Collinson, 1990). The type of diversity that prevails in different areas depends on both climatic and socioeconomic conditions and farmers' response to these. In the dry areas of West Asia and North Africa, for example, barley is often the only feasible rainfed crop, and the cultivars that are grown at present, and that have been grown for centuries, are genetically heterogeneous (Ceccarelli et al., 1987; Weltzien and Fischbeck, 1990).

3. This diversity, which is typical of resource poor farming, is in marked contrast with the uniformity pursued by formal breeding (Wolfe, 1991) and production practices in most crops grown in favorable environments, and is one of the causes for a different mechanism of seed supply (Grisley, 1993).

4. Whereas in high-input agriculture the seed market is the main source of seed supply (particularly for grain crops), in resource-poor agriculture, seed is usually produced on the farm, after some form of selection done by the farmer, or it is purchased from neighboring farmers (Almekinders et al., 1994). Formal breeding not only replaces diversity with uniformity, but it also tries to reach farmers with new cultivar seed through mechanisms and institutions that are not

efficient, are not responsive to the needs of resource-poor farmers, and often are not trusted by them.

Breeding for unfavorable environments, based on selection (not merely testing) in the target environments, is more complex than selection for favorable environments, largely because of the large genotype × year within location interactions. The barley breeding methodology for dry areas at ICARDA is the following:

1. Breeding material (including parental material and segregating populations) is evaluated in the target environments using farmers' agronomic practices, including rotations. In the driest site (200–250 mm of total annual rainfall), this means no use of fertilizers, pesticides, or weed control. Concurrently, the material is evaluated at the main experiment station (long-term average rainfall of 373 mm) with a level of inputs commonly used in moderately favorable areas. In all the "experiment sites," the material is evaluated strictly under rainfed conditions.

2. Experimental designs have evolved from the randomized block design to the lattice design, to α-lattice design, and more recently to row and column with spatial analysis (Singh et al., 2003). This has progressively improved the control of environmental variability.

3. Segregating populations are evaluated as bulks for three years, taking advantage of the large year-to-year variation in total rainfall, rainfall distribution, and temperature patterns. Individual plant selection is done only within the selected bulks.

4. Selection is practiced for high grain yield at each of the experiment sites, regardless of the performance in other experiment sites. This promotes breeding material with specific adaptation without sacrificing wide adaptation.

5. In addition to grain yield, traits used as selection criteria in the driest sites are plant height, tillering, straw softness, and disease resistance; in the wettest sites, earliness, lodging, and disease resistance are selection traits.

Using this methodology, we discovered that locally adapted landraces and *H. spontaneum* (Grando et al., 2001) were useful sources of breeding material that would have been missed, had the evaluation taken place only in high-yielding environments. The presence of useful diversity within landraces has been documented in many crops. The diversity within barley landraces collected in Syria and Jordan has been documented by Ceccarelli et al. (1987), van Leur et al. (1989), Weltzien (1988, 1989), Weltzien and Fischbeck (1990), and Ceccarelli et al. (1995).

To exploit specific adaptation fully and make positive use at the international level of GE interactions, ICARDA's barley breeding program developed a model of decentralized breeding (Ceccarelli et al., 1994), in which ICARDA's barley breeding program produces targeted F_2 segregating populations (often based on crosses designed by national programs), selection between F_2 populations is conducted by national programs in different agro-ecological environments within each country under conditions as similar to farmers' fields as possible, and lines selected from superior F_2 populations are advanced at ICARDA and then yield-tested in different locations within each country.

However, decentralization of the selection component to national programs will not create a response to the needs of resource-poor farmers if decentralization is only from one experiment station to another. This problem can be solved by what may be considered the most extreme decentralization and possibly the most effective way of exploiting specific adaptation, i.e., farmers' participation in selection in their specific environment.

Participatory breeding

Participatory plant breeding (PPB) complements both Mendelian and molecular breeding and, being based on the same genetic principles as formal breeding, incorporates and takes full advantage of modern biotechnological techniques. Three common characteristics of most agricultural research that help explain its limited impact in marginal areas are:

1. The research agenda is usually decided unilaterally by scientists and is not discussed with the users.
2. Agricultural research seldom uses an integrated approach and is instead typically compartmentalized into disciplines and/or commodities. As a consequence there is a continuous emphasis on "interdisciplinary research;" this contrasts with the integration existing at farm level.
3. There is a disproportion between the large amount of technologies generated by agricultural scientists and the relatively small number of them actually adopted and used by the farmers.

PPB has been proposed as a way to address these three points by "turning upside down" the delivery phase of a plant breeding program: in a conventional breeding program, the most promising lines are released as varieties, the certified seed is produced, and only then do farmers decide whether or not to adopt them. In a participatory program, the process is driven by the adoption that takes place during the final stages of selection, and therefore adoption rates are higher, and risks are minimized. Last but not the least, the investment in seed production is nearly always paid off by farmers' adoption.

The program is based on the following concepts (Ceccarelli and Grando, 2007; Ceccarelli et al., 2007):

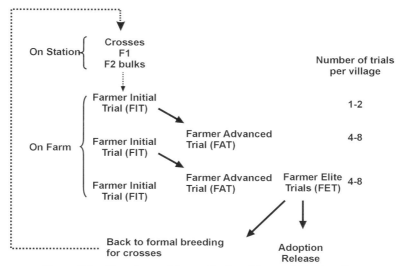

Figure. 7.1.   The scheme of decentralized-participatory barley breeding.

1. The trials are grown in farmers' fields using the host farmer's agronomic practices.
2. Selection is conducted by farmers in their own fields, so that farmers are the key decisions makers.
3. The traditional linear sequence "Scientist–Extension–Farmer" is replaced by a team approach with scientists, extension staff, and farmers participating in all major steps of variety development.

The two major justifications for PPB are the large effects of GE interactions on selection response and the possibility of large differences between breeders' and farmers' selection criteria. An example of different selection criteria between farmers and breeders can be found in relation to the use of barley as animal feed. Breeders often use grain yield as the sole selection criterion, which usually brings with it a high harvest index and lodging resistance. However, in unfavorable environments, lodging is often not a problem because of moisture stress, and farmers are interested not only in grain yield, but also in forage yield and in the palatability of both grain and straw. This often results in on-farm selection of genotypes that are highly susceptible to lodging on-station, and thus these genotypes are not selected for development in formal breeding programs. The low quality of straw of the newly released varieties is one of the reasons for the low adoption rates in most of North Africa and West Asia.

The model of plant breeding we use in Syria and in a number of other countries is a bulk-pedigree system, in which the crosses are made on-station, where the F_1 and the F_2 are also grown, while in the farmers' fields, the bulks are yield-tested across a period of three years (Fig. 7.1) starting from the F_3. The activities in farmers' fields begin with the yield testing of bulks (three years after making a cross), in trials called Farmers Initial Trials (FIT), which are unreplicated trials.

The breeding materials selected by farmers from the FIT are yield-tested for a second year in the Farmer Advanced Trials (FAT), which are replicated trials grown by 4–8 farmers in each village. While within a village, the FAT contain the same entries, the type and the number of entries and checks varies from village to village. The number of FAT in each village depends on how many farmers are willing to grow this type of trial. Each farmer decides the rotation, the seed rate, the soil type, the amount and the time of fertilizer application. Therefore, the FAT are planted under a variety of conditions and managements. During selection, farmers exchange information about the agronomic management of the trials, and rely greatly on this information before deciding which lines to select. Therefore, one of the advantages of the program is that the lines start to be characterized for their responses to environmental or agronomic factors at an early stage of the selection process.

The entries selected from the FAT are yield-tested for a third year in the Farmer Elite Trials (FET) grown by 4–8 farmers in each village. These entries are also used on-station as parents in the crossing program. The three types of trials are planted by scientists using plot drills and are then entirely managed by the farmers. Recently, we added a fourth year testing called large scale (LS), in which farmers plant in large areas about 1–3 entries that they believe could become varieties.

During selection, some farmers are assisted by a researcher to record both quantitative and qualitative data. Some farmers practice the selection at various stages, but the majority do selection when the crop is close to full maturity, and, using a scoring method from 0 = discarded to 4 = the most desirable, farmers express their opinion on each individual entry.

In each trial, the scientists record the following data: plant height, spike length, grain yield, total biomass and straw yield, harvest index, and 1000-kernel weight. The data are subjected to different types of analysis, some of which were developed at ICARDA, such as the spatial analysis of un-replicated or replicated trials (Singh et al., 2003). The environmental-standardized Best Linear Unbiased Predictors (BLUPs) obtained from the GENSTAT programs are then used to analyze genotype × environment interactions using the GGEbiplot software (Yan, 2001). Therefore, the PPB trials generate the same quantity and quality of data as generated by the MET in a conventional breeding program, but the additional information on farmers' preferences is usually not available in the MET. As a consequence, varieties produced by PPB are eligible to be submitted to the process of official variety release that in several countries, including many in the developing world, is the legal prerequisite for their commercialization. The PPB programs, based on the method described above (with minor modifications), are currently implemented in Syria, Jordan, Egypt, Eritrea, and Algeria.

The method described here is only one of the possible methods that could be eventually used in the PPB program. Many other methods are possible, with scientists and farmers playing variable roles.

Four areas of impact have been observed so far with PPB:

1. Variety development: new varieties were spontaneously disseminated from farmer to farmer as early as three years after starting the program. In Syria alone, 12 varieties have been adopted since the inception of the program in 2000, including areas where conventional breeding did not have any impact.
2. Institutional attitudes: in several countries, PPB has generated considerable changes in the attitude of policy makers and scientists toward the benefits of participatory research and has generated changes in national breeding programs.
3. Farmers' skills and empowerment: the cyclic nature of the PPB programs has considerably enriched farmers' knowledge, improved their negotiation capability, and enhanced their self-esteem. By the same token, scientists (breeders) have been enriched by the farmers' indigenous

knowledge of the crops they grow and the environments in which they grow them.

4. Enhancement of biodiversity: different varieties have been selected in different areas within the same country, in response to different environmental constraints and users' needs.

Molecular breeding

The advent of molecular markers has revolutionized the genetic analysis of crop plants and provided valuable new tools with which to identify chromosome regions influencing resistance to biotic and abiotic stresses. The use of DNA molecular marker techniques can considerably reduce the complexity of combining a number of desirable traits in the same line. These techniques make it feasible to develop linkage maps for barley (Graner et al., 1991; Ramsay et al., 2000; Rostoks et al., 2005). Together with statistical techniques, these linkage maps are used to locate and estimate phenotypic effects of QTL (Backes et al., 1995; Teulat et al., 2002, 2003; Ellis et al., 2000, 2002; Matus et al., 2003; Forster et al., 2004b; Diab et al., 2004). The QTL analysis is a powerful tool to locate genes on chromosomes. Therefore, the identification of molecular markers closely linked to QTL of interest to the breeder will allow the use of marker-assisted selection and thus increase the efficiency of selection in the exploitation of desirable germplasm. In addition, QTL mapping is a first step toward unraveling the molecular basis of agronomic traits i.e., by map-based cloning, with the ultimate goal of assigning functions to genes.

In this overview on barley biotechnology, we will first review the marker technologies that have been employed in both genetic analysis and molecular breeding of barley, in particular, Simple Sequence Repeat (SSR) and single nucleotide polymorphism (SNP) markers. These marker technologies are still very much in use and will, for a number of years to come, be practical tools for plant breeders to use to introgress biotic and abiotic resistance genes.

The DNA markers are often used for QTL mapping studies to locate genes and QTL and to derive markers that, after validation in different genetic backgrounds, can be used for marker-assisted selection. As an example of this type of study, QTL for many agronomic traits were identified in the cross Arta/*H. spontaneum* 41-1. For some of these traits, markers can be derived, and if validated, they can be used for marker-assisted selection.

Marker-assisted selection (MAS) is an indirect selection tool that can often help the plant breeder introgress traits faster and more precisely than in conventional plant breeding. The traits for which MAS is currently being used are mostly biotic stresses, but current research is focusing also on abiotic stresses. While marker systems for cold tolerance are already feasible, markers for drought tolerance or components of drought tolerance are still being investigated.

DNA marker technologies

Various types of molecular markers have been used for genetic diversity studies and molecular breeding in barley. However, the data generated were seldom directly comparable between different marker systems or studies. This was particularly true for the marker assays, such as random amplification of polymorphic DNA (RAPD) (Williams et al., 1990) and amplified fragment length polymorphism (AFLP) (Vos et al., 1995), but is also an issue for marker systems such as restriction fragment length polymorphisms (RFLPs), sequence tagged sites (STS), cleaved amplified polymorphic sites (CAPS), sequence characterized amplified regions (SCARs), etc. In recent years, SSR or microsatellite markers have been developed for barley (Liu et al., 1996; Powell et al., 1996; Ramsay et al., 2000; Varshney et al., 2005) as well as in other plant species. SSRs can be found in both coding and noncoding DNA sequences of all higher organisms examined to date. There are indications that the microsatellite frequency is higher in transcribed regions, especially in the untranslated portions (Morgante et al., 2002). The SSRs have become the markers of choice for many applications. Their abundance, high level of repeat-number polymorphism (manifested as the occurrence of a large number of alleles per locus), and codominant inheritance have facilitated their extensive use in genome mapping, phylogenetic inference, and population genetics (Kota et al., 2001; Maestri et al., 2002; Matus and Hayes, 2002; Russell et al., 1997, 2003, 2004). The SSRs have been developed either from genomic libraries (Liu et al., 1996; Ramsay et al., 2000) or have been mined from libraries of expressed-sequence tags (ESTs) (Kantey et al., 2002; Varshney et al., 2005). These genic or functional markers (Andersen and Lübberstedt, 2003) are the ideal markers for MAS, especially if they are derived from the gene that is selected for (e.g., *Yd2*). (Collins et al., 1996). Chloroplast (cp) microsatellites have also been developed and used to elucidate species relationships within the genus *Hordeum* (Provan et al., 1999). However, it is the use of the same SSR markers (e.g., Liu et al., 1996; Ramsay et al., 2000) that has made it possible to compare studies of genetic diversity and to assign markers to bins in the barley genome.

The single nucleotide polymorphisms (SNPs) are bi-allelic markers and represent the smallest units of genetic variation in genomes (Rafalski, 2002). However, the extraordinary abundance of SNPs in the genome largely offsets their disadvantage of being biallelic and makes them the most attractive molecular marker system developed so far. Although the development of SNP markers is still underway in crop plant

species, these markers have already been successfully used for genetic diversity studies (Kota et al., 2001; Kanazin et al., 2002; Bundock and Henry, 2004; Bundock et al., 2006). The allelic frequencies of a given SNP may vary in different populations. Kanazin et al. (2002) evaluated the prevalence of SNP polymorphism at 54 loci across five genotypes and found 38 SNP loci, which revealed the occurrence of one SNP per 189 bases. Russell et al. (2004) studied the frequency and distribution of nucleotide diversity within 23 genes in three germplasm groups representing European cultivars, landraces, and wild accessions; they identified one SNP in every 78 base pairs (bp) and one insertion-deletion in every 680 bp. Rostoks et al. (2005) developed more than 2,000 genome-wide SNPs from eight diverse accessions. The genes used for SNP discovery were selected based on their transcriptional response to a variety of abiotic stresses, such as high salt, drought, low temperature, low nitrate, and water logging. More than 300 SNPs were found in three mapping populations (Oregon Wolfe Dominant × Oregon Wolf Recessive, Steptoe × Morex, and Lina × *H. spontaneum* 92) in an integrated linkage map incorporating a large number of RFLP, AFLP, and SSR markers. The integrated map showed good agreement with the published maps relative to marker order and known abiotic-stress QTL mapped in relevant crosses.

QTL mapping in balanced populations

The QTL analysis provides a powerful tool for locating genes on chromosomes and estimating phenotypic effects of QTL. Yield performance under drought is a particularly complex phenomenon because yield itself is a quantitative trait, and plants exhibit a diverse range of genetically complex mechanisms for drought resistance, including mechanisms of drought escape, drought avoidance, and drought tolerance (e.g., osmotic adjustment). This complexity may explain why most QTL studies of field-grown barley have so far been performed under favorable conditions, whereas only a few studies have been conducted under drought stress (Teulat et al., 2001; Baum et al., 2003; Talamé et al., 2004; Forster et al., 2004a,b). These studies have shown that developmental genes, notably those involved in flowering time and plant stature, have pleiotropic effects on abiotic stress resistance and ultimately determine yield potential.

Agronomic performance in balanced populations was tested under drought-prone conditions of West Asia, North Africa, and Australia. Baum et al. (2003) and Grando et al. (2005) used 194 recombinant inbred lines (RILs), randomly selected from a population of 494 RILs, and tested them in two years and two locations in Syria, with rainfall ranging from 433.7–227.4 mm. The map extended across 890 centimorgan (cM) and contained 189 marker loci, including one morphological marker locus (*btr* = brittle rachis), 158 AFLP loci, and 30 SSR loci. Teulat et al. (2001, 2002, 2003), Forster et al. (2004b), and Diab et al. (2004) used a progeny of 167 two-row barley RILs, and the two parents 'Tadmor' (selected by ICARDA from Arabi Aswad, a Syrian landrace) and 'Er/Apm' (a selected line released in Tunisia), which were grown in three Mediterranean sites with rainfall ranging from 458 mm–54.6 mm. One hundred and thirty-three markers covered 1,500 cM of the genome. Diab et al. (2004) added more candidate genes and ESTs to the available map.

Heading date is a developmental trait, and is controlled by three genetically independent mechanisms of the photothermal response, i.e., day length (photoperiod), plant response to a sum of temperature across a period (earliness per se), and response to a period of low temperature at the initial stages of plant development (vernalization) (Ellis et al., 1990). The major genes for photoperiodic response are the *Eam1* or *Ppd-H1* located on chromosome 2H (bin 4) (Laurie et al., 1995) and *Ppd-H2* on chromosome 1H (bin 14) (Laurie et al., 1995). Earliness per se (*eps*) genes are minor genes that influence onset of flowering independent of conditions of temperature or light. They have been identified as *eps2S* on 2H (bin 6), *eps3L* on 3H (bin 13), *eps4L* on 4HL, *eps5L* on 5H (bin 6), *eps6L.1* and *eps6L.2* on 6H (bin 7, 13), *eps7S* on 7H (bin 3) and *eps7L* on 7H (bin 12) (Laurie et al., 1995). As for vernalization, *Vrn-H1* on chromosome 5H (bin 11), *Vrn-H2* on 4H (bin 13), and *Vrn-H3* on 1H (bin 13) determine the vernalization requirement in barley (Laurie et al., 1995; Dubcovsky et al., 1998).

For days to heading, a few major locations have been repeatedly identified in the above-mentioned populations. Quantitative trait loci have been identified on chromosomes 7H (bin 2-3) (Baum et al., 2003), 3H (bin 12) (Baum et al., 2003; Teulat et al., 2001), 2H (bin 3-7) (*Ppd-H1,* Baum et al., 2003) 2H (bin 11-13) (Baum et al., 2003; Teulat et al., 2001), and 5H (bin 6) (Baum et al., 2003). Under better-than-normal rainfall conditions, the location on 3H (bin 12), 2H (bin 3-7) (*Ppd-H1*), and 2H (bin 13) were more important in the Arta/*H. spontaneum* 41-1 population, whereas under drier conditions, the 7H (bin 3) location was more important (Baum et al., 2003). Here *H. spontaneum* 41-1 contributed the drought escape allele. In the Tadmor/Er/Apm population, 2H (bin 13), 3H (bin 12), 5H (bin 11), and 7H (bin 9) are the important loci. Forster et al. (2004a) pointed out that for the Tadmor// Er/Apm populations, flowering time variation in the RILs under non-vernalization conditions might be due to the *Vrn-H2* locus on 4H, and this might account for a cluster of QTL in this region on chromosome 4H, even though no QTL for heading date was among the QTL. Tadmor, however, has vernalization requirements that might not always be met in North African sites.

One of the most important characters for the extreme drought conditions with annual rainfall less than 200 mm is plant height (under drought stress). Here, alleles from *H. spontaneum* 41-1 become very useful for increasing plant height. A few QTL with major effects have been identified by Baum et al. (2003) on chromosomes 2H (bin 13), 3H (bin 12), and 7H (bin 3). These might be more specific to *H. spontaneum* 41-1 background. In the Tadmor//Er/Apm population, major QTL for plant height were located on 2H (bin 13), 3H (bin 6), 4H (bin 13), and 6H (bin 10) (Teulat et al., 2001).

In the Arta/*H. spontaneum* 41-1 population, QTL for grain yield and biological yield have been identified and located at the *btr* locus on 3H (bin 3), and the *sdw1* (bin 12) locus in the part of the population with non-brittle lines. In both locations, pleiotropic effects for several other characters were identified. Other grain yield QTL were identified on 4H (bin 13), 5H (bin 11), and 7H (bin 3) (Baum et al., 2003). Teulat et al., 2001) identified QTL for grain yield on chromosome 4H (bin 13) and on chromosome 7H (bin 9).

Cold and frost tolerance is a necessary requirement for barley varieties grown in the Fertile Crescent. In the Arta/*H. spontaneum* 41-1 population, eight QTL for cold damage were identified. For the QTL with minor effects (2H, 4H, and 6H), the allele from the *H. spontaneum* line showed the better protection. For all other QTL, the allele of 'Arta' showed better ability to protect the plant from cold damage. Three of the four alleles with a large effect on cold damage were localized on chromosome 5H at the known locations for frost tolerance QTL (Pan et al., 1994; Reinheimer et al., 2004; Francia et al., 2004). Additionally, a QTL on 7H was identified. This is in agreement with the observation of Karsai et al. (2004), whose survey of *spontaneum* accessions of Middle East origin showed some degree of vernalization response.

Numerous QTL studies have been conducted using balanced populations. The QTL studies with and without *H. spontaneum* have shown that developmental genes, notably those involved in flowering time and plant stature, showed pleiotropic effects on abiotic stress resistance and ultimately determined yield. Important loci in this regard are the semi-dwarf gene *sdw3* (bin 7 on 2H), and the flowering loci *Ppd-H1* (bin 4 on 2H) and *Ppd-H2* (bin 10–12 on 1H), the vernalization requirement locus *Vrn-H2* (bin 12-13 on 4H), the brittle rachis locus *btr1* (bin 3 on 3H), and the 2-row/6-row locus *vrs* (bin 10 on 2H). To better understand which genes are involved in drought and other abiotic stresses in the dry areas, two approaches seem logical. The generation of introgression lines (mainly used in relation to introgressions of *H. spontaneum* into *H. vulgare*) and their evaluation in dry areas would give more insight into the effect of certain favorable alleles. Secondly, a larger population needs to be tested to be able to identify confounding effects through developmental traits.

Marker-assisted selection

Genes and QTL related to agronomic traits and associated to both biotic and abiotic stress tolerance have been identified in barley and tagged using molecular markers. Some of these tagged markers are sufficiently close and robust to allow MAS in different genetic backgrounds. Marker-assisted selection allows indirect selection of plants with positive alleles in the seedling stage. The selected plants can then be used for the next crossing cycle, e.g., in marker-assisted backcrossing schemes. The breeding process becomes more precise and faster. Until now, mainly biotic stresses have been tagged and are targets for MAS. Despite the availability of numerous tagged markers, only a limited number are currently being used in different breeding programs.

Many of the known genes for scald resistance have proven to be alleles at a complex locus on the long arm of chromosome 3H (*Rrs1;* Dyck and Schaller, 1961; Barua et al., 1993; Graner and Tekauz, 1996; Genger et al., 2003). Others have been mapped to the short arm of chromosome 7H, (*Rrs2;* Schweizer et al., 1995) and to chromosome 4H (*Rrs3;* Bockelman et al., 1977). Many of the resistant lines in the Mediterranean region carry the chromosome 3H resistance and additional resistances on other chromosomes (Sayed et al., 2004). The chromosome 3H resistance can be selected with closely linked SSR markers. Closely linked AFLP and SCAR markers are also available (Sayed et al., 2004; Genger et al., 2003). A number of markers are available to select for the *mlo* resistance (Büschges et al., 1997). The semi-dominant *Mlg* resistance locus on barley chromosome 4H (Görg et al., 1993) can be selected with a number of CAPS markers (Kurth et al., 2001) or AFLP markers (Sayed et al., 2004). A number of SCAR markers are available to select for the *Yd2* gene controlling BYDV (Jefferies et al., 2003). In addition to *Yd2*, several QTL for tolerance to BYDV have been mapped in some cultivars. The *Yd3* gene on chromosome 6H (Niks et al., 2004) may be another important resistance gene against BYDV. The *Yd3* co-segregates with several PCR-based molecular markers that may be used for MAS.

Examples of successful MAS can also be found for leaf rust (Brunner et al., 2000), stripe rust (Toojinda et al., 2000), cereal cyst nematode resistance (Kretschmer et al., 1997; Jefferies et al., 2003), boron toxicity, and malting quality and others.

Conclusion

The SSR markers are the most common markers currently being employed in barley genetic diversity analysis, genome mapping, and molecular breeding. Numerous linkage maps have been developed and QTL mapping studies have been performed to locate genes and QTL. It is expected that SSRs will continue to be used for that purpose for the foreseeable future. However, SNP markers will most likely follow as the next important marker technology. A number of individual marker assays are being developed, but full genome scan technologies will also become available.

Numerous QTL studies for agronomic performance have been conducted in favorable environments using balanced populations. Although drought tolerance has been recognized as one of the prime breeding goals to increase food production in arid areas, only a few QTL studies are available that analyzed agronomic performance in drought-stressed environments. The QTL studies with and without *H. spontaneum* have shown that developmental genes, notably those involved in flowering time and plant stature, show pleiotropic effects on abiotic stress resistance and ultimately determine yield. Important loci in this regard are the semi-dwarf gene *sdw3* (bin 7 on 2H), the flowering locus *Ppd-H1* (2H, bin 4) and the vernalization requirement locus *Vrn-H2* (4H, bin 13). It was shown that *H. spontaneum* offers novel allelic variance at these loci. These alleles from wild barley can be exploited for the improvement of drought tolerance in elite germplasm by selecting those exotic alleles with the most beneficial pleiotropic effects on drought resistance.

The aim of many QTL mapping studies is the development of marker-assisted selection assays. After validation of marker-trait association in different genetic backgrounds, such markers can be used for indirect selection. Numerous markers for MAS—especially for major genes—are already being used in barley breeding programs. Markers for QTL and minor genes are still being developed and will follow.

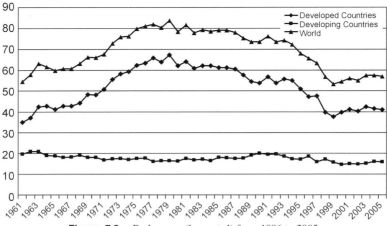

Figure. 7.2.   Barley area (harvested) from 1996 to 2005.

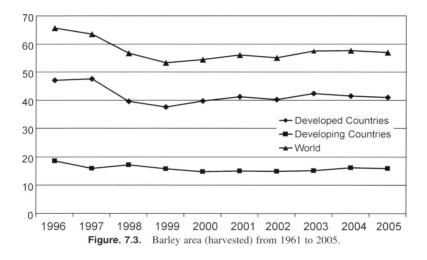

Figure. 7.3.   Barley area (harvested) from 1961 to 2005.

Prospects

As shown in Figure 7.2, global barley area reached a maximum at the end of the 1970s and the beginning of the 1980s, with more than 80 million ha. The area had decreased to a little more than 50 million ha in 1999 and 2000, with a similar pattern of reduction in both developed and developing countries. During the last five years, there has been an increase of 2.5 million ha in the total barley area (Fig 7.3), of which 1.3 million is in developed and 1.2 million is in developing countries.

We anticipate that this trend will probably continue in both developed and developing countries for a number of reasons. Barley continues to be an important feed for livestock, but it is increasingly becoming more popular as human food due to the range of health benefits it offers. Another reason for anticipating a rebound in the area planted to barley is its high level of adaptation to unfavorable conditions and low-input farming systems. If, as predicted, the amount of water available for agricultural uses should decrease or become more expensive, crops that are more "water productive" (a term that seems to be supplanting "drought resistant"), such as barley, are likely to become more important. This is already happening in a number of countries in the areas that are between the typical adaptation zone of wheat and barley.

Eventually, the renowned adaptation of barley to stressful conditions and the health benefits that it offers will make barley an ideal crop for organic farming.

Barley is one of the oldest—if not the oldest—crops, and has been a model plant in classical genetics and in breeding. Despite its relative large genome size (compared to rice), with the advent of molecular genetics, barley is becoming one of the crops more often used in molecular studies because of its very good adaptation to abiotic stresses and particularly to drought.

References

Åberg, E. 1938. *Hordeum agriocrithon nova* sp., a wild six-rowed barley. Ann Agric Coll Sweden 6:159–212.

Åberg, E. 1940. The taxonomy and phylogeny of *Hordeum* L. Sect. Cerealia Ands. Symbolae Bot Upsalienses 4:1–156.

Almekinders, C.J.M., Louwaars, N.P. and de Bruijn, G.H. 1994. Local seed systems and their importance for an improved seed supply in developing countries. Euphytica 78:207–216.

Amri, A., Ouammou, L. and Nassif, F. 2005. Barley-Based Food in Southern Morocco. In *Food Barley: Importance, Uses, and Local Knowledge* (S. Grando and H. Gomez Macpherson, eds.), ICARDA, Aleppo, Syria.

Anderson, J.R. and Lübberstedt, T. 2003. Functional markers in plants. Trends Plant Sci. 8:554–560.

Andersen, M.K. and Reinbergs, E. 1985. Barley breeding. pp. 231–268 In *Barley* (D.C. Rasmusson ed.), Agronomy Monograph No. 26, American Society for Agronomy, Madison, Wisconsin, USA.

Backes, G., Graner, A., Foroughi-Wehr, B., Fischbeck, G., Wenzel, G. and Jahoor, A. 1995. Localization of quantitative trait loci (QTL) for agronomic important characters by the use of a RFLP map in barley (*Hordeum vulgare* L.). Theor. Appl. Genet. 90:294–302.

Bakhteev, F.C. 1960. Systematics of cultivated barleys (in Russian). Publ. Acad. Sci. SSSR, Moscow.

Barua, U.M., Chalmers, K.J., Thomas, W.T.B., Hackett, C.A., Lea, V., Jack, P., Forster, B.P., Waugh, R. and Powel, W. 1993. Molecular mapping of genes determining height, time to heading, and growth habit in barley (*Hordeum vulgare*). Genome 36:1080–1087.

Baum, M., Grando, S., Backes, G., Jahoor, A., Sabbagh, A., and Ceccarelli, S. 2003. QTLs for agronomic traits in the Mediterranean environment identified in recombinant inbred lines of the cross Arta × *H. spontaneum* 41–1. Theor. Appl. Genet. 107:1215–1225.

Bekele, B. 1983. A differential rate of regional distribution of barley flavonoid pattern in Ethiopia, and a view on the center of origin of barley. Hereditas 98:269–280.

Bekele, B., Alemayehu, F. and Lakew, B. 2005. Food barley in Ethiopia. In *Food Barley: Importance, Uses, and Local Knowledge* (S. Grando and H. Gomez Macpherson, eds.), ICARDA, Aleppo, Syria, pp. 53–82.

Belhadafa, H., Bentasil, A., Chafaai, E.A. and Mekkaoui, M. 1992. Evolution de la production et de la consummation de trois principales céréales au Maroc, au cours des dernié

Bockelman, H.E., Sharp, E.L. and Eslick, R.F. 1977. Trisomic analysis of several scald and net blotch resistance genes. Barley Genet. Newsl. 7:11–15.

Bothmer, R. von. 1992a. The wild species of *Hordeum*: Relationships and potential use for improvement of cultivated barley In *Barley: Genetics, Biochemistry, Molecular Biology and Biotechnology* (P. Shewry ed.) Biotechnology in Agriculture No. 5, CAB International, Wallingford, UK, pp. 3–18.

Bothmer, R. von. 1992b. The genepool of barley and preservation of wild species of *Hordeum*. pp. 32–35 In *International Crop Network Series no. 9*. International Board for Plant Genetic Resources, Rome, Italy.

Bothmer, R. von, Jacobsen, N., Baden, C., Jørgensen, R.B. and Linde-Laursen, I. 1991. An Ecogeographical Study of the Genus *Hordeum*. Systematic and Ecogeographic Studies on Crop Genepools no. 7. Internationational Board for Plant Genetic Resources, Rome, Italy.

Brennan, C.S. and Cleary, L.J. 2005. The potential use of cereal (1→3, 1→4)—β-D-glucans as functional food ingredients. Journal of Cereal Science, 42:1–13.

ßres cinquante annés. Internal paper. Direction de Production Végétale ou Ministère de l'Agriculture et la Réforme Agraire. Morocco.

Brunner, S., Keller, B. and Feuillet, C. 2000. Molecular mapping of the Rph7.g leaf rust resistance gene in barley (*Hordeum vulgare* L.). Theor. Appl. Genet. 101:783–788.

Brush, S.B. 1991. A farmer-based approach to conserving crop germplasm. Economic Botany 45:153–161.

Bundock, P.C. and Henry, R.J. 2004. Single nucleotide polymorphism, haplotype diversity and recombination in the isa gene of barley. Theor. Appl. Genet. 109:543–551.

Bundock, P.C., Cross, M.J., Shapter, F.M. and Henry, R.J. 2006. Robust allele-specific polymerase chain reaction markers developed for single nucleotide polymorphisms in expressed barley sequences. Theor. Appl. Genet. 112:358–365.

Burnham, C.R. and Hagberg, A. 1956. Cytogenetic notes on chromosomal interchanges in barley. Hereditas 42:467–482.

Büschges, R., Hollricher, K., Panstruga, R., Simons, G., Wolter, M. Frijters, A., Van Daelen, R., Van der Lee, T., Diergaarde, P., Groenendijk, J., Töpsch, S., Vos, P., Salamini, F. and Schulze-Lefert, P. 1997. The barley *mlo* gene: A novel control element of plant pathogen resistance. Cell 88:695–705.

Capettini, F., Rasmusson, D.C., Dill-Macky, R., Schiefelbein, E. and Elakkad, A. 2003. Inheritance of resistance to Fusarium head blight in four populations of barley. Crop Sci. 43(6):1960–1966.

Capettini, F. 2005. Barley in Latin America. In *Food Barley: Importance, Uses, and Local Knowledge* (S. Grando and H. Gomez Macpherson, eds.), ICARDA, Aleppo, Syria, pp. 121–126.

Ceccarelli, S. 1994. Specific adaptation and breeding for marginal conditions. Euphytica 77:205–219.

Ceccarelli, S. 1996. Adaptation to low/high input cultivation. Euphytica 92:203–214.

Ceccarelli, S. and Grando, S. 2007. Decentralized-participatory plant breeding: An example of demand driven research. Euphytica 155:349–360.

Ceccarelli, S., Grando, S. and Baum, M. 2007. Participatory Plant Breeding in Water-Limited Environment Experimental Agriculture (in press).

Ceccarelli, S. and Grando, S. 2005. Decentralized-participatory plant breeding. pp. 145–156 In Tuberosa R., Phillips R.L. and Gale M. (eds.), *Proceedings of the International Congress: In the Wake of the Double Helix: From the Green Revolution to the Gene Revolution*, 27–31 May 2003, Bologna, Italy.

Ceccarelli, S. and Grando, S. 2002. Plant breeding with farmers requires testing the assumptions of conventional plant breeding: Lessons from the ICARDA barley program. In *Farmers, Scientists and Plant Breeding: Integrating Knowledge and Practice* (Cleveland, David A. and Soleri, D. eds.) CABI Publishing International. Wallingford, Oxon, UK, pp. 297–332.

Ceccarelli, S., Erskine, W., Grando, S. and Hamblin, J. 1994. Genotype × environment interaction and international breeding programmes. Expl. Agric. 30:177–187.

Ceccarelli, S., Grando, S. and van Leur, J.A.G. 1987. Genetic diversity in barley landraces from Syria and Jordan. Euphytica 36:389–405.

Ceccarelli, S., Grando, S. and van Leur, J.A.G. 1995. Understanding landraces: The Fertile Crescent's barley provides lesson to plant breeders. Diversity 11:112–113.

Chapman C.G.D. 1987. Barley genetic resources: The status of collecting and conservation. pp. 43–49 In *Barley Genetics V, Proceedings of the Fifth International Barley Genetics Symposium,* Okayama, Japan.

Chen, F. and Hayes, P.M. 1989. A comparison of Hordeum bulbosum-mediated haploid production efficiency in barley using in vitro floret and tiller culture. Theor. Appl. Genet. 77:701–704.

Collins, N.C., Paltridge, N.G., Ford, C.M. and Symons, R.H. 1996. The Yd2 gene for barley yellow dwarf virus resistance maps close to the centromere on the long arm of barley chromosome 3. Theor. Appl. Genet. 92:858–864.

Collins, H.M., Panozzo, J.F., Logue, S.J., Jefferies, S.P. and Barr, A.R. 2003. Mapping and validation of chromosome regions associated with high malt extract in barley (*Hordeum vulgare* L.). Australian Journal of Agricultural Research 2003:1223–1240.

Coventry, S.J., Collins, H.M., Barr, A.R., Jefferies, S.P., Chalmers, K.J., Logue, S.J. and Langridge, P. 2003. Use of putative QTLs and structural genes in marker-assisted selection for diastatic power in malting barley (*Hordeum vulgare* L.). Australian Journal of Agricultural Research 20.

Diab, A.A., Teulat-Merah, B., This, D., Ozturk, N.Z., Benscher, D. and Sorrells, M.E. 2004. Identification of drought-inducible genes and differentially expressed sequence tags in barley. Theor. Appl. Genet. 109:1417–1425.

Dubcovsky, J., Lijavetzky, D., Appendino, L. and Tranquilli, G. 1998. Comparative RFLP mapping of Triticum monococcum genes controlling vernalization requirement. Theor. Appl. Genet. 97:968–975.

Dyck, P.L. and Schaller, C.W. 1961. Inheritance of resistance in barley to several physiologic races of the scald fungus. Can. J. Genet. Cytol. 3:153–164.

Ellis, R.H., Hadley, P., Roberts, E.H. and Summerfield, R.J. 1990. Quantitative relationship between temperature and crop development and growth. In *Climatic Change and Plant Genetic Resources* (Jackson, M., Ford-Lloyd, B.V. and Parry, M.L., eds.) Belhaven Press, London and New York, pp. 85–115.

Ellis, R.P., Forster, B.P., Robinson, D., Handley, L.L., Gordon, D.C., Russell, J.R. and Powell, W. 2000. Wild barley: A source of genes for crop improvement in the 21st century? J. Exp. Bot. 51:9–17.

Ellis, R.P., Forster, B.P., Gordon, D.C., Handley, L.L., Keith, R.P., Lawrence, P., Meyer, R., Powell, W.,

Robinson, D., Scrimgeour, C.M., Young, G. and Thomas, W.T. 2002. Phenotype/genotype associations for yield and salt tolerance in a barley mapping population segregating for two dwarfing genes. J. Exp. Bot. 53:1163–1176.

Ellis, R.H., Roberts, E.H., Summerfield, R.J. and Cooper, J.P. 1988. Environmental control of flowering in barley (*Hordeum vulgare* L.). II. Rate of development as a function of temperature and photoperiod and its modification by low-temperature vernalization. Annals of Botany 62:145–158.

Ellis, R.H., Summerfield, R.J., Roberts, E.H. and Cooper, J.P. 1989. Environmental control of flowering in barley. (*Hordeum vulgare* L.). III. Analysis of potential vernalization responses, and methods of screening germplasm for sensitivity to photoperiod and temperature. Annals of Botany 63:687–704.

FAO. 1997. The State of the World's Plant Genetic Resources for Food and Agriculture. FAO, Rome, Italy.

FAOSTAT data 2005. Available at: http://faostat.fao.org/faostat/agriculture.

Forster, B.P., Ellis, R.P., Moir, J., Talamé, V., Sanguineti, M.C., Tuberosa, R., This, D., Teulat-Merah, B., Ahmed, I., Mariy, S.A.E.E., Bahri, H., El Ouahabi, M., Zoumarou-Wallis, N., El-Fellah, M. and Ben Salem, M. 2004a. Genotype and phenotype associations with drought tolerance in barley tested in North Africa. Annals of Applied Biology 144:157–168.

Forster, B.P., Ellis, R.P., Thomas, W.T.B., Newton, A.C., Tuberosa, R., This, D., El-Enein, R.A., Bahri, M.H., Ben Salem, M. 2004b. the development and application of molecular markers for abiotic stress tolerance in barley. J. Exp. Bot. 51:19–27.

Fox, G.P., Panozzo, J.F., Li, C.D., Lance, R.C.M., Inkerman, P.A. and Henry, R.J. 2003. Molecular basis of barley quality. Australian Journal of Agricultural Research, 2003:1081–1101.

Francia, E., Rizza, F., Cattivelli, L., Stanca, A.M., Galiba, G., Tóth, B., Hayes, P.M., Skinner, J.S. and Pecchioni, N. 2004. Two loci on chromosome 5H determine low-temperature tolerance in a 'Nure' (winter) × 'Tremois'(spring) barley map. Theor. Appl. Genet. 108:670–680.

Francis, D.G. 1995. Agricultural structure and sustainability: Family production around the world. Journal of Sustainable Agriculture 6:5–15.

Freialdenhoven, A., Peterhänsel, C., Kurth, J., Kreuzaler, F. and Schulze-Lefert, P. 1996. Identification of genes required for the function of non-race-specific *mlo* resistance to powdery mildew in barley. Plant Cell 8:5–14.

Gebhardt, D.J., Rasmusson, D.C. and Wilcoxson, R.D. 1992. Cyclic breeding used to incorporate kernel discoloration resistance into malting barley. Crop. Sci. 32:352–356.

Genger, R.K., Brown, A.H.D., Knogge, W., Nesbitt, K. and Burdon, J.J. 2003. Development of SCAR markers linked to a scald resistance gene derived from wild barley. Euphytica 134:149–159.

Goodchild, A.V. 1997. Effects of rainfall and temperature on the feeding value of barley straw in a semi-arid Mediterranean environment. Journal of Agricultural Science, Cambridge 129:353–366.

Görg, R., Hollricher, K. and Schulze-Lefert, P. 1993. Functional analysis and RFLP-mediated mapping of the Mlg resistance locus in barley. Plant Jour. 3:857–866.

Grando, S., von Bothmer, R. and Ceccarelli, S. 2001. Genetic diversity of barley: Use of locally adapted germplasm to enhance yield and yield stability of barley in dry areas. In *Broadening the Genetic Base of Crop Production* (Cooper, H.D., Spillane, C. and Hodgink, T. eds.). CABI, New York (USA)/FAO, Rome (Italy)/IPRI, Rome (Italy), pp. 351–372.

Grando, S. and Gomez Macpherson, H. (eds.) 2005. Food Barley: Importance, Uses and Local Knowledge. pp. 14–17 In *Proceedings: International Workshop on Food Barley Improvement* Jan 2002, Hammamet (Tunisia) ICARDA, Aleppo, Syria.

Grando, S., Baum, M., Ceccarelli, S., Goodchild, A., Jaby El-Haramein, F., Jahoor, A. and Backes, G. 2005. QTLs for straw quality characteristics identified in recombinant inbred lines of a *Hordeum vulgare* × *H. spontaneum* cross in a Mediterranean environment. Theor. Appl. Genet. 110:688–695.

Graner, A., Jahoor, A., Schondelmaier, J., Siedler, H., Pillen, K., Fischbeck, G., Wenzel, G. and Herrmann, R.G. 1991. Construction of an RFLP map of barley. Theor. Appl. Genet. 83:250–256.

Graner, A. and Tekauz, A. 1996. RFLP mapping in barley of a dominant gene conferring to scald (*Rhynchosporium secalis*). Theor. Appl. Genet. 93:421–425.

Grisley, W. 1993. Seed for bean production in sub-Saharan Africa, issues, problems, and possible solutions. Agricultural Systems 43:19–33.

Harlan, H.V. and Martini, M.L. 1929. A composite hybrid. Am. Soc. Agron. J. 21:487–490.

Harlan, H.V., Martini, M.L. and Stevens, H.A. 1940. A study of methods in barley breeding. USDA Tech. Bull. 720:1–26.

Harlan, J.R. 1991. On the origin of barley: A second look. Barley Genetics II:45–50.

Harlan, J.R. 1992. *Crops & Man*. Second edition. American Society of Agronomy, Crop Science Society of America, Madison, Wisconsin, USA.

Harlan, J.R. and de Wett, J.M.J. 1971. Towards a rational classification of cultivated plants. Taxon 20:509–517.

Haugerud, A. and Collinson, M.P. 1990. Plants, genes and people: Improving the relevance of ant breeding in Africa. Expl. Agric. 26:341–362.

Hockett, E.A. and Nilan, R.A. 1985. Genetics. pp. 187–230 In *Barley* (D.C. Rasmusson ed.),

Agronomy Monograph No. 26, American Society for Agronomy, Madison, Wisconsin, USA.

Inagaki, M.N., Al-Ek, W. and Tahir, M. 1991. A comparison of haploid production frequencies in barley crossed with maize and *Hordeum bulbosum* L. Cereal Research Comm. 19:385–390.

Jadhav, S.J., Lutz, S.E., Ghorpade, V.M. and Salunkhe, D.K. 1998. Barley: Chemistry and value–added processing. Critical Review in Food Science 38:123–171.

Jähne, A., Lazzeri, P.A., Jaeger-Gussen, M. and Lörz, H. 1991. Plant regeneration from embryogenic cell suspensions derived from anther cultures of barley (*Hordeum vulgare* L.). Theor. Appl. Genet. 82:74–80.

Jähne, A. and Lörz, H. 1995. Cereal microspore culture. Plant Science 109:1–12.

Jefferies, S.P., King, B.J., Barr, A.R., Warner, P., Logue, S.J. and Langridge, P. 2003. Marker-assisted backcross introgression of the Yd2 gene conferring resistance to barley yellow dwarf virus in barley. Plant. Breed. 122:52–56.

Jørgensen, J.H. 1994. Genetics of powdery mildew resistance in barley. Crit. Rev. Plant Sci. 13:97–119.

Kanazin, V., Talbert, H., See, D., DeCamp, P., Nevo, E. and Blake, T. 2002. Discovery and assay of single-nucleotide polymorphisms in barley (*Hordeum vulgare*). Plant Mol. Biol. 48:529–537.

Kantey, R.V., La Rota, M., Matthews, D.E., Sorrells, M.E. 2002. Data mining for simple sequence repeats in expressed sequence tags from barley, maize, rice, sorghum and wheat. Plant Mol. Biol. 48:501–510.

Karsai, I., Hayes, P.M., Kling, J., Matus, I.A., Meszaros, K., Lang, L., Bedo, Z. and Sato, H. 2004. Genetic variation in component traits of heading date in *Hordeum vulgare* subsp. *spontaneum* accessions characterized in controlled environments. Crop Sci. 44:1622–1632.

Kota, R., Varshney, R.K., Thiel, T., Dehmer, K.J. and Graner, A. 2001. Generation and comparison of EST-derived SSRs and SNPs in barley (*Hordeum vulgare* L.). Hereditas 135:145–151.

Kretschmer, J.M., Chalmers, J., Manning, S., Karakousis, A., Barr, A., Islam, K.M.R., Logue, S.J., Choe, Y.W., Barker, S.J., Lance, R.C.M. and Langridge, P. 1997. RFLP mapping of the Ha 2 cereal cyst nematode resistance gene in barley. Theor. Appl. Genet. 94:1060–1064.

Kurth, J., Kolsch, R., Simons, V. and Schulze-Lefert, P. 2001. A high resolution genetic map and a diagnostic RFLP marker for the Mlg resistance locus to powdery mildew in barley. Theor. Appl. Genet. 102:53–60.

Laurie, D.A., Pratchett, N., Bezant, J.H. and Snape, J.W. 1995. RFLP mapping of 5 major genes and 8 quantitative trait loci controlling flowering time in a winter × spring barley (*Hordeum vulgare* L.) cross. Genome 38:575–585.

Leur, J.A.G. van, Ceccarelli, S. and Grando, S. 1989. Diversity for disease resistance in barley landraces from Syria and Jordan. Plant Breeding 103:324–335.

Liu, Z.W., Biyashev, R.M. and Saghai Maroof, M.A. 1996. Development of simple sequence repeat DNA markers and their integration into a barley linkage map. Theor. Appl. Genet. 93:869–876.

Liu, X.M., Smith, C.M. and Gill, B.S. 2002. Identification of microsatellite markers linked to Russian wheat aphid resistance genes Dn4 and Dn6. Theor. Appl. Genet. 104:1042–1048.

Lundqvist, U., Franckowiak, J.D. and Konishi, T. 1997. New and revised descriptions of barley genes. Barley Genetic Newsl. 26:22–252.

Lutf Saeed, A. 2005. Barley Production and Research in Yemen. In *Food Barley: Importance, Uses, and Local Knowledge* (S. Grando and H. Gomez Macpherson, eds.), ICARDA, Aleppo, Syria.

Maestri, E., Malcevschi, A., Massari, A. and Marmiroli, N. 2002. Genomic analysis of cultivated barley (*Hordeum vulgare*) using sequence-tagged molecular markers. Estimates of divergence based on RFLP and PCR markers derived from stress-responsive genes, and simple-sequence repeats (SSRs). Mol. Genet. Genomics 267:186–201.

Martin, G.B. and Adams, M.W. 1987. Landraces of Phaseolus vulgaris (Fabaceae) in Northern Malawi. II. Generation and maintenance of variability. Economic Botany 41:204–215.

Matus I.A. and Hayes P.M. 2002. Genetic diversity in three groups of barley germplasm assessed by simple sequence repeats. Genome 45:1095–1106.

Matus, I., Corey, A., Filchkin, T., Hayes, P.M., Vales, M.I., Kling, J., Riera-Lizarazu, O., Sato, K., Powell, W. and Waugh, R. 2003. Development and characterization of recombinant chromosome substitution lines (RCSLs) using *Hordeum vulgare* subsp. *spontaneum* as a source of donor alleles in a *Hordeum vulgare* subsp. *vulgare* background. Genome 46:1010–1023.

Molina-Cano, J.L., Fra-Mon, P., Salcedo, G., Aragoncillo, C., Roca de Togores, F. and Garcia-Olmedo, F. 1987. Morocco as a possible domestication center for barley: biochemical and agromorphological evidence. Theor. Appl. Genet. 73:531–536.

Molina-Cano, J.L., Russell, J.R., Moraleja, M.A., Escacena, J.L., Arias, G. and Powell, W. 2005. Chloroplast DNA microsatellite analysis supports a polyphyletic origin for barley. Theor. Appl. Genet. 110:613–619.

Morgante, M., Hanafey, M. and Powell, W. 2002. Microsatellites are preferentially associated with non repetitive DNA in plant genomes. Nat. Genet. 30:194–200.

Mornhinweg, D.W., Porter, D.R. and Webster, J.A. 2002. Inheritance of Russian Wheat Aphid Resistance in Spring Barley Germplasm Line STARS-9577B. Crop Sci. 42:1891–1893.

Murphy, P.J. and Witcombe, J.R. 1981. Variation in Himalayan barley and the concept of centers of diversity. pp. 26–36 In *Barley Genetics IV* (M.J.C. Asher, R.P. Ellis and A.M. Hayter eds.) Edinburgh, UK.

Negassa, M. 1985. Patterns of phenotypic diversity in an Ethiopean barley collection, and the Arussi-Bale Highland as a centre of origin of barley. Hereditas 102:139–150.

Newman, R.K. 1985. A new market for barley: Food products. Montana AgResearch 2 : 25.

Niks, N.E., Habekuß, A., Bekele, B. and Ordon, F. 2004. A novel major gene on chromosome 6H for resistance of barley against the barley yellow dwarf virus. Theor. Appl. Genet. 109:1536–1543.

Pan, A., Hayes, P.M., Chen, F., Blake, T.H.H., Wright, T.K.S., Karsai, I. and Bedö, Z. 1994. Genetic analysis of the components of winter hardiness in barley (*Hordeum vulgare* L.). Theor. Appl. Genet. 89: 900–910.

Picard, E., Crambes, E., Liu, G.S. and Mihamou-Ziyyat, A. 1995. Evolution des méthodes d'haplodiploïdisation et perspectives pour l'amélioration des plantes. Comptes-Rendus de la Société de Biologie. C.R. Soc. Biol., 1994, 188:107–139.

Pickering, R.A. and Devaux, P. 1992. Haploid production: Approaches and use in plant breeding. pp 519–547. In *Biotechnology in Agriculture No. 5, Barley: Genetics, Biochemistry, Molecular Biology and Biotechnology* (P. Shewry ed.) CAB International, Wallingford, Oxon, UK.

Pickering, P., Johnston, P.A., and Ruge, B. 2004. Importance of secondary and tertiary genepools in barley genetics and breeding. pp. 227–234 In *Cytogenetics and Molecular Analysis: Proceedings of the 9th International Barley Genetics Symposium*, Brno, Czech Republic, 20–26 June 2004.

Pimbert, M.P. 1994. The need for another research paradigm. Seedling, 20–25.

Plucknett, D.L., Smith, N.J.H., Williams, J.T. and Murthi-Anishetty, N. 1987. *Genebanks and the World's Food.* Princeton University Press, Princeton, New Jersey, USA.

Poehlman, J.M. 1986. Adaptation and distribution. pp. 1–17 In *Barley* (D.C. Rasmusson ed.) Agronomy Monograph No. 26, American Society for Agronomy, Madison, Wisconsin, USA.

Powell, W., Morgante, M., Andre, C., Hanafey, M., Vogel, J., Tingey, S. and Rafalski, A. 1996. The comparison of RFLP, RAPD, AFLP and SSR (microsatellite) markers for germplasm analysis. Mol. Breed. 2:225–238.

Provan, J., Russell, J.R., Booth, A. and Powell, W. 1999. Polymorphic chloroplast simple sequence repeat primers for systematic and population studies in the genus *Hordeum*. Mol. Ecol. 8(3):505–511.

Olsen, F.L. 1991. Isolation and cultivation of embryogenic microspores from barley (*Hordeum vulgare* L.). Hereditas 115:255–266.

Rafalski, A. 2002. Application 11 of single nucleotide polymorphsims in crop genetics. Curr. Opin. Plant. Biol. 5:94–100.

Ramage, R.T. 1975. Techniques for producing hybrid barley. Barley Newsl. 18:62–65.

Ramsay, L., Macaulay, M., degli Ivanissevich, S., Maclean, K., Cardle, L., Fuller, J., Edwards, K.J., Tuvesson, S., Morgante, M., Massari, A., Maestri, E., Marmiroli, N., Sjakste, T., Ganal, M., Powell, W. and Waugh, R. 2000. a simple sequence repeat-based linkage map of barley. Genetics 156:1997–2005.

Reinheimer, J.L., Barr, A.R. and Eglinton, J.K. 2004. QTL mapping of chromosomal regions conferring reproductive frost tolerance in barley (*Hordeum vulgare* L). Theor. Appl. Genet. 109:1267–1274.

Roberts, E.H., Summerfield, R.J., Cooper, J.P. and Ellis, R.H. 1988. Environmental control of flowering in barley (*Hordeum vulgare* L.). I. Photoperiod limits to long-day responses, photoperiod-insensitive phases and effects of low-temperature and short-day vernalization. Annals of Botany 62:127–144.

Rostoks, N., Mudie, S., Cardle, L., Russell, J., Ramsay, L., Booth, A., Svensson, J.T., Wanamaker, S.I., Walia, H., Rodriguez, E.M., Hedley, P.E., Liu, H., Morris, J., Close, T.J., Marshall, D.F. and Waugh, R. 2005. Genome-wide SNP discovery and linkage analysis in barley based on genes responsive to abiotic stress. Mol. Gen. Genomics 274:515–527.

Russell, J.R., Fuller, J.D., Macaulay, M., Hatz, B.G., Jahoor, A., Powell, W. and Waugh, R. 1997. Direct comparison of levels of genetic variation among barley accessions detected by RFLPs, AFLPs, SSRs and RAPDs. Theor. Appl. Genet. 95:714–722.

Russell, J.R., Booth, A., Fuller, J.D., Baum, M., Ceccarelli, S., Grando, S. and Powell, W. 2003. Patterns of polymorphism detected in the chloroplast and nuclear genomes of barley landraces sampled from Syria and Jordan. Theor. Appl. Genet. 107:413–421.

Russell, J., Booth, A., Fuller, J., Harrower, B., Hedley, P., Machray, G. and Powell, W. 2004. A comparison of sequence-based polymorphism and haplotype content in transcribed and anonymous regions of the barley. Genome 47:389–398.

Sayed, H., Backes, G., Kayyal, H., Yahyaoui, A., Ceccarelli, S., Grando, S., Jahoor, A. and Baum, M. 2004. New molecular markers linked to qualitative and quantitative powdery mildew and scald resistance genes in barley for dry areas. Euphytica 135:225–228.

Schweizer, G.F., Baumer, M., Daniel, G., Rugel, H. and Röder. 1995. RFLP markers linked to scald (*Rhynchosporium secalis*) resistance gene *Rh2* in barley. Theor. Appl. Genet. 90:920–924.

Shao, Q. and Li, A. 1987. Unity of genetic population for wild barley and cultivated barley in Himalaya area. pp. 35–41 In *Barley Genetics V. Proceeding of the Fifth International Barley Genetics Symposium,* Okayama, Japan.

Sharp, E.L. 1985. Breeding for pest resistance. pp. 313–333 In *Barley* (D.C. Rasmusson ed.), Agronomy Monograph No. 26, American Society for Agronomy, Madison,Wisconsin, USA.

Singh, M., Malhotra, R.S., Ceccarelli, S., Sarker, A., Grando, S. and Erskine, W. 2003. Spatial variability models to improve dryland field trials. Exp. Agric. 39:1–10.

Suneson, C.A. and Stevens, H. 1953. Studies with bulked hybrid populations of barley. U.S. Dept. of Agriculture. Tech Bull. 1067, pp. 1–14.

Takahashi, R. 1955. The origin and evolution of cultivated barely. In: Advances in Genetics 7. (M. Demerec, ed.) Academic, New York, pp. 227–266.

Takahashi, R. and Yasuda, S. 1970. Genetics of earliness and growth habit of barley. *Barley Genetics II,* pp. 388–408.

Talamé, V., Sanguineti, M.C., Chiapparino, E., Bahri, H., Ben Salem, M., Forster, B.P., Ellis, R.P., Rhouma, S., Zoumarou, W., Waugh, R. and Tuberosa, R. 2004. Identification of *Hordeum spontaneum* QTL alleles improving field performance of barley grown under rainfed conditions. Annals of Applied Biology 144(3):309–319.

Tashi, N. 2005. Food Preparation from hulless barley in tibet. In *Food Barley: Importance, Uses, and Local Knowledge* (S. Grando and H. Gomez Macpherson, eds.), ICARDA, Aleppo, Syria.

Teulat, B., Merah, O., Souyris, I. and This, D. 2001. QTLs for agronomic traits from Mediterranean barley progeny grown in several environments. Theor. Appl. Genet. 103:774–787.

Teulat, B., Merah, O., Sirault, X., Borries, C., Waugh, R. and This, D. 2002. QTLs for grain carbon isotope discrimination in field-grown barley. Theor. Appl. Genet. 106:118–126.

Teulat, B., Zoumarou-Wallis, N., Rotter, B., Ben Salem, M., Bahri, H. and This, D. 2003. QTL for relative water content in field-grown barley and their stability across Mediterranean environments. Theor. Appl. Genet. 108:181–188.

Thomson, E.F. and Ceccarelli, S. 1990. Progress and future direction of applied research on cereal straw quality at ICARDA. pp. 249–264 In *Production and Utilization of Lignocellulosics: Plant Refinery and Breeding, Analysis, Feeding to Herbivores and Economic Aspects.* COST 84 Workshop, Bologna. Elsevier, London and New York.

Toojinda, T., Broers, L.H., Chen, X.M., Hayes, P.M., Kleinhofs, A., Korte, J., Kudrna, D., Leung, H, Line, R.F., Powell, W., Ramsay, L., Vivar, H. and Waugh, R. 2000. Mapping quantitative and qualitatitve resistance genes in a doubled haploid population of barley (*Hordeum vulgare*) Theor. Appl. Genet. 101:580–589.

Traxler, G. and Byerlee, D. 1993. A joint-product analysis of the adoption of modern cereal varieties in developing countries. American Journal of Agricultural Economics 75:981–989.

Varshney, R.K., Graner, A. and Sorrells, M.E. 2005. Genic microsatellite markers: Features and applications. Trends in Biotechnology 23:48–55.

Vos, P., Hogers, R., Bleeker, M., Reijans, M., Van de Lee, T., Hornes, M., Fritjers, A., Pot, J., Peleman, J., Kupier, M. and Zabeau, M. 1995. AFLP: A new technique for DNA fingerprinting. Nucl. Acids Res. 23:4407–4414.

Wang, L., Xue, Q., Newman, R.K. and Newman, C.W. 1993. Enrichment of tocopherol, tocotrienol, and oil in barley by milling and pearling. Cereal Chemistry 70(5):499–501.

Wei, F., Gobelman-Werner, K., Morroll, S.M., Kurth, J., Long, M., Wing, R., Leister, D., Schulze-Lefert, P. and Wise, R.P. 1999. The Mla (powdery mildew) resistance cluster is associated with three NBS-LRR gene families and suppressed recombination within a 300-kb DNA interval on chromosome 5S (1HS) of barley. Genetics 153:1929–1948.

Weltzien, E. 1988. Evaluation of barley (*Hordeum vulgare* L.) landraces populations originating from different growing regions in the Near East. Plant Breeding 101:95–106.

Weltzien, E. 1989. Differentiation among barley landrace populations from the Near East. Euphytica 43:29–39.

Weltzien, E. and Fischbeck, G. 1990. Performance and variability of local barley landraces in Near-Eastern environments. Plant Breeding 104:58–67.

Wiebe, G.A. 1979. Barley breeding. pp. 117–127. In *Barley,* U.S. Dept. of Agric. Agric. Handbook 338.

Williams, K.J. 2003. The molecular genetics of disease resistance in barley. Australian Journal of Agricultural Research, 54:1065–1079.

Williams, J.G., Kubelik, A.R., Livak, J., Rafalski, J.A. and Tingey, A.V. 1990. DNA polymorphisms amplified by arbitrary primers useful as genetic markers. Nucleic Acid Research 18:6531–6535.

Wolfe, M.S. 1991. Barley diseases: Maintaining the value of our varieties. Barley Genetics VI:1055–1067.

Xu, T.W. 1982. Origin and evolution of cultivated barley in China. Acta Geneti. Sin. 9:440–446.

Yan, W. 2001. GGEBiplot—A Windows application for graphical analysis of multi-environment trial data and other types of two-way data. Agron. J. 93:1111–1118.

Yasuda, S., Hayashi, J. and Moriya, I. 1993. Genetic constitution for spring growth habit and some other characters in barley cultivars in the Mediterranean coastal regions. Euphytica 70:77–83.

Yirga, C., Alemayehu, F. and Sinebo, W. (eds.) 1998. *Barley-Based Farming Systems in the Highlands of Ethiopia,* Ethiopian Agricultural Research Organization (EARO), Addis Abeba, Ethiopia.

Ziauddin, A., Simion, E. and Kasha, K.J. 1990. Improved plant regeneration from shed microspore culture in barley (*Hordeum vulgare* L.) cv. Igri. Plant Cell Rep. 9:69–72.

Zohary, D. 1959. Is *Hordeum agriocrithon* Åberg the ancestor of six-rowed barley? Evolution 13:279–280.

Zohary, D. and Hopf, M. 1988. Domestication of plants in the Old World. Clarendon Press, Oxford, UK.

Chapter 8

Corn Breeding in the Twenty-first Century

G. Richard Johnson

Introduction

Origin of corn

Corn or maize (*Zea mays* L.), belonging to the grass family Poaceae and tribe Maydeae, originated 5,000 to 10,000 years ago (Hallauer, 1997; Paliwal and Smith, 2002). Corn is the only cultivated species of its tribe and is believed to have originated in the mid-altitude regions of Mexico and Guatemala or Mesoamerica (Paliwal and Smith, 2002; Delgado-Salinas et al., 2004). Archaeobotanical records from Mesoamerican highland sites indicate pollen remains of corn from ca. 6,000 years ago (Matsuoka et al., 2002; Delgado-Salinas et al., 2004). Teosinte (*Zea mexicana*) is the closest wild relative of corn (Paliwal and Smith, 2002).

World corn production

The course of development that occurred in North America may not be relevant or appropriate for other parts of the world. Local environmental, economic, and political factors may lead to other paths of development. The North American experience is a model, however, of corn improvement through a breeding process that was, and is, quite successful. Although early systematic corn breeding began in the United States in the 20th century, corn has become an important crop worldwide. The potential of corn for food and feed was recognized around the globe, and it is now produced in almost all of the countries of the world. The U.S. routinely produces the most corn in the world. Table 8.1 contains information on corn area, total production, and average yield in 2005 for the top 47 corn-producing countries. The U.S. has been ahead of other countries in rapid adoption of hybrids. However, while the U.S. has the highest corn production, it does not have the highest productivity (Table 8.1). Thus, there is room for improvement in U.S. yield per unit of land.

Contributions of East, Shull, and Jones to hybrid corn breeding

Edward Murray East began his studies of inbreeding in about 1904 at the University of Illinois. At the time, he was responsible for the analysis of the kernel chemical composition of inbred strains that Cyril G. Hopkins was selecting for high protein and oil concentration by means of an ear-to-row pedigree method (Crabb, 1947). East's interest in inbreeding was motivated by his observations that the highest protein strains traced back to a single original ear, that the highest oil strains traced back to a very few ears, and that yields of all strains decreased as selection progressed. East hoped to discover a method by which high protein and oil concentrations could be achieved without reduction in yield. East continued his inbreeding studies when he moved to the

Table 8.1. Corn area, total production, and productivity (yield) in 2005 for top 47 countries.

Country	Area (×1000 ha)	Production (×1000 Mt)	Yield (Mt ha^{-1})
USA	30,395	282,260	9.29
China	26,200	135,000	5.15
Brazil	11,469	34,860	3.04
Mexico	8,000	20,500	2.56
Argentina	2,740	19,500	7.12
India	7,400	14,500	1.96
France	1,663	13,712	8.25
Indonesia	3,504	12,014	3.43
South Africa	3,343	11,996	3.59
Italy	1,119	10,510	9.39
Romania	2,662	9,965	3.74
Hungary	1,196	9,017	7.54
Canada	1,084	8,392	7.74
Ukraine	1,700	7,100	4.18
Egypt	840	6,800	8.10
Serbia & Montenegro	1,220	6,600	5.41
Philippines	2,500	5,250	2.10
Nigeria	4,466	4,779	1.07
Thailand	1,150	4,180	3.63
Germany	443	4,083	9.21
Spain	422	3,951	9.36
Pakistan	1,022	3,636	3.48
Vietnam	995	3,500	3.52
Turkey	800	3,500	4.38
Tanzania	2,000	3,230	1.62
Russia Federation	885	3,179	3.59
Ethiopia	1,400	2,740	1.96
Greece	247	2,534	10.26
Kenya	1,700	2,200	1.29
Croatia	410	2,100	5.12
Venezuela	600	2,050	3.42
Poland	343	1,917	5.59
Malawi	1,550	1,750	1.13
Nepal	850	1,716	2.02
Moldova	450	1,695	3.77
Austria	190	1,605	8.46
Korea, Dem People's Rep	500	1,600	3.20
Bulgaria	299	1,586	5.31
Iran	205	1,500	7.32
Chile	134	1,508	11.23
Mozambique	1,300	1,450	1.12
Colombia	606	1,442	2.38
Uganda	750	1,350	1.80
Peru	474	1,243	2.62
Ghana	733	1,158	1.56
Congo, Dem Rep of	1,483	1,155	0.78
Guatemala	603	1,072	1.78

Source: http://faostat.fao.org/site/567/DesktopDefault.aspx?PageID=567.

Connecticut Agricultural Experiment Station a few years later, producing near-homozygous lines by self pollinating in the open-pollinated variety 'Leaming.' East also observed the restoration of vigor and yield from cross-pollination between the inbreds. However, though he could not have failed to appreciate the power of inbreeding to isolate reproducibly stable traits and the restoration of hybrid vigor upon crossing, he apparently did not see that the joining of the two complementary processes, inbreeding and hybridization of inbreds, was the solution to his original problem. Illumination of the fundamental problem of corn breeding was left to George Harrison Shull.

Contemporary with East, Shull was conducting his own experiments on inbreeding and cross breeding at Cold Spring Harbor on Long Island. Shull realized that each plant in an open-pollinated variety was a unique hybrid and that an array of homozygous inbred lines derived by selfing out of the variety essentially represented a sample of duplicated gametes from the variety that were distinct, faithfully reproducible pure lines, and that, most importantly, crosses among the pure lines would produce distinct, faithfully reproducible, pure-line hybrids (Shull, 1908, 1909, 1910). East immediately agreed with Shull's conclusions, but demurred on their practicality, as did most all corn breeders of the time, because of the very low yields and general weaknesses of the then-current inbred lines.

Donald F. Jones solved the entire problem with the invention of the double-cross in 1917 (Crabb, 1947), setting the stage for the development of the hybrid seed industry. Jones realized that the double-cross was still a first-generation cross: just a mixture of four possible non-parental single crosses among four inbred lines. Jones' equally important contribution at the same time was the formulation of a Mendelian theory of hybrid vigor; viz., vigor of hybrids was due to a favorable combination of linked dominant genes from the two parents (Hayes, 1963).

First generation inbred lines

Soon after 1910, a number of investigators began to develop inbred lines by self-pollination and selection within open-pollinated varieties (Hayes, 1963). Among these early breeders were Henry A. Wallace in Iowa, F.D. Richey of the USDA, Lester Pfister, C.L. Gunn, Raymond Baker, and T.A. Kiesselbach. Crabb (1947) paid special attention to the efforts of James Holbert working in conjunction with the seedsman Eugene Funk at Bloomington, Illinois.

Selection and selfing to near-homozygosity was carried on by what was basically a pedigree method. However, in the end, selection was mainly just for survival (Baker, 1984), as the first generation inbreds suffered drastic inbreeding depression with regard to vigor and general health. Most of the surviving inbreds came out of widely grown open-pollinated varieties, such as Reid's Yellow Dent, Krug, Lancaster Sure-Crop, and Funk's 176A, which had been formed by seedsmen through blending various strains and propagating the mixtures under open-pollination with mass selection for vigorous plants and sound ears. Wallace and Brown (1988) give special credit to Robert and James Reid, George Krug, C.E. Troyer, and Isaac Hershey for the development of successful and widely grown open-pollinated varieties.

G.S. Carter produced the first corn hybrid, 'Burr-Leaming' double cross, in 1921, followed by H.A. Wallace's 'Copper Cross' hybrid that was sold in Iowa in 1924 (Troyer, 2004). During 1960–1980, single-cross hybrids replaced double-cross hybrids (Troyer, 2004). With the replacement of open-pollinated varieties with double-cross hybrids, between 1935 and 1965, average U.S. corn yields doubled. The average yields doubled again between 1965 and 1995 as a result of breeding efforts shifting from double-cross hybrids to single-cross hybrids (Hallauer, 2004).

The remarkable production increases that followed from adoption of hybrids derive from breeders' ability to isolate and reproduce superior hybrids identified after crossing homozygous inbred lines. The term "heterosis" was first coined to describe the restoration of hybrid vigor following crossing of inbred lines. Now, heterosis is usually a numeric value indicating the degree to which an F_1 hybrid is superior to either the mid-parent value or the high parent in a cross (Hallauer, 2004; Lamkey and Edwards, 2004). According to Lamkey and Edwards (2004), the terms "hybrid" and "heterosis" are not synonymous because it is possible to produce hybrids that exhibit no heterosis. However, without producing a hybrid, consideration of heterosis is contradictory to reason. Whether the genetic basis of heterosis is dominance or overdominance remains a controversial issue. Nevertheless, a heterotic response occurs whenever there is a difference in gene frequencies between the parental lines of a cross and some degree of dominance at one or more loci, as indicated by the following two-allele model equation (Kang, 1994): Heterosis = $(p - r)^2 \times au$, where p and r represent frequency of favorable allele in parent 1 and parent 2, respectively, and au represents dominance or genetic value of a heterozygote. In a specific cross between two homozygous lines, both p and r can take values of either 0 or 1. If p = r, no heterosis occurs. The equation highlights the fact that the degree of allelic diversity between the parents of a cross is positively correlated with the degree of heterosis observed.

Though heterosis might be a fascinating subject for some, it is not really fundamental to a corn breeder's objectives regarding yield. Corn is a naturally cross-pollinated species, and the goal of the breeder is to isolate and reproduce the best hybrid (Shull, 1908). This task is accomplished by producing homozygous inbred lines, crossing them, testing the hybrids, and selecting the best. Because dominance is involved, crossing of related lines is not apt to produce highly productive hybrids, and so in practice it is avoided. The point is that selection is directed to the productivity of the F_1 hybrids, not the differential productivity of the hybrid versus the parents. Breeders give little thought to heterosis. That hybrid vigor is restored upon crossing unrelated inbred lines is a pleasing phenomenon, but one that is hardly surprising. The dark side of heterosis, i.e., inbreeding depression, is the other side of the coin. The suffering of inbreeding depression in derivation of homozygous inbred lines necessary for the identification and reproduction of superior hybrids is an unavoidable price to pay. Breeders must, of course, select inbred lines that are economically viable in hybrid production. In doing so, however, the breeder gives little thought to inbreeding depression and simply selects for productive lines. In reality, selection pressure is overall much greater for hybrid productivity. Production personnel are usually relied upon to devise the means of producing the most productive hybrids, no matter the characteristics of the inbred parents. Heterosis and inbreeding depression are merely descriptive terms describing the relative productivities of hybrids and inbred parents, and are not emblematic of any distinct developmental or physiological process. The accelerated gain from selection accompanied by the shift from double to single crosses was due to the greater genetic variance afforded by single-cross populations. The concept of heterosis probably has more relevance to breeders who are attempting to convert naturally self-pollinated crops to hybrids. Here the degree of superiority of the F_1 hybrid to the parent lines is economically crucial. Hence, in this case, heterosis may well be an attribute of direct selection.

Beginning of the hybrid corn seed industry

Once a sizeable number of inbred lines had been extracted, selection among double-cross combinations commenced, marking the beginning of the hybrid seed corn industry. Both public and private agencies engaged in this activity, while the production, sale, and distribution of hybrid seed was carried on almost exclusively by the private sector. Henry A. Wallace founded the Pioneer Hi-Bred Company in 1926. The Dekalb Agricultural Association (Tom Roberts, Sr.), Pfister Hybrids (Lester Pfister), Funks Hybrids (Funk Brothers), and many other companies were started shortly thereafter. The first widespread sales of double-cross seed started in 1934 and quickly gained momentum. Within 10 years, nearly 80% of the corn acreage in the Midwest was planted to hybrids (Wallace and Brown, 1988).

At the time that hybrids were introduced to the farming community, another singular event occurred, which had a profound effect on future methods for line and hybrid improvement. George Sprague put together a synthetic variety comprised of 16 inbred lines that were chosen by various breeders for resistance to stalk breakage (Hallauer and Miranda, 1981). Sprague named the synthetic variety 'Iowa Stiff Stalk Synthetic.'

Breeding methods for line and hybrid improvement

From the beginning, pedigree selection has been the most widely used method for development of inbred lines for use as parents in hybrids (Hallauer et al., 1988). The general procedure is to start with F_2 or first-backcross populations and to combine selfing with visual selection for desirable individual plants, with no recombination among selections, the goal being to extract near-homozygous desirable inbred lines as

rapidly as possible so as to quickly initiate crossing and selection among crosses for hybrid performance. Other methods, briefly explored but quickly discarded were convergent improvement (Richey, 1927), featuring complementary backcrossing of two lines with selection for desirable features of each, and gamete selection (Stadler, 1944), essentially a method of sampling gametes from a heterogeneous source, such as an open-pollinated variety, in order to improve the cross-bred performance of an established line.

Missing from these schemes was any recombination among selections to gain a new, improved level of general performance. If Jones' hypothesis of hybrid vigor was correct, then recurrent selection and recombination should lead to continuous improvement of both inbred lines and hybrids. Hayes and Garber (1919) and East and Jones (1920) are given credit for being the first to use recurrent selection in breeding programs designed to raise grain protein content. Jenkins (1935) realized that the combining ability of a line was evident very early in the pedigree method inbreeding process, that it remained relatively stable as selfing continued down to near-homozygosity, and that the best combiners could be selected early in inbreeding, allowing for intense visual selection of the survivors for the remainder of the process. Jenkins (1940) then proposed an implementation of recurrent selection based on recombination of lines selected in early-generation test crosses. Jenkins' early observations concerning early-generation testing were later confirmed in a number of studies (Sprague, 1939, 1946; Jensen et al., 1983).

Though, by 1940, hybrid corn was a tremendous success, corn breeders were perplexed and uncertain about the direction to go next. Attempts to derive improved lines from current hybrids were unsuccessful. Moreover, an alternative hypothesis of hybrid vigor challenging Jones' explanation was receiving more consideration by breeders and geneticists. The alternate explanation was the over-dominance hypothesis (Crow, 1948). If the over-dominance hypothesis was indeed true, further continuous improvement of hybrid corn was doubtful. New hybrids could be produced by continued sampling of the open-pollinated populations, but attempts at raising the base performance of the breeding population by recurrent selection schemes, such as Jenkins', would be futile.

About the same time, Sprague and Tatum (1942) introduced the concepts of general and specific combining ability. General combining ability (the mean performance of a line averaged across all crosses) is determined primarily by additive genetic effects, which can be captured and retained by recombination of selected lines. On the other hand, specific combining ability is the deviation of the observed performance of specific single-cross hybrids from the average general combining abilities of the two parental lines. Upon recombination of selected lines, the specific combining ability deviations, comprised of a residue of various dominance and epistatic effects, are mostly dissipated and not passed on to the progeny. If Jones' hypothesis of hybrid vigor was correct, then gene action in hybrids would be comprised primarily of the additive effects of completely or partially dominant alleles, making Jenkins' recurrent selection scheme credible. However, many felt that dominance alone was insufficient to explain the degree of hybrid vigor in corn and that over-dominant gene action was required to account for the observed results.

If the over-dominance hypothesis held, then prospects for continued hybrid improvement were poor, and it was likely that productivity had reached a plateau. This eventuality was especially disheartening to many in the private sector because, if a plateau had been reached, then soon all hybrids would be more or less equal, and

hybrid seed would devolve into being a commodity item with no distinguishing levels of performance among hybrids through which the best could garner an increased market share (Roberts, Jr., 1977; Sprague, 1983a).

Fred Hull (1945) stepped into the breach to suggest a methodology that he called recurrent selection for specific combining ability. The method consisted of repeated test-cross evaluation of segregated individuals or lines on a homozygous tester coupled with recombination of selections. Progress would be made by improving a cross to a specific inbred line. Obviously though, the potential for long-term improvement with this approach would be limited. If recombination of superior selections failed to provide additional gain, then any improvement could come only from the resampling of a static population.

However, experiments initiated in the 1940s, principally at Iowa State and North Carolina State universities, began to show that additive gene action was predominant, justifying Jones' hypothesis and providing assurance that continued improvement of hybrid corn was possible. More direct proof was supplied by the immediate commercial success of the released inbreds B14 and B37, which were a product of recurrent selection for general combining ability within Stiff Stalk Synthetic.

The final development instrumental to the future of corn breeding occurred in the late 1940s when Comstock, Robinson, and Harvey (1949) outlined the reciprocal recurrent selection procedure. The authors devised the procedure to make maximum use of both general and specific combining ability. The procedure consisted of establishing two unrelated populations, test-crossing individuals, or derived lines, from each population to a number of individuals, or derived lines, in the other. Lines or individuals within each population were selected on the basis of the performance of their half-sib family progeny.

The selected lines or individuals were then intermated within each population. No matings among individuals or lines from different populations occured. The Mendelian authenticity of the procedure rested on the realization that dominance gene action to some degree was required for the expression of hybrid vigor, along with differences in gene frequencies between the two parental populations at the loci responsible, the populations being homozygous lines, synthetic varieties, open-pollinated varieties, or whatever. In effect, each population served as a female tester for lines or individuals in the other population. Selection on a half-sib basis would ensure that additive effects were being captured, assuring continued improvement of each population. If derived lines were used as female testers, recourse to remnant seed would provide the starting point for derivation of pure line hybrids from line pairs with large favorable specific effects.

Some years later, experimental evidence emerged from a study devised by Sprague that supported Jones' hypothesis and derogated the role of overdominance in hybrid vigor (Sprague, 1983b). In designing the experiment, Sprague borrowed from the outline of Hull's method. Two populations were selected and recombined for five generations for improved test-cross performance with a homozygous inbred tester. Both populations showed increased yield, and crosses between the two also produced increased yield in successive generations. Had overdominance been a decisive factor in selection of crosses to the inbred tester, gene frequencies in the two populations would have become increasingly similar as selection progressed, leading to inbreeding depression and decreasing yields when the two populations were crossed. The conclusion was that selection on additive effects of dominant or partially dominant genes was central in the improvement of the two populations.

Fisher (1949) provided a critical analysis of the factors making the hybrid corn breeding process a success. Fisher described a cycle of operations that included choice of parent stock, inbreeding to produce homozygous or near-homozygous lines, and crossing of chosen lines. Fisher noted that selection must take place within the cycle in order for progress to be achieved and argued that it would be most effective if applied to crosses of a large number of homozygous lines tested with great precision across environments using modern experimental designs. Fisher's conclusion remarkably predicted the actual course of the future development of corn breeding.

Theory and practice of corn breeding were now aligned to a notable degree. Shull had illuminated the fundamental problem of corn breeding. Comstock, Robinson, and Harvey had provided the conceptual framework for continued hybrid improvement. Fisher had provided additional theoretical support. The early success of B14 and B37 virtually fixed everyone's choice for one of the reciprocal recurrent populations: Stiff Stalk Synthetic. The opposite reciprocal population was anybody's choice of whatever combined well with Stiff Stalk Synthetic lines. Until this time, corn breeders had given little thought to breeding or "heterotic" groups. Acting on a suggestion in 1947 by Glenn Stringfield, the Committee on Grouping of Inbred Lines for Breeding Purposes of the North Central Corn Improvement Conference arbitrarily divided all existing public lines into two groups (Tracy and Chandler, 2006). The object of the division was ostensibly to maintain genetic diversity and avoid consanguinity among lines that might later be used in the production of hybrids. The division was entirely arbitrary and reflected no general recognition of diversity in origins between the two groups. Indeed, a recent molecular marker study of the original open-pollinated varieties available to the early breeders revealed no pre-existing relationships among the open-pollinated varieties that would have predicted the present population structure (LaBate et al., 2003; Duvick, 2004). Both breeding groups in the current division were drawn from the same pool, and breeders' decisions determined the grouping. The present breeding structure closely resembles the optimum reciprocal recurrent selection framework conducive to maximum long-term gain that was suggested by both Griffing (1963) and Cress (1967), based on theoretical grounds and computer simulation, respectively. Both authors advocated arbitrary division into two populations before initiation of selection. Drift would provide for an initial divergence in gene frequencies, and subsequent selection would drive these frequencies even farther apart. Initial diversity between the two groups is neither necessary nor desirable. Continued selection creates the subsequent diversity. Duvick (2004) also noted that the breeding system in place has operated over the years as virtually a closed system with very little infusion of outside germplasm. Yet progress has continued with no evidence of a plateau.

Industrialization of corn breeding

Corn breeding progressed from a mostly individualistic art and science into what now may fairly be called an industrial process, dominated by a few large corporations, supported and sustained by a sophisticated technological infra-structure. In 1930, there were thousands of corn seed companies; there are fewer than 300 today (Troyer, 2004).

Though industrialization took place during the second half of the 20th century, the stage was set by developments that occurred following the rediscovery of Mendel's laws of inheritance. By the 1960s, the stage was set for the industrialization of hybrid corn breeding. Shull, Comstock, Robinson, Harvey, and Fisher had provided the framework built on a sound Mendelian basis. We knew how to breed corn. All that was left to do, was to do it. Fisher, Haldane, and Wright had developed a mathematical basis for evolutionary theory that reconciled the results of natural selection with Mendelism. Since artificial selection is a manifest counterpart of natural selection, the mathematical theory lent an imprimatur of scientific orthodoxy to corn breeding, making breeders confident that they were on the right track. The genetic foundation for natural selection had provided a complete model for artificial selection.

The industrialization of corn breeding originated in the adoption of large scale, wide-area testing. In addition to plot-to-plot environmental variation, breeders were acutely aware that genotype × environment interaction contributed greatly to imprecision in estimation of breeding value and true hybrid genotypic value. Breeders had at least a very good intuitive appreciation of the law of large numbers: the variance of the sample mean is equal to the population variance divided by the sample size (Kendall and Stuart, 1977). Breeders realized that the only way to get at breeding or genotypic value was through observation of the phenotype, and that the only way to reduce the error component of the phenotype associated with genotype × environment interaction was to sample across as many locations and years as possible. In addition, it was realized that recombination of inbred lines developed under various environmental conditions and then re-evaluated across the original range of environments would lead to selection of lines with relatively stable performance across a broad range of environments, enhancing the chances for a concomitant stability of hybrids across an equally broad range of environments.

Extensive wide-area testing required specialized planting and harvesting equipment, which in turn required capital investment for development and specialized skills for operation. Increased testing required efficient data collection, processing, and distribution. The requirement for development of inbreds in diverse environments necessitated the institution of far-flung breeding stations and a management system to coordinate germplasm exchange among programs, testing, and sharing of winter nurseries. Standardized methods of writing pedigrees and nomenclature were required.

In addition to breeders, pathologists and entomologists were hired, not only to conduct breeding programs for resistance to specific pests, but also to produce and distribute inoculum for application in line development nurseries of less specialized breeders. Progressively, the breeding process was split up into increasingly specialized compartments. Specialization has now progressed to the point where few breeders participate in all phases of hybrid development from initiation of inbreeding through release of a hybrid. In most major companies, breeders have a role either as inbred line developers or as hybrid testers and evaluators, not both.

Virtually all programs practice early testing for combining ability at the S_2 or S_3 stage and develop lines by the pedigree method. The standard breeding system operates as an extended-cycle reciprocal recurrent selection program, with little recombination until the homozygous line stage after extensive testing in hybrids. Most visual selection during inbred development is for disease symptoms or insect damage following artificial inoculation or infestation, or for symptoms of stress induced by planting at extremely high plant densities. Progressively less attention is being devoted to visual selection. The emphasis is on using only the best F_1s, comprised of very elite lines that have been thoroughly tested in hybrid combinations, as breeding starts. The most intensive evaluation of inbred material is in hybrid combination. Some breeding programs now utilize doubled-haploids to advance lines to homozygosity as rapidly as possible so as to accelerate hybrid testing and development. Within this system, the vigor and productivity of inbred lines progressed very rapidly, to the point that by 1980, virtually all hybrids released were single crosses.

The system has proven to be quite efficient. Wide-area testing and selection for wide-adaptation provided accrual of economy-of-scale advantages in production, management of inventory, marketing, and distribution of seed. These advantages in turn induced greater selection pressure for even more widely adapted and stable hybrids, since any slip in stable performance of a relatively few widely marketed hybrids would be disastrous compared with errant performance of just a few small-volume hybrids. Thus, wide-area testing and selection for wide-area adaptation set up a dynamic with positive feedback in which the very success of the process of specialization enabling wide-area testing and industrialization induced even greater specialization and industrialization.

Agronomic, farming, economic, social, and political factors

The development of the hybrid corn industry did not take place in isolation, of course, but within the economic, social, and political milieu of the U.S. in the 20th century. Much of the research and education of the farming public was conducted by the land-grant colleges and universities founded in the preceding century. Hybrid corn was introduced on a wide scale in the midst of the Great Depression of the 1930s. Contributing substantially to the success of the introduction were policies of the Agricultural Adjustment Administration, such as price supports, that ensured fair market prices and profitable return on investment from purchase of hybrid seed (Wallace and Brown, 1988). From the 1930s onward, the U.S. government has maintained policies promoting an ensured supply of cheap and abundant food in high-protein diets, favoring development of a robust feed grain industry. Policies encouraging general economic and industrial growth further ensured the presence of a growing affluent population that could afford a diet high in dairy and animal products.

Policy and social change have affected the history of hybrid corn development, sometimes in surprising ways. For example, the uniformity of hybrid corn when first adopted by farmers spurred interest in investment in mechanical harvesting equipment. As corn harvesting became increasingly mechanized, greater and greater pressure was brought to bear in both inbred and hybrid development on selection for resistance to stalk and root lodging. Because of generally rising economic conditions following the depression and World War II, many rural inhabitants left the farms for better opportunities in the cities, which led in turn to increases in farm size. Increased farm size led to demand for bigger, faster, and more expensive harvesting implements, provoking a concomitant increase in selection pressure for standability. Thus, a dynamic positive feedback mechanism emerged with profound effects on corn breeding.

Another intriguing example of breeding affected by government policy is the wide-scale introduction and adoption of anhydrous ammonia (Arthur, 1960). With the outbreak of the Korean War in 1950, the Truman administration was faced with a dilemma. There was a distinct possibility that the Soviet Union, miscalculating that the U.S. would be too distracted by Korea to respond, might move to seize West Berlin, thus precipitating a two-front war for the U.S. The question put to Secretary of Agriculture Brannan was whether the U.S. farm economy could produce enough food to support and win a two-front war. Brannan's answer was that it could not, but that the solution lay in increasing feed-grain production. The question then became the ways and means by which corn (the principal feed-grain) could be adequately increased. Nitrogen, after water, was judged to be the limiting factor in corn production. Hence Brannan, ostensibly influenced by the advice of Roswell Garst (Garst, 1974), advocated policies promoting wide-scale production and distribution of anhydrous ammonia. These policies consisted of government-mandated priorities for construction materials and accelerated depreciation on plant and equipment. Natural gas is a primary ingredient in one process for the production of ammonia. At the time, natural gas was literally regarded as a waste by-product by the petroleum industry. Consequently, synthesis of anhydrous ammonia was unbelievably cheap. The product had been used to a limited but effective degree as a fertilizer in California. Brannan's policies were put in place, and an industry for the production, distribution, and application of anhydrous ammonia fertilizer quickly developed. The price of nitrogen delivered to the farmer was less than five cents per pound. Adoption by farmers was instantaneous. The Russians did not attempt to grab West Berlin, but nonetheless, after 1952, the Eisenhower administration continued the policy of the previous administration, and farmers continued applying nitrogen at increasingly higher rates. Higher rates of nitrogen application induced greater growth and taller plants that had an increased propensity to lodge, again putting more pressure on selection for standability. Concomitantly, selection pressure for shorter plants arose. Shorter, smaller plants permitted, and in some ways necessitated, higher planting rates and narrower rows. With higher planting rates came selection pressure against the negative effects of shading and moisture stress induced by increased numbers of plants per unit area, providing yet another example of a dynamic wherein a single cultural change induces a subsequent chain of new selection pressures.

Improved planting implements with more uniform depth placement and seed distribution have, along with herbicides and new tillage equipment, enabled adoption of minimum tillage and no-till in corn production. Minimum tillage and no-till facilitated earlier planting, which, in turn, induced selection pressure for higher germination and greater early vigor under cold conditions.

Other factors may vary, but two demands of growers have remained: higher yield and earlier maturity. The first demand is obvious; the second is a bit more subtle. Growers apparently are never completely at ease while the crop is still in the field. Nature is full of surprises, many of which are unpleasant. The longer the crop remains in the field, the greater the opportunity for an unanticipated event, more likely than not, to be bad. Customer preference for early harvest with little sacrifice in yield induced a steady dual selection pressure for both high yield and early maturity over the entire course of hybrid corn breeding.

A major factor contributing to the success of hybrid corn breeding was the adoption of experimental designs and statistical methods, such as the analysis of variance, developed by practitioners including R.A. Fisher and Frank Yates at the Rothamsted Experimental Station in Great Britain and disseminated in the U.S. by George Snedecor, William Cochran, Gertrude Cox, Oscar Kempthorne, and others. Before breeding experiments with adequate designs were instituted, many faulty estimates of breeding value were obtained due to intrinsic experimental bias and lack of precision in performance trials (Sprague, 1978).

Molecular markers

Just after Werner Aber, Daniel Nathans, and Hamilton O. Smith were awarded the Nobel Prize in Medicine or Physiology in 1978 for the discovery of restriction enzymes and their application to problems in molecular genetics, Botstein et al. (1980) suggested the construction of human genetic linkage maps using restriction fragment length polymorphisms as molecular markers for the purpose of mapping traits associated with disease to specific chromosomal sites. Shortly thereafter, plant breeders and geneticists were applying the same technology to map variation in quantitative traits (Stuber et al., 1987; Paterson et al., 1988). Advances in genomic research and marker technology have yielded newer mapping technologies or platforms based, for example, on simple sequence repeats or single nucleotide polymorphisms. No matter the marker platform utilized, however, markers have had four major uses: (1) selection for recurrent parent background genotype during backcross introgression of transgenes (Johnson and Mumm, 1996; Crosbie et al., 2006); (2) introgression of chromosome segments containing major genes with favorable effects but low to moderate penetration from exotic sources into adapted elite lines (Tanksley and Hewitt, 1988); (3) identification of chromosome segments containing candidate genes for positional cloning (Martin et al., 1993; Tanksley et al., 1995; Salvi et al., 2002); and (4) as correlated traits to aid in the improvement of quantitative traits through selection (Lande and Thompson, 1990). Application (1) has been widely used, applications (2) and (3) have had little or no impact, while (4) has been seriously pursued in some segments of the hybrid seed corn industry (Edwards and Johnson, 1994; Johnson and Mumm, 1996; Johnson, 2004; Eathington, 2006).

The use of markers as aids to selection is an adaptation of a well-known breeding technique (Falconer, 1960); viz., selection on a trait that is of no economic value but is highly heritable and genotypically highly correlated with a trait of economic value. Following the procedure of Lande and

Thompson (1990), in a population of individuals or lines subject to selection, a marker score for each individual or line evaluated is calculated by the regression of trait value on the number of allelic copies at each marker locus. The score is then combined in a selection index with the directly observed trait value to produce a predicted breeding value for each individual or line. The relative weight given to marker score in the index is a function of additive genetic variance attributable to marker score versus the heritability of the trait under selection. If heritability of the trait under selection is sufficiently low and/or the percent of additive genetic variance attributable to variation in marker score is sufficiently high, progress from selection via the index will be greater than selection on the trait alone. Dense coverage of the genome with informative markers will ensure that a substantial amount of the additive genetic variance will be correlated with variation among marker genotypes. Hence progress from selection on quantitative traits with low heritability may be substantially enhanced by including markers in the selection process. Moreover, selection on markers alone, in off-season nurseries, for example, can provide a benefit in reduction of cycle time with still some selection gain per cycle.

The paradigm used in a pair of proof of concept experiments (Johnson and Mumm, 1996) represented the studies by Sprague (1946) and Jensen et al. (1983) that validated the early-generation testing procedure. The primary trait under evaluation was grain yield. The design of both experiments featured estimation of breeding value of S_2 lines crossed to two testers as a predictor of S_4 line performance. The S_4 within S_2 generation lines were derived by direct selfing from an array of S_2 lines. Both the S_2 and S_4 lines were crossed to the same inbred line testers, and both the S_2 and S_4 lines were genotyped with numerous restriction length polymorphism markers scattered across the genome.

The test crosses were evaluated in yield trials in two consecutive years. Genotypic effects at each marker locus were calculated in the S_2 generation test-crosses. Significant marker effect estimates were then applied to marker genotypes of S_4 lines, and a marker score was calculated for each S_4 line. Regression analysis using the "extra sum of squares" principle (Draper and Smith, 1966) was used to analyze the data. In this type of analysis, the independent variables are entered sequentially into the model, and an F statistic is calculated at each step to test if the model with the current variable added accounts for a significant portion of variation in the dependent variable over and above that attributable to the model with only the preceding variables entered. In effect, the F statistic at each step tests the null hypothesis that the current variable has no effect that is independent of the preceding variables. The sums of squares extracted are often called type I sums of squares. Regression analysis of observed S_4 yield on S_2 progenitor yield and S_4 marker scores revealed that S_4 marker scores remained highly significant after accounting for S_2 progenitor yield. The reverse, however, was not true. This result held no matter if the S_2 yields and S_4 marker scores were taken from the same or different years. This clearly indicated that marker scores significantly enhanced predictive value of early-generation testcrosses. The same result was obtained for each experiment, and, in addition, by Eathington et al. (1997) shortly thereafter. Parallel to Jenkins' (1940) reasoning in his advocacy of recurrent selection due to his positive results with early-generation testing, it then followed that if markers were effective in identifying superior lines in early-generation testcrosses, then they would be valuable aids in recurrent selection. Subsequent selection experiments bore out this conclusion (Crosbie et al., 2006). Johnson and Mumm (1996) also described a marker-aided reciprocal recurrent selection system and presented results of increase in yield due to selection.

It was apparent, however, that to achieve success with marker-aided breeding, half-way measures were inadequate. Marker-aided breeding needed to be integrated with all phases of hybrid and inbred line development; or perhaps, to put it another way, line and hybrid development needed to be integrated with markers. The result was not really marker-aided breeding, but rather marker-directed breeding. Achieving the integration, however, required massive investment in genotyping capabilities, including tissue sampling, DNA extraction, robotics, automated marker scoring, and information management systems for collection, integration, analysis, and distribution of data from performance trials, nurseries, and laboratories. That the process could function and be managed by other than a large industrial corporation is inconceivable (Eathington 2, 2006). Successful implementation of a technology amenable to success only in a large, industrial corporation virtually guarantees a trajectory toward further industrialization. Even though new inventions and technologies may emerge that can be adopted and applied to advantage by small businesses and entrepreneurs, only the landscape of the industry would be altered. The fact of industrialization will still hold.

Impact of biotechnology

Biotechnology fully entered corn breeding 10 years before the end of the 20th century when a DeKalb Genetics Corporation laboratory in Mystic, Connecticut, fittingly not too far from the site where East conducted his inbreeding experiments eighty-some years before, demonstrated stable transformation of a monocotyledonous species (Gordon-Kamm et al., 1990). Cells from embryogenic corn suspension cultures were transformed using microprojectile bombardment with the bacterial gene "bar" that encodes the enzyme phosphinothricin acetyltransferase (PAT), which confers resistance to the herbicide bialaphos. Transformed calli were identified by treating the suspension cultures with bialaphos. Fertile transformed corn plants were regenerated, and of 53 progeny tested, 29 expressed PAT activity, with all 29 containing the bar gene. Localized application of bialaphos to leaves of bar-transformed plants induced no necrosis, confirming functional activity of PAT in the transgenic plants. Subsequently, DeKalb obtained a patent on the method (US Patent 5,489,520), and shortly thereafter, Japan Tobacco Corporation obtained a patent for *Agrobacterium*-mediated transformation of monocots (US Patent 5,591,616 A). *Agrobacterium*-mediated transformation is now the generally preferred method because of lower copy integration, relatively more precise mode of DNA transfer, more efficient transformation, greater capability to transfer relatively larger pieces of DNA, and lower cost (Hiei et al., 1997; Mohanty et al., 1999).

Currently, corn hybrids transformed with genes for resistance to herbicides, European corn borer, fall armyworm, and corn rootworm have been released for sale to the public. Among transgenes in the development pipelines of major corn breeding companies are genes conferring tolerance to drought and better utilization of nitrogen. Research has been initiated on identifying and developing yield-enhancing genes through genetic engineering.

When discussing transformation, it may be beneficial to define the terms "transgene," "event," and "trait," which are often used in a confusing way by many authors and speakers. The term "transgene" may often be inaccurately descriptive and possibly obsolete. What the term now usually implies is a construct, or cassette, consisting of the structural portions of one or more genes, along with controlling elements, such as promoters and enhancers, that is introduced into a host plant by biolistic or *Agrobacterium*-mediated transformation. The term "event" implies the insertion of the construct into a specific site (or sites) in the host genome (for a number of reasons, single-site insertion is preferred). The event may be thought of as a compound metaphor alluding to construct insertion at a specific locus. Under the most agreeable circumstances, the event segregates normally with linked loci and is completely heritable; i.e., fully expressed (constituently or otherwise) and regularly and normally transmitted to offspring in a manner consistent with typical endogenous genes. "Trait" and "traits" are simply the functional products of the event; i.e., usually the proteins produced as a result of the expression of the structural genes encoded within the event. A single event may be responsible for the expression of one or more traits. The proteins, or traits, resulting from expression of the event exert an effect on the phenotype of a definable characteristic of the plant (e.g., rootworm resistance), which, unfortunately, in the vernacular argot of breeders and quantitative geneticists, is also often referred to as a trait.

The event is the critical element in the transformation process; it provides added value to the product and the element that must gain regulatory approval for release. Until authorization, the event must be maintained under conditions of strict containment and control. Consequently, transformation is usually done on a hybrid-by-hybrid basis, either by direct transformation of one or both of the parental lines or by backcrossing from a transformed donor. Under these conditions, transformation has little direct impact on corn breeding per se, but it does have a considerable effect on management of the pipeline for hybrid advancement and commercial release. Because of the time required for either direct transformation or backcross modification of parents, hybrids chosen for transformation must be deserving of consideration but not too close to

obsolescence and imminent displacement by newer, superior hybrids. On the other hand, selection of hybrids with too little testing incurs the cost of wasted resources devoted to ultimately non-commercial candidates. Multiple gene stacks also complicate the problem since in a stack, one gene may become obsolete and require replacement by an improved version while the companions in the stack remain viable. Quality control to assure presence of the event and absence of contamination during foundation increase or production also becomes very important, and of course management of inventory becomes more difficult.

All transgenic events authorized to date have displayed little or no interaction with endogenous genes. Hence, programs integrating events into selected hybrids could be easily maintained distinct from the standard breeding programs. Continuation of this separation may not be possible in the future. Events conferring stress tolerance or yield enhancements may well interact in biochemical networks with endogenous genes. Consequently, interactions with genetic background may become quite important and demand evaluation in the conventional breeding context. Regulatory agencies may, in fact, demand that this be done. Regardless of the breadth of distribution of transgenic events across the breeding enterprise, rigorous event stewardship is required (Mumm, 2006), entailing standardization and meticulous recording of pedigrees. Adequate event stewardship essentially requires that the pedigree and critical characteristics of every plant in every row in every nursery and evaluation plot be recorded and maintained in a central database. Management of such a system obviously requires a level of specialization and skill attainable probably only in a large corporation with considerable industrial management capability.

Without interaction of transgenic events with endogenous background, the effects of event and background are additive. Even so, the possibility of transformation exerts a selection pressure for higher yield through traditional breeding. No matter the added value increment of the transgenic event, total yield still determines hybrid value. In a competitive marketing environment, the high cost of transgene development can be profitably absorbed only if the highest yielding, widely adapted, and consequently widely sold, hybrids are transformed. The licensee of an event is accordingly under at least as much pressure to develop high-yielding hybrids to justify licensing costs.

Johnson (2004) broadly defined genomics as the science and technology involved in the study of the fundamental organization of the genome and suggested that, more narrowly, genomics is concerned with the physical characterization of genomic DNA; i.e., determination of nucleotide sequence. Functional genomics connotes the relationship of a genomic component to the expression of any element downstream of the gene involved in the development of a measurable phenotypic attribute of the organism. Phenotypic attributes may include, for example, abundance of messenger RNA, small nuclear RNAs, proteins, metabolites, or whole plant characteristics, such as yield or disease resistance. Functional genomics, then, comprises the link between DNA sequence and the development of whole plant traits, allowing simultaneous mapping of any intermediary phenotypes affecting the development of the whole plant phenotype (Walsh and Henderson, 2004; Gibson and Weir, 2005). More important to breeding, the concept of marker-aided selection can be extended to include the intermediary functional genomic phenotypes as correlated traits within a selection index. If the intermediary phenotypes have high heritability, coupled with high correlation with the economically valuable primary trait under selection, gain from selection will be enhanced. In an experiment investigating favorable response to selection of net photosynthetic rate (NPR) under drought conditions, inclusion of glutamate dehydrogenase (GDH) mRNA expression in an index, along with NPR, resulted in an 2.5-fold increase in estimated selection gain relative to selection on NPR alone (Johnson, 2004).

Sequencing of the entire genome along with gene expression studies will enable discovery of synonymous structural gene sequence variation and allele-specific variation within functionally important genes. Allele-specific variation can then be correlated with variability among hybrid individuals in synthetic populations to extend marker-aided breeding to broad-based outbred populations rather than focusing piecemeal on separate F_2-derived or backcross populations.

All in all, adequate return on investment required for the adoption of biotechnology and genomics depends on an industrialized platform adequate to support the technologies necessary for the implementation.

Conclusion

This chapter represents an endeavor to trace the evolution of corn breeding in North America into the industrial process that it is today, trying to note the important developments, notable individuals involved, and the social and political climate that fostered the development. In saying that corn breeding has become industrialized, no judgment as to the rectitude or inevitability of the outcome is implied. This is simply what happened as conditioned by the scientific, economic, political, and cultural climate of the 20th century in North America. The outcome on balance was good. The industry supported a feed-grain production economy that ensured a cheap and abundant feed supply for a growing urban population with no environmental degradation. The environment may actually have been improved since the level of productivity achieved permitted removal of unsuitable acreages from intense cultivation. The rise in productivity also enabled an orderly rural to urban population shift, releasing farm labor to be absorbed by a growing and increasingly productive urban economy that was also stimulated by many of the same economic, political, and cultural factors affecting agriculture.

That continuous increase in productivity was achieved with virtually no input of germplasm from outside sources is notable and somewhat perplexing. In searching for an answer, one may wish to view as a model the Illinois long-term selection experiments for kernel oil and protein concentration (Dudley and Lambert, 2004). These experiments started with a very narrow germplasm base. Yet no evidence for a selection plateau for high oil or protein has been observed after more than 100 cycles of selection. It seems unlikely that migration is an important factor, and the strong directional trend seems to rule out drift alone as a decisive factor. Mutation may be a factor, but it is difficult to reconcile the observed results as being due to mutation alone. Two intriguing explanations have been offered by Dudley (2006), both having been suggested earlier by authors considering long-term selection in chickens. Eitan and Soller (2004) attributed the long-term success as due to (1) shifting of genotype × environment effects induced by changes in husbandry and management, and (2) emergence of selection-induced epistatic variation due to shifts in genetic background. In each environmental shift, genes that were previously neutral assume strong positive or negative effects due to interaction with the new environment. In other words, environmental shifts expose new sets of genes to selection. Likewise, continued selection drives some sets of genes to near-fixation, altering the epistatic landscape of the genome to expose allelic variability at previously neutral loci to continued directional selection. The results of Carlborg et al. (2006) provided empirical support for the selection-induced variation hypothesis. It is easy to see how mutational and drift events would harmonize with epistatic and genotype × environment variation to magnify variability released during selection. Darwin, of course, long ago commented on the release of variation during the domestication of plants and animals (Winther, 2000), underscoring the fact that perhaps there really is nothing new under the sun.

References

Arthur, I.W. 1960. Classroom lecture. Iowa State Univ., Ames.

Baker, R.F. 1984. Some of the open pollinated varieties that contributed the most to modern hybrid corn. p. 1–19. Ann. Ill. Corn Breeders School. Univ. Ill., Urbana.

Botstein, D., R.L. White, M. Skolnick, and R.W. Davis. 1980. Construction of a genetic linkage map in man using restriction length polymorphisms. Am. J. Hum. Genet. 32:314–331.

Comstock, R.E., H.F. Robinson, and P.H. Harvey. 1949. A breeding procedure designed to make maximum use of both general and specific combining ability. Agron. J. 41:360–367.

Carlborg, O., L. Jacobsson, P. Ahgren, P. Seigel, and L. Anderson. 2006. Epistasis and the release of genetic variation during long-term selection. Nat. Genet. 38:418–420.

Crabb, A.R. 1947. *The hybrid-corn makers*. Rutgers Univ. Press, New Brunswick, NJ.

Cress, C.E. 1967. Reciprocal recurrent selection and modifications in simulated populations. Crop Sci. 7:561–567.

Crosbie, T.M., S.R. Eathington, G.R. Johnson, M. Edwards, R. Reiter, S. Stark, R.G. Mohanty, M. Oyervides, R.E. Buehler, A.K. Walker, R. Dobert, X. Delannay, J.C. Pershing, M.A. Hall, and K.R. Lamkey. 2006. Plant breeding: Past, present, and future. pp. 3–50. In K.R. Lamkey and M. Lee (eds.) *Plant Breeding: The Arnel R. Hallauer Int. Symp.* Blackwell Publ., Ames, IA.

Crow, J.F. 1948. Alternative hypothesis of hybrid vigor. Genetics 33:477–487.

Delgado-Salinas, J. Caballero and A. Casas. 2004. Crop domestication in Mesoamerica. pp. 310–313. In R. M. Goodman (ed.) *Encyclopedia of Plant and Crop Science*. Marcel Dekker, Inc., New York.

Draper, N.R. and H. Smith. 1966. *Applied Regression Analysis*. John Wiley & Sons, Inc., New York.

Dudley, J.W. 2006. From means to QTL: The Illinois long-term selection experiment as a case study in quantitative genetics. Presentation given at the International Plant Breeding Symposium. Mexico City. Aug 20–25.

Dudley, J.W. and R.J. Lambert. 2004. 100 generations of selection for oil and protein in corn. Plant Breed. Rev. 24(Part 1):79–110.

Duvick, D.N., J.S.C. Smith, and M. Cooper. 2004. Long-term selection in a commercial hybrid maize breeding program. Plant Breed. Rev. 24(Part2):109–151.

East, E.M. and D.F. Jones. 1920. Genetic studies on the protein content of maize. Genetics 5:543–610.

Eathington, S.R. 2006. Practical uses of molecular markers in a commercial breeding program. Presentation given at the International Plant Breeding Symposium. Mexico City. Aug 20–25.

Eathington, S.R., J.W. Dudley, and G.K. Rufener II. 1997. Usefulness of marker-QTL associations in early generation selection. Crop Sci. 37:1686–1693.

Edwards, M.D. and L. Johnson. 1994. RFLPs for rapid recurrent selection. Proc. Joint Plant Breed. Symp. Series. Am. Soc. Hort. and Crop Sci. Soc. Am., Corvallis, OR.

Eitan, Y. and M. Soller. 2004. Selection induced variation. pp. 153–176. In S.P. Wasser (ed.) *Evolutionary Theory and Processes*. Kluwer Acad. Publ., Netherlands.

Falconer, D.S. 1960. Introduction to quantitative genetics. Ronald Press Co., New York.

Fisher, R.A. 1949. *The Theory of Inbreeding*. Hafner Publ. Co., Inc., New York.

Garst, R. 1974. Statement for the record. pp. 340–347. In *Fertilizer Supply, Demand, and Prices*. U.S. Senate Committee on Agriculture and Forestry. Govt. Printing Office, Washington. Gibson, G. and B. Weir. 2005. The quantitative genetics of transcription. Trends Genet. 21:616–623.

Gordon-Kamm, W.J., T.M. Spencer, M.L. Mangano, T.R. Adams, R.J. Daines, W.G. Start, J.V. O'Brien, S.A. Chambers, W.R. Adams Jr., N.G. Willets, T.B. Rice, C.J. Mackey, R.W. Krueger, A.P. Kausch, and P.G. Lemaux. 1990. Transformation of maize cells and regeneration of fertile transgenic plants. Plant Cell 2:603–618.

Griffing, B. 1963. Comparisons of potentials for general combining ability selection methods utilizing one or two random mating populations. Aust. J. Biol. Sci. 16:838–862.

Hallauer, A.R. 1997. Maize improvement. pp. 15–27. In M.S. Kang (ed.) *Crop Improvement for the 21ˢᵗ Century*. Research Signpost, Trivandrum, India.

Hallauer, A.R. 2004. Breeding hybrids. pp. 186–188. In R.M. Goodman (ed.) *Encyclopedia of Plant and Crop Science*. Marcel Dekker, Inc., New York.

Hallauer, A.R. and J.B. Miranda, Fo. 1981. Quantitative genetics in maize breeding. Iowa State Univ. Press, Ames, IA.

Hallauer, A.R., W.A. Russell, and K.R. Lamkey. 1988. Corn breeding. pp. 463–564. In G.F. Sprague and J.W. Dudley (eds.) *Corn and Corn Improvement*, 3ʳᵈ ed. Crop Sci. Soc. Am., Madison, WI.

Hayes, H.K. 1963. *A Professor's Story of Hybrid Corn*. Burgess Publ. Co., Minneapolis, MN.

Hayes, H.K. and R.J. Garber. 1919. Synthetic production of high protein corn in relation to breeding. J. Am. Soc. Agron. 11:308–318.

Hiei, Y., T. Komari, and T. Kubo. 1997. Transformation of rice mediated by *Agrobacterium tumefaciens*. Plant Mol. Biol. 35:205–218.

Hull, F.H. 1945. Recurrent selection and specific combining ability in corn. J. Am. Soc. Agron. 37:134–145.

Jenkins, M.T. 1935. The effect of inbreeding and of selection within inbred lines of corn upon hybrids made after successive generations of selfing. Iowa State Coll. J. Sci. 9:429–450.

Jenkins, M.T. 1940. The segregation of genes affecting yield of grain in maize. J. Am. Soc. Agron. 32:55–63.

Jensen, S.D., W.E. Kuhn, and R.L. McConnell. 1983. Combining ability studies in elite U.S. maize germplasm. Proc. Ann. Corn & Sorghum Ind. Res. Conf. 38:87–96.

Johnson, G.R. and R.H. Mumm. 1996. Marker assisted maize breeding. Proc. Ann. Corn & Sorghum Ind. Res. Conf. 51:75–84.

Johnson, R. 2004. Marker-assisted selection. Plant Breed. Rev. 24(Part 1):293–309.

Kang, M.S. 1994. *Applied Quantitative Genetics*. M.S. Kang, Publisher, Baton Rouge, LA.

Kendall, M. and A. Stuart. 1977. *The Advanced Theory of Statistics*. Vol. 1., 4ᵗʰ ed. MacMillan Publ. Co. Inc., New York. Labate, J.A., K.R. Lamkey, S.E. Mitchell, S. Kresovich, H. Sullivan, and J.S.C. Smith. 2003. Molecular and historical aspects of corn belt dent diversity. Crop Sci. 43:80–91.

Lamkey, K.R. and J.W. Edwards. 2004. Breeding plants for heterosis. pp. 189–192. In R.M. Goodman (ed.) *Encyclopedia of Plant and Crop Science*. Marcel Dekker, Inc., New York.

Lande, R. and R. Thompson. 1990. Efficiency of marker-assisted selection in improvement of quantitative traits. Genetics 124:743–746.

Martin, G.B., S.H. Brommenschenkel, J. Chunwonse, A. Frary, M.W. Ganal, R. Spivey, T. Wu, E.D. Earle, and S.D. Tanksley. 1993. Map-based cloning of a protein kinase gene conferring disease resistance in tomato. Science 262:1432–1436.

Matsuoka, Y., Y. Vigouroux, M.M. Goodman, G.J. Sanchez, E. Buckler, and J. Doebley. 2002. A single domestication of maize shown by multilocus microsatellite genotyping. Proc. Natl. Acad. Sci. USA 99(9):6080–6084.

Mohanty, A., N.P. Sarma, and A.K. Tyagi. 1999. Agrobacterium-mediated high frequency transformation of an elite indica rice variety Pusa Basmati 1 and transmission of the transgene to R2 progeny. Plant Sci. 147:127–137.

Mumm, R. 2006. Practical considerations driving the balance between backcross and forward breeding in the development of transgenic maize hybrids. pp. 93–100. Ann. Ill. Corn Breeders School. Univ. Ill., Urbana.

Paliwal, R.L. and M.E. Smith. 2002. Tropical maize: Innovative approaches for sustainable productivity and production increases. pp. 43–73. In M.S. Kang (ed.) *Crop Improvement: Challenges in the Twenty-first Century*. Food Products Press, an imprint of Haworth Press, Binghamton, New York.

Paterson, A.H., E.S. Lander, J.D. Hewitt, S. Peterson, S.E. Lincoln, and S.D. Tanksley. 1988. Resolution of quantitative traits into Mendelian factors using a complete RFLP linkage map. Nature 335:721–726.

Richey, F.D. 1927. The convergent improvement of selfed lines of corn. Am. Nat. 61:430–449.

Roberts, T.H. Jr. 1977. Personal communication.

Salvi, S., R. Tuberosa, E. Chiapparino, M. Maccaferri, S. Veillet, L. van Beuningen, P. Isaac, K. Edwards, and R.L. Phillips. 2002. Toward positional cloning of Vgt1, a QTL controlling the transition from the vegetative to the reproductive stage of maize. Plant Mol. Biol. 48:601–613.

Shull, G.H. 1908. The composition of a field of maize. Am. Breed. Assoc. Rep. 4:296–301.

Shull, G.H. 1909. A pure-line method in corn breeding. Am. Breed. Assoc. Rep. 5:51–59.

Shull, G.H. 1910. Hybridization methods in corn breeding. Am. Breed. Mag. 1:98–107.

Sprague, G.F. 1939. An estimation of the number of top-crossed plants required for adequate representation of a corn variety. J. Am. Soc. Agron. 31:11–16.

Sprague, G.F. 1946. Early testing of inbred lines. J. Am. Soc. Agron. 38:108–117.

Sprague, G.F. 1978. Personal communication.

Sprague, G.F. 1983a. Personal communication.

Sprague, G.F. 1983b. Heterosis in maize: Theory and practice. pp. 47–70. In R. Frankel (ed.) *Heterosis*. Springer-Verlag, Berlin.

Sprague, G.F. and L.A. Tatum. 1942. General vs. specific combining ability in single crosses of corn. J. Am. Soc. Agron. 34:923–934.

Stadler, L.J. 1944. Gamete selection in corn breeding. J. Am. Soc. Agron. 36:988–989.

Stuber, C.W., M.D. Edwards, and J.F. Wendell. 1987. Molecular marker-facilitated investigations of quantitative-trait loci in maize. II. Factors influencing yield and its component traits. Crop Sci. 27:639–648.

Tanksley, S.D., M.W. Ganal, and G.B. Martin. 1995. Chromosome landing: A paradigm for map-based cloning in plants with large genomes. Trends Genet. 11:63–68.

Tanksley, S.D. and J. Hewitt. 1988. Use of molecular markers in breeding for soluble solids content in tomato—a re-examination. Theor. Appl. Genet. 75:811–823.

Tracy, W.F. and M.A. Chandler. 2006. The historical and biological basis of the concept of heterotic patterns in corn belt dent maize. pp. 219–233. In K.R. Lamkey and M. Lee (eds.) *Plant Breeding: The Arnel R. Hallauer Int. Symp.* Blackwell Publ., Ames, IA.

Troyer, A.F. 2004. Breeding widely adapted cultivars: Examples from maize. pp. 211–214. In R.M. Goodman (ed.) *Encyclopedia of Plant and Crop Science*. Marcel Dekker, Inc., New York.

Wallace, H.A. and W.L. Brown. 1988. *Corn and its Early Fathers*. Rev. ed. Iowa State Univ. Press, Ames, IA.

Walsh, B. and D. Henderson. 2004. Microarrays and beyond. J. Anim. Sci. 82(E. Suppl.):E292–E299.

Winther, R.G. 2000. Darwin on variation and heredity. J. Hist. Biol. 33:425–455.

Chapter 9

Soybean Breeding Achievements and Challenges

Silvia R. Cianzio

Introduction

World distribution

Soybean, *Glycine max* (L.) Merr., a crop first mentioned in the year 664 B.C. (Cianzio, 1997), has evolved thru intervention by growers and breeders to be, at present, a valuable crop commodity both in food and industrial uses. It originated in Northern China and has traveled across the globe to become one of the world's most important sources of protein feed for livestock (Boerma and Specht, 2004). It is also one of the most important world providers of vegetable oil. Currently, it has been adopted as a crop in about 50 countries, and it has emerged as a major commodity traded in world markets (Wilcox, 2004). The United States, Brazil, Argentina, and the People's Republic of China are the major soybean-producing countries and are also the main exporters of soybean.

As a leading crop for animal feedstock and human use, soybeans have been subjected to intense breeding efforts to make use of the intrinsic traits of the species and fulfill the nutritional requirements of animals and humans. The objective of this chapter is to review some recent breeding achievements and discuss future challenges. Numerous books, reviews, and research papers have been published on soybeans. The reader is therefore encouraged to review these earlier publications, which discuss in detail work conducted during the past century.

Growth and development

Soybean is an annual plant that may be sparsely or densely branched, depending on genetic and environmental factors and the interaction among them, i.e., daylength, intra- and inter-row plant spacing, soil fertility and genotype (Lersten and Carlson, 2004). Seeds of cultivated soybean rapidly imbibe water after planting. However, some genotypes, in particular the wild types, may have a large proportion of hard seeds that are slow to take up water and thus germinate unevenly. After cotyledons emerge, the two primary leaves of the seedling expand opposite to each other, each leaf having an ovate shape. All other subsequent leaves are trifoliolate, alternately arranged in two opposite rows on the stem.

The root system has been described as diffuse. Nodulation is visible about 10 d after planting and begins when rhizobia (*Rhizobium japonicum*) attach themselves to epidermal cells on the root. Roots have also been shown to establish mycorrhizal associations.

Depending on environment and genotype, soybean plants enter reproductive phase by forming axillary buds that develop clusters of flowers, of 2 to 35 each (Carlson and Lersten, 2004). Flowers are perfect, with the male and female organs enclosed within the corolla. They open early in the morning, shedding pollen shortly before or after the flower opens (Poehlman, 1979).

Soybean is normally self-pollinated, although pollination by insects can occur, and it is usually less than 1%.

There are two types of stem growth habit related to flower initiation in soybean (Carlson and Lersten, 2004). One type is the indeterminate stem, with inflorescences in axillary racemes and the terminal bud in the stem growing vegetatively during most of the growing season. Pod development and stem elongation begin at the bottom of an indeterminate plant and progress toward the top as new nodes form; however, all seeds mature simultaneously (Fehr, 1987). Other genotypes have a determinate growth habit in which flowers are borne in both axillary and terminal racemes, with stem elongation ceasing upon differentiation of the terminal bud (Poehlman, 1979). Pod and seed development occur simultaneously throughout the length of the stem in a determinate plant. A semi-determinate growth habit, intermediate between the two, has also been identified.

Soybean seed normally contains approximately 40% protein and 20% oil on a dry weight basis. However, armed with new knowledge about genes governing seed composition, crop scientists have been able to greatly alter the seed composition of some cultivars. Specialty types have been developed that meet human health and special market needs (Wilson, 2004).

Speciation

The genus *Glycine* (Willd.) belongs to the family Fabaceae, subfamily Papilionoideae, tribe Phaseoleae, subtribe Glycininae. It is composed of two subgenera, *Glycine* (perennials) and *Soja* (annuals) (Palmer and Hymowitz, 2004). The subgenus *Soja* contains two species, *G. max*, *G. soja*, and a third form identified as *G. gracilis*, not definitively established as a separate species. Crossability barriers have not been observed in hybrids among the three *Glycine* species (Hymowitz, 2004).

Glycine max (L.) Merr., the cultivated soybean, is a true domesticate or cultigen, which would not exist in the absence of human intervention. Genomic relationships among diploid species of *Glycine* established by cytogenetic, biochemical, and molecular analyses (Hymowitz, 2004), have determined that *Glycine max* (L.) Merr. is a diploid species with 2n = 40, to which the genome symbol GG was assigned.

Breeding achievements

Breeding emerged as an organized and planned activity when growers realized that they could select their preferred cultivated types to plant in the following season. Soybean is no exception to this. The first soybean cultivar, 'Ogemaw,' released in 1902, originated as a selection made among progenies of controlled hybridizations (Palmer and Hymowitz, 2004).

In the U.S., it was the USDA and state agricultural experiment stations that initially conducted cultivar development. This arrangement may have also been similar in other soybean producing countries. Currently in the U.S., and probably in other parts of the world, private companies develop and release most of the cultivars; nevertheless, the public sector still continues to be an important contributor to soybean breeding in terms of basic research, training of future plant breeders, and public distribution of germplasm.

The general approach to breeding for any trait in soybean is to first create genetic variability by artificially crossing cultivars selected as parents. The next step is to derive inbred lines using any of the breeding methods available. The third step is to conduct extensive field-testing across a wide range of locations and years, which are collectively termed as environments. In general, all soybean breeders use some sort of recurrent selection in which superior genotypes, identified and released as cultivars or germplasm lines, are crossed to form the next cycle population, from which new selections are made. Over the years as work is accomplished, and through the emergence of new techniques and tools, which are incorporated into this universal recurrent cycle, superior lines are developed and recombined to meet the breeding objectives of each individual program.

Seed yield

In soybean, seed yield was, and continues to be, the most important trait to breed for because growers derive their income from it. Any other trait of importance in soybean breeding is subservient to breeding for yield and must be incorporated into high-yielding genotypes if it is to reach growers and become of general use. However, since yield is the final expression of plant life span, increased yield and yield stability also benefit from breeding for other traits, such as disease and pest resistance, physiological traits, agronomic adaptation, and standability, among many others. According to Wilcox (2001), public soybean breeders in the northern region of the U.S. have increased yield at a rate of about 10% annually, which would be about 60% during the past 60 years.

For seed yield as a breeding objective, perhaps one of the most important factors to consider is selection of parents that will form the next population. Yield is a quantitative trait, controlled by many genes and strongly influenced by the environment (Brim, 1973). Yield heritability in soybean ranges from 3% to 58%. Of all the economically important traits in soybeans, yield is the characteristic that has the lowest heritability estimate. It also has the widest variation of all traits (Cianzio, 1997). Parents used in crossing for population development can be elite adapted soybean lines or exotic germplasm, and both have been used to increase seed yield.

Selection of parents can be done on the basis of comparative evaluation per se, or by testcross evaluation (Orf et al., 2004). In the U.S., if existing material is to be used, data are available from many of the variety tests conducted throughout the soybean production region. If new material is needed, several methods have been proposed and evaluated for the identification of parental germplasm with favorable alleles not present in existing cultivars (Cerna et al., 1997; Kenworthy, 1980; Panter and Allen, 1995a, 1995b; St Martin et al., 1996).

The method proposed by St. Martin et al. (1996) requires development of all possible combinations among cultivars and germplasm lines to capture the largest numbers of favorable alleles, and the calculations of several statistical measures that may reflect allele contribution. This method proved useful in parent selection, particularly when the testcross statistic was compared with the F_1 mid-parent heterosis. The best linear unbiased prediction method, proposed by Panter and Allen (1995a, 1995b), was reported to be effective in identifying a higher percentage of superior crosses when compared with the mid-parent heterosis value. The method proposed by Kenworthy (1980) also involved the production of testcrosses; it was devised for selecting exotic parents on the basis of their phenotype per se or in testcrosses. Cerna et al. (1997) compared measures of genetic divergence, determined by restriction fragment length polymorphism (RFLP) and isozyme constitution, with the expression of heterosis for seed yield in soybean. The underlying assumption was that if a relationship were to exist between these diversity measurements, then diversity at the molecular or isozyme level could be used as selection criterion for crossing parents. Using 21 soybean genotypes of maturity groups (MG) I, II, and III, adapted to Iowa in the U.S., the authors developed four sets of populations on the basis of seed yield, different geographic origin, and diversity in RFLP markers and isozyme loci. Their results suggested that parent selection on the basis of RFLP and isozyme loci to exploit heterosis in seed yield may not be a feasible alternative.

None of the methods proposed for parent selection proved to be completely effective and efficient. The production of a set of test-crosses prior to the identification of parents for population development might add additional time, effort, and expense to a breeding program, and might not always guarantee that the best parents could be identified. It could also be argued that additional resources applied to the search of promising parents could be directly used to actual field test of advanced lines. Selection of parents continues to be a key and challenging issue in breeding high-yielding soybean cultivars. As molecular (Smalley et al., 2004) and genomic knowledge becomes more extensive in depth and scope, it might contribute to facilitating parent selection in breeding for this trait.

A somewhat related topic in the parent selection arena is the always-present concern about the narrow genetic base of soybean and whether it could be expanded by the use of exotic germplasm in breeding efforts. Early studies indicated that soybean cultivars in the U.S. and Canada had a limited genetic base (Specht and Williams, 1984). Pedigree analysis of 258 cultivars released from 1947–88 showed that 80 ancestors accounted for 99% of the parentage in cultivars, whereas only 26 ancestors accounted for nearly 90% of the total ancestry (Gizlice et al., 1994). Since these early studies, and due to concerns about the narrow genetic base making the crop vulnerable to pest outbreaks, concerted efforts in the U.S. have been made to explore the National Soybean Germplasm Collection (USDA-ARS, University of Illinois, Urbana, Illinois) in search of novel genes (Carter et al., 2004). Germplasm lines possessing yield genes contributed by plant introductions (PI) have been released, along with populations with different levels of PI in their parentage (Fehr and Cianzio, 1981). The populations developed and released by Fehr and Cianzio (1981) are

an early example of how important genetic diversity was considered to be. The authors developed five populations varying in PI parentage; one had 0% (AP14), one had 100% PI parentage (AP10); the other three varying in PI parentage in increments of 25%. No increments in seed yield were obtained on the populations which had some PI percentages in their genetic constitution. The best yielding population was AP14, which had 0% PI in its genetic makeup. However, the use of PIs in developing the populations effectively increased the genetic variability.

Similar results have also been obtained by other researchers, suggesting that seed yield increase attributed to the use of PIs in a breeding program is still a debatable issue in soybean. Ininda et al. (1996) compared seed yield and genetic improvement in five populations with different percentages of PI parentage, to which three cycles of recurrent selection had been applied. Their results indicated that use of PIs to increase rate of genetic improvement for seed yield was not effective and that selections obtained from populations developed with elite parents continued to be the most effective method for immediate-term development of high-yielding cultivars. These observations are not meant to deter breeders from using exotic germplasm in breeding programs to search for yield genes. They do indicate, however, that use of PIs with the objective to increase seed yield in the immediate term may not be a realistic expectation. This quote from Lawrence and Frey (1975) is still current: "Introgression of exotic germplasm into a plant breeding program frequently is suggested as a procedure for increasing genetic variability available to the plant breeder. However, the problem that confronts the practical breeder is to identify useful genes from the exotic materials and incorporate them into adapted genotypes without disrupting the superior gene complexes of the adapted genotypes." In spite of the limited genetic base of cultivars in the U.S., yield gains continue to be made

(Wilcox, 2001), and according to St. Martin (2001), the best current cultivars are not yet even close to their potential limit in yield.

Agronomic traits

Efforts to make improvements in various agronomic traits undoubtedly also contribute to realizing genetic gains in yield. Perhaps one of the most important agronomic traits in soybean is maturity group (MG) (differential photoperiod sensitivity) because it determines the area of adaptation for a cultivar, which in turn determines whether or not it will have a chance to reach full yield potential. This is one of the reasons why, in the U.S. and Canada, most breeding is conducted both within a maturity group and for a specific area of adaptation.

As a method to describe response to daylength, 10 maturity groups have been established for identifying the region of adaptation of soybean cultivars in the U.S. and Canada (Hartwig, 1973). Groups 00, 0, and I are adapted to longer days in the northern areas of the U.S. and southern regions of Canada. Succeeding groups are adapted to regions farther south, with the MG VIII cultivars being the maturity group that can be planted in the southern U.S. MG IX and X are mainly adapted to tropical regions; however, all MG can be planted in tropical locations (Cianzio et al., 1991). In general, within a maturity group there is an approximate range of 8–10 days among the earliest and the latest cultivar of the group, and this criterion is used to assign experimental lines to different MG for regional testing in the U.S.

Lodging resistance is, for obvious reasons, another desirable agronomic trait that contributes to yield. Presently, in most programs this is not a breeding objective per se, because most of the existing adapted germplasm possess, in general, good standability. It could however, be used as a general criterion for discarding lines at early stages of development, if lines from some population show an unusual amount of lodging. Plant height is another desirable trait, although like lodging, it is not used as a breeding objective per se. Seed quality, seed shattering, and the appearance of green stems at harvest time are similar examples. If any of these traits are noticed at early stages of line development, they become general selection criteria for discarding undesirable material.

Seed size, seed composition, pest resistance, and stress-resistance are agronomic traits that are specifically bred for. Each trait, however, needs to be incorporated into genotypes of superior yield if they are to reach public acceptance and use.

Breeding for seed size and composition

During the past 30 years, soybean evolved from being a world supplier of protein and oil to being a complex and sophisticated end product. In the process, the two traditional seed composition traits (size and oil content) were expanded to include many other seed nutritional components. Newly developed attention to human and animal nutrition triggered these changes, and it is the breeders who should be credited with responding in a timely manner.

Seed size

Differences in seed size are of importance mainly in the market for direct human consumption. In the U.S. in general, and at Iowa State University (ISU) in particular, research to develop lines with either small or large seed sizes has been conducted (Bravo et al., 1980 and 1981; LeRoy et al. 1991a, 1991b) and numerous germplasm lines and cultivars have been released (Fehr et al., 1990a, 1990b, 1990c, 1990d). A current list of germplasm released by Iowa State University for both large and small seed size is available at www.iastate.edu/isurf. The ISU group also conducted research on breeding methods to modify seed size in soybean, both small- and large-seeded types (Johnson et al., 2001), and on development of molecular markers for the selection of seed size (Hoeck et al., 2003).

Large-seeded soybeans are developed at ISU with high protein content for the tofu-production market and also for direct human consumption. New lines and cultivars have been developed, and the efforts continue in that direction (Fehr et al., 1990a and 1990b). Research has also been conducted to address particular concerns about developing large-seeded soybeans and how to select for the trait (Cianzio et al., 1982).

Small-seeded soybean cultivars are also developed at ISU for the specialty market of natto production and for direct human consumption as soybean sprouts. Along with cultivar releases (Fehr et al., 1990c, 1990d), research has also been conducted to facilitate breeding for small seed size (Bravo et al., 1980; Cianzio and Fehr, 1987; LeRoy et al., 1991a).

Protein and oil composition

Soybean storage protein is composed primarily of three fractions defined by sedimentation value: 2S (α-conglycinin), 7S (β-conglycinin), and 11S (conglycinin) (Wilson, 2004). The 2S fraction contains protein such as protease inhibitors (Werner and Wemmer, 1991, cited by Wilson 2004); the 7S contains trimers of α, α', and β subunits (Maruyama et al., 2001, cited by Wilson, 2004), and the 11S is composed of hexamers (Nielsen et al., 2001, cited by Wilson, 2004).

Among all vegetable sources, soybean protein may provide the most complete amino acid balance for food and feed, however, additional improvements in protein content should still benefit the soybean industry (Wilson, 2004). There is continued interest in increasing total protein content in soybean seed to contribute to the feed industry. The problem in this area arises because of a negative correlation that exists between

protein and oil content (Brim and Burton, 1979), and a negative association between protein and seed yield (Wehrmann et al., 1987). Still, Iowa State University (Fehr et al. 1990e, 1990f, 1990g, 1990h), and other institutions (Burton et al., 1994) have released several promising germplasm lines.

A number of storage protein genes have been isolated, sequenced, and expressed in transgenic plants, making evident the potential of genetic engineering for modifying protein content and composition (Parrot and Clemente, 2004). There are, however, problems related to the control of gene copy numbers, the site of transgene insertion, and the effects of changing the native primary structure of polypeptides that may impede progress toward improvements in protein quality (Wilson, 2004). Using traditional breeding techniques, the ISU group has released numerous cultivars and germplasm lines with modified protein content (Fehr et al., 1990e, 1990f, 1990g, 1990h). Extensive research on the presence of genetic variability among low- and high-protein content lines, on the expression of protein content in temperate and tropical locations, and on the transfer of high-protein content was also conducted at ISU to facilitate the genetic manipulation of soybean protein (Cianzio and Fehr, 1982a; Cianzio et al., 1985; Wehrmann et al., 1987) and the release of lines for public use.

Crude soybean oil contains various glycerolipids, primarily phospholipids, diacylglycerol, and triacylglycerol (Wilson et al., 1980). Each glycerolipid class is composed of molecular species formed by various combinations of the five fatty acids, 16:0 palmitic acid, 18:0 stearic acid, 18:1 oleic acid, 18:2 linoleic acid, and 18:3 linolenic acid.

Classical genetic studies on oil composition, conducted during the 1980s and mid-1990s, greatly contributed to the understanding of inheritance of fatty acids and how these components could be genetically manipulated to alter fatty acid composition (Erickson et al., 1988; Fehr et al., 1992; Graef et al., 1985; Hartmann et al., 1996; Hawkins et al., 1983; Ndzana et al., 1994; Pantalone et al., 2002; Rahman et al., 1997; Ross et al., 2000; Schnebly et al., 1994; Stojsin et al., 1998; Stoltzfus et al., 2000a, 2000b; Streit et al., 2000; Walker et al., 1998; Wilcox and Cavins, 1985). Recent advances in molecular technology have also contributed to the understanding of fatty acid synthesis and provided molecular markers to alter fatty acid composition in soybean oil (Wilson, 2004).

Presently, there is an array of mutant genes that may be used to alter fatty acid composition in soybean, the majority of which were developed through mutagenesis and/or recombination. The genes, arranged by their phenotypic effect, along with the literature reference in which they were first reported, are listed in Table 9.1. The genes

Table 9.1. Genes affecting oil composition in soybean, their phenotype and the literature reference in which the information has been reported.

Phenotype	Gene	Reference
Low 16:0		Burton et al. (1994)
		Erickson et al. (1988)
		Schnebly et al. (1994)
		Stojsin et al. (1998)
High 16:0		Erickson et al. (1988)
		Schnebly et al. (1994)
		Schnebly et al. (1994)
		Stoltzfus et al. (2000b)
		Narvel et al. (2000)
		Stoltzfus et al. (2000a)
High 18:0		Pantalone et al. (2002)
		Hammond and Fehr (1983)
		Graef et al. (1985)
		Graef et al. (1985)
		Rahman et al. (1997)
		Rahman et al. (1997)
High 18:1		Hitz et al. (1995)
Low 18:3	*fan*	Wilcox and Cavins (1985)
	fan₁	Hawkins et al. (1983)
	fan₂	Fehr et al. (1992)
	fan₃	Fehr et al. (1992)

have been incorporated into high-yielding cultivars that have been released, as is the case of the low linolenic lines released at Iowa State University currently being commercially grown.

Pest resistance and abiotic stress-related factors

As in any crop species, numerous diseases, pests, and abiotic stress-related factors, which compromise plant health and productivity, threaten soybeans. Boethel (2004), Grau et al. (2004), Niblack et al. (2004), and Tolin and Lacy (2004) have provided detailed and comprehensive treatment of this subject.

In the U.S. alone, more than 100 pathogens are known to affect soybeans, of which about 35 cause yield loss of economic importance. Disease severity and prevalence, and agronomic loss are closely associated with the environment, crop management practices, and the reaction of soybean cultivars to plant pathogens (Grau et al., 2004). Frequently, some diseases are geographically limited, whereas others are widely distributed.

Breeding research in a particular disease is done at the place where the disease is a recurrent economic problem. For soybean cultivars planted in the northern regions of the U.S. and particularly in the state of Iowa, the most important pathogens and pests are Phytophthora root rot, brown stem rot, sudden death syndrome, and soybean cyst nematode. Of the stress-related factors, iron deficiency chlorosis is the most important, which also affects soybean yield. The following sections focus on these diseases and abiotic stress factors.

Phytophthora root rot (PRR)

This pathogen was first observed in Indiana in 1948 and in Ohio in 1951, although the causal agent was not described until 1958

(Kaufmann and Gerdemann, 1958). It is now widespread into the major soybean production areas of the U.S., as well as other parts of the world (Grau et al., 2004). *Phytophthora sojae* (Kaufmann and Gerdemann, 1958) is the causal organism. It produces root and stem rot of susceptible soybean cultivars when soils are saturated for prolonged periods of time. *P. sojae* populations in the U.S. are comprised of many physiologic races with virulence to overcome many of the *Rps* genes currently deployed (Grau et al., 2004). *P. sojae* causes seed and seedling pre-emergence damping off and stem rot, which results in reduced stands. It also causes post-emergence damping off when young seedlings are infected at the hypocotyl, cotyledon, or root. The extent of PRR symptoms depends on the resistance level of the cultivar, and its expression ranges from no symptoms to stunted, chlorotic, and wilting plants, to dead plants. Wilcox and St. Martin (1998) compared resistant near-isogenic lines and blends of soybean with *Rps* genes with lines that were susceptible to predominant *P. sojae* populations. They observed that susceptible isogenic lines yielded 65% to 93% of the yield of resistant lines. Tooley and Grau (1984) reported yield loss mainly attributed to reductions in plant stands, although smaller seed weights and reduced number of pods per plant were also observed.

PRR has been successfully managed with host resistance, which may be of three types (Grau et al., 2004): (1) single dominant *Rps* genes detected with the hypocotyl inoculation test; (2) single dominant gene *Rps2* that has an intermediate or partial kill in the hypocotyl inoculation test but a hypersensitive response with root inoculation (Kilen et al., 1974); and (3) partial resistance, which, following root inoculation, is expressed as fewer rotted roots, and the disease progresses at a much slower rate (Schmitthenner, 1985). Eight *Rps* genes have been identified, of them *Rps1* and *Rps3*

have multiple alleles. Gene nomenclature, original source where resistance was identified, and references for the work have been provided by Grau et al. (2004).

In Iowa, PRR causes yield loss mainly due to reduced plant populations and is widely spread throughout the state. In a survey conducted in 1994 (Yang et al., 1996), the most prevalent races identified in the state could be controlled by using the *Rps1-k* gene in breeding for resistant genotypes, which confers resistance to PRR races 1–11, 13–15, 17–18, 22, and 24. A more recent survey conducted in 2000 and 2001 indicated a change in the race composition in Iowa, with Race 25 becoming more prevalent (Niu et al., 2003). Race 25 defeats the *Rps1-k* gene, which is forcing the use of a different resistance gene—the *Rps6* gene—in breeding resistant cultivars. Periodic surveys are conducted throughout the state to provide directions about which genes need to be deployed. Recently, Cianzio and Bhattacharyya (unpublished results) have begun to use the newest identified gene *Rps8* (Burnham et al., 2003), which confers resistance to most of the known PRR races. To facilitate cultivar and germplasm development with the use of molecular marker-assisted selection, Cianzio developed populations that were used for mapping the location of the *Rps8* gene at the Bhattacharyya lab. Results of this work have indicated that the gene *Rps8* maps closely to the *Rps3* region (Sandhu et al., 2005). Work conducted at the Bhattacharyya lab at ISU has also demonstrated that a deletion at the molecular level associated with the *Rps4* gene was responsible for the lack of resistance in soybean cultivars (Sandhu et al., 2004).

Breeding for PRR resistance at ISU is done by using the backcross method of breeding. Earlier work conducted at ISU in which different strategies were compared indicated that the backcross method would be the most efficient way to transfer a gene into a high-yielding superior cultivar (Wehrmann et al., 1988). A superior, susceptible cultivar is used as the recurrent parent, and crossed to a donor parent, which is the source of the desired resistant gene (Cianzio et al., 2005). Usually four backcross generations are obtained and considered sufficient to recover the recurrent parent genotype for release. At every backcross generation, progeny tests done using the hypocotyl inoculation method confirm resistance/susceptible reactions of genotypes. As more molecular markers become available, these will be used more often in marker-assisted selection, as is the case with the *Rps8* gene (Cianzio and Bhattacharyya, unpublished results).

A number of resistant germplasm lines and cultivars have been released by the program at ISU. The most recent is the release of line AR1, which possess resistance to soybean cyst nematode and to PRR conferred by the *Rps1-k* gene (Cianzio et al., 2005). As race distribution and composition change in Iowa, it might be necessary to deploy different genes to protect cultivars. Periodic race surveys conducted in the state will help soybean breeders decide which genes to deploy.

Brown Stem Rot (BSR)

BSR was first observed in Illinois in 1944 (Allington and Chamberlain, 1948), and it is currently prevalent in the north central region of the U.S.; it is present in approximately 69% to 73% of the soybean fields in Illinois, Iowa, and Minnesota (Workneh et al., 1999). BSR, caused by the vascular fungus *Phialophora gregata* (Allington and Chamberlain, 1948) W. Gams f.sp. *sojae Kobayasi, Yamamoto, Negish et Ogoshi* (Gray and Grau, 1999), is ranked as the fifth most important disease of soybean in the U.S. (Wrather et al., 2001).

Genetic variation among isolates has been identified and classified into two groups: Type I-defoliating, which causes both foliar and internal stem symptoms; and Type II-nondefoliating, which only causes internal stem symptoms (Gray, 1971). This phenotypic concept of pathotypes is also supported by DNA-based characterization. Chen et al. (2000) identified a region of variability among isolates of *P. gregata* and developed species-specific primers capable of separating isolates by polymerase chain reaction (PCR) into two distinct genotypes: A (1020 bp) and B (830 bp). Genotype A isolates are capable of causing both foliar symptoms and internal stem discoloration; the B isolates cause only internal stem symptoms (Hughes et al., 2002).

Mild internal stem symptoms can be observed during the vegetative stages of soybean as early as the V4 stage of development (Fehr et al., 1971). However, internal symptoms intensify as the plant enters reproductive stages. The onset of foliar symptoms does not occur until growth stages R4 and R5, and foliar symptoms peak at R7 (Gray and Grau, 1999; Mengistu and Grau, 1987).

Yield losses of 10% to 30% are commonly caused by BSR (Bachman et al., 1997), and losses of up to 44% have been reported (Dunleavy and Weber, 1967). Greatest yield loss occurs in seasons with a cool period in early August followed by hot, dry weather during late pod-fill (Mengistu and Grau, 1987; Mengistu et al., 1987). Two-thirds of the yield losses have been attributed to a reduction in seed numbers and one-third to a reduction in seed size (Weber et al., 1966). Severely diseased plants lodge extensively, which results in an even greater mechanical harvest loss in addition to physiological yield loss.

Three loci have been reported to control resistance to BSR: *Rbs1* locus that traces to PI 84946-2, *Rbs2* that traces to PI 437483, and *Rbs3* locus that traces to both PI 84946-2 and PI 437970 (Bachman and Nickell, 2000; Eathington, et al., 1995; Hanson et al.,

1988; Wilmot and Nickell, 1989). The three resistance genes have been mapped to a region near the simple sequence repeat (SSR) markers Satt431 and Satt244 on LGJ (Bachman and Nickell, 2000; Bachman et al., 2001; Cregan et al., 1999; Klos et al., 2000; Lewers et al., 1999). A desirable interaction between resistance to BSR and to the soybean cyst nematode (SCN) has been observed, particularly when the SCN resistance traces to PI 88788 (Tabor et al., 2003a). Soybean cultivars with SCN resistance tracing to PI 88788 are frequently observed to express none to mild symptom severity when infected by *P. gregata* (Grau et al., 2004).

In Iowa, the disease is widely present throughout the state, and it has even been hypothesized that *P. gregata* may be a native inhabitant of the soils in Iowa. The currently recommended management strategy is use of BSR-resistant soybean cultivars combined with rotation to non-host crops.

Breeding for resistance to BSR has been and continues to be an important objective at ISU. Several cultivars have been released as a result of these efforts, an example of which is cultivar 'BSR101,' which is widely planted in the state and used in research (Tachibana et al., 1987). Another recent release by the ISU program is IA2050 (Cianzio and Fehr, 2000). Cultivars and germplasm lines will be available in the future from the combined use of the three different screening techniques (Cianzio, unpublished results; Tabor et al., 2003b and 2006). The three screening techniques used for development of these cultivars and germplasm lines are field plantings on infested soil, use of the colonization method developed by Tabor et al. (2003b, 2006), and use of molecular markers (Lewers et al., 1999). The experimental lines that will be released are obtained from populations derived from two-parent crosses in which one of the parents provides the resistance genes, and the second parent provides the yield genes.

Selection and breeding for BSR-resistant cultivars had relied heavily on field screenings conducted on soils naturally infested with BSR, such field screenings are the classic screening/selection tools to identify resistant genotypes. Identification of quantitative trait loci (QTL) and molecular markers (Klos et al., 2000; Lewers et al., 1999) have made possible the use of marker-assisted selection to identify resistant genotypes by tracing to known sources of resistance (Tabor et al., 2003b). In addition, a new screening protocol has been recently developed by Tabor et al. (2003b, 2006) which allows screening of genotypes irrespective of the source of resistance. The screening technique, named the "colonization method" is based on differential internal growth of *P. gregata* within susceptible and resistant genotypes. Growth of the fungus in the susceptible genotype is concurrent with stem growth of the plant (Tabor et al., 2006). Conversely, in the resistant genotypes, internal fungal growth is slowed down compared with growth rate of the healthy stem. By cutting small portions of the stem as the plant grows, plating the stem portions in agar, and observing if there is mycelium formation on the plate, it is possible to determine the extent of internal fungus growth inside the stem.

The three screening methods, field tests, molecular markers, and the colonization method are highly correlated, and they provide reliable results to prove resistance. In the breeding program at ISU, both the colonization method and molecular markers are used to identify resistant genotypes.

Sudden Death Syndrome (SDS)

SDS was first observed in the U.S. in Arkansas in 1971, and it is now widespread in the major soybean-producing areas of the U.S. It is caused by the soil-borne fungus *Fusarium solani* f.sp. *glycines* (Roy, 1997), which has been reported to have low genetic variation (Achenbach et al., 1997). SDS symptoms generally appear during reproductive stages of the soybean, producing interveinal chlorotic spots on leaves, that later expand into interveinal chlorotic streaks with the leaves becoming necrotic. The symptomatic leaflets can abscise and drop off.

Yield loss due to SDS may not always be noticeable, however, losses can approach 100% in areas where the disease is severe (Hartman et al., 1995). Yield is affected thru reduction of seed size caused by leaf area reductions due to foliar necrosis and premature defoliation. Seed number reductions have also been observed, resulting from flower and pod abortion (Rupe et al., 1993).

Inheritance of resistance has been identified as being qualitative (Stephens et al., 1993) and also quantitative (Njiti et al., 2002). Depending on the resistance source, seven or eight quantitative trait loci (QTL) have been identified, which forces consideration of the trait for breeding purposes as a quantitative trait. Important progress has been made in understanding the genetic control of this disease, both at the plant and the molecular level (Njiti et al., 1996, 2001, 2002; Schultz et al., 2006). Two genes, *QRfs1* and *QRfs2*, for resistance to SDS have been identified (Triwitayakorn et al., 2005). An extensive molecular analysis is now available for this disease (Shultz et al., 2006), and interactions between the *Dt* loci (genes affecting stem termination in soybean; Palmer et al., 2004) with SDS resistance have also been established (Njiti and Lightfoot, 2006). As is the case with other diseases, interactions between different diseases have been identified. In the case of SDS, an interaction has been established in the expression of SDS in soybean between *Fusarium solani* f.sp. *glycines*, the causal agent of SDS, and *Heterodera glycines*, the soybean cyst nematode (Xing and Westphal, 2006). Genotypes resistant to SCN appear to have an added advantage in being resistance

to SDS. Li et al. (2004) have also published information on the expressed sequence tags of *Fusarium solani* during infection of soybean.

In Iowa, SDS was first observed during the 2000 season and since then breeding efforts have been in place to develop resistant cultivars (Cianzio, unpublished information). This is the most effective approach to the control of SDS. In the breeding program at ISU, resistance sources obtained from Southern Illinois State University in Carbondale, Illinois, are used in two- and three-way crosses to develop populations. Transferring the resistance genes from southern material to soybeans of the earlier maturity groups, such as those adapted to Iowa, poses some practical problems. Large population sizes are necessary to obtain lines of appropriate maturity and with the appropriate number of QTL that confer resistance to SDS. Population development to conduct studies to determine individual QTL contribution to final expression of resistance is underway (Cianzio, unpublished results).

Screening for SDS resistance also poses challenges to breeding efforts. The screening has been done in fields that have a history of SDS occurrence, and even when the correct trial locations have been selected, development of symptoms to the level required for genotype evaluation only occurs about half the time. This lack of consistency of field symptoms has been a major limitation in SDS field research and in cultivar development. In addition to field testing, greenhouse evaluation methods have also been developed with somewhat inconsistent results, although they do provide supportive data for the characterization of varieties regarding their level of SDS resistance (Navi and Yang, 2003, 2005; Njiti et al., 2001; Yang and Navi, 2003). Currently, there are a number of greenhouse methods employed at different institutions in which correlations between field and greenhouse results have proven to be adequate. In the breeding program at ISU, both field and greenhouse screenings are conducted for line evaluation, and currently several advanced experimental lines possessing adequate resistance and superior yield performance have been obtained (Cianzio, unpublished results).

Soybean Cyst Nematode (SCN)

SCN is the most important soybean pathogenic-nematode, and it is distributed worldwide. In the U.S., it was first found in North Carolina in 1954, and it is now distributed in all major soybean production areas in the country. Numerous nematodes attack soybeans, but undoubtedly this is the one that causes the most economic damage (Niblack et al., 2004). The SCN (*Heterodera glycines*) is a vermiform organism which lives in the soil and feeds on soybean roots. The second-stage juveniles are the only life stage of the nematode capable of infecting and establishing a feeding relationship with the soybean plant.

Typical symptoms for the disease include plant stunting and leaf chlorosis in oval patches elongated in the direction of tillage. However, *H. glycines* can cause soybean yield loss even in the complete absence of symptoms. In these cases yield losses may reach 15% or more of the yield that could have been obtained in the absence of the nematode. In plantings on high-fertility soil that may be considered high-yield environments, yield losses in the absence of visual symptoms may reach up to 30%. The primary effect of the nematode on the plant is to reduce the number of pods and seeds.

One of the most difficult issues with nematodes is the variation among populations and their ability to develop and reproduce on resistant soybean lines (Niblack et al., 2004). Even highly resistant lines allow a few females to develop and reproduce, which results in selection of populations able to attack the resistant lines. Resistance genes have been obtained primarily from plant introductions, i.e., PI 88788, Peking, PI 90673, PI 209332 and, more recently, PI437654 (Diers et al., 1997). PI 88788 is the predominant source of resistance in cultivars of the midwestern U.S. (Diers and Arelli, 1999), with a few cultivars released from PI 90763 (Cianzio et al., 2002; Hartwig and Young, 1990), PI 437654 (Anand, 1992a), and PI 209332 (Anand, 1992b; Orf and MacDonald, 1995).

Inheritance of resistance in soybean is conditioned by at least four different genes, which makes breeding for the trait a challenging endeavor. In Peking, three recessive genes (*rhg1*, *rhg2*, *rhg3*) were reported in 1960; these conditioned resistance to an SCN-field population later classified as the Race 1 phenotype. An additional dominant gene (*Rhg4*) was identified in Peking that controlled resistance to the SCN Race 3 phenotype. Resistance in Peking to a cultured, homogeneous SCN population of the Race 3 phenotype was later verified as conditioned by only three resistance genes, *rhg1*, *rhg2*, and *Rhg4* (Arelli et al., 1992). Additionally, Qiu et al. (1997) reported that resistance in PI 88788 was conditioned by one recessive (*rhg1* or *rhg2*) and two dominant genes (*Rgh4*, *Rhg*). PI 437654 was found to have resistance genes in common with Peking (Webb et al., 1995). Inheritance of resistance to Race 3 in PI 209332 is conditioned by one dominant and one recessive gene.

Molecular-marker studies have shown that the *G. max* sources of SCN resistance have genes in common (Concibido et al., 2004; Diers et al, 1997). PI 437654 (Webb et al., 1995), PI 209332, PI 88788, PI 90763, PI 89772, and Peking (Concibido et al., 1996; Chang et al., 1997; Yue et al., 2001) all have the major SCN resistance gene *rhg1* on linkage group G (Cregan et al., 1999). This locus controls a large portion of the total variation for resistance and is effective against several SCN races (Concibido et al., 1996, 1997, 2004). In addition, Peking

(Mahaligan and Skorupska, 1995; Chang et al., 1997), PI 209332 (Concibido et al., 1994), and PI 437654 (Webb et al., 1995) have the resistance gene *Rhg4* that maps near the I locus (black seed coat pigmentation) on linkage group A2 (Cregan et al., 1999). *Glycine soja* PI 468916 has been shown to possess resistance to SCN conditioned by the *rhg1* gene, and has been suggested as an additional source of resistance to use in breeding to increase genetic diversity (Kabelka et al., 2006).

All together, 62 marker-QTL associations have been reported in 17 papers for resistance to SCN races 1, 2, 3, 5, 6, and/or 14 in a total of 13 soybean accessions (nine resistance sources) (Concibido et al., 2004; Glover et al., 2004). In a study using meta-analysis of QTL locations, QTL on linkage groups G, A2, B1, E, and J were confirmed and can be candidates for fine-mapping and cloning (Guo et al., 2006). Guo et al. also identified two clusters of QTL on linkage group G, one of them being *rhg1*. These results provide guidance on which candidate genes could be fine-mapped, and they could eventually open up new avenues for more efficient breeding approaches.

In spite of difficulties in dealing with several genes at once and the complexities of the nematode populations in the soil, numerous cultivars have been released with resistance to SCN, both in the public and private sector; cultivars are available in all maturity groups grown in the U.S. Some recent public release lines are AR1 (Cianzio et al., 2005), A95-684043 (Cianzio et al., 2002), and LDX01-1-65 (Diers et al., 2005).

The breeding program at ISU develops populations for the selection of SCN-resistant cultivars and germplasm lines with adequate yield. The populations are formed by crossing resistant parents to high-yielding genotypes and then developing inbred lines that are classified by maturity and evaluated on their resistance to SCN at early stages.

Evaluations are conducted in greenhouse conditions. Soybean seed are planted in plastic containers with the shape of a cone (usually identified as "cone-tainers") filled with SCN-infested soil (Cianzio, unpublished results). The cone-tainers are placed inside a waterproof box used as a heated-water bath facility to maintain soil temperature in the cone-tainers at an adequate level for cyst nematode population development. It has been determined that for soybean cyst nematodes, the thermal optimum for embryogenesis and egg hatch with low mortality is 24°C (Altson and Schmitt, 1988). The resistant lines are tested for yield and superior lines are advanced for further yield testing.

When a superior line is identified for both yield and SCN resistance traits, molecular marker analysis and greenhouse tests are conducted to determine the type of SCN resistance the genotype has, as indicated by the resistant nematode phenotypic race and the H.G. type. The letters "H.G." stand for *Heterodera glycines*, and are used to remind the breeder and grower that a field nematode population is a collection of nematodes with variable genetic constitution that can infect several different soybean plant genotypes. Since 2002 (Niblack et al., 2002), resistance to SCN is defined not only using the traditional concept of phenotypic race, but also determining the H.G. type. The H.G. type information reports the level of cyst infection a nematode population has on the individual genotypes of soybean differential lines traditionally used to determine the race.

Screening for cyst nematode resistance has evolved from the classical SCN-infested soil evaluation to the use of molecular markers and evaluation of the H.G. types under controlled conditions. Early breeding efforts relied heavily on plantings on infested soil, both in the field and in greenhouse conditions by transporting/using soil from infested sites. The greenhouse screening is the screening test to determine resistance to

SCN which is conducted at ISU during the winter months of the northern hemisphere (January–March). The test is performed at the early breeding stages of line development to discard any susceptible material prior to the first year yield tests planted during the summer.

The new classification system to determine the H.G. type was devised by Niblack et al. (2002) to cope with the heterogeneous and heterozygote nature of the cysts in natural populations. To control some of the inherent variation of the nematode population, Arelli et al. (1992) developed nematode inbred lines to conduct greenhouse tests that could render repeatable results. The molecular markers identified also contribute to increasing efficiency of breeding for this pest, with the phenotyping conducted in greenhouse tests, using mainly inbred lines of the nematode. Presently, in the ISU breeding program, the three screening techniques are used to confirm as much as possible the resistant nature of lines prior to release (Cianzio et al., 2005).

Iron-deficiency chlorosis (IDC)

The problem with IDC was first noted in Iowa when the cultivar 'Corsoy' became widely planted in the state (deMooy, 1972). This genotype, when planted on calcareous soil with a relatively high pH, showed chlorosis symptoms in the leaves, later identified as caused by iron (Fe) deficiency in the soil. Fe-deficiency in soybean may occur when susceptible genotypes are planted on high-pH calcareous soils. The frequency of these soils is estimated at 25% to 30% worldwide, with the largest portion occurring in the U.S. (Cianzio, 1993, 1999). Cultivars unable to acquire and utilize Fe efficiently may develop foliar chlorosis leading to economically important yield loss (Froehlich and Fehr, 1981; Niebur and Fehr, 1981). Of the two physiological strategies developed by plants to cope with Fe-deficiency problems,

soybeans possess Strategy I, which involves: (1) initial excretion of protons to acidify the surrounding rhizosphere and increase solubility; (2) reduction of Fe^{+3} to Fe^{+2} completed by a Fe^{+3}-chelate reductase, and (3) transport of Fe^{+2} across the plasmalemma of root epidermal cells (Hell and Stepahn, 2003; Fox and Guerinot, 1998, both cited by Cianzio et al., 2006).

Chlorosis in genotypes susceptible to Fe deficiency occurs in the interveinal tissue of young leaves, while the veins themselves remain green (Cianzio, 1993; Cianzio et al., 1979). Under severe deficiency, all leaves are affected and plants remain stunted, shedding most of their leaves at later stages. Under less severe conditions, only young leaves may be discolored. Although plants may recover during the growing season, yield is reduced whenever any yellowing occurs (Niebur and Fehr, 1981).

The first report on the inheritance of the trait indicated that differences in Fe efficiency were conditioned by a single gene (Weiss, 1943) and that the genotype of the rootstock was the controlling factor in the use of Fe (Brown et al., 1958, cited by Cianzio, 1993). Continual variation in the expression of the trait prompted additional genetic studies, which determined that the trait was conditioned by a single major gene and modifiers (Cianzio and Fehr, 1980). Additional research conducted with a different set of parents indicated inheritance to be population-dependent, in some cases quantitative with additive gene action (Cianzio and Fehr, 1982b).

The best method to prevent IDC is to plant IDC-resistant cultivars. In northern Iowa, the soil association Clarion-Nicollet-Webster commonly occurs and requires planting of IDC-resistant genotypes if growing soybeans is to be economically feasible for growers. The complex inheritance of the trait and the genotype × environment interaction, however, poses an interesting challenge in breeding for IDC resistance

(Lin et al., 1997; Charlson et al., 2003 and 2005). In spite of these difficulties, cultivars and germplasm lines have been released with adequate resistance to IDC, both in the public (Cianzio et al., 2006a, 2006b; Fehr and Cianzio, 1980; Fehr et al., 1984; Jessen et al., 1988), and in the private sector.

In breeding for IDC, populations are formed via artificial hybridization, lines are developed via inbreeding, and screened for IDC prior to field tests to evaluate for yield and agronomic traits. In breeding for the trait, different strategies for population development are used, such as backcross, single-cross, and three- and four-way crosses. Research that compared different population types has indicated that it is possible to obtain Fe-efficient genotypes by using any of the different population strategies (Cianzio and Voss, 1994; Hintz et al., 1987).

Molecular characterization of Fe-deficiency chlorosis in one population has identified two QTL in linkage group B2, and one in each of linkage group G and N (Lin et al., 1997). In a second population, one QTL each in linkage groups A1 and N was identified (Lin et al., 1997). The lines were evaluated in the field on calcareous soil, and QTL were also identified for chlorophyll concentration. The location of QTL in the molecular linkage groups, however, varied somewhat depending if phenotyping was done using visual chlorosis scores in field tests on calcareous soils or if phenotyping was done by measuring chlorophyll concentration. The same group of genotypes were phenotyped using the two different methods: field evaluation and determination of chlorophyll concentrations. R^2 values calculated for each of the QTL suggested a typical quantitative inheritance in one population, whereas in the second population, there was evidence of a major gene controlling the trait. The R^2 value in this second population indicated that 73% of the total variation was

accounted for in the major gene controlling the trait. More recent studies conducted using a third population failed to confirm the location of QTL in linkage groups previously identified by Lin et al. (1997). An attempt to identify markers for use in marker-assisted selection indicated that Satt481 was an informative marker, however, no flanking marker could be identified in the population (Charlson et al., 2005).

Fe-deficiency chlorosis is a trait that is highly influenced by the environment, with a significant genotype × environment interaction in all statistical analyses (Lin et al., 1997). This interaction is one of the factors that causes difficulties in the identification of QTL. Another confounding issue is that, by all indications, the trait seems to be truly quantitative, with many genes interacting and each contributing a small effect to final expression of the trait. QTL identified by the use of nutrient solution tests confirmed QTL that were identified when the phenotyping was conducted in the field, providing some indication that the use of nutrient solution tests could be used for the selection of lines that will also perform adequately in field conditions (Lin et al., 1998). A limiting factor in using nutrient solution screening is that selection would be based on individual plants; also, the fewer lines may be screened than in field evaluations, due to space restrictions in greenhouse conditions.

Screening is conducted in field plantings on calcareous soils (Cianzio, 1993), in the greenhouse using nutrient solution tests (Coulombe et al., 1984), and more recently using molecular markers (Charlson et al., 2003 and 2005). Reliable data are obtained from replicated field tests conducted in several locations and in different years; however, this extensive testing adds time and resources to cultivar development.

Nutrient solution tests have been effective in the selection of resistant genotypes, with the added advantage that selection can be conducted during winter, increasing efficiency of breeding by diminishing the number of genotypes that will be planted in the field during summer. However, selection is based on individual plant performance, and space may be a limiting factor when large population sizes need to be evaluated.

Use of molecular markers as a selection tool might be a practical alternative (Lin et al., 1997, and 1998; Charlson et al., 2005) in the sense that it can be done during the winter, and a large number of individuals may be characterized. However, to date, molecular markers are parent-population-dependent; if different genotypes are used to develop populations, the markers may be neither informative nor useful in the selection of resistant genotypes. Currently, the breeding program at ISU screens genotypes in field tests planted on calcareous soils, and, at later stages, applies molecular markers depending on the sources of IDC resistance used to develop the populations.

Other diseases, pests, and abiotic stress-related factors

Depending on the region in which soybeans are planted, numerous other diseases, pests, and abiotic stress-related factors are of importance. Many accomplishments in basic research and in the release of resistant cultivars have been achieved in areas that concentrate on these factors, but it is beyond the scope of this chapter to cover them all. Numerous publications and books have been written that cover these areas in great detail and depth.

Also not detailed here are the interactions between the pathogens already discussed and the many other different pests and abiotic stress factors that affect soybeans. These interactions can cause greater yield reductions in soybean. They can also complicate breeding efforts, by forcing researchers to work with more than one disease simultaneously. Interactions have been established between BSR and SCN (Tabor et al., 2001), between iron deficiency chlorosis and SCN (Charlson et al., 2005), and between SDS and SCN (Njiti et al., 1996), to name a few. Other possible interactions between diseases, pests, and abiotic factors have also been reported (Boerma and Specht, 2004).

The soybean breeding program for disease resistance at ISU examines some of these interactions. The approach has been to breed first for the disease considered to be the main target. Once this has been incorporated into a high-yielding superior genotype, genes are added for the second disease or trait of importance. An example of this approach is the recent release of the germplasm line AR1 (Cianzio et al., 2005), which possesses resistance to SCN and PRR. Resistance to PRR was incorporated into the superior SCN-resistant genotype, which was the first targeted pest. The second step was to transfer the *Rps1-k* gene for resistance to PRR into the line via the backcross method of breeding.

Breeding methods and tools

Soybean is a highly self-pollinated species and all commercially grown cultivars are either pure lines (inbred lines) or mixtures of pure lines. The great majority of cultivars presently released have resulted from sexual hybridization followed by selection. Breeding methods include pedigree, bulk, mass selection, single-seed descent, backcross, and diverse forms of recurrent selection. These methods have been in development since the early days of plant breeding and may be regarded as the classic tools. Recently developed tools are used in conjunction with these methods. Presently, all the tools, both classic and recently developed, coexist in breeding programs, with the breeder deciding which approach is more suitable for the trait being improved. As more becomes known about genes, their sequence, distribution throughout the genome, and location, the classic tools will undoubtedly be modified, and the precision of breeding for a certain trait should increase.

Classic tools

The majority of the genetic progress achieved in soybean seed yield, pest resistance, nutritional quality, and industrial uses has been attained by use of classical breeding methods. Commercial cultivars planted by growers are mostly pure lines or blends of pure lines, generally developed by (1) selecting parents to form segregating populations, (2) development of inbred lines, and (3) yield testing. Different breeding methods may be applied during inbred line development, and different strategies may be used to develop populations, with methods and strategies varying depending on the breeding program, objectives, and available resources. Perhaps the crucial point for any breeding program and trait under selection is, as mentioned before, selection of parents to form the segregating populations. Second in importance may be the methods and/or tools used to phenotype individuals, and thirdly, how the testing of lines is done to evaluate the selected material.

A recent survey on the breeding methods currently used by both public and private soybean breeders was conducted by Dr. Lori Scott, from the University of Minnesota (personal communication); it provides an overview of how soybean breeding is currently done in the U.S. (Table 9.2).

Populations are formed, in a majority of cases, by using elite soybean lines as parents, although the use of plant introductions is becoming more common. Two-parent crosses continue to be the preferred approach in population development, along with artificial hybridizations used in crossing. The non-emasculation technique is the most widely used tool for crossing, as it saves time and successfully delivers hybrid seed. The technique, developed and evaluated by Walker et al. (1979), is now widely used. Single-seed descent is the preferred strategy for inbred line development, and the F_4 generation is the prevalent generation in which selections are done.

Field testing is the environment mostly used for evaluations, and the majority of breeders use row plots. In general, evaluation of advanced experimental lines is done in regional tests, either in the public sector or within seed companies. The number of years required for cultivar development ranges from 3 to 10; the shorter periods are the results of making use of off-season nurseries. Because use of off-season nurseries is so prevalent, this may be why the single-seed descent method is so frequently used.

Off-season nurseries have greatly facilitated efficiency of breeding in soybean. Soybean nurseries may be located either in tropical environments or in temperate zones (Cianzio, 1985), the latter located in the hemisphere opposite to which the breeding program is housed. Use of off-season nurseries allows the planting of more than one generation in a year, with all generations planted in field conditions.

The breeding program for disease resistance at Iowa State University uses off-season plantings in Puerto Rico, where it is possible to obtain two generations during the winter months of the northern hemisphere. Depending on the requirements of the work, the lines are planted continuously in Puerto Rico, until yield tests and agronomic evaluations make it necessary to conduct plantings in the adapted location. As examples, Tables 9.3 and 9.4 contain information showing how the backcross breeding method and inbred line development are conducted using off-season nurseries facilities by the breeding for disease resistance program at ISU.

Table 9.2. Different breeding methods used by public and private soybean breeders in the USA as described in a survey conducted in February 2006 (source: L. Scott, Univ of Minnesota, personal communication). Answers to suvery questions are expressed in percentages.

	Soybean breeders	
	Private sector	Public sector
Germplasm sources		
Plant introductions	47	73
Own elite lines	94	97
Others' elite lines	71	77
Public cultivars/lines	82	97
Private cultivars with permission	76	73
Exotic germplasm	65	77
Lines from genetic transformation	53	50
Lines from mutagenic treatments	12	37
Population types		
Two-parent	100	97
Multi-parent	88	67
Backcross	65	87
Hybridization techniques		
Hand pollination with emasculation	24	57
Hand pollination without emasculation	88	50
Male sterility/natural pollinators	12	13
Recombinant DNA	12	7
Inbred line development method		
Bulk	41	20
Pedigree	35	60
Single-seed descent	82	90
Generation of selection		
F_2	18	47
F_3	41	50
F_4	88	83
F_5	41	73
F_6	24	37
Test environment		
Field—summer	100	93
Field—off season	82	63
Greenhouse	82	73
Growth chamber	29	30
Wet lab chemistry	35	47
NIRS*	65	73
Marker-assisted selection	88	57
Plot types		
Spaced-plants	35	33
Hill plots	41	20
Row plots	94	93
Multi-row plots	94	93
Regional tests		
Public tests	24	96
Private tests	53	37
Own trials	6	3
	Average-range	Average-range
Years from crossing to cultivar release	6.3 3–9.5	8.1 5–10

*NIRS = Near infrared analysis.

Table 9.3.   Description of the backcross breeding method as it is conducted by the breeding program for disease resistance at ISU using the two research sites: the off-season nursery in Puerto Rico, and the adapted environment in Iowa.

Year	Season	Work conducted	Generation planted
		Puerto Rico site	
1	1	Cross	Parents
	2	BC_1F_1 seed	F_1 plants
	3	BC_2F_1 seed	BC_1F_1 plants
2	1	BC_3F_1 seed	BC_2F_2 plants
	2	BC_4F_1 seed	BC_3F_1 plants
	3	BC_4F_1 plants	BC_4F_2 plants
3	1	BC_4F_2 seed	BC_4F_2-derived lines in F_3
	2	$BC_4F_{2:3}$ derived lines	BC_4F_2-derived lines in F_4
	3	Increase of homozygous resistant BC_4F_4 derived lines	BC_4F_4-derived lines in F_5
		Iowa site	
4	Summer	1) BC_4F_4-derived lines in replicated yield test	$BC_4F_{4:6}$
		2) Purification rows	BC_4F_6

Table 9.4.   Description of the strategy followed by the breeding program for disease resistance at ISU to develop inbred lines using the two research sites: the off-season nursery in Puerto Rico, and the adapted environment in Iowa.

Year	Season	Work conducted	Generation planted
		Puerto Rico site	
1	1	Cross	Parents
	2	F_1 seed	F_1 plants
	3	F_2 seed	F_2 plants
2	1	F_3 seed	F_3 plants
		Iowa site	
	Summer	F_4 plants, maturity classification	F_4-derived lines
3	Summer	First yield test	F_4-derived lines in F_6
4	Summer	Second yield test	F_4-derived lines in F_7
5	Summer	1) Regional tests	F_4-derived lines in F_8
		2) Purification rows	

Recently developed tools

During the 1990s, two techniques already in use in other species, molecular genetic mapping or genomics, and plant transformation emerged as new tools for use in soybean breeding and genetics. Both are now implemented in breeding programs, and their incorporation in breeding is expected to increase even more in the near future.

Soybean genomics

Development of molecular genetic maps based upon DNA sequence polymorphisms was triggered by the suggestion that restriction fragment polymorphisms (RFLP) could serve as an approach for the development of numerous DNA markers (Botstein et al., 1985, cited by Shoemaker et al., 2004). Subsequently, development of numerous additional classes of DNA markers opened up new possibilities for soybean genomic work.

The soybean genome is of average size, and the haploid size is about 1.1×10^9 base pairs (Arumuganathan and Earle, 1991). Its size is about seven and one-half times larger than the genome of *Arabidopsis*, and two and one-half times larger than that of rice (*Oryza sativa* L.), but less than half the size of the corn (*Zea mays L.*) genome. Approximately 40% to 60% of the soybean genome sequence can be defined as repetitive (Gurley et al, 1979; Goldberg, 1978, cited by Shoemaker et al., 2004), a fact that has slowed down progress in mapping and gene identification.

The complete karyotype analysis was reported by Singh and Hymowitz (1988) using pachytene analysis. The analysis of the pachytene chromosomes has shown that more than 35% of the soybean genome is made up of heterochromatin.

The diploid number of soybean is 40 (2n = 40). It has been suggested that *Glycine* was probably derived from a diploid ancestor (n = 11) which underwent aneuploid loss to n = 10 and subsequent polyploidization to the present 2n = 40 (Lackey, 1980). In spite of being a polyploid, the genome for the most part behaves genetically like a diploid. The possible ploidy level of soybean may be one explanation of why there are many examples of duplicate genes independently controlling the same trait, a fact that has been confirmed in inheritance studies using as parents different accessions in the National Soybean Germplasm Collection.

Several types of markers have been developed and used in mapping the soybean genome, i.e., restriction fragment length polymorphism (RFLP), simple sequence repeat (SSR), restriction fragment length polymorphisms and DNA amplification fingerprinting markers (RAPD), amplified fragment length polymorphism markers (AFLP), and single nucleotide polymorphism markers (SNP) (Shoemaker et al., 2004). An extensive treatment of this area and a description of technologies used for DNA marker analysis has been provided by Shoemaker et al. (2004).

Breeding programs are currently using some of the markers to aid in marker-assisted selection (Charlson et al., 2005; Njiti et al., 2002; Tabor et el., 2006), which also helps breeders understand the concept of genetic diversity at the molecular level (Concibido et al., 2004; Shultz et al., 2006). Gene identification, the next step, will greatly increase breeding efficiency. Random sequencing of gene transcripts is a simple and efficient method of identification of many expressed genes in an organism (Putney et al., 1983, cited by Shoemaker et al., 2004). The sequences, known as expressed sequence tags (EST) are a valuable and efficient method for gene discovery and genome sequencing (Shoemaker et al., 2004). Work in this area is currently underway for soybean and will provide structural and functional genomic information that will be very valuable for research and breeding in soybean.

Use of molecular markers in marker-assisted selection in soybean breeding is becoming more widely used as the numbers of markers identified increases and their consistent phenotypic association is confirmed across several different soybean mapping populations. Numerous markers have been developed that are of use in soybean breeding programs and contribute to precision in the selection process (Concibido et al., 2004; Wilson, 2004). Markers have also been used as tools to confirm QTL presence in recent germplasm releases (Cianzio et al., 2005, 2006a, 2006b). The Soybase Database (www.soybase.ncgr.org) contains the most up-to-date information on identified molecular markers and the literature references related to the marker.

Soybean plant transformation

Genetic transformation of soybean has become firmly established as a viable breeding and research tool. This was possibly due to technical advances in cell and molecular biology that were achieved in the 1980s, which made possible development of the Roundup Ready (RR) soybean by the Monsanto Company (St. Louis, MO) (Parrot and Clemente, 2004).

So far, genetic engineering has rendered two commercial products that have been widely accepted by soybean producers and the general public, the Round-Up Ready (RR) soybean, and the High Oleic (HO) soybean (Parrot and Clemente, 2004). There are numerous other traits derived from genetic engineering of soybean that have been reported in the literature, however, none of them is yet commercially available. According to Parrot and Clemente (2004), the engineered traits can be classified as (1) agronomic, (2) protein and oil quality, (3) specialty oils, (4) removal of allergenic proteins, and (5) pharmaceuticals. In spite of the many traits in which genetic engineering has been practiced to develop genetically modified organism (GMO), only the Round-Up Ready (RR) soybean and the High Oleic (HO) soybean have reached the commercial level. That is because in order for new transgenic cultivars to be commercially viable, they must have sufficient value added to recover the costs of research and development, compensate all owners of intellectual property involved in development of the GMO, and recover all regulatory costs, not to mention the agronomic production costs (Parrot and Clemente, 2004). This has not been feasible for most of the traits. However, the advent of functional genomics will increase the number of genes available for transformation that will meet the aforementioned criteria. Until then, plant transformation remains an important research tool, with some application to practical breeding programs.

Challenges in soybean breeding

In spite of progress made in soybean breeding and production, challenges to both researchers and producers will always be present. In an ever-changing world, nature and needs are in constant motion, evolving and driving each other, at times in consonance, and at other times in opposite directions.

One of the most important challenges in soybean breeding will be to continue the quest for higher and more stable yields across the wide range of environments in which soybeans are planted. Seed yield drives the economics of the commodity. A growing world population's food needs, coupled with serious problems in food distribution, are a constant concern for researchers and producers in reference to seed yield per se and yield stability. The breeding work conducted to-date indicates that yield limits have not been reached. The continuous progress made on the trait, and the efforts in bringing new genetic materials into the gene pool, bode well for future yield improvements.

In the research field, a constant challenge is resource management and the establishment of research priorities. As a new economic order seems to be establishing itself, appropriate use of resources is basic to defining objectives and accomplishments. Each program must conduct its own exercise and analysis to determine how to use resources and define priorities. This is a challenge that all researchers might be facing simultaneously, however, individually, each must make his/her adjustments and decisions to better serve the grower.

A topic related to resource management and priority identification is the time that passes between population development and the release of a cultivar or germplasm. Research conducted at off-season nurseries has greatly contributed to shortening the time necessary to develop and release cultivars and germplasm lines (Ortiz et al., 1986, 1988; Cianzio and Ortiz 1993). However, if other ways could be found or could be added into the scheme to shorten this time even more, it would be of great benefit to all involved in soybean breeding and production. Research has also been conducted on how to increase the efficiency of using off-season sites in tropical regions by using artificial lights to facilitate artificial hybridization among genotypes of widely different maturity groups (Cianzio and Muniz, 1997).

The dynamic picture of diseases and pests poses new challenges to breeding work and soybean production. These changes may be of different types, i.e., changes in race prevalence and the emergence of new races and phenotypes that may result from gene recombination and selection pressure exerted by resistant soybean cultivars; the interaction among races; and the appearance of new diseases and pests. This latter may occur through at least two different mechanisms. One occurs when organisms evolve to survive in new environments. The other is the introduction of foreign diseases and pests, often due to deficient controls in sanitary laws and regulations, or to environmental factors that affect the incidence and distribution of the pests. Two examples that directly impact soybean production in the U.S. have been the recent appearance of soybean aphids and the introduction of soybean rust from the northern region of South America to the continental U.S. Depending on the economic importance and impact the new pests or races, they may redefine breeding objectives and determine new resource allocation. As an example, at the breeding program for pest and disease resistance at ISU, work has already begun on breeding for soybean aphid resistance as well as for soybean rust. The decision to

implement the work was a direct consequence of the economic impact that soybean aphids already have had on soybean production and the possible impact that soybean rust may have in the future.

Progress in areas of genomics and plant transformation will greatly impact breeding efficiency and will open up new possibilities for breeders. It is expected that gene discoveries will facilitate soybean breeding, making it a more exact science. It is also expected that plant transformation will facilitate gene transfers. It might be possible that in the near future, the laborious and time-consuming task of artificial hybridizations may not be required anymore, and that genes may be placed in the desired genotypes through genetic engineering and plant transformation. Without question, the perfected use of these techniques will open up new possibilities and add new dimensions to the breeding work, some that are yet to be identified.

One challenging aspect of soybean breeding that has not been discussed is the human factor. The complexity of the work and the depth and quantity of knowledge in each area requires team and multidisciplinary work if breeders are to accomplish their objectives.

Summary

Soybean is one of the most important world providers of protein feed and vegetable oil, and it is widely planted across the world. It is an annual self-pollinated crop, planted commercially as pure lines or mixtures of pure lines. Breeders have developed soybean cultivars and germplasm lines that meet market and consumer demands. Soybean breeding is conducted by both the public and the private sector in most of the countries in which the crop is grown. The most important breeding objective to address is seed yield; important contributors to high and stable seed yield are agronomic traits, pests, and abiotic resistances. Every program adjusts the particular breeding objectives to requirements of the target soybean production region. Work in genomics and plant transformation is progressing rapidly, and it is expected to greatly change traditional soybean breeding. However, effective human interaction among multidisciplinary teams will be required if breeders are to be successful in meeting the future challenges that are sure to come.

References

Achenbach, L.A., J.A. Patrick, and L.E. Gray. 1997. Genetic homogeneity among isolates of Fusarium solani that cause soybean sudden death syndrome. Theor. Appl. Genet. 95:474–478.

Allington, W.B., and D.W. Chamberlain. 1948. Brown stem rot of soybean. Phytophathology 38:793–802.

Alston, D.G., and D.P. Schmitt. 1988. Development of *Heterodera glycines* life stages as influenced by temperature. J. Nematol. 20:366–372.

Anand, S.C. 1992a. Registration of 'Hartwig' soybean. Crop Sci. 32:1069–1070.

Anand, S.C. 1992b. Registration of 'Delsoy 4710' soybean. Crop Sci. 32:1294.

Arelli, P.R., S.C. Anand, and J.A. Wrather. 1992. Soybean resistance to SCN Race 3 is conditioned by an additional dominant gene. Crop Sci. 32:862–864.

Arumuganathan, K., and E.D. Earle. 1991. Nuclear DNA content of some important plant species. Plant Mol. Biol. Rep. 9:208–219.

Bachman, M.S., and C.D. Nickell. 2000. Investigating the genetic model for brown stem rot resistance in soybean. J. Hered. 91:316–321.

Bachman, M.S., J.P. Tamulonis, C.D. Nickell, and A. F. Bent. 2001. Molecular markers linked to brown stem rot resistance genes Rbs1 and Rbs2 in soybean. Crop Sci. 41:527–535.

Bachman, M.S., C.D. Nickell, P.A. Stephens, A.D. Nickell, and L.E. Gray. 1997. The effect of Rbs2 on yield of soybean. Crop Sci. 37:1148–1151.

Boerma, H.R., and J.E. Specht. 2004. Preface. In Soybeans: Improvement, Production, and Uses, edited by Boerma, H.R. and J.E. Specht, pp. xiii–xv. 3rd edition, ASA, CSSA, SSSA, Madison, Wisconsin, USA.

Boethel, D.J. 2004. Integrated management of soybean insects. In Soybeans: Improvement, Production, and Uses, edited by Boerma, H.R. and J.E. Specht, pp. 853–882. 3rd edition, ASA, CSSA, SSSA, Madison, Wisconsin, USA.

Bravo, J.A., W.R. Fehr, and S.R. Cianzio. 1980. Use of pod width for indirect selection of seed weight in soybeans. Crop Sci. 20:507–510.

Bravo, J.A., W.R. Fehr, and S.R. Cianzio. 1981. Use of small-seeded soybean parents for the improvement of large-seeded cultivars. Crop Sci. 21:430–432.

Brim, C.A. 1973. Quantitative genetics and breeding. In Soybeans: Improvement, Production and Uses, edited by Caldwell, B.E., pp. 155–176. ASA, CSSA, and SSSA, Madison, Wisconsin, USA.

Brim, C.A., and J.W. Burton. 1979. Recurrent selection in soybeans. II. Selection for increased protein content. Crop Sci. 19:494–497.

Burnham, K.D., A.E. Dorrance, D.M. Francis, R.J. Fioritto, and S.K. St. Martin. 2003. Rps8, a new locus in soybean for resistance to Phytophthora sojae. Crop Sci. 43:101–105.

Burton, J.W., R.F. Wilson, and C.A. Brim. 1994. Registration of N79-077-12 and N87-2122-4, two soybean germplasm lines with reduced palmitic acid in seed oil. Crop Sci. 34:313.

Carlson, J.B., and N.R. Lersten. 2004. In Soybeans: Improvement, Production, and Uses, edited by Boerma, H.R. and J.E. Specht, pp. 59–97. 3rd edition, ASA, CSSA, SSSA, Madison, Wisconsin, USA.

Carter, T.E. Jr., R.L. Nelson, C.H. Sneller, and Z. Cui. 2004. In Soybeans: Improvement, Production, and Uses, edited by Boerma, H.R. and J.E. Specht, pp. 303–416. 3rd edition, ASA, CSSA, SSSA, Madison, Wisconsin, USA.

Cerna, F.J., S.R. Cianzio, A. Rafalski, S. Tingey, and D. Dyer. 1997. Relationship between seed yield heterosis and molecular marker heterozygosity in soybean. Theor. Appl. Genet. 95:460–467.

Chang, S.J.C., T.W. Doubler, V.Y. Kilo, J. Abu Thredeih, R. Prabhu, V. Freire, R. Stuttner, J. Klein, M.E. Schmidt, P.T. Gibson, and D.A. Lightfoot. 1997. Association of loci underlying field resistance to sudden death syndrome (SDS) and cyst nematode (SCN) race 3. Crop Sci. 37:965–971.

Charlson, D.V., T.B. Bailey, S.R. Cianzio, and R.C. Shoemaker. 2003. Breeding soybean for resistance to iron-deficiency chlorosis and soybean cyst nematode. Soil Sci. Plant Nutr. 50:1055–1062.

Charlson, D.V., T.B. Bailey, S.R. Cianzio, and R.C. Shoemaker. 2005. Molecular marker Satt481 is associated with iron-deficiency chlorosis resistance in a soybean breeding population. Crop Sci. 45:2394–2399.

Chen, W., C.R. Grau, E.A. Adee, and X. Meng. 2000. A molecular marker identifying subspecific populations of the brown stem rot pathogen, Phialophora gregata. Phytopathology 90:875–883.

Cianzio, S. R. 1985. Off-season nurseries enhance soybean breeding and genetics programs. In Proc. World Soybean Research Conference III, edited by Shibles, R.M., pp. 329–336. Westview Press, Boulder and London.

Cianzio, S.R. de. 1993. Case study with soybeans: Iron efficiency evaluation in field tests compared with controlled conditions. In Iron Chelation in Plants and Soil Microorganisms, edited by Barton, L.L. and Hemming, B.C. pp. 387–397. Academic Press, Inc., New York, NY.

Cianzio, S.R. 1997. Soybean breeding and improvement. In Crop Improvement for the 21ˢᵗ Century, edited by Kang, M.S., pp. 1–13. Research Signpost, T.C. 36/248(1), Trivandrum-695 008, India.

Cianzio, S.R. 1999. Breeding crops for improved nutrient efficiency: Soybean and wheat as case studies. In Mineral Nutrition of Crops, edited by Rengel, Z., pp. 267–287. Haworth Press, Inc., New York, London, Oxford.

Cianzio, S.R., and W.R. Fehr. 1980. Genetic control of iron deficiency chlorosis in soybeans. Iowa State J. of Res. 54:367–375.

Cianzio, S.R., and W.R. Fehr. 1982a. Genetic variability for soybean seed composition in crosses between high- and low-protein parents. J. Agric. Univ. Puerto Rico 66:123–129.

Cianzio, S.R., and W.R. Fehr. 1982b. Variation in the inheritance of resistance to iron deficiency chlorosis in soybeans. Crop Sci. 22:433–434.

Cianzio, S.R., and W.R. Fehr. 1987. Inheritance of agronomic and seed traits in Glycine max × Glycine soja crosses. J. Agric. Univ. Puerto Rico 71:53–64.

Cianzio, S.R., and W.R. Fehr. 2000. Soybean variety IA2050. ISURF #02674.

Cianzio, S.R., and R. González-Muñiz. 1997. Synchronous flowering of soya bean plantings of Maturity Groups III–VI under lighted photoperiods in Puerto Rico. Trop. Agric. (Trinidad) 74:203–209.

Cianzio, S.R., and C.E. Ortiz. 1993. A visual indicator for harvest of immature viable seed of indeterminate soybean genotypes. J. Agric. Univ. Puerto Rico 77:33–44.

Cianzio, S.R., and B.K. Voss. 1994. Three strategies for population development in breeding high-yielding soybean cultivars with improved iron efficiency. Crop Sci. 34:355–360.

Cianzio, S.R., J.F. Cavins, and W.R. Fehr. 1985. Protein and oil percentage of temperate soybean genotypes evaluated in tropical environments. Crop Sci. 25:602–606.

Cianzio, S.R., W.R. Fehr, and I.C. Anderson. 1979. Genotypic evaluation for iron deficiency chlorosis by visual scores and chlorophyll concentration. Crop Sci. 19:644–646.

Cianzio, S.R., S.J. Frank, and W.R. Fehr. 1982. Seed width to pod width ratio for the identification of green soybean pods that have attained maximum length and width. Crop Sci. 22:463–466.

Cianzio, S.R., E.C. Schroder, and C.T. Ramirez. 1991. Response of soybeans of different maturity groups to sowing date in tropical locations in Puerto Rico. Trop. Agric. (Trinidad) 68(4): 306–312.

Cianzio, S.R., R.C. Shoemaker, and D.V. Charlson. 2006. Genomic resources of agronomic crops.

In Iron Nutrition in Plants and Rizhospheric Micro-organisms, edited by Barton, L.L. and J. Abadia, pp. 449–466. Springer, The Netherlands.

Cianzio, S.R., P. Arelli, M. Uphoff, L. Mansur, S. Schultz, and R. Ruff. 2002. Soybean line A95-684043. ISURF #02975.

Cianzio, S.R., D.A. Lightfoot, T.L. Niblack, N. Rivera-Velez, P. Lundeen, and G. Gebhart. 2005. Soybean line AR1. ISURF #03376.

Cianzio, S.R., R.C. Shoemaker, D. Charlson, G. Gebhart, P. Lundeen, and N. Rivera-Velez. 2006a. Soybean line AR2. ISURF #03381.

Cianzio, S.R., R.C. Shoemaker, D. Charlson, G. Gebhart, P. Lundeen, and N. Rivera-Velez. 2006b. Soybean line AR3. ISURF #03380.

Concibido, V.C, B.W. Diers, and P.R. Arelli. 2004. A decade of QTL mapping for cyst nematode resistance in soybean. Crop Sci. 44:1121–1131.

Concibido, V.C., R.L. Denny, D.A. Lange, J.H. Orf, and N.D. Young. 1997. Genome mapping on soybean cyst nematode resistance genes in 'Peking', PI 90673, and PI 88788 using DNA markers. Crop Sci. 37:258–264.

Concibido, V.C., R.L. Denny, S.R. Boutin, R. Hautea, J.H. Orf, and N.D, Young. 1994. DNA marker analysis of loci underlying resistance to soybean cyst nematode (Heterodera glycines Ichinohe). Crop Sci. 34:240–246.

Concibido, V.C., N.D, Young, D.A. Lange, R.L. Denny, and J.H. Orf. 1996. RFLP mapping and marker-assisted selection of soybean cyst nematode resistance in PI 209332. Crop Sci. 36:1643–1650.

Coulombe, B.A., R.L. Chaney, and W.J. Weibold. 1984. Use of bicarbonate in screening soybeans for resistance to iron chlorosis. J. Plant Nutr. 7:411–425.

Cregan, P.B., T. Jarvik, A.L. Bush, R.C. Shoemaker, K.G. Lark, A.L. Khaler, N. Kaya, T.T. VanToai, D.G. Lohnes, J. Chung, and E.J. Specht. 1999. An integrated genetic linkage map of the soybean. Crop Sci. 39:1464–1490.

Diers, B.W., and P.R. Arelli. 1999. Management of parasitic nematodes of soybean through genetic resistance. In Proceedings World Soybean Research Conference VI, edited by Kauffmann, H.E., pp. 300–306. Superior Printing, Champagne, Illinois, USA

Diers, B.W., H.T. Skorupska, P.R. Arelli, and S.R. Cianzio. 1997. Genetic relationship among soybean plant introductions with resistance to soybean cyst nematode. Crop Sci. 37:1966–1972.

Diers, B.W., P.R. Arelli, S.R. Carlson, W.R. Fehr, E.A. Kabelka, R.C. Shoemaker, and D. Wang. 2005. Registration of LDX01-1-65 soybean germplasm with soybean cyst nematode resistance derived from Glycine soja. Crop Sci. 45:1671–1672.

deMooy, C.J. 1972. Iron deficiency chlorosis in soybeans—What can be done about it? Coop. Ext.

Serv., Iowa State University, Pub. 531, Ames, Iowa, USA.

Dunleavy, J.M., and C.R. Weber. 1967. Control of brown stem rot of soybean with corn-soybean rotations. Phytopathology 57:114–117.

Eathington, S.R., C.D. Nickell, and L.E. Gray. 1995. Inheritance of brown stem rot resistance in soybean cultivar BSR101. J. Hered. 86:55–60.

Erickson. E.A., J.R. Wilcox, and J.F. Cavins. 1988. Inheritance of altered palmitic acid percentage in two soybean mutants. J. Hered. 28:465–468.

Fehr, W.R. 1987. Soybean. In Principles of Cultivar Development, edited by Fehr, W.R., pp. 533–557. Vol. 2, Crop Species, Macmillan Publishing Co., New York, NY, USA.

Fehr, W.R., and S.R. Cianzio. 1980. Registration of AP9(S1)C2 soybean germplasm. Crop Sci. 20:677.

Fehr, W.R., and S.R. Cianzio. 1981. Registration of soybean germplasm populations AP10-AP14. Crop Sci. 21:477–478.

Fehr, W.R., G.A. Welke, E.G. Hammond, D.N. Duvick, and S.R. Cianzio. 1992. Inheritance of reduced linolenic acid content in soybean genotypes A16 and A17. Crop Sci. 32:903–906.

Fehr, W.R., S.R. Cianzio, and G.A. Welke. 1990a. Registration of 'LS201' soybean. Crop Sci. 30:1163.

Fehr, W.R., S.R. Cianzio, and G.A. Welke. 1990b. Registration of 'LS301' soybean. Crop Sci. 30:1163–1164.

Fehr, W.R., S.R. Cianzio, G.A. Welke, and A.R. LeRoy. 1990c. Registration of 'SS201' soybean. Crop Sci. 30:1161.

Fehr, W.R., S.R. Cianzio, and G.A. Welke. 1990d. Registration of 'SS202' soybean. Crop Sci. 30:1161.

Fehr, W.R., S.R. Cianzio, and G.A. Welke. 1990e. Registration of 'HP201' soybean. Crop Sci. 30:1361–1362.

Fehr, W.R., S.R. Cianzio, and G.A. Welke. 1990f. Registration of 'HP202' soybean. Crop Sci. 30:1362.

Fehr, W.R., S.R. Cianzio, and G.A. Welke. 1990g. Registration of 'HP203' soybean. Crop Sci. 30:1362–1363.

Fehr, W.R., S.R. Cianzio, and G.A. Welke. 1990h. Registration of 'HP204' soybean. Crop Sci. 30:1362–1363.

Fehr, W.R., B.K. Voss, and S.R. Cianzio. 1984. Registration of a germplasm line of soybean A7. Crop Sci. 24:390–391.

Fehr, W.R., C.E. Caviness, D.T. Burmood, and J.S. Pennington. 1971. Stage of development descriptions for soybeans, Glycine max (L.) Merrill. Crop Sci. 11:929–931.

Froehlich, D.M., and W.R. Fehr. 1981. Agronomic performance of soybeans with different levels of iron deficiency chlorosis on calcareous soil. Crop Sci. 21:438–441.

Graef, G.L., L.A. Miller, W.R. Fehr, and E.G. Hammond. 1985. Fatty acid development in a soybean mutant with high stearic acid. J. Am. Oil Chem. Soc. 62:773–775.

Gray, L.E. 1971. Variation in pathogenicity of Cephalosporium gregatum isolates. Phytophathology 61:1410–1411.

Gray, L.E., and C.R. Grau. 1999. Brown stem rot. In Compendium of Soybean Diseases, edited by Hartman, G.L. pp. 28–29. 4th edition, APS Press, St. Paul, Minnesota, USA.

Grau, C.R., A.E. Dorrance, J. Bond, and J.S. Russin. 2004. Fungal diseases. In Soybeans: Improvement, Production, and Uses, edited by Boerma, H.R. and J.E. Specht, pp. 679–764. 3rd edition, ASA, CSSA, SSSA, Madison, WI, USA.

Gizlice, Z., T.E. Carter Jr., and W.J. Burton. 1994. Genetic base for North American public soybean cultivars released between 1947–1988. Crop Sci. 34:1143–1151.

Glover, K.D., D. Wang, P.R. Arelli, S.R. Carlson, S.R. Cianzio, and B.W. Diers. 2004. Near isogenic lines confirm a soybean cyst nematode resistance gene from PI 88788 on linkage group J. Crop Sci. 44:936–941.

Guo, B., D.A. Sleper, P. Lu, J.G. Shannon, H.T. Nguyen, and P.R. Arelli. 2006. QTLs associated with resistance to soybean cyst nematode in soybean: Meta-analysis of QTL locations. Crop Sci. 46:595–602.

Hammond, E.G., and W.R. Fehr. 1983. Registration of A6 germplasm line of soybean. Crop Sci. 23:192–193.

Hanson, P.M., C.D. Nickell, L.E. Gray, and S.A. Sebastian. 1988. Identification of two dominant genes conditioning brown stem rot resistance in soybean. Crop Sci. 28:41–43.

Harrington, T.C., and D.L. McNew. 2003. Phylogenetic analysis places the phialophora-like anamorph genus Cadophora in the Helotiales. Mycotaxon 87:141–151.

Hartman, G.L., G.R. Noel, and L.E. Gray. 1995. Occurrence of soybean sudden death syndrome in east-central Illinois and associated yield losses. Plant Dis. 79:314–318.

Hartmann, R.B., W.R. Fehr, G.A. Welke, E.G. Hammond, D.N. Duvick, and S.R. Cianzio. 1996. Association of elevated stearate with agronomic and seed traits of soybean Crop Sci. 37:124–127.

Hartwig, E.E. 1973. Varietal development. In Soybean: Improvement, Production and Uses, edited by Caldwell, B.E., pp. 187–210. ASA 1973, Madison, Wisconsin, USA.

Hartwig, E.E., and L.D. Young. 1990. Registration of 'Cordell' soybean. Crop Sci. 30:231–232.

Hawkins, S.E., W.R. Fehr, and E.G. Hammond. 1983. Resource allocation in breeding for fatty acid composition of soybean oil. Crop Sci. 23:900–904.

Hintz, R.W., W.R. Fehr, and S.R. Cianzio. 1987. Population development for the selection of high-yielding soybean cultivars with resistance to iron-deficiency chlorosis. Crop Sci. 27:707–710.

Hitz, W.D., N.S. Yadav, R.S. Reiter, C.J. Mauvais, and A.J. Kinney. 1995. Reducing polyunsaturation in oils of transgenic canola and soybean. In Plant Lipid Metabolism, edited by Kader, J.C., pp. 506–508. Kluwer Academic Publication, Dordrecht, the Netherlands.

Hoeck, J.A., W.R. Fehr, R.C. Shoemaker, G.A. Welke, S.L. Johnson, and S.R. Cianzio. 2003. Molecular marker analysis of seed size in soybean. Crop Sci. 43: 68–74.

Hughes, T.J., W. Chen, and C.R. Grau. 2002. Pathogenic characterization of genotypes A and B of Phialophora gregata f.sp. sojae. Plant Dis. 86:729–735.

Hymowitz, T. 2004. Speciation and cytogenetics. In Soybeans: Improvement, Production, and Uses, edited by Boerma, H.R. and J.E. Specht, pp. 97–136. 3rd edition, ASA, CSSA, SSSA, Madison, Wisconsin, USA.

Ininda, J., W.R. Fehr, S.R. Cianzio, and S.R. Schnebly. 1996. Genetic gain in soybean populations with different percentages of plant introduction parentage. Crop Sci. 36:1470–1472.

Jessen, H.J., W.R. Fehr, and S.R. Cianzio. 1988. Registration of germplasm lines of soybean A11, A12, A13, A14, and A15. Crop Sci. 28:204.

Johnson, S.L., W.R. Fehr, G.A. Welke, and S.R. Cianzio. 2001. Genetic variability for seed size of two- and three-parent soybean populations. Crop Sci. 41:1029–1033.

Kabelka, E.A., S.R. Carlson, and B.W. Diers. 2006. Glycine soja PI 468916 SCN resistance loci's associated effects on soybean seed yield and other agronomic traits. Crop Sci. 46:622–629.

Kaufmann, M.J., and J.W. Gedemann. 1958. Root and stem rot of soybean caused by Phytophthora sojae n. sp. Phytophathology 48:201–208.

Kenworthy, W.J. 1980. Strategies for introgressing germplasm in breeding programs. In Proceedings World Soybean Research Conference II, edited by Corbin, F.T., pp.217–225. N.C. Westview Press, Boulder, Colorado, USA.

Kilen, T.C., E.E. Hartwig, and B.L. Keeling. 1974. Inheritance of a second major gene for resistance to Phytophthora root rot in soybeans. Crop Sci. 14:260–262.

Klos, K.L.E., M.M. Paz, L.F. Marek, P.B. Cregan, and R.C Shoemaker. 2000. Molecular markers useful for detecting resistance to brown stem rot in soybean. Crop Sci. 40:1445–1452.

Lackey, J. 1980. Chromosome numbers in the Phaseoleae (Fabaceae:Faboideae) and their relation to taxonomy. Am J. Bot. 67:595–602.

Lawrence, P.K., and K.J. Frey. 1975. Backcross variability for grain yield in oat species crosses (Avena sativa L. × A. sterilis L.). Euphytica 24:77–85.

LeRoy, A.R., S.R. Cianzio, and W.R. Fehr. 1991a. Direct and indirect selection for small seed of soybean in temperate and tropical environments. Crop Sci. 31:697–699.

LeRoy, A.R., W.R. Fehr, and S.R. Cianzio. 1991b. Introgression of genes for small seed size from Glycine soja into G. max. Crop Sci. 31:693–697.

Lersten, N.R., and J.B. Carlson. 2004. Vegetative morphology. In Soybeans: Improvement, Production, and Uses, edited by Boerma, H.R. and J.E. Specht, pp. 15–58. 3rd edition, ASA, CSSA, SSSA, Madison, Wisconsin, USA.

Lewers, K.S., E.H. Crane, C.R. Bronson, J.M. Schupp, P. Keim, and R.C. Shoemaker. 1999. Detection of linked QTL for soybean brown stem rot resistance in 'BSR 101' as expressed in a growth chamber environment. Molecular Breeding 5:33–42.

Li, S., A.G. Hernandez, L. Liu, X. Zeng, G.L. Hartman, and L.L. Domier. 2004. Analysis of expressed sequence tags of Fusarium solani f.sp. glycines during infection of soybean. Phytophathology 94: S159.

Lin, S., S. Cianzio, and R. Shoemaker. 1997. Mapping genetic loci for iron deficiency chlorosis in soybean. Molecular Breeding 3:219–229.

Lin, S., J.S. Baumer, D. Ivers, S.R. Cianzio, and R.C. Shoemaker. 1998. Field and nutrient solution tests measure similar mechanisms controlling iron deficiency chlorosis in soybean. Crop Sci. 38:254–259.

Mahaligan, R., and H.T. Skorupska. 1995. DNA markers for resistance to Heterodera glycines I. Race 3 in soybean cultivar Peking. Breed. Sci. 45:435–443.

Mengistu, A., and C.R. Grau. 1987. Seasonal progress of brown stem rot and its impact on soybean productivity. Phytopathology 77:1521–1529.

Mengistu, A., H. Tachibana, A.H. Epstein, K.G. Bidne, and J.D. Hatfield. 1987. Use of leaf temperature to measure the effect of brown stem rot and soil moisture stress and its relation to yields of soybeans. Plant Dis. 71:632–634.

Narvel, J.M., W.R. Fehr, J. Ininda, G.A. Welke, E.G. Hammond, D.N. Duvick, and S.R. Cianzio. 2000. Inheritance of elevated palmitate in soybean seed oil. Crop Sci. 40:635–639.

Navi, S.S., and Yang X.B. 2003. Dip inoculation technique to Identify resistance to soybean sudden death syndrome Phytopathology 93:S65.

Navi, S.S., and Yang, X.B. 2005. Seedling inoculation screening technique to identify resistance to soybean sudden death syndrome caused by Fusarium solani f.sp. glycines. In Proceedings of the Global Conference II on Plant Health—Global Wealth, November 25–29, 2005, pp. 252–253. Dept of Plant. Pathology, Rajasthan College of Agriculture, Maharana Pratap. University of Agriculture & Technology, Udaipur, India 313 001.

Ndzana, X., W.R. Fehr, G.A. Welke, E.G. Hammond, D.N. Duvick, and S.R. Cianzio. 1994. Influence of reduced palmitate content on agronomy and seed traits of soybean. Crop Sci. 34:646–649.

Niblack, T.L., G.L. Tylka, and R.D. Riggs. 2004. Nematode pathogens of soybean. In Soybeans: Improvement, Production, and Uses, edited by Boerma, H. R. and J.E. Specht, pp. 821–852. 3rd edition, ASA, CSSA, SSSA, Madison, Wisconsin, USA.

Niblack, T.L., P.R. Arelli, G.R. Noel, C.H. Opperman, J. Orf, D.P. Schmitt, J.G. Shannon, and G.L. Tylka. 2002. A revised classification scheme for genetically diverse populations of Heterodera glycines. J. Nematol. 34:270–288.

Niebur, W.S., and W.R. Fehr. 1981. Agronomic evaluation of soybean genotypes resistant to iron deficiency chlorosis. Crop Sci. 21:551–554.

Niu, X.F., X.B. Yang, P. Lundeen, M.D. Uphoff, and S.R. Cianzio. 2003. Changes in Phytophthora sojae races in Iowa soybean fields. Phytopathology 93: S66.

Njiti, V.N., and D.A. Lightfoot. 2006. Genetic analysis infers Dt loci underlie resistance to SDS caused by Fusarium virguliforme in indeterminate soybeans. Can. J. Plant Science 41:83–89.

Njiti, V.N., M.A. Shenaut, R.J. Suttner, M.E. Schmidt, and P.T. Gibson. 1996. Soybean response to sudden death syndrome: Inheritance influenced by cyst nematode resistance in Pyramid × Douglas progenies. Crop Sci. 36:1165–1170.

Njiti, V.N., J.E. Johnson, G.A. Torto, L. Gray, Y. Luo, P.T. Gibson, and D.A. Lightfoot. 2001. Inoculum rate influences selection for field resistance to sudden death syndrome in the greenhouse. Crop Sci. 41:1726–1723.

Njiti, V.N., K. Meksem, M.J. Iqbal, J.E. Johnson, K.F. Zobrist, V.Y. Kilo, and D.A. Lightfoot. 2002. Common loci underlie field resistance to soybean sudden death syndrome in Forrest, Pyramid, Essex, and Douglas. Theor. Appl. Genet. 104:294–300.

Orf, J.H., and D.H. MacDonald. 1995. Registration of 'Faribault' soybean. Crop Sci. 35:1227.

Orf, J.H., B.W. Diers, and H.R. Boerma. 2004. Genetic improvement: Conventional and molecular-based strategies. In Soybeans: Improvement, Production, and Uses, edited by Boerma, H.R. and J.E. Specht, pp. 417–444. 3rd edition, ASA, CSSA, SSSA, Madison, Wisconsin, USA.

Ortiz, C., S.R. Cianzio, and P.R. Hepperly. 1988. Fungi and insect infestations of soybean seeds harvested at immature stages in tropical environments. J. Agric. Univ. Puerto Rico 72:73–79.

Ortiz, C., S.R. Cianzio, P.R. Hepperly, and W.R. Fehr. 1986. Premature harvest of soybeans for rapid

generation advance. J. Agric. Univ. Puerto Rico 70:159–167.

Palmer, R.G., and T. Hymowitz. 2004. Soybean: Germplasm, Breeding, and Genetics. In Encyclopedia of Grain Science, edited by Wrigley, C., H. Corke, and C. Walker, pp. 136–146. Elsevier Science Ltd., London, UK.

Pantalone, V.R., R.F. Wilson, W.P. Novitzky, and J.W. Burton. 2002. Genetic regulation of elevated stearic acid concentration in soybean oil. J. Am. Oil Chem. Soc. 79:549–553.

Panter, D.M., and F.L. Allen. 1995a. Using best linear unbiased predictions to enhance breeding for yield in soybean. I: Choosing parents. Crop Sci. 35:397–405.

Panter, D.M., and F.L. Allen. 1995b. Using best linear unbiased predictions to enhance breeding for yield in soybean. II: Selection of superior crosses from a limited number of yield trials. Crop Sci. 35:405–410.

Parrot, W.A., and T.E. Clemente. 2004. Transgenic soybean. In Soybeans: Improvement, Production, and Uses, edited by Boerma, H.R. and J.E. Specht, pp. 265–302. 3rd edition, ASA, CSSA, SSSA, Madison, Wisconsin, USA.

Poehlman, J.M. 1979. Breeding Field Crops. Avi Publishing Co., Inc., Westport, Connecticut, USA.

Qiu, B.X., D.A. Sleper, and P.R. Arelli. 1997. Genetic and molecular characterization of resistance to Heterodera glycines race isolates 1, 3, and 5 in Peking. Euphytica 96:225–231.

Rahman, S.M., Y. Takagi, and T. Kinoshita. 1997. Genetic control of high stearic acid content in seed oil of two soybean mutants. Theor. Appl. Genet. 95:772–776.

Roy, K.W. 1997. Fusarium solani on soybean roots: Nomenclature of the causal agent of sudden death syndrome and identity and relevance of F. solani form B. Plant Dis. 81:259–266.

Ross, A.J., W.R. Fehr, G.A. Welke, and S.R. Cianzio. 2000. Agronomic and seed traits of 1%-linolenate soybean genotypes. Crop Sci. 40:383–386.

Rupe, J.C., W.E. Sabbe, R.T. Robbins, and E.E. Gbur, Jr. 1993. Soil and plant factors associated with sudden death syndrome of soybean. J. Prod. Agric. 6:218–221.

Sandhu D., H. Gao, S.R. Cianzio, and M.K. Bhattacharyya. 2004. Deletion of a disease resistance nucleotide-binding-site leucine-rich-repeat-like sequence is associated with the loss of the Phytophthora resistance gene Rps4 in soybean. Genetics 168:2157–2167.

Sandhu, D., K.G. Schallock, N. Rivera-Velez, P. Lundeen, S. Cianzio, and M.K. Bhattacharyya. 2005. Soybean Phytophthora resistance gene Rps8 maps closely to the Rps3 region. Journal of Heredity 2005 96(5): 536 (originally published online on June 15, 2005, doi:10.1093/jhered/esi081).

Schnebly, S.R., W.R. Fehr, G.A. Welke, E.G. Hammond, and D.N. Duvick. 1994. Inheritance of reduced and elevated palmitate in mutant lines of soybean. Crop Sci. 34:829–833.

Schmitthenner, A.F. 1985. Problems and progress in control of Phytophthora root rot of soybean. Plant Dis. 69:362–368.

Shoemaker, R.C., P.B. Cregan, and L.O. Vodkin. 2004. Soybean genomics. In Soybeans: Improvement, Production, and Uses, edited by Boerma, H.R. and J.E. Specht, pp. 235–264. 3rd edition, ASA, CSSA, SSSA, Madison, Wisconsin, USA.

Shultz, J.L., D. Jayaraman, K.L. Shopinski, M.J. Iqbal, S. Kazi, K. Zobrist, R. Bashir, S. Yaegashi, N. Lavu, A.J. Afzal, C.R. Yesudas, M.A. Kassem, C. Wu, H.B. Zhang, C.D. Town, K. Meksem, and D.A. Lightfoot. 2006. The Soybean Genome Database (SoyGD): A browser for display of duplicated, polyploid, regions and sequence tagged sites on the integrated physical and genetic maps of Glycine max. Nucleic Acid Research 34:D1–D8.

Singh, R.J., and T. Hymowitz. 1988. The genomic relationship between Glycine max (L.) Merr. and G. soja Sieb. and Zucc. as revealed by pachytene chromosomal analysis. Theor. Appl. Genet. 76:705–711.

Smalley, M.D., W.R. Fehr, S.R. Cianzio, F. Han, S.A. Sebastian, and L.G. Streit. 2004. Quantitative trait loci for soybean seed yield in elite and plant introduction germplasm. Crop Sci. 44:436–442.

Specht, J.E., and J.H. Williams. 1984. Contribution of genetic technology to soybean productivity retrospect and prospect. In Genetic Contributions to Grain Yields of Five Major Crop Plants, edited by Fehr, W.R., pp. 49–74. CSSA Special Publication 7. CSSA and ASA, Madison, Wisconsin, USA.

Stephens, P.A., C.D. Nickell, and F.L. Kolb. 1993. Genetic analysis of resistance to Fusarium solani in soybean. Crop Sci. 33:929–930.

St. Martin, S.K. 2001. Selection limits—How close are we? Soybean Genetics Newsletter, Available at: http://www.soygenetics.org/articles/sgn2001–003.htm.

St. Martin, S.K., K.S. Lewers, R.G. Palmer, and B.R. Hedges. 1996. A testcross procedure for selecting exotic strains to improve pure-line cultivars in predominantly self-pollinating species. Theor. Appl. Genet. 92:78–82.

Stojsin, D., G.R. Ablett, B.M. Luzzi, and J.W. Tanner. 1998. Use of gene substitution values to quantify partial dominance in low palmitic acid soybean. Crop Sci. 38:1437–1441.

Stoltzfus, D.L., W.R. Fehr, G.A. Welke, E.G. Hammond, and S.R. Cianzio. 2000a. A fap7 allele for elevated palmitate in soybean. Crop Sci. 40:1538–1542.

Stoltzfus, D.L., W.R. Fehr, G.A. Welke, E.G. Hammond, and S.R. Cianzio. 2000b. A fap5 allele for elevated palmitate in soybean. Crop Sci. 40:647–650.

Streit, L.G., W.R. Fehr, G.A. Welke, E.G. Hammond, and S.R. Cianzio. 2000. Family and line selection for reduced palmitate, saturates, and linolenate of soybean. Crop Sci. 41:63–67.

Tabor, G.M., G.L. Tylka, and C.R. Bronson. 2001. Heterodera glycines increases susceptibility of soybeans to Phialophora gregata. Phytopathology 91(suppl.): S87.

Tabor, G.M., G.L. Tylka, J.E. Behm, and C.R. Bronson. 2003a. Heterodera glycines infection increases incidence and severity of brown stem rot in both resistant and susceptible soybean. Plant Dis. 87:655–661.

Tabor, G.M., Tylka, G.L, Cianzio, S.C., and Bronson, C.R. 2003b. Resistance to Phialophora gregata is expressed in the stems of resistant soybeans. Plant Dis. 87: 970–976.

Tabor, G.M., S.R. Cianzio, G.L. Tylka, R. Roorda, and C.R. Bronson. 2006. A new greenhouse method to assay soybean resistance to brown stem rot. Plant Dis. (in press).

Tachibana, H., B.K. Voss, and W.R. Fehr. 1987. Registration of 'BSR101' soybean. Crop Sci. 27:612.

Tolin, S.A., and G.H. Lacy. 2004. Viral, bacterial, and phytoplasmal diseases of soybean. In Soybeans: Improvement, Production, and Uses, edited by Boerma, H.R. and J.E. Specht, pp. 765–820. 3rd edition, ASA, CSSA, SSSA, Madison, Wisconsin, USA.

Tooley, P.W., and C.R. Grau. 1984. The relationship between rate-reducing resistance to Phytophthora megasperma f.sp. glycinea and yield of soybean. Phytopathology 74:1209–1216.

Triwitayakorn, K., V.N. Njiti, M.J. Iqbal, S. Yaegashi, C. Town, and D.A. Lightfoot. 2005. Genomic analysis of a region encompassing QRfs1 and QRfs2: Ggenes that underlie soybean resistance to sudden death syndrome. Genome/Génome 48: 125–138.

Walker, A.K., S.R. Cianzio, J.A. Bravo, and W.R. Fehr. 1979. Comparison of emasculation and nonemasculation for artificial hybridization of soybeans. Crop Sci. 19:285–286.

Walker, J.B., W.R. Fehr, G.A. Welke, E.G. Hammond, D.N. Duvick, and S.R. Cianzio. 1998. Reduced-linolenate content associations with agronomic and seed traits of soybean. Crop. Sci. 38:352–355.

Wang, J., T.L. Niblack, J.N. Tremaine, W.J. Wiebold, G.L. Tylka, C.R. Marrett, G.R. Noel, O. Myers, and M.E. Schmidt. 2002. The soybean cyst nematode reduces soybean yield without causing obvious symptoms. Plant Dis. 87:623–628.

Webb, D.M., B.M. Baltazar, P.R. Arelli, J. Schupp, K. Clayton, P. Keim, and W.D. Beavis. 1995. Genetic mapping of soybean cyst nematode race 3 resistance loci in the soybean PI 437654. Theor. Appl. Genet. 91:574–581.

Weber, C.R., J.M. Dunleavy, and W.R. Fehr. 1966. Influence of brown stem rot on agronomic performance of soybeans. Agron. J. 258:519–520.

Wehrmann, V.K., W.R. Fehr, and S.R. Cianzio. 1988. Analysis of strategies for transfer of an allele for resistance to Phytophthora in soybean. Crop Sci. 28:248–250.

Wehrmann, V.K., W.R. Fehr, S.R. Cianzio, and J.F. Cavins. 1987. Transfer of high seed protein to high-yielding soybean cultivars. Crop Sci. 27:297–299.

Weiss, M.G. 1943. Inheritance and physiology of efficiency in iron utilization in soybeans. Genetics 28:253–268.

Wilcox, J.R. 2001. Sixty years of improvement in publicly developed elite soybean lines. Crop Sci. 41:1711–1716.

Wilcox, J.R. 2004. World distribution and trade of soybean. In Soybeans: Improvement, Production, and Uses, edited by Boerma, H.R. and J.E. Specht, pp. 1–14. 3rd edition, ASA, CSSA, SSSA, Madison, Wisconsin, USA.

Wilcox, J.R., and J.F. Cavins. 1985. Inheritance of low linolenic acid content of the seed oil of a mutant in Glycine max. Theor. Appl. Genet. 71:74–78.

Wilcox, J.R., and S.K. St. Martin. 1998. Soybean genotypes resistant to Phytophthora sojae and compensation for yield losses of susceptible isolines. Plant Dis. 82:303–306.

Wilson, R.F. 2004. Seed composition. In Soybeans: Improvement, Production, and Uses, edited by Boerma, H.R. and J.E. Specht, pp. 621–678. 3rd edition, ASA, CSSA, SSSA, Madison, Wisconsin, USA.

Wilson, R.F., H.H. Weissinger, J.A. Buck, and G.D. Faulkner. 1980. Involvement of phospholipids in polyunsaturated fatty acid synthesis in developing soybean cotyledons. Plant Physiol. 66:545–549.

Wilmot, D.B., and C.D. Nickell. 1989. Genetic analysis of brown stem rot resistance in soybean. Crop Sci. 29:672–674.

Wrather, J.A., W.C. Stienstra, and S.R. Koenning. 2001. Soybean disease loss estimates for the United States from 1996 to 1998. Can. J. Plant Pathol. 23:122–131.

Workneh, F., G.L. Tylka, X.B. Yang, J. Faghihi, and J.M. Ferris. 1999. Regional assessment of soybean brown stem rot, Phytophthora sojae, and Heterodera glycines using area frame sampling: Prevalence and effects of tillage. Phytopathology 89:204–211.

Xing, L.-J., and A. Westphal. 2006. Interaction of Fusarium solani f.sp. glycines and Heterodera glycines in sudden death syndrome of soybean. Phytopathology (in press).

Yang X.B., and Navi, S.S. 2003. Fungal colonization in phloem/xylem tissues of taproots in relation to foliar symptoms expression of soybean sudden death syndrome. Phytopathology 93:S92.

Yang, X.B., R.L. Ruff, X.Q. Meng, and F. Workneh. 1996. Races of Phytophthora sojae in Iowa soybean fields. Plant Dis. 80:1418–1420.

Yue, P., D.A. Sleper, and P.R. Arelli. 2001. Mapping resistance to multiple races of Heterodera glycines in soybean PI 89772. Crop Sci. 41:1589–1598.

Part 3

Carbohydrate Suppliers: Root Crops and Banana

Chapter 10

Breeding Potato as a Major Staple Crop

John E. Bradshaw

Introduction

Domestication and evolution of the modern potato crop

The 219 wild tuber-bearing *Solanum* species recognized by Hawkes (1990), and grouped into 19 series, are distributed from the south-western United States (38°N) to central Argentina and adjacent Chile (41°S) and cover a great ecogeographical range (Spooner and Hijmans, 2001). They form a polyploid series from diploid (*2n = 2x = 24*) to hexa-ploid (*2n = 6x = 72*), in which nearly all of the diploid species are self-incompatible out-breeders and the tetraploids and hexaploids are mostly self-compatible allopolyploids that display disomic inheritance (Hawkes, 1990). But where and when did domestica-tion occur and which species were involved in the process? Simmonds (1995) concluded that just a few closely related diploid species in series Tuberosa (e.g., *Solanum brevicaule, S. leptophyes,* and *S. canasense*) were domes-ticated in the Andes of southern Peru and northern Bolivia more than 7,000 years ago. More recently, Spooner et al. (2005a) have provided molecular taxonomic evidence for a single domestication in Peru from the northern group of members of the *S. brevi-caule* complex of species. The result of domestication was the diploid species *S. ste-notomum* from which six other cultivated species were derived, including *S. tuberosum,* which became the most widely grown species in South America. *S. tuberosum* is a tetra-ploid (*2n = 4x = 48*) species that displays tet-rasomic inheritance. The short-day adapted landrace populations of the Andes and the long-day adapted ones of coastal Chile are genetically distinct groups (Raker and Spooner, 2002) that have been classified as separate subspecies (*S. tuberosum* subsp. *andigena* and subsp. *tuberosum*) and are also referred to as Andigena and Chilean Tuberosum potatoes (Fig. 10.1).

There is still much debate over which group of potatoes was introduced into Europe at the end of the 16th century and subse-quently to the rest of the world from the 17th century onward (Pandey and Kaushik, 2003). Hawkes and Francisco-Ortega (1993) have argued that Andigena potatoes were intro-duced into the Canary Isles around 1562 and from there to mainland Europe in the 1570s. Then, as the growing of potatoes spread northeastward across Europe, the potato became adapted to the long summer days of northern Europe and again evolved sufficiently to be classified as subspecies *tuberosum,* albeit with Andigena cytoplasm. Molecular analyses of old Japanese cultivars revealed that they were derived from sub-species *andigena* through early European potatoes (Hosaka et al., 1994). In contrast, recent molecular analyses have clustered all Indian potato varieties, including putatively remnant Andean populations, with Chilean Tuberosum (Spooner et al., 2005b). Hence it now seems safest to assume that the early

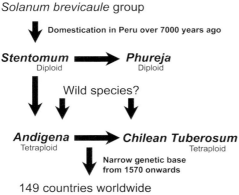

Solanum brevicaule group

Fig. 10.1.   Domestication of cultivated potatoes.

introductions of cultivated potatoes to Europe came from both the Andes and Chile, but were few in number and hence captured only some of the biodiversity present in the cultivated potatoes of South America.

The potato is now the fourth most important food crop after wheat, maize, and rice, with 311 million tons produced from 19 million hectares at a mean fresh-weight yield of 16.4 t/ha in 2003 (FAO statistics). As well as being a staple food, the potato is grown as a vegetable for table use, is processed into french fries and crisps (chips), and is used for dried products and starch production. A study in 1995 (CIP/FAO, 1995) found that processing was the fastest growing sector of the world potato economy, and today processors are building factories in countries where the potato is primarily grown as a staple food. In some countries the potato is still fed to animals, but this use is decreasing.

Potatoes are grown in 149 countries from latitudes 65°N to 50°S and at altitudes from sea level to 4,000 m (Hijmans, 2001); they are grown, in fact, wherever it is neither too hot nor too cold and there is adequate water from rain or irrigation. In practice this means that they are grown as a summer crop in the tropical highlands of Bolivia, Peru, and Mexico; year round in parts of China, Brazil, the equatorial highlands of South America

(e.g., Ecuador and Colombia), and East Africa (e.g., Kenya and Uganda); as a winter crop they are grown in the lowland subtropics (e.g., northern India and southern China); as spring and autumn crops in the Mediterranean (e.g., North Africa); and in summer in the lowland temperate regions of the world (North America, western and eastern Europe, and northern China). The growing season can be as short as 75 days in the lowland subtropics and as long as 180 days in the high Andes. As potatoes cannot be grown year round in most parts of the world, it is normal to have to store both seed tubers for planting the next crop and ware tubers for consumption.

Need for new cultivars

Modern potato breeding began in 1807 in England when Knight made the first deliberate hybridizations between varieties by artificial pollination (Knight, 1807). It flourished in Europe and North America during the second half of the 19th century when many new cultivars were produced by farmers, hobby breeders, and seedsmen. There were exchanges of germplasm between Europe and North America, which included the descendents of 'Rough Purple Chili,' a Chilean Tuberosum cultivar that had been introduced into the U.S. in 1851 (Goodrich, 1863). These descendents were widely employed as female parents in crosses with European Tuberosum at the end of the 19th century and, as a consequence, Chilean Tuberosum cytoplasm predominates in modern cultivars. This uniformity of cytoplasm has been a cause for concern ever since the epidemics of southern corn leaf blight on maize hybrids with T (Texas) cytoplasm in 1970–71 (Ullstrup, 1972). Modern potato breeding started later, in the 1930s, in China and India, but these countries are now two of the leading potato producers in the world (Gaur and Pandey, 2000; Jin et al., 2004). The extent of progress since 1807 can be judged by the latest *World Catalogue of Potato Varieties* (Hils and Pieterse, 2005), which lists more than 4,000 cultivars from more than 100 countries. This is a remarkable achievement since outside of Latin America the genetic base of potato breeding is a relatively small sample of potatoes from the Andes and coastal Chile. Some introgressions of disease resistance genes and some base broadening during the 20th century do not alter this conclusion (Bradshaw and Mackay, 1994a). They have, however, resulted in the use for breeding of *S. phureja,* which is the second most widely cultivated species in South America (see Fig. 10.1).

Despite the large number of cultivars currently available, there is a continuing need for new ones. In Asia (China and India) and Africa there is a need for increased and stable potato production to meet increasing demands for food resulting from human population growth during a period of environmental (including climate) change. New cultivars must deliver higher yields under low inputs, disease and pest attacks, and environmental stresses such as heat, cold, drought, and salinity. If possible, they should also have improved nutritional and health properties, but the overriding need is for increased and stable yields. Such food security is high on the agenda of the International Potato Center (CIP) in Lima, Peru. The CIP was founded in 1970 as an international agricultural research center (IACR), and is now a Future Harvest Center; since 1971, CIP has been supported by the Consultative Group on International Agricultural Research (CGIAR), whose aim is the eradication of human hunger and poverty through research.

Once food security has been achieved, breeding objectives change, as can be seen in countries such as those in the European Union where the potato industry is trying to increase potato usage in an economically and environmentally sustainable way. New cultivars must give economic benefits

through more yield of saleable product at less cost of production, whether the potatoes are for processing or table use. They must have built-in resistances to pests and diseases that give environmental benefits through reduced use of pesticides and fungicides. Increased water and mineral use efficiencies are also desirable for better use of water and fertilizers, both nitrogen and phosphate. Finally, they must help meet consumer demands for convenience, nutritional and health benefits, improved flavor, and novel products. Today, in meeting the need for new cultivars, breeders have a tremendous opportunity to utilize all of the biodiversity available in the tuber-bearing *Solanum* species and to apply recent advances in potato genomics to potato breeding (Bonierbale et al., 2003). These topics will therefore be considered before returning to breeding objectives and strategies.

Germplasm collections and taxonomy of tuber-bearing *Solanum* species

Germplasm collections of wild species

Recognition of Central and South America as the centers of origin and diversity of tuber-bearing *Solanum* species resulted in numerous collecting expeditions (Fig. 10.2), from those pioneered by the Russians in the 1920s (Hawkes, 1990) to the more recent ones of the 1990s (Spooner and Hijmans, 2001). Reference books have been written on the potatoes of Argentina, Brazil, Paraguay, and Uruguay (Hawkes and Hjerting, 1969), Bolivia (Hawkes and Hjerting 1989; Ochoa, 1990), Peru (Ochoa, 2004), and North and Central America (Spooner et al., 2004). The collecting expeditions led to the establishment of a number of potato germplasm collections. The world collection is held at the CIP in Lima, Peru. The other main germplasm collections are the

Fig. 10.2. Collecting wild tuber-bearing species near Cochabamba in Bolivia in 1963.

Commonwealth Potato Collection (CPC, Dundee, Scotland), the Dutch-German Potato Collection (CGN, Wageningen, The Netherlands), the Groß Lusewitz Potato Collection (GLKS, IPK, Groß Lusewitz, Germany), the Potato Collection of the Vavilov Institute (VIR, St. Petersburg, Russia), the U.S. Potato Genebank (NRSP-6, Sturgeon Bay, Wisconsin, U.S.) and Potato Collections in Argentina, Bolivia, and Peru. Together they comprise the Association for Potato Intergenebank Collaboration and have established an Inter-genebank Potato Database (IPD) that can be accessed through the internet (www.potgenebank. org). The IPD contains 7,112 different accessions of 188 taxa (species, subspecies, varieties, and forms) out of the 247 tuber-bearing wild potato taxa recognized by Hawkes (Huaman et al., 2000a). Accessions are normally held in true seed form.

When assessing wild potatoes for plant-breeding purposes it is important to appreciate their wide geographical distribution and great range of ecological adaptation (Hawkes, 1994). In the southwestern U.S. and in Central America, wild species generally occur at medium to fairly high altitudes. In South America they are found along the Andes from Venezuela to northwest Argentina and also in the lowlands of Chile, Argentina, Uruguay, Paraguay, and southeastern Brazil. The adaptive range among the different species is very great; it includes the high Andean regions from 3,000 to 4,500 m where frosts are common, dry semi-desert conditions and scrub and cactus deserts, cool temperate pine and rain forests, woodlands, and coastal plains. Wild species have also developed resistances to a wide range of pests and diseases. Data are available through the IPD links to individual collections for more than 33,000 evaluations of wild potato accessions covering 55 traits (dry matter, starch, reducing sugar and glycoalkaloid content, and resistances to fungi, bacteria, viruses, viroids, insects, and environmental stresses such as frost and heat/drought).

The screening of germplasm collections for desirable traits will no doubt continue and so will genetic studies on their inheritance, followed by introgression of desirable alleles into Tuberosum potatoes as described by Bradshaw and Ramsay (2005) for the CPC. These approaches will be supplemented by an increasing amount of molecular-marker and DNA-sequence data that will aid in the selection of core collections for research purposes, establish the integrity of genebank accessions, and help breeders in their search for desirable novelty. The success of this latter objective may depend on whether or not allele frequencies and distributions can be associated with features of the natural habitats of the accessions.

CIP's collection of cultivated species

CIP assembled a collection of more than 15,000 accessions of native potato cultivars from nine countries in Latin America (Argentina, Bolivia, Chile, Colombia, Ecuador, Guatemala, Mexico, Peru, and Venezuela). The cultivars belong to seven species of which *S. tuberosum* (tetraploid) is the most widely cultivated. *S. stenotomum* (diploid) and *S. chaucha* (triploid) are grown in the central Andes of Peru and Bolivia, whereas frost-tolerant *S. ajanhuiri* (diploid), frost-resistant *S. curtilobum* (pentaploid), and frost-resistant *S. juzepczukii* (triploid) are grown at higher altitudes (up to 4,500 m for *S. juzepczukii*) in these same countries. *S. phureja* (diploid) was selected from *S. stenotomum* by Andean farmers for lack of tuber dormancy and faster tuber development so that they could grow up to three crops per year in the lower, warmer, eastern valleys of the Andes. *S. phureja* was therefore able to spread into northern Ecuador, Colombia, and Venezuela and is the second most widely cultivated species in South America.

It is worth mentioning how the other species were derived from *S. stenotomum*. *S. ajanhuiri* is the hybrid of *S. stenotomum* with the wild frost-resistant species *S. megistacrolobum*; *S. juzepczukii* is the hybrid with the wild frost-resistant species *S. acaule*;

and *S. chaucha* is the hybrid with *S. tuberosum* subsp. *andigena*. *S. curtilobum* is the hybrid between an unreduced gamete of *S. juzepczukii* and a normal gamete of *S. tuberosum* subsp. *andigena*. Finally, the origins of *S. tuberosum* subsp. *andigena* and subsp. *tuberosum* are still a matter of debate. The former could be a hybrid between *S. stenotomum* and the wild weed species *S. sparsipilum* either through unreduced gametes or normal gametes and subsequent chromosome doubling (Ortiz and Ehlenfeldt, 1992; Hawkes, 1994). However, molecular evidence has raised doubts about the role of *S. sparsipilum,* and Grun (1990) refers simply to an unidentified wild species with actinomorphic calyces. The hypothesis that subsp. *andigena* is a simple autotetraploid of *S. stenotomum* is unlikely to be true despite its cytological behavior and tetrasomic inheritance (Hawkes, 1994). Sukhotu and Hosaka (2006) have concluded from chloroplast and nuclear DNA markers that subsp. *andigena* arose from *S. stenotomum* through sexual polyploidization from unreduced gametes many times at many places in the fields of *S. stenotomum* and that these tetraploids were subsequently modified by occasional and unintentional selection of natural hybrids with neighboring wild species to give present-day subsp. *andigena*. Hosaka (2004) has suggested that Chilean Tuberosum cytoplasm is derived from the wild species *S. tarijense* so that subsp. *tuberosum* is not simply subsp. *andigena* that has been selected to tuber in long days.

The CIP identified duplicate accessions in their potato collection and reduced the number of individual cultivars to 3,527; 552 of these were diploids, 128 triploids, 2836 tetraploids (2,644 subsp. *andigena*, 144 subsp. *tuberosum* and 48 hybrids), and 11 pentaploids (Huaman et al., 1997). The collection is maintained by clonal propagation. By 1997 researchers at CIP had already conducted 46,124 evaluations on the collection for the reactions of cultivars to biotic and abiotic stresses and for the presence of various desirable traits. To make the data more useful, Huaman et al. (2000b) used cluster analysis of morphological data to establish a core set of 306 accessions from the 2,379 cultivars of subsp. *andigena* still held at CIP. The set was chosen to represent the widest morphological diversity and to maximize geographical representation. Evaluation data were taken into account in choosing the representative accession of each cluster. Isozyme analysis of the entire collection confirmed that the core collection had captured a representative sample of the alleles at nine allozyme loci, with only the loss of alleles whose frequency was less than 0.05% (Huaman et al., 2000c). A simulation study revealed, however, that a core collection size of 600 would be required to adequately represent allele frequencies and locus heterozygosity (Chandra et al., 2002). The CIP core collection should nevertheless be a valuable base for detailed evaluation and future breeding both in South America and worldwide. The other genebanks mentioned above also have collections of cultivated species, but the one at CIP is recognized as the world collection.

The International Treaty on Plant Genetic Resources

Some of the biological issues involved in the conservation of potato genetic resources have recently been discussed by Bamberg and del Rio (2005) along with the complicated issue of germplasm ownership. The International Treaty on Plant Genetic Resources for Food and Agriculture came into force on June 29, 2004 and was ratified by 55 countries (www.fao.org). It offers a multilateral system for easy access and exchange of germplasm in return for fair and equitable sharing of the benefits. Potato (*S. tuberosum*) is one of the 35 food crops in the initial list covered by the multilateral system.

It is too early to assess the impact of the treaty on the utilization of tuber-bearing *Solanum* species in potato breeding.

Taxonomy of wild and cultivated species

The taxonomy of tuber-bearing *Solanum* species is complicated and under revision. Hawkes (1990) recognized 219 wild and 7 cultivated species and arranged them into 19 series of subsection *Potatoe* of section *Petota* of subgenus *Potatoe* of genus *Solanum*. He grouped series I to IX in superseries *Stellata* and series X to XIX in superseries *Rotata*. He considered the sequence of subsections, superseries, and series to reflect an approximate evolutionary sequence. Based on this sequence, he suggested a possible scenario for the evolution of wild potato species, even as he acknowledged that modification may be required as a result of continuing experimental work, particularly with molecular markers. Hawkes also recognized a further nine closely related non-tuber-bearing species that he grouped into two series of subsection Estolonifera. A partial list of species appears in Table 10.1. The species form a polyploid series from diploid ($2n = 2x = 24$) to hexaploid ($2n = 6x = 72$) in which homologous genomes have been classified into five groups, A, B, C, D, and E by Matsubayashi (1991). Spooner et al. (2004) summarized the putative genome compositions of the polyploid species, but it is clear that further research is required to resolve the origins of the genomes in polyploid potato species. A particularly promising approach is the comparison of DNA sequences in single-copy gene segments from the different genomes contained in diploid and polyploid taxa (Bryan and Ramsay, 2005). The results should help potato breeders in their choice of germplasm with different genomes.

Spooner and Hijmans (2001) reviewed accepted species based on a literature survey,

Table 10.1. *Solanum* species.

Series	Species and abbreviations	Ploidy	EBN	Distribution
Subsection Estolonifera				
I Etuberosa	*S. brevidens* (brd)	2*x*	1	Argentina, Chile
II Juglandifolia				
Subsection Potatoe				
I Morelliformia				
II Bulbocastana	*S. bulbocastanum* (blb)	2*x*	1	Mexico
III Pinnatisecta	*S. pinnatisectum* (pnt)	2*x*	1	Mexico
IV Polyadenia	*S. polyadenium* (pld)	2*x*	1	Mexico
V Commersoniana	*S. commersonii* (cmm)	2*x*	1	Argentina
VI Circaeifolia				
VII Lignicaulia				
VIII Olmosiana				
IX Yungasensa	*S. chacoense* (chc)	2*x*	2	Argentina, Bolivia
	S. tarijense (tar)	2*x*	2	Argentina, Bolivia
	S. yungasense (yun)	2*x*	2	Bolivia
X Megistracroloba	*S. megistacrolobum* (mga)	2*x*	2	Peru, Bolivia
XI Cuneoalata				
XII Conicibaccata				
XIII Piurana				
XIV Ingifolia				
XV Maglia				
XVI Tuberosa (wild)	*S. berthaultii* (ber)	2*x*	2	Bolivia
	S. brevicaule (brc)	2*x*	2	Bolivia
	S. bukasovii (buk)	2*x*	2	Peru
	S. canasense (can)	2*x*	2	Peru
	S. gourlayi (grl)	2*x*	2	Argentina
	S. leptophyes (lph)	2*x*	2	Bolivia, Peru
	S. mochiquense (mcq)	2*x*	1	Peru
	S. sparsipilum (spl)	2*x*	2	Bolivia, Peru
	S. spegazzinii (spg)	2*x*	2	Argentina
	S. vernei (vrn)	2*x*	2	Argentina
	S. verrucosum (ver)	2*x*	2	Mexico
XVI Tuberosa (cultivated)	*S. ajanhuiri* (ajh)	2*x*	2	Bolivia, Peru
	S. chaucha (cha)	3*x*	3	Bolivia, Peru
	S. curtilobum (cur)	5*x*	4	Bolivia, Peru
	S. juzepczukii (juz)	3*x*	2	Bolivia, Peru
	S. phureja (phu)	2*x*	2	Venezuela to Bolivia
	S. stenotomum (stn)	2*x*	2	Bolivia, Peru
	S. tuberosum (tbr)	4*x*	4	Worldwide
	subsp. *tuberosum* (tbr)	4*x*	4	Worldwide
	subsp. *andigena* (adg)	4*x*	4	Venezuela to Argentina
XVII Acaulia	*S. acaule* (acl)	4*x*	2	Bolivia, Peru
XVIII Longipedicellata	*S. fendleri* (fen)	4*x*	2	USA, Mexico
	S. stoloniferum (sto)	4*x*	2	Mexico
XIX Demissa	*S. demissum* (dms)	6*x*	4	Mexico
	S. hougasii (hou)	6*x*	4	Mexico

including new species described and names placed in synonymy since Hawkes' treatment. They listed 203 tuber-bearing species. Three species in Table 10.1 (*S. brevidens, S. canasense,* and *S. gourlayi*) will not be found in this new list but are mentioned under their original names because these are still in common usage. Huaman and Spooner (2002) then reviewed the seven cultivated potatoes with reference to the International Code of Nomenclature of Cultivated Plants (Trehane et al., 1995) and concluded that they belonged to one taxon, thus reducing the number of species to 197. Further changes in the delimitation of species are being reported as molecular-marker and DNA-sequence data are used to clarify species relationships. Two examples are the proposed reductions of species in series Longipedicellata (van den Berg et al., 2002) and in the *S. brevicaule* complex of species where Spooner et al. (2005a) concluded that the northern group of members is poorly defined and may be reduced to a single species of which *S. bukasovii* is the earliest valid name. The number of species is therefore likely to be further reduced in the future, but of more interest to potato breeders is the crossability of groups of species.

Reproductive biology

Crossability of groups of species

The crossability of species has been determined through artificial pollinations done over many years (Jansky, 2006). The results can be explained primarily, but not exclusively, in terms of endosperm balance number (EBN) that can be regarded as the effective rather than the actual ploidy of the species (Johnston et al., 1980). In crosses between species with the same EBN, the hybrids have a normal endosperm for nourishing the hybrid embryo, whereas in crosses between species with different EBNs the endosperm degenerates. The endosperm develops following the fusion of a sperm nucleus from the male parent with two polar nuclei from the female parent to give a triple fusion nucleus, and it is the genetic composition of this nucleus that is important. Under the EBN hypothesis, the endosperm is normal when the three nuclei have the same EBN and hence a two maternal to one paternal ratio of endosperm balance factors. Attempts have been made to understand the genetic and biological basis of the EBN concept, but EBNs are determined

experimentally starting with the primitive diploid species in series I to VII (and possibly VIII) which have an EBN of 1 (Hermsen, 1994). There are five main groups of species: diploid 1 EBN; diploid 2 EBN; tetraploid 2 EBN; tetraploid 4 EBN (including *S. tuberosum*); and hexaploid 4 EBN (Hawkes and Jackson, 1992). Species within these crossability groups have evolved by means of geographical and ecological isolation rather than by genetic incompatibility, so hybridizations within groups are usually successful.

Today breeders can usually achieve sexual hybridization between *S. tuberosum* and wild species by manipulation of ploidy with due regard to EBN (Ortiz, 1998, 2001; Jansky, 2006). EBN can be doubled meiotically through the unreduced $2n$ gametes produced from several naturally occurring recessive meiotic mutations that are common in *Solanum* species (Tai, 1994). It can also be doubled mitotically through the use of colchicine for chromosome doubling, although this doubling can also occur naturally during callus culture. EBN can be halved by haploidization using in vitro androgenesis (anther culture) or more commonly in *S. tuberosum* by parthenogenesis using pollinations with particular clones of diploid *S. phureja* (Wenzel, 1994; Veilleux, 2005). Furthermore embryo rescue can be used to secure a hybrid when embryo abortion is due to a defective endosperm (Hermsen, 1994; Jansky, 2006). As the largest compatibility group is the 2 EBN group, it is now common for potato breeders to secure tetraploid hybrids from $4x$ (*S. tuberosum*) × $2x$ ($2x$ *S. tuberosum* × wild species) crosses in which an unbalanced endosperm prevents the development of triploid embryos. There are, however, other barriers to hybridization, such as interspecific pollen-pistil incompatibility and nuclear-cytoplasmic male sterility, and these can present breeders with problems. Unilateral incompatibility is known to occur when a self-incompatible (SI) species is pollinated by a self-compatible (SC) one. For example, *S. verrucosum* (SC female) × *S. phureja* (SI male) is successful, but the reciprocal cross fails (Hermsen, 1994; Jansky, 2006). Sometimes incompatible pollen can be helped to achieve fertilization through a second pollination with compatible pollen, a technique known as mentor pollination (Hermsen, 1994; Jansky, 2006). These phenomena have been reviewed by Camadro et al. (2004) in the context of how sympatric species maintain their integrity. From time to time, potato breeders will have unexpected successes and failures when attempting to overcome barriers to hybridization.

Potato breeders can also use somatic (protoplast) fusion to achieve difficult or impossible sexual hybridizations (Wenzel, 1994; Thieme and Thieme, 2005; Veilleux, 2005). Protoplast fusion can be induced chemically by the polycation polyethyleneglycol (PEG) or via electrofusion, which is now the preferred method. Somatic hybrids derived from the same fusion combination do show genotypic and phenotypic variation and hence require screening for the desired product. Somatic fusion has allowed the production of fertile hexaploid hybrids between tetraploid 4 EBN *S. tuberosum* and diploid 1 EBN species, such as the non-tuber-bearing species *S. brevidens,* which has tuber soft rot and early blight resistances (Tek et al., 2004), and *S. bulbocastanum,* which has a major gene for broad spectrum resistance to late blight (Naess et al., 2000). Somatic fusion has also allowed the production of tetraploid hybrids from dihaploids (the haploids produced from 4*x S. tuberosum*), including male-sterile ones (Thieme and Thieme, 2005). Dominant resistance genes as well as quantitative traits can be combined in the same way as when gametes fuse in sexual hybridization. Furthermore, Wenzel et al. (1979) proposed the use of haploidy, somatic fusion, and classical breeding steps for combining several traits

of potato. Homozygous dihaploids are produced by spontaneous or induced doubling of monohaploids that have been produced from tetraploid clones by two successive haploidization steps. Breeding at the diploid level then takes place before using somatic fusion to return to the tetraploid level and maximum heterozygosity.

Sexual reproduction

The reproductive biology of potato is ideal for creating and maintaining variation. Potatoes flower and set true seed in berries following natural pollination by insects, particularly bumblebees. Outcrossing is enforced in cultivated (and most wild) diploid species by a single S-locus, multiallelic, gametophytic self-incompatibility system (Dodds, 1965). Cross-pollination between field plots of *S. phureja* has been estimated to decline from 5.1% at 10 meters to 0.2% at 80 meters based on a pollen donor possessing a dominant marker (Schittenhelm and Hoekstra, 1995). This information is useful in planning isolation distances for natural true-seed multiplication of genebank accessions and cultivars propagated by this method.

While self-incompatibility does not operate in tetraploid *S. tuberosum,* 40% (range 21% to 74%) natural cross-pollination was estimated to occur in subsp. *andigena* in the Andes (Brown, 1993) and 20% (range 14% to 30%) in an artificially constructed Andigena population (Glendinning, 1976). This sexual reproduction creates an abundance of diversity by recombining the variants of genes that arose by mutation. As a consequence, potatoes are highly heterozygous individuals that display inbreeding depression on selfing. The genetically unique seedlings that grow from true seeds produce tubers that can be replanted as seed tubers and hence distinct clones can be established and maintained by asexual (vegetative) reproduction. No doubt many

Andean cultivars were produced by farmer selection from naturally occurring variation. Domestication must have involved selection of less bitter and hence less toxic tubers, and Andean farmers certainly retained a much wider variety of tuber shapes and skin and flesh colors than seen in wild species (Simmonds, 1995). Subsequent selection for early maturity, appropriate dormancy, and resistance to abiotic and biotic stresses must have occurred in many environments. Interestingly, naturally occurring variation was still exploited by breeders after they started to make deliberate hybridizations. North America's most popular potato cultivar 'Russet Burbank' was derived from Rough Purple Chili by three generations of open-pollination, with selection and release in 1914 (Ortiz, 2001).

Today most cultivars come from deliberate artificial hybridizations. The floral characteristics of potatoes and methods of artificial hybridization and self-pollination have been described by Plaisted (1980). Details can also be found in Caligari (1992), Douches and Jastrzebski (1993), and in the textbook *Breeding Field Crops* by Poehlman and Sleper (1995). A temperature of 19°C and 16 hours of daylength are recommended for crosses involving *S. tuberosum* subsp. *Tuberosum,* so an extension of natural daylength is required for crossing in the tropics and subtropics. In contrast, subsp. *andigena* is, in effect, day-neutral for flowering, but not for tuberization. Not all desired crosses are successful, as problems can arise from clones failing to flower, buds and flowers dropping either before or after fertilization, low pollen production, and failure to produce viable pollen (male sterility). Breeders usually encourage flowering by the periodic removal of daughter tubers, and sometimes breeders graft young potato shoots onto tomato or other compatible Solanaceous plants. Pollinations can also be done on flowers attached to stems that have been cut and placed in jars of water with an anti-bacterial agent to reduce contamination (Peloquin and Hougas, 1959).

Asexual reproduction and seed certification

Asexual reproduction through seed (daughter) tubers allows a genetically unique seedling to be multiplied and maintained as a new cultivar. Such vegetative reproduction has, however, a number of disadvantages that can cause problems in countries where the potato is a staple food. Viral and other diseases can be transmitted through seed tubers and accumulate across clonal generations. Seed tubers can be produced in areas unfavorable for disease development, but it is expensive to transport bulky and heavy tubers to the fields where they are to be grown for food. Premature sprouting of tubers can also be a problem, particularly where farmers need to save tubers for planting the next season but do not have cold storage facilities. Indeed, Lang (2001) has explained how farmers in northern India turned to cold storage during the 1970s in order to try to keep seed firm and healthy.

Potato yields and quality are certainly best when crops are planted with high-quality disease-free seed tubers; experience around the world has shown that widespread availability of such tubers is most likely to be achieved through statutory seed certification schemes operating in areas where potatoes are only grown for seed.

Meristem culture and micropropagation have now become routine in seed certification schemes and have been reviewed by Wenzel (1994) and Veilleux (2005). Meristem-tip culture can be used in combination with thermotherapy and chemotherapy to eliminate viruses. Rooted plantlets are indexed for the presence of viruses, and the virus-free plants are increased by clonal multiplication, starting with in vitro axillary node and shoot production. The resulting axillary plants can be used to produce nodal cuttings that can be used for rooting in soil and transplanting to the field, greenhouse production of minitubers, and in vitro production of microtubers. Multiplication of field generations needs to be kept to a minimum to maintain high seed health. It is common practice for the certifying authority to hold in vitro pathogen-free nuclear stocks of cultivars under multiplication. It is important to point out that strict quarantine procedures are required when potatoes are transferred from one country to another to prevent the introduction of diseases, particularly non-indigenous ones. Advances with in vitro techniques are proving useful in this regard. Lang (2001) has described how CIP supplies its new cultivars to farmers in East Africa through a seed multiplication scheme that starts in Kenya with nodal cuttings being taken from virus-free sprouts. These are supplied in vitro from CIP headquarters in Lima, Peru to the quarantine station in Kenya.

In the future, cryopreservation may become an important component in the maintenance of clonal germplasm collections, which would reduce the labor needed to do routine subculture of in vitro stocks. Much research in this area has been done in recent years with some encouraging results that have been reviewed by Veilleux (2005).

Genetic transformation

Potato transformation using *Agrobacterium*-mediated systems was developed during the 1980s, first with *A. rhizogenes* (Ooms et al., 1986) and then more successfully with *A. tumefaciens* (Stiekema et al., 1988). This system has opened up new possibilities for potato crop improvement, which is discussed in later sections. Today, genetic transformation is most often done using *A. tumefaciens* Ti plasmid-mediated gene transfer; this is a relatively straightforward approach, but regeneration can be a problem with some

genotypes (Dale and Hampson, 1995). Hence the easily transformed cultivar 'Desiree' is still a frequent choice for research purposes, particularly as proof of function of cloned genes.

To obtain this proof, the gene of interest is incorporated into the bacterial plasmid along with a promoter and selectable marker, such as resistance to an antibiotic or herbicide, the bacterium is co-cultured with freshly cut tuber discs or leaf or internode explants of the potato, and regeneration of shoots with the selectable marker takes place in plant tissue culture in the presence of the selectable agent. The choice of promoter is important for gene expression. The CaMV 35S promoter has frequently been used for constitutive gene expression, but others have been developed for higher constitutive expression and for leaf or tuber expression (Douches and Grafius, 2005). Recently, both Chang et al. (2002) and Morris et al. (2006) have been able to cotransform potatoes simultaneously with two transgenes using a single selectable marker. The transgene constructs were multiplied separately in *A. tumefaciens* clones and cultures of the *Agrobacterium* were mixed and incubated with the potato internode explants. From 300 explants, Morris et al. (2006) generated 38 independent transformants of which 4 contained both transgenes. Clearly, this increases the speed and efficiency of transformation as a breeding method and makes it a more attractive proposition.

Somaclonal variation and mutation breeding

Somaclonal variation can occur when plant regeneration and multiplication involves tissue culture, particularly when there is a callus phase. After such variation was first reported in protoclones of the cultivar Russet Burbank (Shepard et al., 1980), it was investigated as a breeding method in its own right. Desirable improvements on parent cultivars were reported and assessed under field conditions, but most somaclonal variation comprised undesirable changes that did not merit further breeding efforts (Kumar, 1994; Veilleux, 2005). Hence, somaclonal variation now tends to be viewed as a source of undesirable variants that need to be screened out of breeding programs that involve plant regeneration.

Mutations are the ultimate source of all genetic variation. The potato is not ideal for deliberate mutation breeding because it is a clonally propagated tetraploid crop. Nevertheless, limited success has been achieved with deliberate mutation breeding, for example, by selecting gamma-ray mutants of the cultivar 'Lemhi Russet' for improved resistance to blackspot bruise and low-temperature sweetening (Love et al., 1993).

Potato genetics

The rediscovery in 1900 of Mendel's published work of 1865 marked the birth of modern genetics and opened the way to crop improvement by scientific breeding methods. Potato genetics, however, proved difficult for most of the 20th century. The methods used and progress made were reviewed by Bradshaw and Mackay (1994b). *S. tuberosum* is a tetraploid that displays tetrasomic inheritance and hence has complex inheritance patterns. Tetrasomic inheritance also means diploid gametes in which diallelic interactions (dominance effects) can be transmitted to the next generation; sister chromatids can occur as a result of double reduction and two alleles can be identical by descent. Furthermore, one generation of random mating does not achieve the single locus equilibrium of genotype frequencies (i.e., the diploid equivalent of Hardy-Weinberg equilibrium), nor does it remove the effects of inbreeding. The potato also has small and relatively numerous chromosomes, which made cytological investigation difficult. Yeh and Peloquin (1965) were able to identify 12 unique

pachytene chromosomes based on the distribution patterns of euchromatin and heterochromatin. Later, Pijnacker and Ferwerda (1984) used Giemsa C-banding and silver nitrate staining to produce an ideogram of the twelve basic chromosomes of potato, which vary in length and centromere position. The two chromosome numbering systems that resulted do not, however, correlate with each other, and they do not correlate with the system now used for the 12 genetic linkage groups.

Some of the problems and complexities of working with a tetraploid were overcome after 1958 with the production of haploids (also called dihaploids) of *S. tuberosum* (Hougas et al., 1958) and genetic studies at the diploid level involving crosses with other diploid *Solanum* species. The dihaploids were, however, usually male sterile, and most dihaploids and diploid species were self-incompatible. Furthermore most economically important traits displayed continuous variation, which required biometrical rather than Mendelian analysis. Hence in potato genetics it was not possible to achieve the same degree of sophistication as in the genetic analysis of crosses between true-breeding inbred lines that display disomic inheritance. Nevertheless, as in other crops, knowledge of quantitative genetics provided the foundation for efficient conventional potato breeding, which is still the main route to new cultivars. The concepts of heritability, additive and non-additive genetic variation, genotype × environment interaction, and population improvement are all important in predicting and improving the response to selection and rate of progress. High-quality mechanized fieldwork and computer-based data capture and analysis are essential in this endeavor.

Genetic knowledge of the potato has increased dramatically since the first molecular-marker map appeared in 1988 (Bonierbale et al., 1988). High-density molecular linkage maps of the potato and tomato genomes were aligned and shown to differ by five major paracentric-like inversions on chromosomes 5, 9, 10, 11, and 12 (Gebhardt et al., 1991; Tanksley et al., 1992); Dong et al. (2000) demonstrated that the 12 potato chromosomes can be identified in a straightforward way by the presence of FISH (fluorescence in situ hybridization) signals from chromosome-specific cytogenetic DNA markers. These were BACs (bacterial artificial chromosomes) that had been screened with genetically mapped RFLP (restriction fragment length polymorphism) markers as probes. Dong et al. (2000) were also able to find the orientations of 11 of the 12 genetic linkage maps in terms of the orientations of the short/long arms of the chromosomes. RFLP-marker probes have also been used to find conserved genetic linkage between loci that share sequence similarity in different species, including potato and the model plant *Arabidopsis thaliana,* for which the entire genome has been sequenced (Gebhardt et al., 2003). Syntenic regions were found that covered 41% of the potato genetic map and 50% of the *Arabidopsis* physical map. Evidence was found for ancient intra- and inter-chromosomal duplications in the potato genome, but partial genome coverage and the redundancy of syntenic blocks limited the use of synteny for comparative functional analysis.

Comparative genomics in the family Solanaceae has progressed since 2004 under the auspices of "The International Solanaceae Genome Initiative" (http://www.sgn. cornell.edu/solanaceae-project/) and should result in the sequencing of the potato (840 million base pairs of DNA) and tomato (950 million base pairs) genomes; the total length of the combined female and male maps was 1,120 centimorgans. For tomato, the plan is to concentrate on the distal portions of each chromosome that comprise gene-rich euchromatin, ignoring the heterochromatin that is concentrated around the centromeres. In contrast, for potato, the Dutch-led consortium plans to produce a complete physical map and sequence the whole genome (www.potatogenome.net). The physical map is being aligned to a Ultra High Density (10,000 AFLP markers) genetic map (http://www.dpw.wageningen-ur.nl/uhd).

Today the genes underlying qualitative traits can be mapped directly onto the dense molecular-marker maps as individuals can be classified into distinct categories for trait and marker and the number of individuals in each category can be counted. Jacobs et al. (1995), for example, used a diploid backcross population to construct a genetic map of the 12 chromosomes of potato, integrating molecular and classical markers, including the self-incompatibility locus (*S*) on chromosome 1 that can cause segregation distortion in its neighborhood (Gebhardt et al., 1991). The total length of the combined female and male maps was 1,120 cM. This tells us that two chiasmata per bivalent is the typical level of intra-chromosomal recombination in potato. Quantitative trait loci (QTL) can be mapped indirectly through associations between trait scores and molecular markers. Of relevance to potatoes are the interval mapping models developed by Schäfer-Pregl et al. (1996) for the offspring of non-inbred diploid parents and by Hackett et al. (2001) for a full-sib family derived under tetrasomic inheritance by crossing two tetraploid parents.

Progress is also being made in understanding gene expression and how gene function at the biochemical level relates to observed phenotypes. Where genetic and biochemical information is available on metabolic pathways, such as those for carbohydrates, anthocyanins, and carotenoids in potatoes, candidate genes are being postulated and sought for improving both qualitative and quantitative traits. Furthermore, advances in metabolite profiling are allowing a systems approach to understanding biochemical pathways and hence the key genes to target for desired phenotypic (trait) changes. By 2006 there were 219,765 potato EST (expressed sequence tags) sequences in the public database GenBank (http://www.ncbi.nlm.nih.gov), and these can be regarded as a catalogue of partially sequenced genes that are expressed in target potato cells and organs of interest for crop improvement.

The value of ESTs will increase as they are located on the genetic and physical maps of potato and as the extent of colinearity of the different Solanaceae genomes is established and exploited in comparative genomics, in which known gene position and function in one genome can be used to make inferences to other genomes (e.g., from tomato to potato, pepper, and eggplant). Finally, potato microarray chips are now becoming available (e.g., through the Potato Oligo Chip Initiative of Wageningen University, the Netherlands) that contain at least 75% of the potato transcriptome (messenger RNAs = expressed genes); these chips should allow a much more complete analysis of gene expression than has previously been possible.

These advances in potato genetics are being accompanied by the development of plant-breeding methods based on genomics, transformation, and marker-assisted selection. These molecular breeding methods use and exploit DNA-based information for crop improvement and will increasingly complement the conventional approaches already mentioned.

Breeding objectives

Breeding objectives all involve improving the adaptation of the potato to specific growing environments, storage regimes, and end-uses. To be of practical benefit, however, objectives need to be translated into improvements over existing cultivars and selection criteria that can be used by breeders. They also raise some strategic questions that need to be answered before considering selection criteria in detail.

Type of cultivar

Most potato cultivars are clonally (vegetatively) propagated through seed (daughter) tubers and are genetically uniform. There are, however, circumstances in which developing cultivars using current methods of true potato seed (TPS) propagation are an attractive proposition, despite such cultivars being genetically variable and hence inferior to the best genotype that could be selected from within the TPS progeny.

Breeders must decide which type of cultivar they are going to produce. The main advantages of TPS are reduced seed costs due to the much smaller amount of planting material required, flexibility of planting time because the farmer does not have to consider the physiological age and condition of seed tubers, and freedom from tuber-borne diseases, such as viruses, despite the existence of a few true seed-borne viroids. The main disadvantages of TPS are difficulties in establishing the crop, later maturation, and less uniformity of end product (Golmirzaie et al., 1994). Simmonds (1997) sums this up as the benefits (i.e., food security or profit) from clean seedlings versus infected clones.

TPS propagation is most attractive in the torrid zones of the lowland tropics and subtropics. In practice, TPS potatoes are established in Bangladesh, China, Egypt, India, Indonesia, Nicaragua, Peru, the Philippines, southern Italy, and Vietnam (Almekinders et al., 1996; Ortiz, 1997; Simmonds, 1997). Chilver et al. (2005) have recently reviewed on-farm profitability and prospects for TPS and concluded that widespread geographic adoption is unlikely in the immediate future, but that investment in a small but sustained TPS breeding effort can be justified in both China and India. While TPS can be propagated by direct drilling and transplanting, preference is now given to rapid multiplication by cuttings or shoot-tip culture to give tuberlets   (first   generation   tubers)   or mini-tubers that can be chitted before planting (Simmonds, 1997). Methods involving multiplication allow some selection to be practiced within a TPS progeny, and hence are a compromise between clonal (vegetative) and true seed multiplication. Breeding strategies are considered later in this chapter.

General or specific adaptation to environments and end-uses

Another important strategic decision concerns breeding for general or specific adaptability because countries in which potatoes are grown as a staple food include many different growing conditions. A cultivar that gives high yields in all growing environments (high and stable yield) is desirable but may not be achievable. Different cultivars may be required for different environments. Furthermore, in many countries the potato is both cooked as a vegetable for direct use and processed into products. Hence, the breeder has to decide on the number of different breeding programs and cultivars required to cover a range of target environments and end-uses. The answer requires an assessment of genotype × environment (G × E) interactions. A brief review of methods and results of relevance to potatoes was done by Bradshaw (1994a).

The complexity or simplicity of G × E interactions can usually be seen from a principal component analysis, such as the one done by Forbes et al. (2005) to determine the stability of resistance of potato cultivars to *Phytophthora infestans*. The same analysis would be appropriate for yield stability across a range of environments. It is important to know if crossover interactions occur, and Baker (1988) explained one technique for determining if a cultivar does consistently better than another in one environment but consistently worse in another environment.

A related question concerns the environments in which to practice selection. It is known that selection in one type of environment can have consequences for performance in different types of environment. Given that the mean global fresh-weight yield of potatoes is low at 16.4 t/ha (FAO statistics for 2003), there is a danger that selection for higher yield might take place in a high-yielding environment at a research station and not result in benefits for farms with poorer environments. In other words, selection might be more effective if done in poorer environments under low inputs of fertilizer and without chemical weed and disease control. Interestingly, for higher yielding environments in Great Britain, Simmonds (1981) analyzed the increase in agricultural yields from 34 t/ha in 1964 to 42.9 t/ha in 1976 and concluded that 4.0 t/ha was environmental (better husbandry and management), 5.5 t/ha was due to the near replacement of cultivars 'King Edward' and 'Majestic' with three modern cultivars 'Pentland Crown,' 'Maris Piper,' and 'Desiree,' and that the difference of 0.6 t/ha was a small but negative G × E effect. In other words, the new cultivars were not more responsive than the old ones to environmental change. Nevertheless, selection experiments with the fungus *Schizophyllum commune* (Jinks and Connolly, 1973, 1975) and with the tobacco *Nicotiana rustica* (Jinks and Pooni, 1982) have confirmed theoretical predictions about the sensitivity of genotypes to environmental change. Upward selection, for yield for example, in above-average environments should result in genotypes with a high sensitivity, whereas upward selection in below-average environments should result in genotypes with a low sensitivity, and the result could be crossover G × E (Fig. 10.3). Ceccarelli et al. (2001) have put forward a strategy to exploit G × E interactions in national and international breeding programs for cereals in unfavorable environments and subsistence agriculture, which is also relevant to potatoes. It involves using farmers in decentralized and

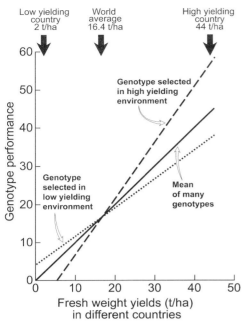

Fig. 10.3.   Crossover G × E.

participatory plant breeding for marginal environments. Participatory plant breeding research has recently been reviewed by Morris and Bellon (2004) in the context of opportunities and challenges for the international crop-improvement system.

The potato's vegetative means of reproduction does lend itself to selection experiments in contrasting environments, but extensive studies have not been done. In practice, the logistics of seed-tuber multiplication mean that potato breeders are likely to select their early generations at local seed and ware sites and then test relatively few potential cultivars in a wider range of environments. Brown et al. (1996) found a greater correlation for total marketable yield between the Scottish ware site where clones were selected and sites in England (r = 0.43 to 0.70) than with sites in the Mediterranean (r = 0.00 to 0.67). Although the very best clones from the Scottish ware site performed reasonably well in the Mediterranean, the results supported the idea that selection

would be optimized by selecting in environments more similar to those in which the cultivars are to be grown. Given the global importance of potato, it would seem sensible for potato-growing countries to have their own breeding programs targeted at local agro-ecological environments, notwithstanding commercial companies wanting to see their new cultivars grown as widely as possible.

Ideotypes

The global average fresh-weight yield of potatoes in 2003 was 16.4 t/ha with a range from 2 to 44 t/ha by country (FAO statistics). In contrast, yields of 120 t/ha have been achieved experimentally in the absence of pests and pathogens and with adequate inputs of water and fertilizers (Mackay, 1996). This could mean that actual yields are well below the potential yields of current cultivars because of poor cultural practices and high incidences of pests and diseases. But it could also mean that there is scope to improve potential yields, for example by better adaptation to local temperatures. Ideotyping is one approach to breeding cultivars for increased yields in particular environments. Haverkort and Grashoff (2004) define an ideotype as the ideal genotype for a particular environment. The ideotype has a growth cycle with a length that matches that of the available growing season, which is characterized by a temperature window (neither too cold nor too hot) as well as amounts of available solar radiation, water, and nutrients.

Haverkort and Grashoff advocate that, before starting a breeding program, an ideotyping exercise be done to guide the breeder on the alterations to characters that might achieve the aims of the program. This is possible through their crop-growth model that considers parameters that represent desirable genetic properties of cultivars:

- Frost tolerance for earlier planting
- Greater sprout growth rate for earlier emergence
- A lower base temperature for crop growth
- Higher initial leaf areas at emergence and/or higher leaf area development rates
- Thinner leaves in temperate climates with less intense solar radiation than in the tropics
- Changes in maximum leaf area index
- Patterns of senescence
- Deeper rooting
- Increased water-use efficiency

The results of simulation runs of ideotyping exercises are encouraging, but need testing in practice. The outcome may be the use of physiological selection criteria for adaptation to target environments in ways that have not proved feasible in the past. A more likely immediate outcome is help in the choice of parents with complementary traits for crossing followed by selection for yield in the target environment and growing season. It will, however, be important to ensure that crops are either planted with healthy (i.e., disease-free) seed tubers at the right physiological stage or that farmers change to TPS, and also that cultivars with better built-in resistance to abiotic and biotic stresses are grown.

Selection criteria

Breeding objectives will vary from country to country, but all programs are likely to involve selection for higher yield, appropriate maturity and dormancy, tuber characteristics that affect quality and suitability for particular end-uses, and resistance to abiotic and biotic stresses. The objectives that follow are considered achievable, but priorities will need to be assigned in any particular program as progress is most noticeable where the focus in on few rather than many traits.

Yield and maturity

There is less interest in dry-matter yield than in high fresh-weight yields with a dry-matter content (usually estimated by specific gravity) appropriate for the desired end-use. The manufacturers of french fries and chips (crisps) require a high dry-matter content because it is associated with a high yield of product and low oil absorption. In contrast, high dry-matter content can result in disintegration on boiling (sloughing) and hence can be undesirable in a cultivar for table use. Yield components may not be important for a staple food, whereas for some markets and uses tuber number and size (weight) are objectives in their own right.

Fast tuber development and early maturity are important where growing seasons are short. Even when they are long, such as in northern temperate latitudes, first and second early cultivars are grown as well as maincrop ones to meet the requirements for fresh produce across many months. Not surprisingly, early cultivars tend to have lower yields than maincrop ones that are left to grow for a full season.

Dormancy

The dormancy of a potato tuber is the obligate period after harvest of nonsprouting under conditions favorable to sprouting. Where one crop is grown annually, tubers are stored both for planting in the next season and to provide continuity of supply for eating and for processing. Long dormancy is required, provided it does not delay the sprouting of seed tubers beyond the planting time for the next season. In contrast, short dormancy is required when multiple crops per year are planted.

Tuber shape, eye depth, and defects

Tuber shape can be described in a series of qualitative terms from round to extra long, oval; quantitatively, shape is described as the length to breadth ratio. Unusual shapes can be found in Andean potato cultivars and have been described by Huaman et al. (1977). Regular shapes with shallow eyes are preferred for processing and table use to reduce wastage but may not be so critical where potatoes are a staple food. Round potatoes are currently preferred for making chips (crisps) and long, oval ones for french fries.

There are a number of tuber defects that breeders will select against by discarding clones in which they see the problem expressed to a marked degree (Dale and Mackay, 1994). External defects are growth cracking, mechanical damage and bruising, and greening. Internal defects require tubers to be cut for detection and include hollow heart, brown center, and internal rust spot.

Nutritional value, pigmentation, and glycoalkaloids

The potato has proved to be a rich source of energy in the form of carbohydrate (starch) and provides significant amounts of good quality protein as well as vitamins, particularly ascorbic acid (vitamin C) and minerals (Dale and Mackay, 1994). The major storage protein is patatin, a lipid acyl hydrolase that accounts for 30% to 40% of the soluble protein in potato tubers (Ganal et al., 1991). While high and stable yields for food security of a staple crop are probably more important than improving nutritional value per se, the latter is worthy of consideration and the following points are worth making.

Yellow or orange flesh is due to carotenoids that are antioxidants with health-promoting attributes. These attributes have been reviewed by Taylor and Ramsay (2005) along with recent advances in understanding carotenoid biosynthesis and the prospects for improving plant food quality. The pink, red, blue, and purple colors in potato tubers are due to anthocyanins whose antioxidant

properties are now appreciated and have been demonstrated in potatoes with red and purple flesh as well as skins (Brown et al., 2003). Hence, breeding potatoes with higher contents of antioxidant pigments should be given serious consideration as a way of improving their nutritional value (Brown, 2005). An educational program may be required to encourage their consumption where people prefer white-fleshed potatoes, but it is reassuring to remember that Andean farmers selected and maintained the wide range of tuber shapes and skin and flesh colors that we see today.

An important new objective for the Future Harvest Centers is improving the health of poor people by breeding staple food crops that are rich in micronutrients, a process called biofortification (http://www.harvestplus.org/biofaqs.html). The biofortification of crops with essential mineral elements has been reviewed by White and Broadley (2005). The aim is to address the problem of micronutrient malnutrition from diets poor in bioavailable vitamins and minerals. The initial emphasis is on higher levels of iron, zinc, and beta-carotene (vitamin A). The CIP is involved in this work; potatoes are one of the crops in phase two of the project. Research is still at the pre-breeding feasibility stage, but higher beta-carotene content is probably a realistic target given recent success in modifying carotenoid biosynthesis through genetic transformation (Ducreux et al., 2005). The topic of genetic transformation is dealt with more thoroughly later in this chapter.

Vitamin C is the main vitamin in potatoes, with a cultivar range of 84 to 145 mg per 100 g dry-weight reported by Augustin (1975), who also reported on variation due to site and storage conditions. Hence, breeding for higher concentrations of vitamin C would appear to be a realistic objective if considered nutritionally desirable. Breeding for efficiency in the uptake of mineral micronutrients from the soil may also increase disease resistance and stress tolerance, but further research is required.

Steroidal glycoalkaloids are a class of potentially toxic compounds with a bitter taste, which are found throughout the family Solanaceae. Domestication must have involved selection of less bitter and, hence, less toxic tubers. Today it is important for breeders to check the content of the parental material they use to breed finished cultivars, particularly when they have used wild species as a source of a desired character. If there are any doubts about the levels in a new cultivar, the breeder should check that total glycoalkaloid content is well below the 20 mg per 100 g fresh weight considered safe for human consumption (Dale and Mackay, 1994).

Cooking and processing quality

Today people want food that is flavorful, healthy, and a pleasure to eat, and this applies to potatoes whether they are simply cooked as a staple food or processed into convenience products. Improved flavor and texture are important breeding goals but still difficult to translate into selection criteria that can be applied to a large number of clones. Flavor involves taste, aroma, and texture, which is more than a simple description of degrees from extremely waxy to extremely floury. These are complex traits that depend upon the interactions of many factors. Because of their complexity, consumer preferences are determined by sensory panel evaluation rather than by analytical measurements. This will not change until more research has been done on the contributions of volatile and non-volatile compounds to flavor as well as the role of starch content and distribution in creating texture.

In contrast, it is relatively easy to select for freedom from after-cooking blackening and enzymic browning and for light colored fry-products both post-harvest and after storage under various temperature regimes (Dale and Mackay, 1994). After-cooking blackening is most noticeable in susceptible cultivars following boiling or steaming but can occur following baking, frying, or dehydration. It has no known detrimental effect on taste or nutritive value but is considered unsightly and hence undesirable. High tuber chlorogenic acid levels compared with citric acid levels tend to favor the production of after-cooking blackening. Enzymic browning is another discoloration of the flesh (black, brown, and red pigments) that can be found in peeled, cut, or bruised potatoes. It can increase costs of production through losses in marketable yield, labor required for removing affected tubers, and chemical control during processing. Enzymic browning results from the oxidation of tyrosine and other orthodihydric phenols by polyphenoloxidase (PPO). Light-colored fry products require low tuber contents of glucose and fructose. At the high temperatures used for frying, these reducing sugars are primarily responsible for the non-enzymic browning that occurs through a typical Maillard reaction and results in dark-colored, bitter-tasting products. Selection for lower levels of reducing sugars out of cold storage (i.e., resistance to cold-induced sweetening) is required to reduce or avoid the use of sprout-suppressant chemicals if potatoes are to be stored below 8 to 10°C for prolonged periods (Dale and Mackay, 1994). Selection for lower levels of reducing sugars should also reduce the potential for acrylamide production in roasted and fried products and, hence, address recent concerns that have been raised about acrylamide and human health (Amrein et al., 2003).

Starch

Starch is the primary storage compound in tubers. In countries with a starch industry it is also the starting material for the preparation of more than 500 different commercial products (Davies, 2002). The physical

Table 10.2. The major pests of potato.

Insects	
Potato tuber moth	*Phthorimaea operculella* Zeller
Colorado potato beetle	*Leptinotarsa decemlineata* Say
Andean potato weevil	*Premnotrypes* spp.
Green peach aphid	*Myzus persicae*
Leafminor fly	*Liriomyza huidobrensis*
Nematodes	
Golden potato cyst nematode	*Globodera rostochiensis* Woll.
White potato cyst nematode	*Globodera pallida* Stone
Southern root-knot nematode	*Meloidogyne incognita*
Columbia root-knot nematode	*Meloidogyne chitwoodi*
Northern root-knot nematode	*Meloidogyne hapla*
Root-knot nematode	*Meloidogyne fallax*

properties of starch vary with plant source and there are considerable opportunities to generate novel starches for use in food and non-food products. We shall see later that genetic engineering has already generated novel potato starches of which the two extremes are high amylopectin starch and high amylose starch.

Resistance to abiotic stresses

If food production is to expand by growing potatoes in a wider range of environments and for longer growing seasons, then resistance to abiotic stresses becomes important. These stresses include drought, heat, cold, mineral deficiency, and salinity, with water stress being the most important one affecting potato production in most areas of the world (Vada, 1994). In setting breeding objectives, it is important to distinguish between drought avoidance (e.g., through early maturity), drought tolerance, and water-use efficiency. In warm climates, intercropping with shade crops does help to reduce heat stress, but heat-tolerant cultivars are desirable. However, until the physiology, biochemistry, and molecular genetics of stress responses are better understood, breeders are likely to continue to select for high yield and quality in stress environments rather than use physiological selection criteria.

Resistance to major pests and diseases

Serious yield losses and reductions in quality can occur when potato plants and tubers are eaten by insects, mites, and nematodes or succumb to fungal, bacterial, and viral diseases. The major pests are listed in Table 10.2 and the major diseases in Table 10.3. A summary of the global distribution of potato pests has been given by Evans et al. (1992) and of potato diseases by Hide and Lapwood (1992). In any particular breeding program, the breeder must be realistic in deciding priorities for improving resistance. For major problems, new and durable forms of resistance will be actively sought and incorporated into new cultivars as quickly as possible. For less important pests and diseases, potential cultivars will simply be screened to avoid extreme susceptibility.

Potato tuber moth (PTM, *Phthorimaea operculella* Zeller) is the most serious insect pest of potatoes worldwide and is particularly troublesome in warm tropical and subtropical climates. The larvae mine the foliage, stems, and tubers in the field and also in storage. Colorado potato beetle (CPB, *Leptinotarsa decemlineata* Say) is a major pest in northern latitudes, particularly in North America, Europe, and Russia, where severe infestations are capable of defoliating

Table 10.3.   The major fungal, bacterial, and viral diseases of potato.

Fungal disease of foliage, stem and tubers	
Late blight (an oomycete)	*Phytophthora infestans* (Mont.) de Bary
Fungal diseases of foliage, stem and tubers in warm climates	
Verticillium wilt	*Verticillium dahliae* Kelb.
Early blight	*Alternaria solani* (Ellis and Martin) Jones & Grout
Fusarium wilt	*Fusarium solani* (Mont.) and *F. oxysporum* Schlecht App. and Wr. var. *eumartii* (Carp.) Wr.
Fungal diseases of tubers	
Wart	*Synchytrium endobioticum* (Schilb.) Perc
Dry rot	*Fusarium coeruleum* (Lib.) ex Sacc. *F. sulphureum* (Schlecht.) *F. sambucinum* (Fuckel) and *F. avenaceum* Sacc. (Corda ex. Fr.)
Black scurf	*Rhizoctonia solani* (Kühn.) & stem canker
Common scab (an actinomycete)	*Streptomyces scabies* (Thaxt.) Waksman & Henrici
Powdery scab	*Spongospora subterranea* (Wallr.) Lagerh
Gangrene	*Phoma foveata* Foister
Skinspot	*Polyscytalum pustulans* (Owen & Wakef.) M.B. Ellis
Silver scurf	*Helminthosporium solani* Dur. & Mont.
Black Dot	*Colletotrichum coccodes* (Wallr.) Hughes
Bacterial diseases of foliage, stem and tubers	
Blackleg and soft rot	*Pectobacterium* (formerly *Erwinia*) *carotovorum* subsp. *atrospecticum* *P. carotovorum* subsp. *carotovorum* *P. chrysanthemi*
Bacterial wilt or brown rot	*Ralstonia* (formerly *Pseudomonas*) *solanacearum* Smith
Bacterial ring rot	*Clavibacter michiganensis* subsp. *sepedonicus* (Spieck and Kotth.) Davis et al.
Viral diseases	
PLRV	*Potato leafroll virus*
PVY	*Potato virus Y*
PVA	*Potato virus A*
PVX	*Potato virus X*
PVM	*Potato virus M*
PVS	*Potato virus S*
TRV	*Tobacco rattle virus*
PMTV	*Potato mop-top virus*
PSTVd (a viroid)	*Potato spindle-tuber viroid*

a crop. Andean potato weevil (APW, *Premnotrypes* spp.), green peach aphid (GPA, *Myzus persicae*), and leafminer flies (LMF, *Liriomyza huidobrensis*) are also major pests in many developing countries (Raman et al., 1994). Plant-parasitic nematodes are small worm-like animals that inhabit the soil and penetrate and feed on roots. The cyst nematodes *Globodera rostochiensis* and *G. pallida* are the most damaging nematodes for potato on a worldwide scale, followed by the root-knot nematodes (*Meloidogyne* spp.). The southern root-knot nematode (SRN, *M. incognita*) is the main problem in warm climates, whereas the Columbia root-knot nematode (CRN, *M. chitwoodi*) and the northern root-knot nematode (NRN, *M. hapla*) are the main nematode pests in temperate regions. *M. chitwoodi* and *M. fallax* are emerging problems in Europe (Kouassi et al., 2006).

Late blight, caused by the oomycete pathogen *Phytophthora infestans* (Mont.) de Bary, is still the most serious disease of potato foliage and tubers worldwide, some 160 years after its first impact outside of Mexico when severe epidemics swept through North America and Europe and resulted in the Irish potato famine (Large, 1940). The CIP estimated annual damage in developing countries at $3 billion, with poor farming communities in highland ecoregions disproportionately affected (CIP, 1997). In 1996, CIP launched a 10-year Global Initiative on Late Blight (GILB) to meet the threat to food security posed by the new populations of *P. infestans* that had been spreading from Mexico to the rest of the world since 1984 (Goodwin and Drenth, 1997). These new populations comprised both the A1 and the A2 mating types and hence had the potential for sexual reproduction. They also contained strains with increased aggressiveness and strains insensitive to the widely used systemic fungicide metalaxyl (GILB, 1999).

Verticillium wilt and early blight are the most common soil-borne and air-borne diseases, respectively, of potato plants in the warm-growing-season areas of potato production (Pavek and Corsini, 1994). They are seen during the later stages of tuber bulking and result in a hastening of maturity and hence early dying, smaller tubers and lower dry-matter content. Some fungal diseases of tubers are caused by persistent soil-borne fungi, such as wart, common scab (strictly speaking caused by an actinomycete), and powdery scab; some fungal diseases are tuber-borne (at least in part): black scurf, gangrene, silver scurf, dry rot, skinspot, and black dot (Wastie, 1994). Some are important causes of spoilage of tubers in storage, such as gangrene in northern Europe and dry rot in warmer climates, while others such as wart and common and powdery scab are disfiguring, blemish-forming diseases of tubers.

Bacterial wilt, or brown rot (*Ralstonia solanacearum,* formerly called *Pseudomonas solanacearum*), is regarded by CIP (1997) as the next most limiting phytopatholgical factor after late blight to potato production in the developing world, particularly in tropical and sub-tropical regions where biovars (a classification based on biochemical properties) 1, 3, and 4 predominate. Stem blackleg early in the growing season and tuber soft rot in storage are serious bacterial diseases caused by *Erwinia* species. *E. carotovora* subsp. *atroseptica* (reclassified as *Pectobacterium carotovorum* subsp. *atrosepticum*) is common in temperate climates, whereas *E. chrysanthemi* (reclassified as *Pectobacterium chrysanthemi*) predominates in warmer areas and *E. carotovora* subsp. *carotovora* (reclassified as *Pectobacterium carotovorum* subsp. *carotovorum*) is well adapted to both climatic regions (Weber, 1990). Bacterial ring rot is a recurring seed-tuber-transmitted disease problem in temperate regions, despite many countries treating it as a quarantine disease and having a zero-tolerance policy for the import of seed potatoes.

The viruses with a worldwide distribution are Potato leafroll virus (PLRV), Potato virus Y (PVY), Potato virus A (PVA), Potato virus X (PVX), and Potato virus S (PVS). The PLRV is probably the most damaging and widespread of the viruses, with PVY next in importance. The PLRV is aphid-transmitted in a persistent manner, whereas PVY is aphid-transmitted in a nonpersistent manner, and hence it is harder to control with aphicides. Potato virus M (PVM) can be a devastating virus in the seed and ware potato production areas of Central and Eastern Europe (Marczewski et al., 2006). Likewise, *Tobacco rattle virus* (TRV) and *Potato mop-top virus* (PMTV) can be locally important (Solomon-Blackburn and Barker, 2001; Barker and Dale, 2006), the former in cooler climates and the latter in light sandy soils (Brown and Mojtahedi, 2005). They are transmitted by Trichodorid nematodes and *Spongospora subterranea,* respectively, and cause spraing symptoms in tubers, which are a particular problem for processors. *Potato spindle-tuber viroid* (PSTVd) is an example of a true seed-transmitted disease and one which is treated as a quarantine disease in countries where it is not endemic.

Introgression of genes from wild species

Past introgressions and durability of disease and pest resistance

In 1908 in Germany and 1909 in Great Britain, the Mexican wild species *S. demissum* was recognized as a source of genes for extreme resistance to late blight, which could be backcrossed into *S. tuberosum* (Muller and Black, 1951). This marked the start of introgression breeding and was followed by genes for resistance to viruses and cyst nematodes, which were also lacking in the European and North American potato. It took from three to seven backcrosses to transfer a major dominant resistance gene from a wild species into a successful cultivar (Ross, 1986). Resistance to late blight was introgressed from *S. demissum* and *S. stoloniferum;* resistance to viruses from these species, together with *S. chacoense* and *S. acaule,* and resistance to potato cyst nematodes was introgressed from *S. vernei* and *S. spegazzinii.* By the end of the 1980s these wild species, together with cultivated *S. tuberosum* subsp. *andigena* and *S. phureja,* had been used extensively in the breeding of successful cultivars in Europe (Ross, 1986). Likewise the species *S. demissum, S. chacoense,* and *S. acaule* dominated among North American cultivars (Plaisted and Hoopes, 1989). Interestingly, *S. chacoense* is also a source of leptines (acetylated leaf glycoalkaloids) that have antifeedant properties and have been shown to confer plant

resistance to the Colorado potato beetle and the potato tuber moth (Douches and Grafius, 2005).

One of the most successful cultivars to be introduced into China by CIP, CIP-24, had *S. acaule, S. demissum,* and *S. stoloniferum* in its pedigree (Ortiz, 2001). Furthermore, the genetic base of modern Indian potato selections can be traced to 49 ancestors of which 10 from the United Kingdom account for 41% of the total genomic constitution (Gopal and Oyama, 2005). The most frequent ancestors were two clones from Great Britain, 2814a1 and 3069d4, which were the parents of the popular Indian cultivar 'Kufri Jyoti.' They trace back to a cross between *S. rybinii* (a variant of *S. phureja*) and *S. demissum,* which was made by Dr. Black in 1937 (Bradshaw and Ramsay, 2005) to introgress late blight resistance from *S. demissum* (Fig. 10.4). In conclusion, the introgression of genes from wild species has been fairly limited in number and hence has had a relatively small impact on the narrow genetic base of modern cultivars.

The durability of resistance is clearly an important issue given the time and resources required for introgression. The major gene

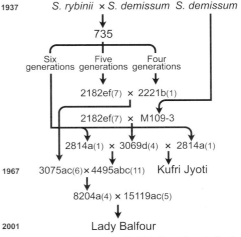

Fig. 10.4. Pedigrees of 'Kufri,' 'Jyoti,' and 'Lady Balfour.'

resistances to PVY have proved durable, and the hybrid parent MPI 61.303/34 from the Max Plank Institute in Köln has been used extensively in European breeding as the source of the *Ry* gene for extreme resistance from *S. stoloniferum*. The major gene resistances to PVX have also proved durable despite the occurrence of resistance-breaking strains (Jones, 1985). The widely used *H1* gene from a CPC accession (CPC 1673) of Andigena potatoes has remained effective against *G. rostochiensis* in Great Britain because Ro1 is still the main pathotype, but its widespread deployment has encouraged the spread of *G. pallida* (Bradshaw and Ramsay, 2005). In contrast, the *S. demissum*-derived R genes failed to provide durable resistance to late blight either singly or in combination due to the evolution of new races of *P. infestans* (Malcolmson, 1969). Breeders therefore switched to selection for high levels of quantitative field resistance, either by using races of *P. infestans* compatible with the R genes present in their material or by creating R gene-free germplasm so that screening could be done with any race (Toxopeus, 1964; Black, 1970; Wastie, 1991; Ortiz, 2001). Interestingly, the two clones 2814a1 and 3069d4 (see Fig. 10.4) are great grandparents of cultivar 'Lady Balfour' that has high levels of field resistance in both its foliage and tubers and was bred at the Scottish Crop Research Institute (SCRI) for organic potato production. Currently there is much debate over whether or not the R genes for blight resistance being found in other wild species will be more durable per se or can be deployed in a more durable way. One suggestion is a multi-line cultivar produced by map-based cloning of different R genes and independent transformation of a popular but susceptible cultivar (Huang, 2005; Smilde et al., 2005). For a multi-line cultivar to provide durable resistance, complex races of *P. infestans* that can overcome more than one of the resistance genes must be less aggres-sive (fit) than simple races that can only overcome one resistance gene.

Molecular marker-assisted introgression and gene cloning

In future breeding for resistance to abiotic and biotic stresses, we can anticipate a greater use of wild species given the wide range of habitats in which they have evolved. By manipulation of ploidy, with due regard to EBN, virtually any potato species can be used for the introgression of desirable genes into *S. tuberosum* (Ortiz, 1998, 2001), with protoplast fusion a further option for otherwise difficult or impossible hybridizations.

Molecular marker-assisted introgression offers the possibility of faster progress than can be achieved by traditional backcrossing (Hermsen, 1994; Barone, 2004; Iovene et al., 2004). This is because one can select genotypically rather than phenotypically against the wild species genome, while selecting either phenotypically or genotypically for the desired gene(s) from the wild species. With adequate molecular marker coverage of all 12 potato chromosomes it is possible to estimate the optimal combination of population sizes and number of backcross generations and to select in a very precise way for the desired products of meiosis in each backcross generation (Hospital, 2003) (Fig. 10.5).

Where introgression is performed at the tetraploid level, the result may not be a genotype with a complete set of 48 chromosomes. This need not affect the performance of the genotype and its vegetative propagation but could affect its fertility and use as a parent for further breeding. The result of an introgression from *S. brevidens* was a high-yielding clone, C75–5+297, with resistances to both tuber soft rot and early blight (Tek et al., 2004). Molecular and cytogenetic analyses revealed that C75–5+297 had 47 chromosomes, including four copies of chromosome 8, three from potato and one

Chromosome with R gene

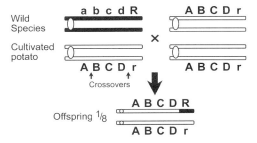

Other eleven chromosomes

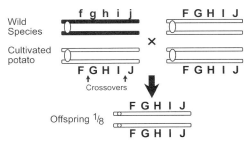

Fig. 10.5. First backcross of diploid wild-cultivated species hybrid to diploid cultivated potato with selection for R-gene and as many whole chromosomes as possible from cultivated potato using molecular markers (A to J). The probability of chromosome with R-gene and at least three other whole chromosomes from cultivated potato is 0.0528.

from *S. brevidens,* which was the only part of the wild species genome present. In contrast, Barone et al. (2001) did obtain 48 chromosomes and evidence of recombination between *S. commersonii* (an EBN1 diploid) and *S. tuberosum* chromosomes in their molecular marker-assisted introgression of tuber soft-rot resistance.

As potatoes are heterozygous outbreeders, use of the same recurrent parent during introgression results in a self of the recurrent parent accompanied by inbreeding depression. This can be avoided by using different Tuberosum parents for each backcross, but that results in an entirely new cultivar, which may or may not be the desired outcome. The only way to introduce a gene into a known cultivar is by the transgenic route, which is discussed in a later section.

Base broadening versus introgression

Peloquin and his coworkers (Hermundstad and Peloquin, 1987; Jansky et al., 1990) proposed a novel breeding strategy designed to introgress specific characteristics and to broaden the genetic base of potato in a way similar to that envisaged by Chase (1963) in his analytic breeding scheme. The hybrids between haploids (dihaploids) of *S. tuberosum* and diploid wild species with an EBN of 2 will often form tubers in long-day growing conditions. Jansky et al. (2004), for example, found that in 125 out of 154 families of haploid-wild species crosses, more than 50% of the plants tuberized in long days and could therefore be selected for tuber characteristics. If they also produce $2n$ gametes by FDR (first division restitution), much of the genetic diversity of the wild species can be efficiently transferred to the tetraploid offspring from $4x \times 2x$ crosses which result in about 25% of the wild species' genes in the final product (Tai, 1994). Tai (1994) concluded, however, that haploid-wild species hybrids need to be improved before they are used in $4x \times 2x$ crosses, for example, through population improvement by recurrent selection (Rousselle-Bourgeois and Rousselle, 1992). The issue that needs to be resolved by experiment is whether or not wild species contain sufficient desirable alleles to warrant base broadening as opposed to the introgression of just a few desirable genes.

Base broadening and population improvement

During the second half of the 20th century, recognition was given to the value of the cultivated species of Latin America for broadening the genetic base of European and

North American breeding programs. Furthermore, CIP recognized the need to make broad-based germplasm from South America available to national programs in developing countries, particularly germplasm with durable resistance to late blight.

Simple mass selection for tubering in long days

Breeding experiments in the U.K., the U.S., the Netherlands, and Canada have all demonstrated that through simple mass selection under northern latitude, long-day summer conditions, subsp. *andigena* will adapt and produce parents suitable for direct incorporation into European and North American potato breeding programs (Bradshaw and Mackay, 1994a). The first experiment was started by Simmonds in the U.K. in 1959 with a gene pool of Andigena potatoes from the CPC of which approximately 45% were Bolivian in origin, 35% south Peruvian, 10% north Peruvian, and 10% Colombian (Simmonds, 1969). He practiced simple phenotypic mass selection on a two-year cycle. In the first year, large populations of seedlings were grown in the field and selected for high yields of tubers of acceptable size, shape, and color. In the second year, the selected tubers were planted in an isolation site and open-pollinated berries were harvested to provide the next generation of seedlings. Within four generations the better Andigena clones were comparable in yield, maturity, and cooking quality to Tuberosum cultivars, but were still inferior in tuber shape to modern cultivars. In the U.S., Plaisted (1987) introduced the deliberate pollination of selected clones with bulk pollen to avoid inbreeding depression from selfing. He achieved more vigorous plants for selection by raising his seedlings in pots to produce seed tubers for planting field trials. This enabled him to screen the seedlings in the glasshouse for resistance to late blight and viruses, but it also increased the

length of each cycle to three years. Maris (1989) used hand pollination between individual selected clones in his scheme. A brief summary of research on Neotuberosum and more original references can be found in Bradshaw and Mackay (1994a).

During the period 1962–1979, Carroll (1982) also employed a mass selection method to produce a population of *S. phureja/S. stenotomum* adapted to long-day North European conditions. The starting material was diploid stocks from the CPC comprising approximately 75% *S. phureja* and 25% *S. stenotomum*. The method relied on natural insect pollination and seed set and included both seedling and tuber populations with a minimum generation cycle of two years. In practice, the generation time was three years, with seedlings raised in a polythene tunnel rather than the field. The self-incompatibility of the diploids was presumed to ensure cross-pollination. The population adapted to tubering in long days after only one to two generations of selection, possibly because *S. phureja* had already been selected in the Andes for faster tuber development. Yield improved across several generations mainly as a result of increase in tuber size without a reduction in tuber numbers. The proportions of oval and long oval, regular-shaped tubers increased and diploid cultivars, such as 'Mayan Gold,' have now been produced from this population, but they are targeted at niche markets because their yield is only two-thirds that of Tuberosum potatoes. Direct hybridization of members of this improved diploid population with tetraploid potato cultivars via unreduced pollen grains ($4x \times 2x$ crosses) resulted in tetraploid hybrids, some of which were superior to standard tetraploid cultivars in both total and marketable yield (Carroll and De Maine, 1989). A similar program was started in 1966 by F. L. Haynes at North Carolina State University from approximately equal proportions of *S. phureja* and *S. stenotomum* (Haynes and Lu 2005). Today the population comprises 72 families, each of which can trace its ancestry to one of the 72 original plant introductions because within-half-sib family selection was practiced. This was done on a two-year cycle for 10 generations, the first five to six for general adaptation, followed by four to five for high specific gravity. The program was then continued at USDA/ARS-Beltsville by K.G. Haynes, but on a four-year cycle in which a seedling generation in the glasshouse was followed by three years of field evaluation in which the number of clones per family was reduced from 100 to 4 to 1. Open-pollinated seed was collected from the 288 clones evaluated in the second year in the field, but only used from the 72 clones selected after the third year in the field. Haynes has recently turned her attention to using the population for recurrent selection for higher levels of resistance to both early and late blight (Haynes and Lu, 2005).

Despite these breeding efforts, relatively few clones of Neotuberosum (long-day subsp. *andigena*) and long-day *S. phureja/S. stenotomum* have been used to any extent in the breeding of modern cultivars. Part of the reason for this is that while adaptation to tubering in long days was quickly achieved, other problems remained. Neotuberosum clones lacked the regularity of tuber shape of intensively selected subsp. *tuberosum* clones and long-day-adapted *S. phureja* clones lacked tuber dormancy. So these populations still need selection for further improvements to achieve the original goal of direct use as parents in breeding finished cultivars for European and North American markets (i.e., contributing 50% of their genes to their offspring, or 25% if the hybrids are backcrossed to Tuberosum). More encouraging for immediate use is the yield heterosis shown by the hybrids. Under long days, Tuberosum × long-day Andigena/Neotuberosum hybrids have shown yield advantages of 13–17% over Tuberosum × Tuberosum hybrids (Glendinning, 1969;

Cubillos and Plaisted, 1976; Tarn and Tai, 1983; Maris, 1989), but it appears that the extent of heterotic advantage is dependent on day length (Cubillos and Plaisted, 1976). Likewise, the tetraploid offspring from Tuberosum (4x) × Tuberosum/Phureja hybrids (2x) have shown yield advantages of 27% over their 4x parents and controls (Tai and De Jong, 1980), but in a more recent study a 15% yield advantage occurred only in the higher yielding of two environments (Buso et al., 1999).

Three strategies can be envisaged for using the cultivated species of Latin America in breeding potatoes adapted to long days:

1. Biodiverse populations of long-day-adapted Andigena and Phureja/Stenotomum can be created and evaluated as sources of desirable alleles for marker-assisted incorporation into Tuberosum and thence into finished cultivars. Using molecular markers and Tuberosum, Neotuberosum, and Phureja germplasm in the Cornell University potato breeding program, Bonierbale et al. (1993) found that specific combinations of individual marker fragments were more important than maximum heterozygosity for tuber yield heterosis.
2. Biodiverse populations can be improved by recurrent selection in target environments and then crossed to adapted Tuberosum to create hybrids that display yield heterosis because of genetic differences between the two groups of germplasm.
3. Heterosis can be deliberately sought in crosses between two groups of germplasm by practising reciprocal recurrent selection in which the selection of parents for maintaining the two groups is based on the performance of their hybrids.

But which is the best strategy? This question cannot be answered at present because it is not clear how inferior a biodiverse population has to be to modern cultivars for targeted gene transfer to be superior to base broadening.

CIP's blight-resistant populations

Probably the best known broad-based germplasms from CIP's breeders are their A and B populations with quantitative resistance to late blight. Population A was developed in the 1980s from germplasm that contained the major R genes that had failed to give durable resistance. In contrast, populations B1 to B3 were developed from 1988 onward from germplasm lacking such R genes. The aim was to develop improved cultivars for a wide range of environments that would have high and stable levels of resistance to late blight and viruses as well as having suitable culinary and processing qualities (Trognitz et al., 2001; Landeo, 2002). In population A, complex races of *P. infestans* were used to overcome the major R genes known to be present and hence allow selection for quantitative field resistance. This proved difficult in practice and the blight resistance of the cultivars distributed to developing countries was often disappointing (Landeo, 2002). As a result of these experiences, the R gene-free B populations were established. Group B1 was a population of native cultivated forms of short-day-adapted *S. tuberosum* subsp. *andigena* that lack R-genes; Group B2 derived from crosses of the subsp. *andigena* clones with subsp. *tuberosum* clones that also lacked R genes; and Group B3, which was derived from population A, comprised subsp. *tuberosum* germplasm with *S. demissum*-derived resistance, subsp. *andigena*, Neotuberosum (Plaisted, 1987) and complex ABPT hybrids of *S. acaule* and *S. bulbocastanum* with *S. phureja* and *S. tuberosum* subsp. *tuberosum* (Hermsen, 1994). Trognitz et al. (2001) provide a brief description of how race-specific resistance was eliminated from the B3 population by progeny testing with crosses to a tester that

did not possess R genes for resistance. The breeding in each group involved recurrent selection for quantitative resistance to late blight and other desirable traits together with selection in the geographical areas where the new cultivars were to be grown, for example, in longer days in Argentina (Trognitz al., 2001) and in East Africa (Mulema et al., 2004/5). Clones with good general combining ability were also identified for use as parents in the local breeding programs of the National Agricultural Research Systems in developing countries. Landeo (2002) has reported in the B populations good progress in selecting for resistance and other traits, large additive genetic variances and high heritability estimates for resistance, and stability of resistance across diverse environments and pathogen populations in tropical environments.

Family selection involving progeny testing

In 1991 an experimental breeding program was set up at SCRI to combine quantitative resistances to late blight and the white potato cyst nematode with acceptable tuber yields and quality. Such resistances were available in *S. tuberosum* germplasm held at SCRI as a result of past breeding and introgression from wild and short-day-adapted cultivated species, but they had not been combined together in widely grown cultivars. It is possible to operate full-sib family selection on a three-year cycle (Bradshaw et al., 2003). Crosses are made in the first year followed by glasshouse seedling progeny tests in the second year and a tuber progeny test of the surviving progenies in the field in the third year. The most visually attractive clones are selected from the tuber progeny tests to provide parents for the next cycle of crosses. Replicated seedling progeny tests are used for assessing the full-sib families for foliage blight, tuber blight, and nematode resistance as well as for breeders' visual assessment (preference) of tubers primarily for yield and regularity of shape. The latter progeny test provides the tubers for the field progeny test in which quality traits, such as fry color, can be assessed. The parents selected for the next cycle of crosses are multiplied at the seed site and assessed in a replicated yield trial at a ware site in the same year as the next set of seedling progeny tests. Mid-parent values for yield and quality traits can complement the results of the seedling progeny tests. Other potential parents can be assessed in the yield trial and used in the next cycle of crosses if superior to current parents. Diallel analysis revealed the presence of large amounts of additive genetic variance for foliage blight, tuber blight, and nematode resistance as well as for breeders' preference and fry color (Bradshaw et al., 1995, 2000a). Good progress from full-sib family selection was predicted and is being achieved in practice (Bradshaw, 2005).

The breeders operating base broadening and population improvement programs have usually done genetic analyses to check that narrow-sense heritabilities were high enough and additive genetic variation sufficient for progress with the traits under recurrent selection. The design of their programs was, however, mainly determined by practical considerations and genetic principles rather than detailed genetic analysis aimed at optimizing the response to selection. Unless more than one crop can be grown in a year and unless good quality transplants can be established with irrigation, the crossing and selection cycle will take three years because there will be a seedling generation grown in a glasshouse or polythene tunnel to provide seed tubers for field evaluation (Fig. 10.6). Family selection intensity will be limited by the minimum number of families that need to be selected to avoid loss of favorable alleles and inbreeding depression (probably around 20) and the maximum number of families that can be assessed in progeny tests (often around 200). Larger numbers of clones can be assessed within families, but there is a limit to the number of traits that can be assessed on single plants or small plots, and heritabilities are likely to be low for many traits because of lack of replication (Bradshaw et al., 1998). A review of the critical points for the effective exploitation of family selection in plant breeding was done by Simmonds (1996).

Breeding cultivars at the tetraploid level for clonal propagation

Parents

Potato breeding worldwide has traditionally involved making crosses between pairs of parents with complementary features, and this is still the main route to new cultivars. Increasingly, parents will have genes introgressed from wild species and from complementary groups of germplasm, such as Neotuberosum and Andigena, to exploit yield heterosis. The aim is to generate genetic variation on which to practice phenotypic selection across a number of vegetative generations for clones with as many desirable characteristics as possible for release as new cultivars. The choice of parents and crosses is all important, as breeding can never simply be a numbers game. The number of possible bi-parental crosses increases from 4,950 from 100 parents to 499,500 from 1,000 parents and on to a staggering 49,995,000 from 10,000 parents. Breeders can now complement phenotypic assessments of potential parents with a genotypic assessment of diversity and content using molecular markers and hence capture allelic diversity in a smaller core set of parents. They can also use genetic distance based on molecular markers (Powell et al., 1991) to complement co-ancestry/pedigree analysis (Tarn et al., 1992; Gopal and Oyama, 2005) to avoid closely related parents yielding inbreeding depression and

Year 1

Crossing in glass-
house to provide
families of true
potato seed.

Year 2

Families raised from
true potato seed in
glasshouse.
Samples of seed
can also be used for
seedling progeny
tests for disease
resistance.

Year 3

Families raised from
seed tubers at seed
site. Selection
between and within
families for next
round of crossing.

Fig. 10.6.   Population improvement on a three-year cycle.

to ensure genetic variation for continued progress. Both analyses are required because clustering based on molecular markers can be different from clustering based on pedigree (Sun et al., 2003). Furthermore, as genetic knowledge accumulates, it will be possible to choose parents known to possess desired major genes and large-effect QTL and to select for these in their offspring.

Later in this chapter we shall see that major genes have been mapped for flesh, skin and flower color, tuber shape and eye depth, and resistances to late blight, nematodes, PVX, PVY, PVA and PLRV, and wart. Large-effect QTL have also been mapped for maturity and resistances to late blight, potato cyst nematodes, and PLRV. In

contrast, many economically important traits still appear to be complex polygenic traits; these include dormancy, dry matter and starch content, fry color, resistance to *Erwinia*, tuberization, and yield. For these traits, breeders will still have to rely on phenotypic data and the concepts of quantitative genetics to determine crossing strategies. For a highly heritable trait like fry color, the midparent value is a good predictor of the mean performance of the offspring, and a few carefully chosen crosses can be made (Bradshaw et al., 2000a). In contrast, with an only moderately heritable trait, such as yield, offspring mean is less predictable, and more crosses need to be made to ensure that they include the best possible cross for the trait.

Early generations

The program at the SCRI before 1982 was typical in its handling of the early generations (Bradshaw and Mackay, 1994a). Visual selection reduced the number of potential cultivars from 100,000 in the seedling generation (SG) in the glasshouse to 1,000 clones in replicated yield trials at a ware site in the third clonal generation (TCG). The first clonal generation (FCG) comprised 40,000 spaced plants at a high-grade seed site with a short growing season, which was rather atypical of normal ware production. The second clonal generation (SCG) of 4,000 unreplicated four-plant plots was also grown at the seed site. Several independent reviews concluded that such intense early-generation visual selection was ineffective, except for skin and flesh color and possibly tuber shape and eye depth (Tai and Young, 1984; Caligari, 1992; Tarn et al., 1992; Bradshaw and Mackay, 1994a). Bradshaw et al. (1998) confirmed that selection by visual preference within crosses in the seedling and first clonal generations is highly ineffective, but that worthwhile progress can be made from selection in the second clonal generation for faster emergence, earlier maturity, higher yield, and greater regularity of tuber shape.

The experimental breeding program set up at SCRI in 1991 provides a solution to the problem of ineffective selection between seedlings in a glasshouse and between spaced plants at a seed site (Bradshaw et al., 2003). Population improvement by full-sib family selection is combined with breeding finished cultivars. The progeny tests are used to discard whole progenies before starting conventional within-progeny selection at the unreplicated small-plot stage. The number of clones on which to practice selection is increased by sowing more true seed of the best progenies but without selection until the small-plot stage. There is no loss of time if these resowings are chosen on the basis of the seedling progeny tests and progressed straight from the glasshouse to small plots at the seed site.

Intermediate and later generations

The SCRI program prior to 1982 was also typical in its handling of the intermediate and later generations (Bradshaw and Mackay, 1994a). Decreasing numbers of potential cultivars were assessed for three years in replicated yield trials at a local ware site before the most promising ones underwent two years of more extensive testing at a number of sites, including sites overseas. One or a few clones would then be entered into official statutory trials (National List Trials) and registered for Plant Breeders' Rights. Multiplication from disease-free stock would start with a view to commercialization. During these intermediate and final stages of selection, the production of seed tubers was separated from the trials that were grown under ware conditions; this practice was designed to resemble those of good commercial practice as much as possible. This separation meant that the ware site used for the intermediate generations did not have to be local, and more than one site could be used once sufficient seed tubers were available for planting. Clones undergoing selection were assessed for their yield and agronomic performance, internal defects, and cooking and processing characteristics, as described by Bradshaw et al. (1998, 2003). They were also assessed for their disease and pest resistance in special tests, which will be discussed in the next section. More recently, the time from seedling to cultivar has been reduced by two years by using progeny tests to discard whole progenies, by starting replicated trials earlier at more than one site, and by using micropropagation to multiply promising clones for more extensive testing (Mackay, 2005). But the whole process can still take nine years, so it is important to use potential cultivars at the earliest stage possible to ensure the fastest progress.

A characteristic of yield trials in the intermediate generations at a single site is the large number of entries and small plots, which often comprise just 5 to 10 plants in single-row plots. Connolly et al. (1993) and Bradshaw (1994b) found inter-plot competition for fresh-weight yield in such trials, which led to some changes in the ranking of entries. Overall, however, competition was thought unlikely to seriously affect a breeder's ability to select clones that are high yielding under monoculture. Plot-to-plot variation can be high in small-plot trials, so efficient trial designs are required, such as those with incomplete block or row and column designs, possibly supplemented with spatial analysis (Mead, 1997).

Bradshaw and Mackay (1994a) have discussed how to use the resources available in the intermediate generations to maximize the response to selection. The theory is relatively straightforward for single-stage, single-trait selection but becomes very complicated for multi-stage, multi-trait selection. A selection index would accommodate multi-trait selection, but at present the use of independent culling levels is normal practice. Furthermore, the selection criteria and testing procedures used by breeders have been largely governed by practical considerations and experience of the reliability of the various tests used rather than by genetic knowledge about heritabilities and genetic correlations between traits. Nevertheless, it might still be worthwhile for breeders to try to develop a more robust decision-making process based on multi-stage, multi-trait selection theory and estimated genetic parameters, or at least to assess the extent to which current practice is sub-optimal.

Once the advanced generations are grown at a number of sites for more than one year, genotype × environment interactions can be assessed along with the effectiveness of the trial system for identifying new cultivars for the target environments and end-uses (Bradshaw, 1994a). While new cultivars are more likely to be adopted if breeders involve end-users in deciding objectives and selecting germplasm, there is currently much interest in involving small-scale farmers in developing countries in the process to increase food production and security. Methods for the quantitative analysis of data from participatory methods in plant breeding were discussed at a workshop in Germany in 2001, and the proceedings have been published and are well worth reading (Bellon and Reeves, 2002). For potatoes, farmer participation in Peru involved the farmers using cards with happy, serious, or sad faces to indicate their preferences.

Selection for disease and pest resistance

Effective selection for disease and pest resistance requires reliable screening using the most appropriate isolate or isolates of the pathogen (Bradshaw et al., 2000b; Jansky, 2000). As pathogens are often differentiated into species, subspecies, biovars, or physiological races before screening for resistance, it is important to determine if host genotype × pathogen isolate interactions occur that are relevant to the program. Complications can arise when different resistance mechanisms appear to occur in different parts of the plant, and separate disease tests are required to guarantee complete resistance, examples being foliage and tuber resistance to late blight, resistance to *Erwinia* blackleg of stems, and soft rot of tubers. Complications can also arise with diseases such as late blight and early blight where resistance is often associated with late maturity, so it is important to score for both traits if one wants to break the association.

For any disease test, there are many components that need to be determined from the literature or by experiment: the best stage of plant growth for inoculation and scoring; inoculum concentration; method of inoculation; and environmental conditions (temperature, day-length, humidity). An extensive set of references for bacteria, fungi, viruses and viroids, insects, and nematodes can be found in the review by Tarn et al. (1992). Screening needs to be integrated with selection for other traits as explained by Mackay (1987) and Jellis (1992). Examples of screening for resistance to an air-borne fungal disease (late blight), a persistent soil-borne fungus (powdery scab), and a tuber-borne disease (gangrene) have been given by Bradshaw et al. (2000b); examples for viruses are given by Solomon-Blackburn and Barker (2001), and for insects by Raman et al. (1994). Jansky (2000) has reviewed publications from the 1990s about screening and breeding for resistance to late blight, early blight, Verticillium wilt, common scab, soft rot and blackleg, ring rot and PLRV, PVX, and PVY.

Breeding for insect resistance provides two examples of indirect selection through traits associated with resistance, namely glandular trichomes and leptine glycoalkaloids (Douches and Grafius, 2005). Three wild species, *S. berthaultii*, *S. polyadenium* and *S. tarijense*, have high densities of glandular trichomes that have been bred into cultivated potatoes. Breeding line NYL235-4 from Cornell University has glandular trichomes derived from *S. berthaultii* and has a positive effect against both the potato tuber moth and the Colorado potato beetle (Douches and Grafius, 2005). Leptines found in the leaves of some accessions of *S. chacoense* have antifeedant properties that have been shown to confer resistance to the same pests (Douches and Grafius, 2005). These leptines are currently being transferred to cultivated potatoes.

Genetic knowledge and molecular marker-assisted selection

As knowledge increases about the number and chromosomal locations of genes affecting important traits, breeders should be able to design better breeding programs. They will be able to choose parents that complement one another for desirable genes and then select for these genes in the offspring. They will be able to determine the seedling population size required for certainty in finding the desired combination of genes or, more realistically, the number of cycles of crossing and selection required before this is achievable in practice in the size of population that they can handle. Efficiency and rate of progress would be hastened with the identification of superior clones with the desired combination of genes as seedlings in the glasshouse and the use of modern methods of rapid multiplication to progress them to commercialization. Such progress will require molecular-marker-assisted selection or, preferably, direct recognition of the desired genes, as has recently been achieved for the *RB* gene for late blight resistance from *S. bulbocastanum* (Colton et al., 2006). Progress to date has been slow, but is expected to accelerate.

Diagnostic markers are now available for five major genes for disease resistance that are of value to breeders. Kasai et al. (2000) have developed sequence characterized amplified region (SCAR) markers for the gene *Ry* (from Andigena, on chromosome 11) that confers extreme resistance to PVY. These SCAR markers should be powerful tools in marker-assisted selection as they showed high accuracy for detection of the *Ry* gene and one marker, RYSC3, was generated only in genotypes carrying *Ry*, namely 14 out of 103 breeding lines and cultivars with diverse genetic backgrounds.

Bakker et al. (2004) have identified an AFLP marker (EM1) that co-segregates with the *H1* gene for resistance to *G. rostochiensis* pathotype Ro1. EM1 and *H1* were present in 19 resistant cultivars and absent from 26 susceptible ones. However, Bakker et al. (2004) recommended conversion to a cleaved amplified polymorphic sequence (CAPS) marker for use in marker-assisted selection because such markers are cheaper and easier to handle. Colton et al. (2006) have developed and thoroughly tested a polymerase chain reaction-based DNA marker for tracking the *RB* gene from *S. bulbocastanum* for resistance to late blight. Finally, Gebhardt et al. (2006) have demonstrated the use of diagnostic markers for *Rx1* (extreme resistance to PVX), *Sen1* (resistance to *Synchytrium endobioticum* pathotype 1, the cause of potato wart), and *Gro1* (resistance to all known pathotypes of *G. rostochiensis*), as well as *Ry*, for pyramiding these major genes for pathogen resistance in potato. The pyramiding of such major genes, together with QTL alleles of large effect for disease resistance, is likely to be the main use of marker-assisted selection because it removes the need for costly, complex, and sometimes inaccurate disease testing.

Inheritance of important traits

The inheritance of all the important traits mentioned in this chapter was thoroughly reviewed in 1994 in *Potato Genetics*, edited by Bradshaw and Mackay (1994b). More recent information can be found in a review of potato breeding via ploidy manipulations (Ortiz, 1998) and in *Genetic Improvement of Solanaceous Crops Volume I: Potato*, edited by Razdan and Mattoo (2005). The main advances since 1994 have been in the mapping of major genes and QTL, and it is these advances that are dealt with in this section.

Advances have not occurred for all traits, the most notable exceptions being resistances to abiotic stresses. In the future, we can anticipate the development of genetic analyses of these and other complex physiological traits such as water- and mineral-use efficiency, but it is too early to report results in this chapter. It is worth mentioning that 10 to 15 copies of the gene for the major storage protein of potato, patatin, have been mapped genetically and physically to a single locus at the end of the long arm of chromosome 8, but the implications of this for potato breeding are not yet clear (Ganal et al., 1991).

Yield, dormancy, tuberization, and dry matter, starch, and sugar content

The inheritance of yield, dormancy, tuberization, and dry matter and starch content has been reviewed by Bradshaw (2006) and starch-sugar metabolism by Solomos and Mattoo (2005). These traits must still be viewed as complex polygenic traits and handled as such in breeding programs. Most published work on yield and its components, tuber number and size, has come from breeders doing combining ability analyses that are specific to their particular population of genotypes. Tetraploid examples for yield can be found in the review by Bradshaw (2006) where differences will be seen in the relative importance of GCA (general combining ability) and SCA (specific combining ability). Neele et al. (1991) suggested that SCA tends to be more important than GCA in crossing schemes involving closely related parents (e.g., Killick, 1977) and Tai (1976) suggested the same for traits such as tuber yield, which have been subjected to continuous directional selection.

The genetic basis of yield heterosis is an issue in breeding tetraploid cultivars for both clonal and TPS propagation. It is a question of whether or not interactions between three or four different alleles are important compared with interactions between two different alleles and, hence, whether they should be sought after. The evidence is inconclusive, but as mentioned earlier, Bonierbale et al. (1993) found that specific combinations of individual molecular-marker fragments were more important than maximum heterozygosity for tuber yield heterosis in diverse cultivated germplasm. Fragments whose segregation patterns showed significant associations with the measured yield characteristics corresponded to 26 mapped loci from seven of the potato linkage groups (1, 2, 4, 5, 7, 9, 12).

The QTL mapping experiments that have been done in recent years hold out the possibility of finding candidate genes of importance for which allelic variants are worth seeking, but they also confirm the need to treat polygenic traits as systems of interacting genes on which to practice recurrent selection.

In a QTL-mapping experiment for tuber-starch content (%) and tuber yield involving two different crosses among dihaploid breeding lines, Schäfer-Pregl et al. (1998) found 18 putative QTL for tuber-starch content, which covered all 12 chromosomes and eight for tuber yield on eight chromosomes. In one of the crosses, 14 candidate genes were postulated for tuber-starch content through correlations between the map positions of the QTL and loci on the molecular-function map developed by Chen et al. (2001) for carbohydrate metabolism and transport. In a diploid population, (*S. tuberosum* × *S. chacoense*) × *S. phureja*, Freyre and Douches (1994) identified 10 putative QTL for specific gravity that were located on chromosomes 1, 2, 3, 5, 7, and 11. In the same population they also identified six QTL for dormancy, one on each of chromosomes 2, 3, 4, 5, 7, and 8 (Freyre et al., 1994). Van den Berg et al. (1996a) studied the genetics of tuber dormancy in reciprocal backcrosses between diploid *S. tuberosum* and *S. berthaultii*, a species that is noted for its long dormancy, as well as its high densities of glandular trichomes that are associated with insect resistance (Douches and Grafius, 2005). They detected QTL on nine linkage groups that affected tuber dormancy at 13°C in the dark, either alone or through epistatic interactions. Van den Berg et al. (1996b) also used the crosses to study tuberization and found 11 distinct loci on seven linkage groups. Most of these loci had small effects, but a QTL on chromosome 5 explained 27% of the variation in tuberization. Later, a QTL affecting leaf gibberellin A1 levels was found in this region of chromosome 5, with high levels associated with late tuberization (Ewing et al., 2004). The backcrosses involving *S. berthaultii* have also been used to map QTL for glandular trichomes and foliar glycoalkaloids (Yencho et al., 1996, 1998).

Another approach to finding candidate genes for potato crop improvement is through molecular-function maps, such as the one referred to above for carbohydrate metabolism and transport. The amount of starch accumulated in mature potato tubers is the net result of photosynthetic carbon fixation, the synthesis of transient starch and its conversion into sucrose in photosynthetically active source leaves, the vascular transport of sucrose from the leaves to the developing sink tuber, and starch synthesis and degradation in the tuber during the growth period. Chen et al. (2001) developed the map by searching public databases for genes of interest from potato, tomato, and other plant species. They used DNA sequence information to develop PCR-based marker assays (mainly CAPs and a few SCARs) that allowed them to locate corresponding potato genes on existing RFLP linkage maps. Potato cDNA and EST clones coding for relevant enzymes were also used as marker probes for RFLP analysis. In total, 69 genes were identified that mapped to 85 genetic loci. In this context it is worth noting that fry color after storage at low temperatures still needs to be treated as a complex polygenic trait despite a number of obvious candidate genes for variation in reducing sugars and cold sweetening. Menendez et al. (2002) identified QTL on chromosomes 1, 3, 7, 8, 9, and 11, each of which explained more than 10% of the variability for reducing sugars. The QTL were linked to genes encoding invertase, sucrose synthase 3, sucrose phosphate synthase, ADP-glucose pyrophosphorylase, sucrose transporter 1, and a putative sucrose sensor. Another candidate gene is the one on chromosome 11 encoding UDP-glucose pyrophosphorylase (Sowokinos, 2001).

Maturity

Maturity is an interesting trait because selection for earlier maturity must have taken place in many environments and has been accompanied by changes in other traits, such as susceptibility to diseases like early and late blight. Combining-ability and regression analyses have both pointed to high heritability and the importance of GCA (Bradshaw, 2006). A QTL of large effect has been located on chromosome 5 in the region of marker GP179 (Collins et al., 1999; Oberhagemann et al., 1999; Visker et al., 2003; Bradshaw et al., 2004). This marker is also associated with QTL for tuberization, tuber yield, tuber-starch content, vigor and susceptibility to late blight (Schäfer-Pregl et al., 1998; Oberhagemann et al., 1999).

Tuber shape, eye depth and pigmentation

Tuber shape, eye depth, and pigmentation are all traits for which major genes have been mapped since 1988 (Bradshaw, 2006). Hence, the desired combinations of alleles at these genetic loci can be sought in breeding programs through the correct choice of parents and visual selection. Round shape is dominant over long and controlled by gene *Ro* on chromosome 10 (van Eck et al., 1994a). This locus is closely linked (4 cM) to one for tuber eye depth, *Eyd*, with deep eye dominant over shallow (Li et al., 2005). There are four major dominant genes, *R*, *P*, *F*, and *I*, which control anthocyanin pigmentation in the flower and tuber skin (van Eck et al., 1993; 1994b). Locus *R* on chromosome 2 and locus *P* on chromosome 11 are involved in the biosynthesis of red and blue

anthocyanins, respectively, with *P* epistatic to *R* in the tubers (*rrP*-purple) but not in the flowers (*rrP*-blue). Loci *F* and *I* determine whether or not *R* and *P* are expressed in the flowers (petals) and tubers (both skin and flesh), respectively, and are closely linked to locus *Ro* on chromosome 10 (*Ro* at 42 cM, *F* at 57 cM, and *I* at 62 cM). A summary can be found in the review by Bradshaw (2006) of the alleles at these loci and other genes affecting the distribution of pigments, for example to the tuber flesh. Jung et al. (2005) have recently confirmed that the dominant allele at the *P* locus codes for activity of flavonoid 3'-, 5'-hydroxylase, whereas the recessive *p* allele lacks this activity. Variation at the *R* locus is thought to affect the specificity of dihydroflavonol 4-reductase (De Jong et al., 2003) while *F* and *I* are thought to encode transcription factors (De Jong et al., 2004). Finally white, yellow, or orange flesh is controlled by alleles at a single locus on chromosome 3 (Bonierbale et al., 1988; Brown et al., 1993) with orange (*Or*) dominant over both yellow (*Y*) and white (*y*), and yellow dominant over white. It is not yet agreed whether this locus codes for phytoene synthase, beta-carotene hydroxylase, or some other enzyme of importance for carotenoid biosynthesis. Modifying genes, tuber size, and to a lesser extent, environmental factors affect the intensity of yellowness (Bradshaw, 2006).

Resistance to pests and diseases

Since 1988 major genes have been mapped for resistance to late blight, wart, viruses (PVY, PVA, PVX, PVM, and PVS), cyst nematodes, and root-knot nematodes. The most important ones are shown in Table 10.4. Many of the genes listed are worth using in breeding for resistance. The *S. demissum*-derived R-genes for blight resistance have, however, failed to provide durable resistance, and the durability of the more recently described major genes from

Table 10.4. Major genes for disease and pest resistance.

Disease or pest	Gene and source	Chromosome	Latest relevant reference
Late blight	*R1 (dms)*	5	Ballvora et al. 2002
	R2 (dms)	4	Li et al. 1998
	R3a, R3b (dms)	11	Huang et al. 2005
	R5-R11 (dms)	11	Huang 2005
	R10, R11 (dms)	11	Bradshaw et al. 2006
	Rpi-ber	10	Rauscher et al. 2006
	RB (blb)	8	Song et al. 2003
	Rpi-blb1	8	van der Vossen et al. 2003
	Rpi-blb2	6	Park et al. 2005a
	Rpi-blb3	4	Park et al. 2005a
	Rpi-abpt (blb)	4	Park et al. 2005b
	R2-like	4	Park et al. 2005c
	Rpi-moc1 (mcq)	9	Smilde et al. 2005
	Rpi-pin1 (pnt)	7	Kuhl et al. 2001
Wart	*Sen1 (tbr?)*	11	Hehl et al. 1999
	Sen1-4 (tbr?)	4	Brugmans et al. 2006
PVY	*Ry (adg)*	11	Hamalainen et al. 1997
	Ry (sto?)	11	Brigneti et al. 1997
	Ry (sto)	12	Song et al. 2005
	Ry-f (sto)	12	Flis et al. 2005
	Ny (tbr)	4	Celebi-Toprak et al. 2002
PVA	*Ra (adg)*	11	Hamalainen et al. 1998
PVX	*Rx1 (adg)*	12	Bendahmane et al. 1999
	Rx2 (acl)	5	Bendahmane et al. 2000
	Nx (phu)	9	Tommiska et al. 1998
	Nb (tbr)	5	Marano et al. 2002
PVM	*Gm (grl)*	9	Marczewski et al. 2006
	Rm (mga)	11	Marczewski et al. 2006
PVS	*Ns (adg)*	8	Marczewski et al. 2002
Golden potato cyst nematode	*Gro1 (spg)*	7	Paal et al. 2004
	H1 (adg)	5	Bakker et al. 2004
	GroV1 (vrn)	5	Jacobs et al. 1996
White potato cyst nematode	*Gpa2 (adg)*	12	Rouppe van der Voort et al. 1997
	Gpa3 (tar)	11	Wolters et al. 1998
Columbia root-knot nematode	*R-Mc1 (blb)*	11	Brown et al. 1996
	R-Mc1-fen	11	Draaistra 2006
	R-Mc1-hou	11	Draaistra 2006
Northern root-knot nematode	*R-Mh-tar*	7	Draaistra 2006
Meloidogyne fallix root-knot nematode	*MfaXIIspl*	12	Kouassi et al. 2006

other wild species is not known. Furthermore the genes for extreme resistance to PYX and PVY are superior to the ones for hypersensitive resistance that tend to be strain-specific.

A number of large-effect QTL has been found when analyzing quantitative resistance to late blight, Verticillium wilt, PLRV, and cyst nematodes. The genomic positions of most of these QTL, together with most of the major genes in Table 10.4, can be seen in the paper by Bormann et al., 2004. Since then, QTL for resistance to Verticillium wilt have been reported on chromosomes 2 (explaining 40% of phenotypic variation), 6, 9, and 12 (Simko et al., 2004) and for resistance to PLRV on chromosome 11 (Marczewski et al., 2004). The latest map should be available on the following website: https://gabi.rzpd.de/projects/Pomamo/. Two QTL are often required, however, to achieve the high level of resistance desired in a new cultivar and sometimes additional resistance is also needed, probably from a number of small-effect QTL.

Quantitative trait loci for foliage resistance to late blight have been detected on all 12 chromosomes, but only about one-tenth of them had a large effect (30%–45%) on trait variation (Simko, 2002; Jones and Simko, 2005). Quantitative trait loci have been repeatedly detected on the distal parts of chromosomes 3, 4, and 5, with the resistance allele on chromosome 5 associated with late maturity (Simko, 2002; Jones and Simko, 2005). Quantitative resistance to *G. pallida* provides three examples of two QTL acting additively to give a high level of resistance, namely QTL on chromosomes 5 and 9 for *S. vernei*-derived resistance (Rouppe van der Voort et al., 2000; Bryan et al., 2002), on chromosomes 4 and 11 for *S. tuberosum* subsp. *Andigena*-derived resistance (Bryan et al., 2004), and on chromosomes 5 and 11 for *S. sparsipilum*-derived resistance (Caromel et al., 2005). In mapping resistance to *G. rostochiensis* derived from *S. spegazzinii*, Kreike et al. (1996) found two large-effect QTL for root development on chromosomes 2 and 6 and an epistatic interaction between them. Root development is important because one does not want to select for apparent resistance due to poor roots. In tetraploids there are two examples of two alleles at a QTL providing more resistance than either alone (partial dominance), namely the QTL for resistance to late blight and *G. pallida* on chromosome 4 (Bradshaw et al., 2004; Bryan et al., 2004).

A QTL analysis of resistance to *Erwinia* confirmed its truly polygenic mode of inheritance (Zimnoch-Guzowska et al., 2000). Markers for resistance were found on all 12 chromosomes in the cross between diploid hybrids involving *S. chacoense* and *S. yungasense*. Hence this is an example of quantitative resistance that must still be handled as a complex polygenic trait in breeding. It remains to be seen if this will also be true of the many other sources of resistance to the many pests and diseases of potato yet to be analyzed.

Breeding cultivars for true potato seed

Breeding cultivars for TPS was started at CIP in 1972 with the aim of high yields and acceptable uniformity. None of the current breeding methods can deliver genetic uniformity and hence all of them involve selection for acceptable uniformity. Methods and progress have been reviewed by Golmirzaie et al. (1994) and Ortiz (1997). Breeding strategy has to take account of the reproductive biology of the potato and certain genetic considerations. Clearly, flower production and berry and seed set are important, but as the transport of true seed is relatively cheap, TPS can be produced in a favorable environment that is distant from where it will be evaluated and grown. Current TPS breeding aims to produce tetraploid cultivars, either from $4x \times 4x$ crosses in which heterosis is exploited between Tuberosum and Andigena (Simmonds, 1997); from $4x \times 2x$ crosses in which the $2x$ parent produces a high frequency of $2n$ gametes by FDR so that 83% of its heterozygosity is transmitted to the offspring (Golmirzaie et al., 1994; Clulow et al., 1995); or from $2x \times 2x$ crosses in which both parents produce $2n$ gametes, again with a very high frequency of $2n$ pollen produced by FDR (otherwise the offspring will contain more diploids than tetraploids) (Ortiz and Peloquin, 1991). Simmonds (1997) argued, however, that diploid TPS cultivars should not be ruled out for the future, and De Maine (1996) produced TPS families of long-day-adapted Phureja that appeared as uniform for tuber size and shape as selected clones. Open-pollination of diploids will result in almost 100% outcrossing because of self-incompatibility, whereas open-pollination of tetraploids will normally result in varying amounts of self-pollination and inbreeding depression. The inbreeding depression of the tetraploids may not be outweighed by their apparent intrinsic yield advantage over diploids. Furthermore the seed fertility (seed yield) of diploids is greater than that of tetraploids. Whether or not diploid inbred lines and true F1 hybrids can be produced in the future for maximum heterosis and uniformity is a matter for further research. The inbreeding depression from selfing also means that hand-pollinated tetraploids are superior to open-pollinated ones, but their production is more expensive because of the cost of emasculation of the flowers of the female parent (Simmonds, 1997). The use of male sterility that does not involve loss of visits by bumblebees is an attractive proposition; for example, the use of protoplast fusion to transfer cytoplasmic male sterility into a TPS parental line (Golmirzaie et al., 1994). Another possibility is the use of a dominant marker in the male parent so that hybrids can be selected from seedlings for transplanting and tuber production (Ortiz, 1997).

Synthetic seed is an alternative to TPS that would avoid the need to develop parents that generate uniform hybrids. Hence, there is currently renewed interest in somatic embryogenesis in potato, and sufficient progress has been made for serious consideration to be given to exploiting synthetic potato seed, as seen in the review by Veilleux (2005).

Genetically modified potatoes

Genetic modification of successful and widely grown potato cultivars offers the possibility for further targeted improvements that can be brought about in three ways. Desirable genes can be found in cultivated potatoes and their cross-compatible wild relatives, cloned, and introduced by genetic transformation. Where the control of biochemical pathways of interest is understood, down-regulation of gene expression using antisense technology has proved useful for achieving some desired modifications. In antisense technology, the potato plant is induced to produce large amounts of antisense RNA capable of trapping the messenger RNA from the gene being transcribed in an untranslatable RNA duplex. This is done by transforming the plant with a genetically engineered construct in which the gene has been attached at its wrong end to a strong promoter. Antisense technology, however, does not always completely silence the gene being targeted. Finally—and more controversially—the use of exotic sources of genes in transformation has the clear value of permitting the introduction of traits not found in cross-compatible relatives, which can produce novel function. The genes used to-date have mainly been those that code for proteins that are toxic to pests and pathogens, those whose expression interferes with virus multiplication in host cells, and those that code for key enzymes in biochemical pathways in other organisms, often, but not always in other plant species.

The products of transformation do require screening to select the best transformants for commercialization, including demonstration of substantial equivalence to the parent cultivar (Davies, 2002). Commercialization also involves the demonstration that it is safe both to grow and eat the genetically modified (GM) potatoes. The gradual elimination of antibiotic-resistance markers will help deal with one of the safety issues that have been raised, but the demonstration of economic, environmental, and health benefits will be crucial in convincing skeptical consumers in some countries of the value of genetically modified potatoes. Leading processing and fast-food outlet companies in North America are currently reluctant to purchase GM potatoes because of consumer concerns over GM potato products, and as a consequence, Monsanto stopped marketing GM potatoes in 2001 (Davies, 2002).

Gene cloning

Gene isolation in potato can be done via map-based (i.e., position-based) cloning using BAC technology and dense AFLP genetic maps, but it is time consuming (De Jong, 2005). The only prerequisites are a phenotype that can be unambiguously scored in the individuals of a segregating population and a sufficiently large population size, typically 1,000 or more plants. De Jong (2005) estimated that, with current tools, about five person-years are needed to isolate a new gene in potato, but that this will be dramatically reduced once the potato genome has been sequenced. Sometimes shortcuts can be taken by using a candidate gene approach, but their success is not guaranteed.

The current priority in a number of laboratories worldwide is the molecular cloning of natural resistance genes and their transfer into well-adapted but susceptible cultivars. The *Rx* (Andigena) gene for PVX resistance in cultivar 'Cara' has been cloned and stably introduced into the susceptible cultivar 'Maris Bard' via *Agrobacterium*-mediated transformation (Bendahmane et al., 1999). The transgenic *Rx*-mediated resistance was indistinguishable from the *Rx*-mediated phenotype in cultivar Cara. The *R1* gene for resistance to late blight has been cloned and introduced into the susceptible cultivar Desiree and shown to give a typical hypersensitive response (HR), similar to the resistant line containing R1 (Ballvora et al., 2002). This gene failed to provide durable resistance, but the more recent cloning of the *R3a* gene in cultivar 'Pentland Ace' (Huang

et al., 2005), along with the corresponding avirulence gene, *Avr3a*, in *P. infestans*, offers a possible transgenic solution. The avirulence gene encodes a 147-amino acid extracellular protein that is recognized in the potato cytoplasm and triggers *R3a*-dependent cell death (Armstrong et al., 2005). The transgenic solution would be to produce a GM potato containing *R3a* in which a pathogen-inducible promoter was linked to *Avr3a* so that cell death would only occur in those cells infected with *P. infestans*. Not surprisingly, the search for pathogen-inducible promoters is currently an active area of research (Gurr and Rushton, 2005). The *RB* and *Rpi-blb1* genes (which are allelic) from *S. bulbocastanum* have been cloned and introduced into the susceptible cultivars 'Katahdin' and 'Impala,' respectively, and have been shown to confer broad-spectrum resistance to late blight (Song et al., 2003; van der Vossen et al., 2003). As the *RB* gene provided a high level of field resistance (14% foliage infection compared with 65% in susceptible controls) rather than complete immunity, there is cautious optimism about its potential durability, as well much debate (Colton et al., 2006). As a final example, Paal et al. (2004) have cloned the *Gro1* gene for resistance to *G. rostochiensis* from *S. spegazzinii* and introduced it into the susceptible cultivar Desiree and shown that it confers resistance to pathotype Ro1.

Antisense technology

Bruise-resistant potatoes have been produced by down-regulation of the gene encoding the enzyme polyphenol oxidase (PPO) so that the reaction leading to pigment production is no longer catalyzed (Bachem et al., 1994).

Progress is being made with a transgenic solution to ensuring that potato tubers have low glycoalkaloid contents. Down-regulation of a gene encoding a sterol alkaloid glycosyltransferase (*Sgt1*) resulted in an almost complete inhibition of alpha-solanine accumulation but was compensated for by elevated levels of alpha-chaconine (McCue et al., 2005). Further transformation is required to inhibit alpha-chaconine accumulation.

The essential amino acid methionine has been increased in tubers by antisense inhibition of threonine synthase (Zeh et al., 2001). Methionine levels were increased by a factor of 30 without a reduction in threonine, another essential amino acid.

High amylopectin starch has been produced by the down-regulation of the granule-bound starch synthase gene that controls amylose synthesis (Visser et al., 1991) and of high amylose starch by concurrently down-regulating two starch branching enzymes, A and B (Schwall et al., 2000). Field testing confirmed the stability of the latter modification and demonstrated an increased tuber yield, reduced starch content, smaller granule size, and an increase in reducing sugars (Hofvander et al., 2004). Potatoes containing starch with a very low degree of branching, such as 0.3%, were not suitable for commercial cultivation due to severe starch yield reduction and other effects. With slightly more branching, however, the effects were much reduced and the modified starch was considered suitable for biodegradable plastics, expanded products, and film-forming operations.

Novel traits

Pest resistance from the bacterium Bacillus thuringiensis

The use of *cry* genes to provide insect control in potato has been attempted by numerous research groups since the initial efforts in the late 1980s and has been reviewed by Douches and Grafius (2005). Transgenic resistance to the potato tuber moth (*Phthorimaea opercu-lella* Zeller) has been provided by the *Bt* proteins encoded by the *cry5* (Mohammed

et al., 2000), *cry1Ac9* (Davidson et al., 2004), and *cry1Ia1* (Douches and Grafius, 2005) genes using the constitutive 35S CaMV promoter. Davidson et al. (2004) demonstrated that their transgenic potato lines exhibited stable resistance to larvae across field seasons, between affected plant organs, and between plant organs of different ages. Douches and Grafius (2005) also demonstrated that expression of their gene in both the foliage and tubers reduced damage in the field and in storage.

Transgenic resistance to the Colorado potato beetle (*Leptinotarsa decemlineata* Say) was provided by gene *cry3A*, which encodes the Cry3A protein from the bacterium *Bacillus thuringiensis* var. *tenebrionis*, using the constitutive 35S CaMV promoter. The first GM cultivars with this *Bt* gene to be commercialized by Monsanto in North America from 1995 onward were 'Russet Burbank,' 'Atlantic,' 'Snowden,' and 'Superior' (Duncan et al., 2002). Monsanto advised growers to plant 20% of their field area with a non-Bt-containing potato as a refuge for susceptible beetles and to rotate crops and fields in which transgenic potatoes were grown to extend the durability of resistance (Duncan et al., 2002). Coombs et al. (2003) were able to demonstrate that *cry3A* transgenic lines provided effective host-plant resistance to Colorado potato beetle.

Potato cyst nematode resistance from cystatins

Urwin et al. (2003) were able to demonstrate that constructs based on a cysteine proteinase inhibitor (cystatin) from sunflower and a variant of a rice cystatin conferred similar levels of resistance to *Globodera* spp. as chicken egg white cystatin (CEWC), under the control of the CaMV35S promoter. In a field trial these levels of resistance were similar to that provided by the natural partial resistance of cultivar 'Sante.' Transformation of Sante and the South American cultivar 'Maria Huanca' with CaMV35S/ CEWC raised the status of both cultivars from partial to full resistance.

Pathogen-derived virus resistance

The most recent reviews of virus resistance breeding to include transgenic resistance are those of Solomon-Blackburn and Barker (2001) and Thieme and Thieme (2005). Examples are given of coat protein-mediated, movement protein-mediated, and polymerase-mediated (replicase) resistance. There is discussion about whether or not resistance is mediated by the RNA transcript or the encoded protein and also about the role of post-transcriptional gene silencing.

Russet Burbank potatoes have been produced in which Colorado potato beetle (CPB) resistance was combined with resistance to PLRV provided by a construct designed to prevent virus replication and using the constitutive Figwort Mosaic Virus Promoter (FMV). Russet Burbank and 'Shepody' potatoes have also been produced with combined CPB and PVY resistance, the latter provided by the PVY coat protein gene, again using the FMV promoter. In both these examples, the process started with about 3,000 original clones in 1991 from which six were finally selected for commercialization in 1998 (Davies, 2002). Trait stability was demonstrated in field trials across a number of years, as was the greatly reduced use of pesticides (Duncan et al., 2002).

Resistance to bacteria and fungi

Genes coding for lytic enzymes from a variety of organisms ranging from bacteriophages to humans are being evaluated in a number of laboratories worldwide as a way to achieve transgenic resistance to a number of bacteria and fungi (Davies, 1996). For example, the gene *chly* encoding the enzyme lysozyme from chicken has been introduced

into cv. Desiree via *Agrobacterium*-mediated transformation and shown to enhance resistance to blackleg and soft rot caused by infection with *E. carotovora* subsp. *atroseptica* (Serrano et al., 2000). Interestingly, a pathogen-derived pectate lyase-mediated soft rot resistance has also been achieved and shown to be a heritable trait in crosses between transgenic lines and non-transgenic cultivars (Wegener, 2003/2004).

One of the most encouraging recent developments in transgenic resistance is the achievement of resistance to more than one pathogen. Expression of a gene for a derivative of the antimicrobial peptide dermaseptin B1, from the arboreal frog *Phyllomedusa bicolor*, has been shown to increase resistance to diseases, such as late blight, dry rot, and pink rot and to markedly extend the shelf life of tubers (Osusky et al., 2005).

Improved nutritional value and flavor

Chakraborty et al. (2000) reported improvements in the nutritive value of potato through transformation with a non-allergenic seed albumin gene (*AmA1*) from *Amaranthus hypochondriacus*. The seed protein produced has a well-balanced amino acid composition with no known allergenic properties. Five- to 10-fold increases in transcript levels in tubers were achieved using the tuber-specific granule-bound starch synthase (GBSS) promoter compared with the CaMV35S promoter. Significant 2- to 4-fold increases were achieved in the lysine, methionine, cysteine, and tyrosine content of the protein amino acids, and a 35% to 45% increase was achieved in total protein content.

The introduction of inulin to potato is another way to improve its nutritional value. Inulin is a mixture of linear fructose-polymers with different chain-lengths and a glucose molecule at each C2-end. Inulin belongs to the fructan group of polysaccharides and serves as a storage carbohydrate in many plant species. Compounds such as inulin reduce the energy density of food and are used to enrich food with dietary fiber or to replace sugar and fat. Hellwege et al. (2000) have developed transgenic potato tubers that synthesize the full range of inulin molecules naturally occurring in globe artichoke (*Cynara scolymus*). High molecular weight inulins have been produced by expressing the sucrose:sucrose 1-fructosyl transferase and the fructan:fructan 1-fructosylhydrolase genes from globe artichoke. Inulin made up 5% of the dry weight of the transgenic tuber.

Di et al. (2003) were able to increase the level of soluble methionine in Russet Burbank potato by the introduction of *Arabidopsis thaliana* CGS (cystathionine γ-synthase) cDNA under transcriptional control of the CaMV35S promoter. Interestingly, the increased levels in the raw tubers resulted in 2.4- to 4.4-fold increases of methional in baked tubers; methional is a primary flavor compound that is lost on processing, which is a problem for the food industry.

Carotenoids

The health benefits of carotenoids were discussed earlier. Ducreux et al. (2005) enhanced the carotenoid content of potato tubers by expressing an *Erwinia uredovora crtB* gene encoding phytoene synthase under the control of the tuber-specific patatin promoter. They were able to increase the carotenoid content of *S. tuberosum* cultivar Desiree (light yellow flesh) from 5.6 to 35.5 μg carotenoid per g dry-weight and the content of *S. phureja* cultivar Mayan Gold (deep yellow flesh) from 20 to 78 μg carotenoid per g dry-weight. Of particular significance, however, was the fact that the best transgenic lines of cultivar Desiree and Mayan Gold produced 10.3 and 6.5 μg, respectively, of beta-carotene per g dry-weight whereas none was present in the untransformed controls. The authors suggest approaches for optimizing beta-carotene

accumulation. Morris et al. (2006) have also genetically engineered cultivars Desiree and Mayan Gold to produce astaxanthin, a ketocarotenoid with known health benefits, which is marketed as a dietary supplement (a nutraceutical). Transgenic lines were produced that expressed an algal (*Haematococcus pluvialis*) *bkt1* gene encoding beta-carotene ketolase and under the control of the tuber-specific patatin promoter. The level of unesterified astaxanthin reached 13.9 μg per g dry-weight in cultivar Mayan Gold compared with none in the untransformed control, but was much lower in cultivar Desiree although enough to cause a distinct red coloration.

Lower reducing sugars out of cold storage

While breeders have been able to select for lower levels of reducing sugars out of cold storage, transgenic approaches are also possible based on an understanding of primary carbohydrate metabolism. Stark et al. (1992) increased tuber-starch content (see below) and lowered the levels of reducing sugars by expressing an *E. coli glgC16* mutant gene that encodes the enzyme ADPGlc pyrophosphorylase and increases the production of ADPglucose, which in turn becomes incorporated into the growing starch granule. Greiner et al. (1999) were able to minimize the conversion of sucrose to glucose and fructose by expressing a putative vacuolar invertase inhibitor protein from tobacco, Nt-inhh, in potato plants under the control of the CaMV35S promoter. More information on the control of sugar accumulation, starch biosynthesis, and potato quality can be found in the review by Davies (1998).

Starch content and yield

Within the amyloplast the conversion of glucose-1-P to ADPglucose by the enzyme ADPGlc pyrophosphorylase (ADPGPP) is the first essentially irreversible reaction in starch synthesis in the plastid. When Stark et al. (1992) expressed an *E. coli glgC16* gene encoding a bacterial ADPGPP in tubers of cv. Russet Burbank, starch content was elevated by up to 60% (35% average), but total tuber yield appeared unaffected. In contrast, yield is affected by the unloading in tubers of sucrose after its transportation from the leaves where it is made. Sonnewald et al. (1997) expressed a gene encoding a yeast invertase in tubers and the invertase protein was targeted to the cell wall. This resulted in a substantial increase in tuber glucose content, a 2-to 3-fold decrease in tuber number per plant and a 30% increase in tuber fresh weight. Davies (1998) concluded that the role of specific sucrose transporters on the plasma membrane in regulating overall unloading rates is an important issue for potato as well as for other sink organs well-known to have apoplastic unloading mechanisms. Thus, the transgenic manipulation of carbohydrate metabolism holds promise for the understanding and improvement of yield (carbon partitioning to tubers) and quality traits (improved dry-matter content and even distribution of starch), but much research remains to be done.

Conclusion

The human-directed evolution of the potato began more than 7,000 years ago with domestication in Peru and continues today. Andean farmers first selected less bitter and hence less toxic tubers and made the potato an edible crop. Today, it is still important for breeders to check the tuber glycoalkaloid content of their parental material and finished cultivars, especially when they have used wild species as a source of a desired characteristic. Andean farmers also selected and maintained simply inherited mutants affecting tuber shapes and skin and flesh colors. It is still important to visually select the appropriate alleles at these loci to meet

different consumer preferences around the world.

Potato cultivation spread throughout Latin America and then much later to Europe and the rest of the world, and the potato became adapted to a wide range of environments and end-uses. The traits of importance in this adaptation were ones now known to display polygenic inheritance, such as yield, maturity, dormancy, tuberization, dry-matter and starch content, and starch-sugar metabolism. They are still important traits, and quantitative genetics has provided the foundation for efficient conventional potato breeding, which remains the route to new cultivars. High-quality field trials and computer-based data capture and analyses are essential aspects of this endeavor. The QTL mapping experiments of recent years hold out the possibility of finding candidate genes of importance for which allelic variants should be sought in germplasm collections, but they also confirm the need to treat polygenic traits as systems of interacting genes on which to practice recurrent selection. In the future, breeders may use molecular genotyping methods to identify the best parents and crosses but then use conventional phenotyping for selection purposes. The same may prove to be true for the complex physiological traits required to expand food production by growing potatoes in a wider range of environments and for longer growing seasons. These traits include resistance to drought, heat, cold, mineral deficiency and salinity; of these, the efficient use of water resources could prove the most important.

Many disease and pest problems were encountered as the potato assumed its role as a major staple food crop from a narrow genetic base. Major genes and QTL of large effect for resistance have been found in cultivated potatoes and their wild relatives and introduced into modern cultivars. This process will continue with genes that fail to provide durable resistance, being replaced with new ones and new strategies for their deployment. Molecular-breeding methods such as gene cloning and molecular marker-assisted introgression will be used to accelerate the process. Furthermore, we can anticipate the use of genetic modification of successful and widely grown cultivars to achieve further targeted improvements in yield, quality, nutritional value, and resistance to abiotic and biotic stresses. The use of exotic sources of genes in such transformations has the clear value of permitting the introduction of traits not found in cross-compatible relatives and of novel function. It should be embraced as a way forward to complement the utilization of the biodiversity available in the tuber-bearing *Solanum* species for achieving increased food production and food security in the potato as a major staple crop.

Acknowledgements

I thank the many colleagues at SCRI and associates around the world who have been kind enough to share their thoughts on potato breeding and genetics across many years. I also thank the Scottish Executive Environment and Rural Affairs Department for funding.

References

Almekinders, C.J.M., Chilver, A.S. and Renia, H.M. 1996. Current status of the TPS technology in the world. Potato Res. 39:289–303.

Amrein, T.M., Bachmann, S., Noti, A., Biedermann, M., Barbosa, M.F., Biedermann-Brem, S., Grob, K., Keiser, A., Realini, P., Escher, F. and Amado, R. 2003. Potential of acrylamide formation, sugars, and free asparagine in potatoes: A comparison of cultivars and farming systems. J. Agric. Food Chem. 51:5556–5560.

Armstrong, M.R., Whisson, S.C., Pritchard, L., Bos, J.I.B., Venter, E., Avrova, A.O., Rehmany, A.P., Bohme, U., Brooks, K., Cherevach, I., Hamlin, N., White, B., Fraser, A., Lord, A., Quail, M.A., Churcher, C., Hall, N., Berriman, M., Huang, S., Kamoun, S., Beynon, J.L. and Birch, P.J.R. 2005. An ancestral oomycete locus contains late blight

avirulence gene *Avr3a*, encoding a protein that is recognized in the host cytoplasm. Proc. Natl. Acad. Sci. 102:7766–7771.

Augustin, J. 1975. Variations in the nutritional composition of fresh potatoes. Journal of Food Science 40:1259–1299.

Bachem, C.W.B., Speckmann, G.J., Van Der Linde, P.C.G., Verheggen, F.T.M., Hunt, M.D., Steffens, J.C. and Zabeau, M. 1994. Antisense expression of polyphenol oxidase genes inhibits enzymic browning in potato tubers. Bio. Technology 12:1101–1105.

Baker, R.J. 1988. Tests for crossover genotype-environmental interactions. Can. J. Plant Sci. 68:405–410.

Bakker, E., Achenbach, U., Bakker, J., van Vliet, J., Peleman, J., Segers, B., van der Heijden, S., van der Linde, P., Graveland, R., Hutten, R., van Eck, H., Coppoolse, E., van der Vossen, E., Bakker J. and Goverse, A. 2004. A high-resolution map of the *H1* locus harbouring resistance to the potato cyst nematode *Globodera rostochiensis*. Theor. Appl. Genet. 109:146–152.

Ballvora, A., Ercolano, M.R., Weiß, J., Meksem, K., Bormann, C.A., Oberhagemann, P., Salamini, F. and Gebhardt, C. 2002. The *R1* gene for potato resistance to late blight (*Phytophthora infestans*) belongs to the leucine zipper/NBS/LRR class of plant resistance genes. Plant J. 30:361–371.

Bamberg, J. and del Rio, A. 2005. Conservation of Potato Genetic Resources. In: Genetic Improvement of Solanaceous Crops Volume I: Potato (M.K. Razdan and A.K. Mattoo, eds.), p. 1–38. Science Publishers, Inc., Enfield, New Hampshire.

Barker, H. and Dale, M.F.B. 2006. Resistance to Viruses in Potato. In: Natural Resistance Mechanisms of Plants to Viruses (G. Loebenstein and J.P. Carr, eds.), p. 341–366. Springer, Dordrecht.

Barone, A. 2004. Molecular marker-assisted selection for potato breeding. Amer. J. Potato Res. 81:111–117.

Barone, A., Sebastiano, A., Carputo, D., della Rocca, F. and Frusciante, L. 2001. Molecular marker-assisted introgression of the wild *Solanum commersonii* genome into the cultivated *S. tuberosum* gene pool. Theor. Appl. Genet. 102:900–907.

Bellon, M.R. and Reeves, J. (eds.). 2002. Quantitative Analysis of Data from Participatory Methods in Plant Breeding. CIMMYT, Mexico.

Bendahmane, A., Kanyuka, K. and Baulcombe, D.C. 1999. The *Rx* gene from potato controls separate virus resistance and cell death responses. Plant Cell 11:781–791.

Bendahmane, A., Querci, M., Kanyuka, K. and Baulcombe, D.C. 2000. *Agrobacterium* transient expression system as a tool for the isolation of disease resistance genes: Application to the *Rx2* locus in potato. Plant J. 21:73–81.

Black, W. 1970. The nature and inheritance of field resistance to late blight (*Phytophthora infestans*) in potatoes. Am. Potato J. 47:279–288.

Bonierbale, M.W., Plaisted, R.L. and Tanksley, S.D. 1988. RFLP maps based on a common set of clones reveal modes of chromosomal evolution in potato and tomato. Genetics 120:1095–1103.

Bonierbale, M.W., Plaisted, R.L. and Tanksley, S.D. 1993. A test of the maximum heterozygosity hypothesis using molecular markers in tetraploid potatoes. Theor. Appl. Genet. 86:481–491.

Bonierbale, M.W., Simon, R., Zhang, D.P., Ghislain, M., Mba, C. and Li, X.-Q. 2003. Genomics and Molecular Breeding for Root and Tuber Crop Improvement. In: Plant Molecular Breeding (H.J. Newbury, ed.), p. 216–253. Blackwell, Oxford.

Bormann, C.A., Rickert, A.M., Ruiz, R.A.C., Paal, J., Lubeck, J., Strahwald, J., Buhr, K. and Gebhardt, C. 2004. Tagging quantitative trait loci for maturity-corrected late blight resistance in tetraploid potato with PCR-based candidate gene markers. MPMI 17:1126–1138.

Bradshaw, J.E. 1994a. Quantitative Genetics Theory for Tetrasomic Inheritance. In: Potato Genetics (J.E. Bradshaw and G.R. Mackay, eds.), p. 71–99. CAB International, Wallingford, UK.

Bradshaw, J.E. 1994b. Assessment of five cultivars of potato (*Solanum tuberosum* L.) in a competition diallel. Ann. Appl. Biol. 125:533–540.

Bradshaw, J.E. 2005. Potato Improvement at Scri by Multitrait Genotypic Recurrent Selection. In: Proc. IX Simposio de Atualizacao em Genetica e Melhoramento de Plantas. 25 and 26 August 2005, Universidade Federal de Lavras, Lavras, Brasil, p. 9–28.

Bradshaw, J.E. 2006. Genetics of Agrihorticultural Traits. In: Handbook of Potato Production, Improvement, and Postharvest Management (J. Gopal and S.M.P. Khurana, eds.). The Haworth Press, New York. (in press).

Bradshaw, J.E. and Mackay, G.R. 1994a. Breeding Strategies for Clonally Propagated Potatoes. In: Potato Genetics (J.E. Bradshaw and G.R. Mackay, eds.), p. 467–497. CAB International, Wallingford, UK.

Bradshaw, J.E. and Mackay, G.R. (eds.). 1994b. Potato Genetics, CAB International, Wallingford, UK.

Bradshaw, J.E. and Ramsay, G. 2005. Utilisation of the Commonwealth Potato Collection in potato breeding. Euphytica 146:9–19.

Bradshaw, J.E., Stewart, H.E., Wastie, R.L., Dale, M.F.B. and Phillips, M.S. 1995. Use of seedling progeny tests for genetical studies as part of a potato (*Solanum tuberosum* subsp. *tuberosum*) breeding programme. Theor. Appl. Genet. 90:899–905.

Bradshaw, J.E., Dale, M.F.B., Swan, G.E.L., Todd, D. and Wilson, R.N. 1998. Early-generation selection between and within pair crosses in a potato (*Solanum* *tuberosum* subsp. *tuberosum*) breeding programme. Theor. Appl. Genet. 97:1331–1339.

Bradshaw, J.E., Todd, D. and Wilson, R.N. 2000a. Use of tuber progeny tests for genetical studies as part of a potato (*Solanum tuberosum* subsp. *tuberosum*) breeding programme. Theor. Appl. Genet. 100:772–781.

Bradshaw, J.E., Lees, A.K. and Stewart, H.E. 2000b. How to breed potatoes for resistance to fungal and bacterial diseases. Plant Breed. Seed Sci. 44:3–20.

Bradshaw, J.E., Dale, M.F.B. and Mackay, G.R. 2003. Use of mid-parent values and progeny tests to increase the efficiency of potato breeding for combined processing quality and disease and pest resistance. Theor. Appl. Genet. 107:36–42.

Bradshaw, J.E., Pande, B., Bryan, G.J., Hackett, C.A., McLean, K., Stewart, H.E. and Waugh, R. 2004. Interval mapping of quantitative trait loci for resistance to late blight [*Phytophthora infestans* (Mont.) de Bary], height and maturity in a tetraploid population of potato (*Solanum tuberosum* subsp. *tuberosum*). Genetics 168:983–995.

Bradshaw, J.E., Bryan, G.J., Lees, A.K., McLean, K. and Solomon-Blackburn, R.M. 2006. Mapping the *R10* and *R11* genes for resistance to late blight (*Phytophthora infestans*) present in the potato (*Solanum tuberosum*) R-gene differentials of Black. Theor. Appl. Genet. 112:744–751.

Brigneti, G., Garcia-Mas, J. and Baulcombe, D.C. 1997. Molecular mapping of the *potato virus Y* resistance gene *Rysto* in potato. Theor. Appl. Genet. 94:198–203.

Brown, C.R. 1993. Outcrossing rate in cultivated autotetraploid potato. Am. Potato J. 70:725–734.

Brown, C.R. 2005. Antioxidants in potato. Am. J. Potato Res. 82:163–172.

Brown, C.R. and Mojtahedi, H. 2005. Breeding for Resistance to *Meloidogyne* Species and Trichodorid-Vectored Virus. In: Genetic Improvement of Solanaceous Crops Volume I: Potato (M.K. Razdan and A.K. Mattoo, eds.), p. 267–292. Science Publishers, Inc., Enfield, New Hampshire.

Brown, C.R., Yang, C.P., Mojtahedi, H., Santo, G.S. and Masuelli, R. 1996. RFLP analysis of resistance to Columbia root-knot nematode derived from *Solanum bulbocastanum* in a BC2 population. Theor. Appl. Genet. 92:572–576.

Brown, C.R., Edwards, C.G., Yang, C.-P. and Dean, B.B. 1993. Orange flesh trait in potato: Inheritance and carotenoid content. J. Am. Soc. Hort. Sci. 118:145–150.

Brown, C.R., Wrolstad, R., Durst, R., Yang, C.-P. and Clevidence, B. 2003. Breeding studies in potatoes containing high concentrations of anthocyanins. Am. J. Potato Res. 80:241–250.

Brown, J., Dale, M.F.B. and Mackay, G.R. 1996. General adaptability of potato genotypes selected in the UK for the Mediterranean region. J. Agric. Sci., Camb. 126:441–448.

Brugmans, B., Hutten, R.G.B., Rookmaker, A.N.O., Visser, R.G.F. and van Eck, H.J. 2006. Exploitation of a marker dense linkage map of potato for positional cloning of a wart disease resistance gene. Theor. Appl. Genet. 112:269–277.

Bryan, G.J. and Ramsay, G. 2005. Origins of genomes in polyploid potato species. Scottish Crop Research Institute Annual Report 2003/2004, p. 148–149.

Bryan, G.J., McLean, K., Bradshaw, J.E., De Jong, W.S., Phillips, M., Castelli, L. and Waugh, R. 2002. Mapping QTLs for resistance to the cyst nematode *Globodera pallida* derived from the wild potato species *Solanum vernei*. Theor. Appl. Genet. 105:68–77.

Bryan, G.J., McLean, K., Pande, B., Purvis, A., Hackett, C.A., Bradshaw, J.E. and Waugh, R. 2004. Genetical dissection of H3-mediated polygenic PCN resistance in a heterozygous autotetraploid potato population. Mol. Breed. 14:105–116.

Buso, J.A., Boiteux, L.S. and Peloquin, S.J. 1999. Comparison of haploid Tuberosum-*Solanum chacoense* versus *Solanum phureja*-haploid Tuberosum hybrids as staminate parents of 4x-2x progenies evaluated under distinct crop management systems. Euphytica 109:191–199.

Caligari, P.D.S. 1992. Breeding New Varieties. In: The Potato Crop, 2ⁿᵈ edn. (P. Harris, ed.), p. 334–372. Chapman & Hall, London.

Camadro, E.L., Carputo, D. and Peloquin, S.J. 2004. Substitutes for genome differentiation in tuber-bearing *Solanum*: Interspecific pollen-pistil incompatibility, nuclear-cytoplasmic male sterility, and endosperm. Theor. Appl. Genet. 109:1369–1376.

Caromel, B., Mugniery, D., Kerlan, M-C., Andrzejewski, S., Palloix, A., Ellisseche, D., Rousselle-Bourgeois, F. and Lefebvre, V. 2005. Resistance quantitative trait loci originating from *Solanum sparsipilum* act independently on the sex ratio of *Globodera pallida* and together for developing a necrotic reaction. MPMI 18:1186–1194.

Carroll, C.P. 1982. A mass-selection method for the acclimatization and improvement of edible diploid potatoes in the United Kingdom. J. Agric. Sci., Camb. 99:631–640.

Carroll, C.P. and De Maine, M.J. 1989. The agronomic value of tetraploid F1 hybrids between potatoes of group Tuberosum and group Phureja/Stenotomum. Potato Res. 32:447–456.

Ceccarelli, S., Grando, S., Amri, A., Asaad, F.A., Benbelkacem, A., Harrabi, M., Maatougui, M., Mekni, M.S., Mimoun, H., El-Einen, R.A., El-Felah, M., El-Sayed, A.F., Shreidi, A.S. and Yahyaoui, A. 2001. Decentralized and Participatory Plant Breeding for Marginal Environments. In: Broadening the Genetic Base of Crop Production (H.D.

Cooper, C. Spillane and T. Hodgkin, eds.), p. 115–135. CAB International, Wallingford, UK.

Celebi-Toprak, F., Slack, S.A. and Jahn, M.M. 2002. A new gene, *Nytbr*, for hypersensitivity to *Potato Virus Y* from *Solanum tuberosum* maps to chromosome IV. Theor. Appl. Genet. 104:669–674.

Chakraborty, S., Chakraborty, N. and Datta, A. 2000. Increased nutritive value of transgenic potato by expressing a non-allergenic seed albumin gene from *Amaranthus hypochondriacus*. Proc. Nat. Acad. Sci. 97:3724–3929.

Chandra, S., Huaman, Z., Hari Krishna, S. and Ortiz, R. 2002. Optimal sampling strategy and core collection size of the Andean tetraploid potato based on isozyme data—a simulation study. Theor. Appl. Genet. 104:1325–1334.

Chang, M.M., Culley, D., Choi, J.J. and Hadwiger, L.A. 2002. *Agrobacterium*-mediated co-transformation of a pea beta-1,3-glucanase and chitinase genes in potato (*Solanum tuberosum* L. c.v. Russet Burbank) using a single selectable marker. Plant Sci. 163:83–89.

Chase, S.S. 1963. Analytic breeding in *Solanum tuberosum* L.—a scheme utilizing parthenotes and other diploid stocks. Can. J. Genet. Cytol. 5:359–363.

Chen, X., Salamini, F. and Gebhardt, C. 2001. A potato molecular-function map for carbohydrate metabolism and transport. Theor. Appl. Genet. 102:284–295.

Chilver, A., Walker, T.S., Khatana, V., Fano, H., Suherman, R. and Rizk, A. 2005. On-Farm Profitability and Prospects for True Potato Seed (TPS). In: Genetic Improvement of Solanaceous Crops Volume I: Potato (M.K. Razdan and A.K. Mattoo, eds.), p. 39–63. Science Publishers, Inc., Enfield, New Hampshire.

CIP/FAO. 1995. Potatoes in the 1990s. FAO, Rome.

CIP. 1997. Medium-Term Plan 1998–2000. International Potato Center, Lima, Peru.

Clulow, S.A., McNicoll, J. and Bradshaw, J.E. 1995. Producing commercially attractive, uniform true potato seed progenies: The influence of breeding scheme and parental genotype. Theor. Appl. Genet. 90:519–525.

Collins, A., Milbourne, D., Ramsay, L., Meyer, R., Chatot-Balandras C., Oberhagemann, P., De Jong, W., Gebhardt, C., Bonnel, E. and Waugh, R. 1999. QTL for field resistance to late blight in potato are strongly correlated with maturity and vigour. Mol. Breeding 5:387–398.

Colton, L.M., Groza, H.I., Wielgus, S.M. and Jiang, J. 2006. Marker-assisted selection for the broad-spectrum potato late blight resistance conferred by gene *RB* derived from a wild potato species. Crop Sci. 46:589–594.

Connolly, T., Currie, I.D., Bradshaw, J.E. and McNicol, J.W. 1993. Inter-plot competition in yield trials of potatoes (*Solanum tuberosum* L.) with single-drill plots. Ann. Appl. Biol. 123:367–377.

Coombs, J.J., Douches, D.S., Li, W., Grafius, E.J. and Pett, W.L. 2003. Field evaluation of natural, engineered, and combined resistance mechanisms in potato (*Solanum tuberosum* L.) for control of Colorado potato beetle (*Leptinotarsa decemlineata* Say). J. Amer. Soc. Hort. Sci. 128:219–224.

Cubillos, A.G. and Plaisted, R.F. 1976. Heterosis for yield in hybrids between *S. tuberosum* sp. *tuberosum* and *S. tuberosum* sp. *andigena*. Am. Potato J. 53:143–150.

Dale, M.F.B. and Mackay, G.R. 1994. Inheritance of Table and Processing Quality. In: Potato Genetics (J.E. Bradshaw and G.R. Mackay, eds.), p. 285–315. CAB International, Wallingford, UK.

Dale, P.J. and Hampson, K.K. 1995. An assessment of morphogenic and transformation efficiencies in a range of varieties of potato (*Solanum tuberosum* L.). Euphytica 85:101–108.

Davidson, M.M., Butler, R.C., Wratten, S.D. and Conner, A.J. 2004. Resistance of potatoes transgenic for a *cry*1Ac9 gene, to *Phthorimaea operculella* (Lepidoptera: Gelechiidae) over field seasons and between plant organs. Ann. Appl. Biol. 145:271–277.

Davies, H.V. 1996. Recent developments in our knowledge of potato transgenic biology. Potato Res. 39:411–427.

Davies, H.V. 1998. Prospects for manipulating carbohydrate metabolism in potato tuber. Aspects of Applied Biology 52:245–254.

Davies, H.V. 2002. Commercial Developments with Transgenic Potato. In: Fruit and Vegetable Biotechnology (V. Valpuesta, ed.), p. 222–249. Woodhead Publishing Limited, Cambridge.

De Jong, W. 2005. Approaches to Gene Isolation in Potato. In: Genetic Improvement of Solanaceous Crops Volume I: Potato (M.K. Razdan and A.K. Mattoo, eds.), p. 165–184. Science Publishers, Inc., Enfield, New Hampshire.

De Jong, W.S., De Jong, D.M., De Jong, H., Kalazich, J. and M. Bodis, M. 2003. An allele of dihydroflavonol 4-reductase associated with the ability to produce red anthocyanin pigments in potato (*Solanum tuberosum* L.). Theor. Appl. Genet. 107:1375–1383.

De Jong, W.S., Eannetta, N.T., De Jong, D.M. and Bodis, M. 2004. Candidate gene analysis of anthocyanin pigmentation loci in the Solanaceae. Theor. Appl. Genet. 108:423–432.

De Maine, M.J. 1996. An assessment of true potato seed families of *Solanum phureja*. Potato Res. 39:323–332.

Di, R., Kim, J., Martin, M.N., Leustek, T., Jhoo, J., Ho, C-T. and Tumer, N.E. 2003. Enhancement of the primary flavour compound methional in potato by increasing the level of soluble methionine. J. Agric. Food Chem. 51:5695–5702.

Dodds, K.S. 1965. The history and Relationships of Cultivated Potatoes. In: Essays in Crop Plant Evolution (J.B. Hutchinson, ed.), p. 123–141. Cambridge University Press, Cambridge.

Dong, F., Song, S., Naess, S.K., Helgeson, J.P., Gebhardt, C. and Jiang, J. 2000. Development and applications of a set of chromosome-specific cytogenetic DNA markers in potato. Theor. Appl. Genet. 101:1001–1007.

Douches, D.S. and Grafius, E.J. 2005. Transformation for Insect Resistance. In: Genetic Improvement of Solanaceous Crops Volume I: Potato (M.K. Razdan and A.K. Mattoo, eds.), p. 235–266. Science Publishers, Inc., Enfield, New Hampshire.

Douches, D.S. and Jastrzebski, K. 1993. Potato. In: Genetic Improvement of Vegetable Crops (G. Kalloo and B.O. Bergh, eds.), p. 605–644. Pergamon Press, Oxford.

Draaistra, J. 2006. Genetic analyses of root-knot nematode resistance in potato. Thesis Wageningen University, The Netherlands.

Ducreux, L.J.M., Morris, W.L., Hedley, P.E., Shepherd, T., Davies, H.V., Millam, S. and Taylor, M.A. 2005. Metabolic engineering of high carotenoid potato tubers containing enhanced levels of β-carotene and lutein. J. Exp. Bot. 56:81–89.

Duncan, D.R., Hammond, D., Zalewski, D.J., Cudnohufsky, J., Kaniewski, W., Thornton, M., Bookout, J.T., Lavrik, P., Rogan, G.J. and Feldman-Riebe, J. 2002. Field performance of transgenic potato, with resistance to colorado potato beetle and viruses. HortScience 37:275–276.

Evans, K., Trudgill, D.L., Raman, K.V. and Radcliffe, E.B. 1992. Pest Aspects of Potato Production. In: The Potato Crop (P. Harris, ed.), p. 438–506. Chapman & Hall, London.

Ewing, E.E., Simko, I., Omer, E.A. and Davies P.J. 2004. Polygene mapping as a tool to study the physiology of potato tuberization and dormancy. Amer. J. Potato Res. 81:281–289.

Flis, B., Hennig, J., Strzelczyk-Żyta, D., Gebhardt, C. and Marczewski, W. 2005. The *Ry-fsto* gene from *Solanum stoloniferum* for extreme resistant to *Potato virus Y* maps to potato chromosome XII and is diagnosed by PCR marker GP122718 in PVY resistant potato cultivars. Mol. Breed. 15:95–101.

Forbes, G.A., Chacon, M.G., Kirk, H.G., Huarte, M.A., Van Damme, M., Distel, S., Mackay, G.R., Stewart, H.E., Lowe, R., Duncan, J.M., Mayton, H.S., Fry, W.E., Andrivon, D., Ellisseche, D., Pelle, R., Platt, H.W., MacKenzie, G., Tarn, T.R., Colon, L.T., Budding, D.J., Lozoya-Saldana, H., Hernandez-Vilchis, A. and Capezio, S. 2005. Stability of resistance to *Phytophthora infestans* in potato: An international evaluation. Plant Path. 54:364–372.

Freyre, R. and Douches, D.S. 1994. Development of a model for marker-assisted selection of specific gravity in diploid potato across environments. Crop Sci.34:1361–1368.

Freyre, R., Warnke, S., Sosinski, B. and Douches, D.S. 1994. Quantitative trait locus analysis of tuber dormancy in diploid potato (*Solanum* spp.). Theor. Appl. Genet. 89:474–480.

Ganal, M.W., Bonierbale, M.W., Roeder, M.S., Park, W.D. and Tanksley, S.D. 1991. Genetic and physical mapping of the patatin genes in potato and tomato. Mol. Gen. Genet. 225:501–509.

Gaur, P.C. and Pandey, S.K. 2000. Potato Improvement in Sub-tropics. In: Potato, Global Research & Development—Volume 1 (S.M.P. Khurana, G.S. Shekhawat, B.P. Singh and S.K. Pandey, eds.), p. 52–63. Indian Potato Association, Shimla, India.

Gebhardt, C., Ritter, E., Barone, A., Debener, T., Walkemeier, B., Schachtschabel, U., Kaufman, H., Thompson, R.D., Bonierbale, M.W., Ganal, M.W., Tanksley, S.D. and Salamini, F. 1991. RFLP maps of potato and their alignment with the homeologous tomato genome. Theor. Appl. Genet. 83:49–57.

Gebhardt, C., Walkemeier, B., Henselewski, H., Barakat, A., Delseny, M. and Stuber, K. 2003. Comparative mapping between potato (*Solanum tuberosum*) and *Arabidopsis thaliana* reveals structurally conserved domains and ancient duplications in the potato genome. Plant J. 34:529–541.

Gebhardt, C., Bellin, D., Henselewski, H., Lehmann, W., Schwarzfischer, J. and Valkonen, J.P.T. 2006. Marker-assisted pyramidization of major genes for pathogen resistance in potato. Theor. Appl. Genet. 112:1458–1464.

GILB. 1999. Late Blight: A Threat to Global Food Security. Proc. of the Global Initiative on Late Blight Conference, March 16–19, 1999, Quito, Ecuador.

Glendinning, D.R. 1969. The performance of progenies obtained by crossing Groups Andigena and Tuberosum of *Solanum tuberosum*. Eur. Potato J. 12:13–19.

Glendinning, D.R. 1976. Neo-Tuberosum: New potato breeding material. 4. The breeding system of Neo-Tuberosum, and the structure and composition of the Neo-Tuberosum gene pool. Potato Res. 19:27–36.

Golmirzaie, A.M., Malagamba, P. and Pallais, N. 1994. Breeding Potatoes Based on True Seed Propagation. In: Potato Genetics (J.E. Bradshaw and G.R. Mackay, eds.), p. 499–513. CAB International, Wallingford, UK.

Goodrich, C.E. 1863. The origination and test culture of seedling potatoes. Trans. N.Y. State Agric. Soc. 23:89–134.

Goodwin, S.B. and Drenth, A. 1997. Origin of the A2 mating type of *Phytophthora infestans* outside Mexico. Phytopathology 87:992–999.

Gopal, J. and Oyama, K. 2005. Genetic base of Indian potato selections as revealed by pedigree analysis. Euphytica 142:23–31.

Greiner, S., Rausch, T., Sonnewald, U. and Herbers, K. 1999. Ectopic expression of a tobacco invertase inhibitor prevents cold induced sweetening of potato tubers. Nature Biotech. 17:708–711.

Grun, P. 1990. The evolution of cultivated potatoes. Economic Bot. 44:39–55.

Gurr, S.J. and Rushton, P.J. 2005. Engineering plants with increased disease resistance: How are we going to express it? Trends Biotechnol. 23:283–290.

Hackett, C.A., Bradshaw, J.E., and McNicol, J.W. 2001. Interval mapping of quantitative trait loci in autotetraploid species. Genetics 159:1819–1832.

Hamalainen, J.H., Watanabe, K.N., Valkonen, J.P.T., Arihara, A., Plaisted, R.L., Pehu, E., Miller, L. and Slack S.A. 1997. Mapping and marker-assisted selection for a gene for extreme resistance to *potato virus Y*. Theor. Appl. Genet. 94:192–197.

Hamalainen, J.H., Sorri, V.A., Watanabe, K.N., Gebhardt, C. and Valkonen, J.P.T. 1998. Molecular examination of a chromosome region that controls resistance to potato Y and A potyviruses in potato. Theor. Appl. Genet. 96:1036–1043.

Haverkort, A.J. and Grashoff, C. 2004. Ideotyping-Potato a Modelling Approach to Genotype Performance. In: Decision Support Systems in Potato Production (D.K.L. MacKerron and A.J. Haverkort, eds.), p. 199–211. Wageningen Academic Publishers, Wageningen, The Netherlands.

Hawkes, J.G. 1990. The Potato: Evolution, Biodiversity & Genetic Resources. Belhaven Press, London.

Hawkes, J.G. 1994. Origins of Cultivated Potatoes and Species Relationships. In: Potato Genetics (J.E. Bradshaw and G.R. Mackay, eds.), p. 3–42. CAB International, Wallingford, UK.

Hawkes, J.G. and Francisco-Ortega, J. 1993. The early history of the potato in Europe. Euphytica 70:1–7.

Hawkes, J.G. and Hjerting, J.P. 1969. The Potatoes of Argentina, Brazil, Paraguay and Uruguay: A Biosystematic Study. Oxford University Press, Oxford.

Hawkes, J.G. and Hjerting, J.P. 1989. The Potatoes of Bolivia: Their Breeding Value and Evolutionary Relationships. Oxford University Press, Oxford.

Hawkes, J.G. and Jackson, M.T. 1992. Taxonomic and evolutionary implications of the endosperm balance number hypothesis in potatoes. Theor. Appl. Genet. 84:180–185.

Haynes, K.G. and Lu, W. 2005. Improvement at the Diploid Species Level. In: Genetic Improvement of Solanaceous Crops Volume I: Potato (M.K. Razdan and A.K. Mattoo, eds.), p. 101–114. Science Publishers, Inc., Enfield, New Hampshire.

Hehl, R., Faurie, E., Hesselbach, J., Salamini, F., Whitham, S., Baker, B. and Gebhardt, C. 1999.

TMV resistance gene *N* homologues are linked to *Synchytrium endobioticum* resistance in potato. Theor. Appl. Genet. 98:379–386.

Hellwege, E.M., Czapla, S., Jahnke, A., Willmitzer, L. and Heyer, A.G. 2000. Transgenic potato (*Solanum tuberosum*) tubers synthesise the full spectrum of inulin molecules naturally occurring in globe artichoke (*Cynara scolymus*). Proc. Nat. Acad. Sci. 97:8699–8704.

Hermsen, J.G.Th. 1994. Introgression of Genes from Wild Species, including Molecular and Cellular Approaches. In: Potato Genetics (J.E. Bradshaw and G.R. Mackay, eds.), p. 515–538. CAB International, Wallingford, UK.

Hermundstad, S.A. and Peloquin, S.J. 1987. Breeding at the 2*x* level and Sexual Polyploidization. In: The Production of New Potato Varieties (G.J. Jellis and D.E. Richardson, eds.), p. 197–210. Cambridge University Press, Cambridge.

Hide, G.A. and Lapwood, D.H. 1992. Disease Aspects of Potato Production. In: The Potato Crop (P. Harris, ed.), p. 403–437. Chapman & Hall, London.

Hijmans, R.J. 2001. Global distribution of the potato crop. Am. J. Potato Res. 78:403–412.

Hils, U. and Pieterse, L. 2005. World Catalogue of Potato Varieties. Agrimedia GmbH, Bergen/Dumme, Germany.

Hofvander, P., Andersson, M., Larsson, C-T. and Larsson, H. 2004. Field performance and starch characteristics of high-amylose potatoes obtained by antisense gene targeting of two branching enzymes. Plant Biotechnol. J. 2:311–320.

Hosaka, K. 2004. Evolutionary pathway of T-type chloroplast DNA in potato. Amer. J. Potato Res. 81:153–158.

Hosaka, K., Mori, M. and Ogawa, K. 1994. Genetic relationships of Japanese potato cultivars assessed by RAPD analysis. Am. Potato J. 71:535–546.

Hospital, F. 2003. Marker-assisted Breeding. In: Plant Molecular Breeding (H.J. Newbury, ed.), p. 30–59. Blackwell, Oxford.

Hougas, R.W., Peloquin, S.J. and Ross, R.W. 1958. Haploids of the common potato. J. Hered. 49:103–107.

Huaman, Z. and Spooner, D.M. 2002. Reclassification of landrace populations of cultivated potatoes (*Solanum* sect. *Petota*). Amer. J. Bot. 89:947–965.

Huaman, Z., Williams, J.T., Salhuana, W. and Vicent, L. 1977. Descriptors for the cultivated potato and for the maintenance and distribution of germplasm collections. International Board for Plant Genetic Resources, Rome, Italy.

Huaman, Z., Golmirzaie, A. and Amoros, W. 1997. The Potato. In: Biodiversity in Trust: Conservation and Use of Plant Genetic Resources in CGIAR Centres (D. Fuccillo, L. Sears and P. Stapleton, eds.), p. 21–28. Cambridge University Press, Cambridge, UK.

Huaman, Z., Hoekstra, R. and Bamberg, J.B. 2000a. The inter-genebank potato database and the dimensions of available wild potato germplasm. Amer. J. Potato Res. 77:353–362.

Huaman, Z., Ortiz, R. and Gomez, R. 2000b. Selecting a *Solanum tuberosum* subsp. *andigena* core collection using morphological, geographical, disease and pest descriptors. Amer. J. Potato Res. 77:183–190.

Huaman, Z., Ortiz, R., Zhang, D. and F. Rodriguez, F. 2000c. Isozyme analysis of entire and core collections of *Solanum tuberosum* subsp. *andigena* potato cultivars. Crop Sci. 40:273–276.

Huang, S. 2005. Discovery and characterization of the major late blight resistance complex in potato. Thesis Wageningen University, The Netherlands.

Huang, S., van der Vossen, E.A.G., Kuang, H., Vleeshouwers, V.G.A.A., Zhang, N., Borm, T.J.A., van Eck, H.J., Baker, B., Jacobsen, E. and Visser, R. 2005. Comparative genomics enabled the cloning of the *R3a* late blight resistance gene in potato. Plant J. 42:251–261.

Iovene, M., Barone, A., Frusciante, L., Monti, L. and Carputo, D. 2004. Selection for aneuploid potato hybrids combining a low wild genome content and resistance traits from *Solanum commersonii*. Theor. Appl. Genet. 109:1139–1146.

Jacobs, J.M.E., van Eck, H.J., Arens, P., Verkerk-Bakker, B., te Lintel Hekkert, B., Bastiaanssen, H.J.M., El-Kharbotly, A., Pereira, A., Jacobsen, E. and Stiekema, W.J. 1995. A genetic map of potato (*Solanum tuberosum*) integrating molecular markers, including transposons, and classical markers. Theor. Appl. Genet. 91:289–300.

Jacobs, J.M.E., van Eck, H.J., Horsman, K., Arens, P.F.P., Verkerk-Bakker, B., Jacobsen, E., Periera, A. and Stiekema, W.J. 1996. Mapping of resistance to the potato cyst nematode *Globodera rostochiensis* from the wild potato species *Solanum vernei*. Mol. Breed. 2:51–60.

Jansky, S. 2000. Breeding for Disease Resistance in Potato. In: Plant Breeding Reviews, Volume 19 (J. Janick, ed.), p. 69–165. John Wiley & Sons, New York.

Jansky, S. 2006. Overcoming hybridization barriers in potato. Plant Breed. 125:1–12.

Jansky, S.H., Yerk, G.L. and Peloquin, S.J. 1990. The use of potato haploids to put 2*x* wild species germplasm into a usable form. Plant Breed. 104:290–294.

Jansky, S.H., Davis, G.L. and Peloquin, S.J. 2004. A genetic model for tuberization in potato haploid-wild species hybrids grown under long-day conditions. Amer. J. Potato Res. 81:335–339.

Jellis, G.J. 1992. Multiple resistance to diseases and pests in potatoes. Euphytica 63:51–58.

Jin, L.P., Qu, D.Y., Xie, K.Y., Bian, C.S. and Duan, S.G. 2004. Potato Germplasm, Breeding Studies in

China. In: Proceedings of the Fifth World Potato Congress, p. 175–178. Kunming, China.

Jinks, J.L. and Connolly, V. 1973. Selection for specific and general response to environmental differences. Heredity 30:33–40.

Jinks, J.L. and Connolly, V. 1975. Determination of the environmental sensitivity of selection lines by the selection environment. Heredity 34:401–406.

Jinks, J.L. and Pooni, H.S. 1982. Determination of the environmental sensitivity of selection lines of *Nicotiana rustica* by the selection environment. Heredity 49:291–294.

Johnston, S.A., der Nijs, T.P.M., Peloquin, S.J. and Hanneman, R.E.Jr. 1980. The significance of genic balance to endosperm development in interspecific crosses. Theor. Appl. Genet. 57:5–9.

Jones, R.A.C. 1985. Further studies on resistance-breaking strains of potato virus X. Plant Pathol. 34:182–189.

Jones, R.W. and Simko, I. 2005. Resistance to Late Blight and Other Fungi. In: Genetic Improvement of Solanaceous Crops Volume I: Potato (M.K. Razdan and A.K. Mattoo, eds.), p. 397–417. Science Publishers, Inc., Enfield, New Hampshire.

Jung, C.S., Griffiths, H.M., De Jong, D.M., Cheng, S., Bodis, M. and De Jong, W.S. 2005. The potato *P* locus codes for flavonoid 3′,5′-hydroxylase. Theor. Appl. Genet. 110:269–275.

Kasai, K., Morikawa, Y., Sorri, V.A., Valkonen, J.P.T., Gebhardt, C. and Watanabe, K.N. 2000. Development of SCAR markers to the PVY resistance gene Ry*adg* based on a common feature of plant disease resistance genes. Genome 43:1–8.

Killick, R.J. 1977. Genetic analysis of several traits in potatoes by means of a diallel cross. Ann. appl. Biol. 86:279–289.

Knight, T.A. 1807. On raising of new and early varieties of the potato (*Solanum tuberosum*). Trans. Hort. Soc. Lond. 1:57–59.

Kouassi, A.B., Kerlan, M-C., Caromel, B., Dantec, J-P., Fouville, D., Manzanares-Dauleux, M., Ellisseche, D. and Mugniery, D. 2006. A major gene mapped on chromosome XII is the main factor of a quantitatively inherited resistance to *Meloidogyne fallax* in *Solanum sparsipilum*. Theor. Appl. Genet. 112:1400.

Kreike, C.M., Kok-Westeneng, A.A., Vinke, J.H. and Stiekema, W.J. 1996. Mapping of QTLs involved in nematode resistance, tuber yield and root development in *Solanum* sp. Theor. Appl. Genet. 92:463–470.

Kuhl, J.C., Hanneman, R.E. Jr. and Havey, M.J. 2001. Characterization and mapping of Rpi1, a late–blight resistance locus from diploid (1EBN) Mexican *Solanum pinnatisectum*. Mol. Genet. Genomics 265:977–985.

Kumar, A. 1994. Somaclonal Variation. In: Potato Genetics (Bradshaw, J.E. and Mackay, G.R.,

eds.) p. 197–212. CAB International, Wallingford, UK.

Landeo, J.A. 2002. Durable resistance: Quantitative/Qualtative Resistance. In: Proceedings of the Global Initiative on Late Blight conference, July 11–13, 2002, Hamburg, Germany (C. Lizarraga, ed.), p. 29–36.

Lang, J. 2001. Notes of a Potato Watcher. Texas A&M University Press, College Station.

Large E.C. 1940. Advance of the Fungi. Jonathan Cape, London.

Li, X., van Eck, H.J., Rouppe van der Voort, J.N.A.M., Huigen, D.-J., Stam, P. and Jacobsen, E. 1998. Autotetraploids and genetic mapping using common AFLP markers: The *R2* allele conferring resistance to *Phytophthora infestans* mapped on potato chromosome *4*. Theor. Appl. Genet. 96:1121–1128.

Li, X-Q., De Jong, H., De Jong, D.M. and De Jong, W.S. 2005. Inheritance and genetic mapping of tuber eye depth in cultivated diploid potatoes. Theor. Appl. Genet. 110:1068–1073.

Love, S.L., Thompson-Johns, A. and Baker, T. 1993. Mutation breeding for resistance to blackspot bruise and low temperature sweetening in the potato cultivar Lemhi Russet. Euphytica 70:69–74.

Mackay, G.R. 1987. Selecting and Breeding for Better Potato Cultivars. In: Improving Vegetatively Propagated Crops (A.J. Abbot and R.K. Atkin, eds.), p. 181–196. Academic Press Limited, London & San Diego.

Mackay, G.R. 1996. An agenda for future potato research. Potato Res. 39:387–394.

Mackay, G.R. 2005. Propagation by Traditional Breeding Methods. In: Genetic Improvement of Solanaceous Crops Volume I: Potato (M.K. Razdan and A.K. Mattoo, eds.), p. 65–81. Science Publishers, Inc., Enfield, New Hampshire.

Malcolmson, J.F. 1969. Races of *Phytophthora infestans* occurring in Great Britain. Trans. Br. mycol. Soc. 53:417–423.

Marano, M.R., Malcuit, I., De Jong, W. and Baulcombe, D.C. 2002. High-resolution genetic map of *Nb*, a gene that confers hypersensitive resistance to potato virus X in *Solanum tuberosum*. Theor. Appl. Genet. 105:192–200.

Marczewski, W., Hennig, J. and Gebhardt, C. 2002. The *Potato virus S* resistance gene *Ns* maps to potato chromosome VIII. Theor. Appl. Genet. 105:564–567.

Marczewski, W., Flis, B., Syller, J., Strzelczyk-Zyta, D., Hennig, J. and Gebhardt, C. 2004. Two allelic or tightly linked genetic factors at the *PLRV.4* locus on potato chromosome XI control resistance to potato leafroll virus accumulation. Theor. Appl. Genet. 109:1604–1609.

Marczewski, W., Strzelczyk-Zyta, D., Hennig, J., Witek, K. and Gebhardt, C. 2006. Potato chromosomes IX and XI carry genes for resistance to potato virus M. Theor. Appl. Genet. 112:1232–1238.

Maris, B. 1989. Analysis of an incomplete diallel cross among three ssp. *tuberosum* varieties and seven long-day adapted ssp. *andigena* clones of the potato (*Solanum tuberosum* L.). Euphytica 41:163–182.

Matsubayashi, M. 1991. Phylogenetic Relationships in The Potato and its Related Species. In: Chromosome Engineering in Plants: Genetics, Breeding and Evolution, Part B (T. Tsuchiya and P.K. Gupta, eds.), p. 93–118. Elsevier, Amsterdam.

McCue, K.F., Shepherd, L.V.T., Allen, P.V., Maccree, M.M., Rockhold, D.R., Corsini, D.L., Davies, H.V. and Belknap, W.R. 2005. Metabolic compensation of steroidal glycoalkaloid biosynthesis in transgenic potato tubers: Using reverse genetics to confirm the in vivo enzyme function of a steroidal alkaloid galactosyltransferase. Plant Science 168:267–273.

Mead, R. 1997. Design of Plant Breeding Trials. In: Statistical Methods for Plant Variety Evaluation (R.A. Kempton and P.N. Fox, eds.), p. 40–67. Chapman & Hall, London.

Menendez, C.M., Ritter, E., Schäfer-Pregl, R., Walkemeier, B., Kalde, A., Salamini, F. and Gebhardt, C. 2002. Cold sweetening in diploid potato: Mapping quantitative trait loci and candidate genes. Genetics 162:1423–1434.

Mohammed, A., Douches, D.S., Pett, W., Grafius, E., Coombs, J., Liswidowati, W., Li, W. and Madkour, M.A. 2000. Evaluation of potato tuber moth (Lepidoptera: Gelechiidae) resistance in tubers of Bt-cry5 transgenic potato lines. J. Econ. Entomol. 93:472–476.

Morris, M.L. and Bellon, M.R. 2004. Participatory plant breeding research: Opportunities and challenges for the international crop improvement system. Euphytica 136:21–35.

Morris, W.L., Ducreux, J.M., Fraser, P.D., Millam, S and Taylor, M.A. 2006. Engineering ketocarotenoid biosynthesis in potato tubers. Metab. Eng. (in press).

Mulema, J.M.K., Olanya, O.M., Adipala, E. and Wagoire, W. 2004/5. Stability of late blight resistance in Population B potato clones. Potato Res. 47:11–24.

Muller, K.O. and Black, W. 1951. Potato breeding for resistance to blight and virus diseases during the last hundred years. Z. Pflanzenzuchtung 31:305–318.

Naess, S.K., Bradeen, J.M., Wielgus, S.M., Haberlach, G.T. McGrath, J.M. and Helgeson, J.P. 2000. Resistance to late blight in *Solanum bulbocastanum* is mapped to chromosome 8. Theor. Appl. Genet. 101:697–704.

Neele, A.E.F., Nab, H.J. and Louwes, K.M. 1991. Identification of superior parents in a potato breeding programme. Theor. Appl. Genet. 82:264–272.

Oberhagemann, P., Chatot-Balandras, C., Schäfer-Pregl, R., Wegener, D., Palomino, C., Salamini, F.,

Bonnel, E. and Gebhardt, C. 1999. A genetic analysis of quantitative resistance to late blight in potato: Towards marker-assisted selection. Molecular Breed. 5:399–415.

Ochoa, C.M. 1990. The Potatoes of South America: Bolivia. Cambridge University Press, Cambridge.

Ochoa, C.M. 2004. The Potatoes of South America: Peru. 1: The Wild Species. International Potato Center, Lima.

Ooms, G., Bossen, M.E., Burrell, M.M. and Carp, A. 1986. Genetic manipulation in potato with *Agrobacterium rhizogenes*. Potato Res. 29:367–379.

Ortiz, R. 1997. Breeding for potato production from true seed. Plant Breeding Abstracts 67:1355–1360.

Ortiz, R. 1998. Potato Breeding via Ploidy Manipulations. In: Plant Breeding Reviews, Volume 16 (J. Janick, ed.), p. 15–86. John Wiley & Sons, New York.

Ortiz, R. 2001. The State of the Use of Potato Genetic Diversity. In: Broadening the Genetic Base of Crop Production (H.D. Cooper, C. Spillane and T. Hodgkin, eds.), p. 181–200. CAB International, Wallingford, UK.

Ortiz, R. and Ehlenfeldt, M.K. 1992. The importance of endosperm balance number in potato breeding and the evolution of tuber-bearing *Solanum* species. Euphytica 60:105–113.

Ortiz, R. and Peloquin, S.J. 1991. A new method of producing inexpensive 4*x* hybrid true potato seed. Euphytica 57:103–108.

Osusky, M., Osuska, L., Kay, W. and Misra, S. 2005. Genetic modification of potato against microbial diseases: In vitro and in planta activity of a dermaseptin B1 derivative, MsrA2. Theor. Appl. Genet. 111:711–722.

Paal, J., Henselewski, H., Muth, J., Meksem, K., Menendez, C.M., Salamini, F., Ballvora, A. and Gebhardt, C. 2004. Molecular cloning of the potato *Gro1-4* gene conferring resistance to pathotype Ro1 of the root cyst nematode *Globodera rostochiensis*, based on a candidate gene approach. Plant J. 38:285–297.

Pandey, S.K. and Kaushik, S.K. 2003. Origin, Evolution, History and Spread of Potato. In: The Potato—Production and Utilization in Sub-Tropics (S.M.P. Khurana, J.S. Minhas and S.K. Pandey, eds.), p. 15–24. Mehta Publishers, New Delhi.

Park, T-O., Gros, J., Sikkema, A., Vleeshouwers, V.G.A.A., Muskens, M., Allefs, S., Jacobsen, E., Visser, R.G.F. and van der Vossen, E.A.G. 2005a. The late blight resistance locus *Rpi-blb3* from *Solanum bulbocastanum* belongs to a major late blight *R* gene cluster on chromosome 4 of potato. MPMI 18:722–729.

Park, T.H., Vleeshouwers, V.G.A.A., Hutten, R.C.B., van Eck, H.J., van der Vossen, E., Jacobsen, E. and Visser, R.G.F. 2005b. High-resolution mapping and analysis of the resistance locus *Rpi-abpt* against

Phytophthora infestans in potato. Mol. Breed. (in press).

Park, T.H., Vleeshouwers, V.G.A.A., Huigen, D.J., van der Vossen, E.A.G., van Eck, H.J. and Visser, R. G.F. 2005c. Characterization and high-resolution mapping of a late blight resistance locus similar to *R2* in potato. Theor. Appl. Genet. 111:591–597.

Pavek, J.J. and Corsini, D.L. 1994. Inheritance of Resistance to Warm-growing-season Fungal Diseases. In: Potato Genetics (J.E. Bradshaw and G.R. Mackay, eds.), p. 403–409. CAB International, Wallingford, UK.

Peloquin, S.J. and Hougas, R.W. 1959. Decapitation and genetic markers as related to haploidy in *Solanum tuberosum*. European Potato J. 2:176–183.

Pijnacker, L.P. and Ferwerda, M.A. 1984. Giemsa C-banding of potato chromosomes. Canadian J. Genet. Cytol. 26:415–419.

Plaisted, R.L. 1980. Potato. In: Hybridization of Crop Plants (W.R. Fehr and H.H. Hadley, eds.), p. 483–494. The American Society of Agronomy, Inc., Madison.

Plaisted, R.L. 1987. Advances and Limitations in the Utilization of Neotuberosum in Potato Breeding. In: The Production of New Potato Varieties (G.J. Jellis and D.E. Richardson, eds.), p. 186–196. Cambridge University Press, Cambridge.

Plaisted, R.L. and Hoopes, R.W. 1989. The past record and future prospects for the use of exotic potato germplasm. Am. Potato J. 66:603–627.

Poehlman, J.M. and Sleper, D.A. 1995. Breeding Field Crops, Fourth Edition, Iowa State University Press, Ames.

Powell, W., Phillips, M.S., McNicol, J.W. and Waugh, R. 1991. The use of DNA markers to estimate the extent and nature of genetic variability in *Solanum tuberosum* cultivars. Ann. Appl. Biol. 118:423–432.

Raker, C.M. and Spooner, D.M. 2002. Chilean tetraploid cultivated potato, *Solanum tuberosum*, is distinct from the Andean populations: Microsatellite data. Crop Sci. 42:1451–1458.

Raman, K.V., Golmirzaie, A.M., Palacios, M. and Tenorio, J. 1994. Inheritance of Resistance to Insects and Mites. In: Potato Genetics (J.E. Bradshaw and G.R. Mackay, eds.), p. 447–463. CAB International, Wallingford, UK.

Rauscher, G.M., Smart, C.D., Simko, I., Bonierbale, M., Mayton, H., Greenland, A. and Fry W.E. 2006. Characterization and mapping of *RPi-ber,* a novel potato late blight resistance gene from *Solanum berthaultii*. Theor. Appl. Genet. 112:674–687.

Razdan, M.K. and Mattoo, A.K. (eds.). 2005. Genetic Improvement of Solanaceous Crops Volume I: Potato. Science Publishers, Inc., Enfield, New Hampshire.

Ross, H. 1986. Potato Breeding—Problems and Perspectives. Advances in Plant Breeding 13, Paul Parey, Berlin and Hamburg.

Rouppe van der Voort, J.N.A.M., Wolters, P., Folkertsma, R., Hutten, R. and van Zandvoort, P. 1997. Mapping of the cyst nematode resistance locus *Gpa2* in potato using a strategy based on comigrating AFLP markers. Theor. Appl. Genet. 95:874–880.

Rouppe van der Voort, J.N.A.M., van der Vossen, E., Bakker, E., Overmars, H., van Zandroort, P., Hutten, R., Lankhorst, R.K. and Bakker, J. 2000. Two additive QTL conferring broad-spectrum resistance in potato to *Globodera pallida* are localized on resistance gene clusters. Theor. Appl. Genet. 101:1122–1130.

Rousselle-Bourgeois, F. and Rousselle, P. 1992. Creation et selection de populations diploide de pomme de terre (*Solanum tuberosum* L.). Agronomie 12:59–67.

Schäfer-Pregl, R., Salamini, F. and Gebhardt, C. 1996. Models for mapping quantitative trait loci (QTLs) in progeny of non-inbred parents and their behaviour in the presence of distorted segregation ratios. Genet. Res. 67:43–54.

Schäfer-Pregl, R., Ritter, E., Concilio, L., Hesselbach, J., Lovatti, L., Walkemeier, B., Thelen, H., Salamini, F. and Gebhardt, C. 1998. Analysis of quantitative trait loci (QTLs) and quantitative trait alleles (QTAs) for potato tuber yield and starch content. Theor. Appl. Genet. 97:834–846.

Schäfer-Pregl, R., Salamini, F. and Gebhardt, C. 2000. QTL analysis of new sources of resistance to *Erwinia carotovora* ssp. *atroseptica* in potato done by AFLP, RFLP and resistance-gene-like markers. Crop Sci. 40:1156–1167.

Schittenhelm, S. and Hoekstra, R. 1995. Recommended isolation distances for the field multiplication of diploid tuber-bearing *Solanum* species. Plant Breed. 114:369–371.

Schwall. G.P., Safford, R., Westcott, R.J., Jeffcoat, R., Tayal, A., Shi, Y.C., Gidley, M.J. and Jobling, S.A. 2000. Production of very high amylose potato starch by inhibition of SBE A and B. Nature Biotech. 18:551–554.

Serrano, C., Arce-Johnson, P., Torres, H., Gebauer, M., Gutierrez, M., Moreno, M., Jordana, X., Venegas, A., Kalazich, J. and Holuigue, L. 2000. Expression of the chicken lysozyme gene in potato enhances resistance to infection by *Erwinia carotovora* subsp. *atroseptica*. Amer. J. Potato Res. 77:191–199.

Shepard, J.F., Bidney, D. and Shahin, E. 1980. Potato protoplasts in crop improvement. Science 208:17–24.

Simko, I. 2002. Comparative analysis of quantitative trait loci for foliage resistance to *Phytophthora infestans* in tuber-bearing *Solanum* species. Am. J. Potato Res. 79:125–132.

Simko, I., Haynes, K.G., Ewing, E.E., Costanzo, S., Christ, B.J. and Jones, R.W. 2004. Mapping genes for resistance to *Verticillium albo-atrum* in tetraploid and diploid potato populations using haplotype association tests and genetic linkage analysis. Mol. Genet. Genomics 271:522–531.

Simmonds, N.W. 1969. Prospects of potato improvement. Scottish Plant Breeding Station Forty-Eighth Annual Report 1968–69, p. 18–38.

Simmonds, N.W. 1981. Genotype (*G*), Environmemt (*E*) and *GE* components of crop yields. Exp. Agr. 17:355–362.

Simmonds, N.W. 1995. Potatoes. In: Evolution of Crop Plants, Second Edition (J. Smartt and N.W. Simmonds, eds.), p. 466–471. Longman Scientific & Technical, Harlow.

Simmonds, N.W. 1996. Family selection in plant breeding. Euphytica 90:201–208.

Simmonds, N.W. 1997. A review of potato propagation by means of seed, as distinct from clonal propagation by tubers. Potato Res. 40:191–214.

Smilde, W.D., Brigneti, G., Jagger, L., Perkins, S. and Jones, J.D.G. 2005. *Solanum mochiquense* chromosome IX carries a novel late blight resistance gene *Rpi-moc1*. Theor. Appl. Genet. 110:252–258.

Solomon-Blackburn, R.M. and Barker, H. 2001. Breeding virus resistant potatoes (*Solanum tuberosum*): A review of traditional and molecular approaches. Heredity 86:17–35.

Solomos, T. and Mattoo, A.K. 2005. Starch-Sugar Metabolism in Potato (*Solanum tuberosum* L.) Tubers in Response to Temperature Variations. In: Genetic Improvement of Solanaceous Crops Volume I: Potato (M.K. Razdan and A.K. Mattoo, eds.), p. 209–234. Science Publishers, Inc., Enfield, New Hampshire.

Song, J., Bradeen, J.M., Naess, S.K., Raasch, J.A., Wielgus, S.M., Haberlach, G.T., Liu, J., Kuang, H., Austin-Phillips, S., Buell, C.B., Helgeson, J.P. and Jiang, J. 2003. Gene *RB* cloned from *Solanum bulbocastanum* confers broad spectrum resistance to potato late blight. Proc. Nat. Acad. Sci. 100:9128–9133.

Song, Y-S., Hepting, L., Schweizer, G., Hartl, L., Wenzel, G. and Schwarzfischer, A. 2005. Mapping of extreme resistance to PVY (*Rysto*) on chromosome XII using anther-culture-derived primary dihaploid potato lines. Theor. Appl. Genet. 111:879–887.

Sonnewald, U., Hajirezaei, M.-H., Kossmann, J., Heyer, A., Trethewey, R.N. and Willmitzer, L. 1997. Increased potato tuber size resulting from apoplastic expression of a yeast invertase. Nature Biotechnology 15:794–797.

Sowokinos, J.R. 2001. Allele and isozyme patterns of UDP-glucose pyrophosphorylase as a marker for cold-sweetening resistance in potatoes. Amer. J. Potato Res. 78:57–64.

Spooner, D.M. and Hijmans, R.J. 2001. Potato systematics and germplasm collecting, 1989–2000. Amer. J. Potato Res. 78:237–268.

Spooner, D.M., van den Berg, R.G., Rodriguez, A., Bamberg, J., Hijmans, R.J. and Cabrera, S.I.L. 2004. Wild Potatoes (Solanum section Petota; Solanaceae) of North and Central America. Systematic Botany Monographs Volume 68.

Spooner, D.M., McLean, K., Ramsay, G., Waugh, R. and Bryan, G. J. 2005a. A single domestication for potato based on multilocus amplified fragment length polymorphism genotyping. Proc. Nat. Acad. Sci. 102:14694–14699.

Spooner, D.M., Nunez, J., Rodriguez, F., Naik, P.S. and Ghislain, M. 2005b. Nuclear and chloroplast DNA reassessment of the origin of Indian potato varieties and its implications for the origin of the early European potato. Theor. Appl. Genet. 110:1020–1026.

Stark, D.M., Timmerman, K.P., Barry, G.F., Preiss, J. and Kishore, G.M. 1992. Regulation of the amount of starch in plant tissues by ADPglucose pyrophosphorylase. Science 258:287–292.

Stiekema, W.J., Heidekamp, F., Louwerse, J.D., Verhoeven, H.A. and Dijkhuis, P. 1988. Introduction of foreign genes into potato cultivars Bintje and Desiree using an *Agrobacterium tumefaciens* binary vector. Plant Cell Rep. 7:47–50.

Sukhotu, T. and Hosaka, K. 2006. Origin and evolution of Andigena potatoes revealed by chloroplast and nuclear DNA markers. Genome 49:636–647.

Sun, G., Wang-Pruski, G., Mayich, M. and Jong, H. 2003. RAPD and pedigree-based genetic diversity estimates in cultivated diploid hybrids. Theor. Appl. Genet. 107:110–115.

Tai, G.C.C. 1976. Estimation of general and specific combining abilities in potato. Canadian Journal of Genetics and Cytology 18:463–470.

Tai, G.C.C. 1994. Use of 2*n* Gametes. In: Potato Genetics (J.E. Bradshaw and G.R. Mackay, eds.), p. 109–132. CAB International, Wallingford, UK.

Tai, G.C.C. and De Jong, H. 1980. Multivariate analyses of potato hybrids. 1. Discrimination between tetraploid-diploid hybrid families and their relationship to cultivars. Can. J. Genet. Cytol. 22:277–235.

Tai, G.C.C. and Young, D.A. 1984. Early generation selection for important agronomic characteristics in a potato breeding population. Am. Potato J. 61:419–434.

Tanksley, S.D., Ganal, M.W., Prince, J.P., de Vincente, M.C., Bonierbale, M.W., Broun, P., Fulton, T.M., Giovannoni, J.J., Grandillo, S., Martin, G.B., Messeguer, R., Miller, J.C., Miller, L., Paterson, A.H., Pineda, O., Roder, M.S., Wing, R.A., Wu. W. and Young, N.D. 1992. High density molecular linkage maps of the tomato and potato genomes. Genetics 132:1141–1160.

Tarn, T.R. and Tai, G.C.C. 1983. Tuberosum × Tuberosum and Tuberosum × Andigena potato hybrids: Comparisons of families and parents, and breeding strategies for Andigena potatoes in long-day temperate environments. Theor. Appl. Genet. 66:87–91.

Tarn, T.R., Tai, G.C.C., De Jong, H., Murphy, A.M. and Seabrook, J.E.A. 1992. Breeding Potatoes for Long-day, Temperate Climates. In: Plant Breeding Reviews 9 (J. Janick, ed.), p. 217–332. John Wiley & Sons, New York.

Taylor, M.A. and Ramsay, G. 2005. Carotenoid biosynthesis in plant storage organs: Recent advances and prospects for improving plant food quality. Physiol. Plant. 124:143–151.

Tek, A.L., Stevensen, W.R., Helgeson, J.P. and Jiang, J. 2004. Transfer of tuber soft rot and early blight resistances from Solanum brevidens into cultivated potato. Theor. Appl. Genet. 109:249–254.

Thieme, T. and Thieme, R. 2005. Resistance to Viruses. In: Genetic Improvement of Solanaceous Crops Volume I: Potato (M.K. Razdan and A.K. Mattoo, eds.), p. 293–337. Science Publishers, Inc., Enfield, New Hampshire.

Tommiska, T.J., Hamalainen, J.H., Watanabe, K.N. and Valkonen, J.P.T. 1998. Mapping of the gene Nxphu that controls hypersensitive resistance to potato virus X in Solanum phureja IvP35. Theor. Appl. Genet. 96:840–843.

Toxopeus, H.J. 1964. Treasure-digging for blight resistance in potatoes. Euphytica 13:206–222.

Trehane, P., Brickell, C.D., Baum, B.R., Hetterscheid, W.L.A., Leslie, A.C., McNeill, J., Spongberg, S.A. and Vrugtman, F. 1995. International Code of nomenclature for cultivated plants. Regnum Veg. 133:1–175.

Trognitz, B.R., Bonierbale, M., Landeo, J.A., Forbes, G., Bradshaw, J.E., Mackay, G.R., Waugh, R., Huarte, M.A. and Colon, L. 2001. Improving Potato Resistance to Disease under the Global Initiative on Late Blight. In: Broadening the Genetic Base of Crop Production (H.D. Cooper, C. Spillane and T. Hodgkin, eds.), p. 385–398. CAB International, Wallingford, UK.

Ullstrup, A.J. 1972. The impacts of the Southern corn leaf blight epidemics of 1970–71. Ann. Rev. Phytopathol. 10:37–50.

Urwin, P.E., Green, J. and Atkinson, H.J. 2003. Expression of a plant cystatin confers partial resistance to Globodera, full resistance is achieved by pyramiding a cystatin with natural resistance. Molecular Breed. 12:263–269.

Vada, M.E. 1994. Environmental Stress and its Impact on Potato Yield. In: Potato Genetics (J.E. Bradshaw and G.R. Mackay, eds.), p. 239–261. CAB International, Wallingford, UK.

Van den Berg, J.H., Ewing, E.E., Plaisted, R.L., McMurry, S. and Bonierbale, M.W. 1996a. QTL analysis of potato tuber dormancy. Theor. Appl. Genet. 93:317–324.

Van den Berg, J.H., Ewing, E.E., Plaisted, R.L., McMurry, S. and Bonierbale, M.W. 1996b. QTL analysis of potato tuberization. Theor. Appl. Genet. 93:307–316.

van der Berg, R.G., Bryan, G.J., del Rio, A. and Spooner, D.M. 2002. Reduction of species in the wild potato Solanum section Petota series Longipedicellata: AFLP, RAPD and chloroplast SSR data. Theor. Appl. Genet. 105:1109–1114.

van der Vossen, E., Sikkema, A., te L. Hekkert, B., Gros, J. Stevens, P., Muskens, M., Wouters, D., Pereira, A., Stiekema, W. and Allefs, S. 2003. An ancient R gene from the wild potato species Solanum bulbocastanum confers broad-spectrum resistance to Phytophthora infestans in cultivated potato and tomato. Plant J. 36:867–882.

van Eck, H.J., Jacobs, J.M.E., van Dijk, J., Stiekema, W.J. and Jacobsen, E. 1993. Identification and mapping of three flower colour loci of potato (S. tuberosum L.) by RFLP analysis. Theor. Appl. Genet. 86:295–300.

van Eck, H.J., Jacobs, J.M.E., Stam, P., Ton, J., Stiekema, W.J. and Jacobsen, E. 1994a. Multiple alleles for tuber shape in diploid potato detected by qualitative and quantitative genetic analysis using RFLPs. Genetics 137:303–309.

van Eck, H.J., Jacobs, J.M.E., van den Berg, P.M.M.M., Stiekema, W.J. and Jacobsen, E. 1994b. The inheritance of anthocyanin pigmentation in potato (Solcnum tuberosum L.) and mapping of tuber skin colour loci using RFLPs. Heredity 73:410–421.

Veilleux R.E. 2005. Cell and Tissue Culture of Potato (Solanaceae). In: Genetic Improvement of Solanaceous Crops Volume I: Potato (M.K. Razdan and A.K. Mattoo, eds.), p. 185–208. Science Publishers, Inc., Enfield, New Hampshire.

Visker, M.H.P.W., Keizer, L.C.P., Van Eck, H.J., Jacobsen, E., Colon, L.T. and Struik, P.C. 2003. Can the QTL for the late blight resistance on potato chromosome 5 be attributed to foliage maturity type? Theor. Appl. Genet. 106:317–325.

Visser, R.G.F., Somhorst, I., Kuipers, G.J., Ruys, N.J., Feenstra, W.J. and Jacobsen, E. 1991. Inhibition of the expression of the gene for granule-bound starch synthase in potato by antisense constructs. Mol. Gen. Genet. 225:289–296.

Wastie, R.L. 1991. Breeding for resistance. Adv. Plant Path. 7:193–224.

Wastie, R.L. 1994. Inheritance of Resistance to Fungal Diseases of Tubers. In: Potato Genetics (J.E. Bradshaw and G.R. Mackay, eds.), p. 411–427. CAB International, Wallingford, UK.

Weber, J. 1990. Erwinia—A Review of Recent Research. In: Proc. 11th Triennial Conference of the EAPR, 8–13 July 1990, Edinburgh, p. 112–121.

Wegener, C.B. 2003/4. Inheritance of a pectate lyase mediated soft rot resistance in crosses between transgenic potato lines of cv. Desiree and *S. tuberosum* cultivars. Potato Res. 46:155–166.

Wenzel, G. 1994. Tissue Culture. In: Potato Genetics (J.E. Bradshaw and G.R. Mackay, eds.), p. 173–195. CAB International, Wallingford, UK.

Wenzel, G., Schieder, O., Przewozny, T., Sopory, S.K. and Melchers, G. 1979. Comparison of single cell culture derived *Solanum tuberosum* L. plants and a model for their application in breeding programs. Theor. Appl. Genet. 55:49–55.

White, P.J. and Broadley, M.R. 2005. Biofortifying crops with essential mineral elements. Trends Plant Sci. 10:586–593.

Wolters, P., Vinke, H., Bontjer, I., Rouppe van der Voort, J., Colon, L., Hoogendoom, C. 1998. Presence of Major Genes for Resistance to *Globodera Pallida* in Wild Tuber-Bearing *Solanum* Species and their Location on the Potato Genome. In: 5th Intl. Symp. Mol. Biol. Pot, August 2–6, Bogensee, Germany.

Yeh, B.P. and Peloquin, S.J. 1965. Pachytene chromosomes of the potato (*Solanum tuberosum* group *andigena*). Am. J. Botany 52:1014–1020.

Yencho, G.C., Bonierbale, M.W., Tingey, W.M., Plaisted, R.L. and Tanksley, S.D. 1996. Molecular markers locate genes for resistance to the Colorado potato beetle, *Leptinotarsa decemlineata,* in hybrid *Solanum tuberosum* × *S. berthaultii* potato progenies. Entomol. Exp. Appl. 81:141–154.

Yencho, G.C., Kowalski, S.P., Kobayashi, R.S., Sinden, S.L., Bonierbale, M.W. and Deahl, K.L. 1998. QTL mapping of foliar glycoalkaloid aglycones in *Solanum tuberosum* × *S. berthaultii* potato progenies: Quantitative variation and plant secondary metabolism. Theor. Appl. Genet. 97:563–574.

Zeh, M., Casazza, A.P., Kreft, O., Roessner, U., Bieberich, K., Willmitzer, L., Hoefgen, R. and Hesse, H. 2001. Antisense inhibition of threonine synthase leads to high methionine content in transgenic potato plants. Plant Physiol. 127:792–802.

Zimnoch-Guzowska, E., Marczewski, W., Lebecka, R., Flis, B., Schäfer-Pregl, R., Salamini, F. and Gebhardt, C. 2000. QTL analysis of new sources of resistance to *Erwinia carotovora* ssp. *atroseptica* in potato done by AFLP, RFLP and resistance-gene-like markers. Crop Sci. 40:1156–1167.

Chapter 11

Breeding of Sweetpotato

S.L. Tan, M. Nakatani, and K. Komaki

Introduction

Like most root crops, sweetpotato (*Ipomoea batatas* [L.] Lam) is a source of carbohydrates. Orange-fleshed varieties are rich in β-carotene, a precursor of vitamin A. Indeed, the β-carotene content of some varieties surpasses even that of carrot. Sweetpotato also contains vitamins C, B complex (niacin, riboflavin, and thiamine), and E, as well as potassium, calcium, and iron. Purple-fleshed varieties contain anthocyanin. Anthocyanin, β-carotene, and vitamins C and E are powerful antioxidants that in recent years have been gaining attention because of their purported ability to fend off certain cancers, help in preventing cardio-vascular disease, and ability to slow down the ravages of aging.

Unlike cassava (*Manihot esculenta* Crantz), the storage root of sweetpotato also has a reasonable amount of protein. With so many nutrients—and the added bonus of dietary fiber—sweetpotato can be considered a particularly healthy food.

Although the shoots and leaves have an even higher protein content than the roots, they are only used in the human diet in Southeast Asia. Sharing the same genus as *kangkong*, or water convolvulus or water spinach (*Ipomoea aquatica = I. reptans*), sweetpotato shoots serve as a green vegetable and may be cooked in a similar fashion. In Japan, recent research shows high phytochemical activity in sweetpotato leaves,

which can benefit human health (Ishiguro and Yoshimoto, 2006; Yoshimoto et al., 2006). Elsewhere, shoots and leaves are considered crop residues, and are sometimes fed to ruminant livestock (Mai, 2006).

Sweetpotato ranks seventh in world food production (totaling around 129.4 million tons per year) after wheat (*Triticum aestivum* L.), rice (*Oryza sativa* L.), maize (*Zea mays* L.), white potato (*Solanum tuberosum* L.), barley (*Hordeum vulgare* L.), and cassava. China is by far the world's largest producer of sweetpotato, accounting for 83% of total production volume, followed by Uganda and Nigeria as distant second and thirds (Table 11.1). While sweetpotato is not strictly a primary staple (except in parts of Polynesia and in Papua New Guinea), and is eaten more as a supplementary food, the per capita consumption in China is 71 kilograms per year. One of the reasons that sweetpotato is not favored as a staple is its inherent sweetness, which deters it from being eaten with other foods.

Origin and distribution

Indisputably, sweetpotato is of American origin (de Candolle, 1886), but there is disagreement on the particular place of origin of sweetpotato within the New World. From a prehistoric site dating 8080 B.C. in Peru, remains of sweetpotato have been excavated (Engel, 1970). Many scientists and archaeologists in the United States and Europe

Table 11.1. World production of sweetpotato.

Continent/Country	Production (tonnes)	% World production
Africa	10,757,486	8.44
Angola	430,000	0.34
Burundi	834,394	0.65
Cameroon	175,000	0.14
Congo, Dem. Republic of	224,450	0.18
Egypt	310,000	0.24
Ethiopia	340,000	0.27
Kenya	520,000	0.41
Madagascar	509,175	0.40
Nigeria	2,150,000	3.13
Rwanda	908,306	0.71
Tanzania, United Rep. of	950,000	0.74
Uganda	2,600,000	2.04
Others	816,061	0.64
Asia	113,411,880	88.93
Bangladesh	320,000	0.25
China	106,197,100	83.27
India	900,000	0.71
Indonesia	1,859,744	1.46
Japan	920,000	0.72
Korea, Democratic People's Rep.	350,000	0.27
Korea, Republic of	316,703	0.25
Laos	194,000	0.15
Malaysia	41,000	0.03
Philippines	530,000	0.42
Vietnam	1,600,000	1.25
Others	213,333	0.17
Latin America & Carribean	1,949,409	1.53
Argentina	315,000	0.25
Brazil	495,000	0.39
Cuba	490,000	0.38
Haiti	175,000	0.14
Paraguay	125,000	0.10
Peru	127,289	0.10
Others	222,120	0.17
Oceania	641,100	0.50
Papua New Guinea	520,000	0.41
Others	121,100	0.09
United States of America	720,900	0.57
Others	54,233	0.04
WORLD TOTAL	127,535,008	100.00

Source: Adapted from FAOSTAT data, 2004.
http://apps.fao.org/faostat/collections?version=ext&hasbulk=0&subset=agriculture.

regard sweetpotato to be of Peruvian origin. However, a Japanese geneticist regarded a wild plant from Mexico (*Ipomoea trifida*; K-123) as related to sweetpotato (Nishiyama et al., 1961). Because K-123 was hexaploid and could cross with sweetpotato, Nishiyama et al. (1975) regarded it as the ancestral species of sweetpotato. From this point of view, sweetpotato is of Mexican origin. Austin (1983), on the other hand, claimed that K-123 was not a wild ancestor but an escapee of domesticated sweetpotato. He proposed the concept of a "Batatas complex," which contains 11 species closely related to sweetpotato, including *I. trifida* (Austin, 1978).

On the matter of ancestral wild species of sweetpotato, there has been a long dispute between scientists in Japan and the U.S. The proposal of Kobayashi (1984) of an "*I. trifida* complex" containing species of section Batatas that can cross with sweetpotato changed the situation. Komaki (2001) showed that variation of morphological characteristics and DNA markers were continuous in the *I. trifida* complex. Onjeda et al. (1990) found, among accessions from Colombia, some diploid lines of *I. trifida* that form non-reductive 2n pollen. When both haploid (x) and diploid (2x) pollen are used in crosses within the *I. trifida* complex, it is possible to produce 2x, 3x, 4x, 5x, and 6x plants. Onjeda et al. (1990) thought that this assortment of hexaploids (6x) led to sweetpotato. Although the distribution of the

I. trifida complex is across Central and South America, it is likely that sweetpotato originated in the northwestern area of South America. A report by Kobayashi (1984) supports this hypothesis, because he found some storage root-forming *I. trifida* from the highlands of Colombia.

Christopher Columbus, the Spaniards, and the Portuguese are believed to have brought sweetpotato to Europe. The Spanish were reported to have introduced it from Mexico to the Philippines (Camote route), whereas the Portuguese introduced it from the Caribbean and South America to Europe, Africa, India, Southeast Asia, and East Asia (Batatas route). Nonetheless, the crop has been in cultivation in New Guinea, the Polynesian Islands, and New Zealand since the pre-Columbian era (Kumara route) (Fig. 11.1). The hypothesis that Native Americans introduced it to Oceania from South America traveling by rafts was based on the Kon-Tiki expedition of Thor Heyerdahl in 1952. Recent thinking is that Polynesians traveled mainly on the Kumara route. However, if the

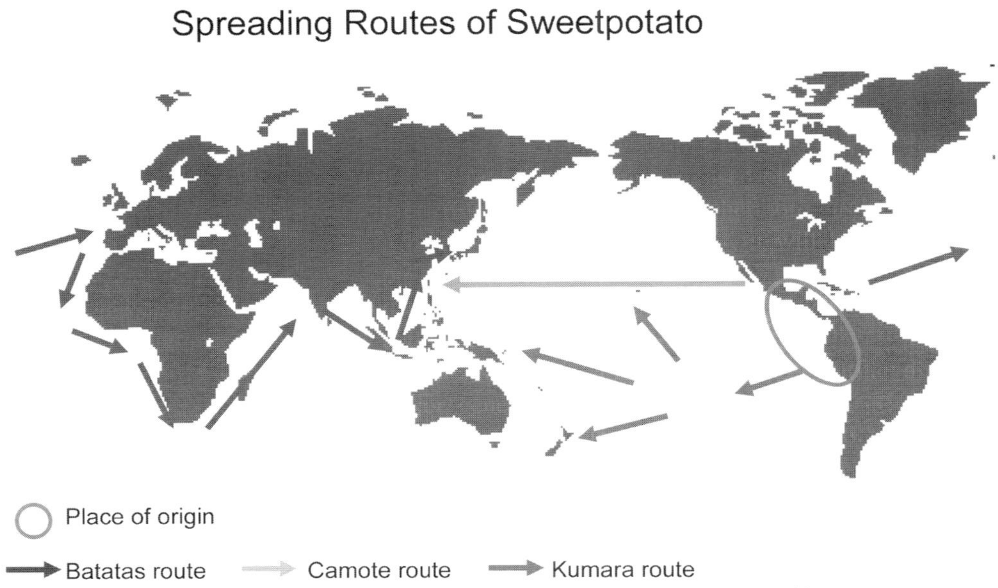

Spreading Routes of Sweetpotato

○ Place of origin

→ Batatas route   → Camote route   → Kumara route

Fig. 11.1. Routes of introduction of sweetpotato from the New World.

introduction of sweetpotato from the American continent to Polynesia was brought about by human trading activity, then why did the traders on rafts not bring other crops of American origin such as maize or cassava as well? It is therefore still obscure whether human activities like trade, or natural phenomenon like true seed transfer by migratory birds (Yen, 1971), disseminated sweetpotato from the American continent to Polynesia.

Bretschneider (cited by Burkill, 1935) claimed that there was mention of sweetpotato in Chinese literature from the third and fourth centuries, but this has not been accepted by the scientific community. More recently, Gavin Menzes (2002) put forth compelling evidence that sweetpotato (among other exotic crops and animals) had been introduced by the Chinese traveling around the world, including to South America, in gigantic junks. These far-reaching expeditions predate Columbus and his "discovery" of the New World.

Another hypothesis considers New Guinea to be a secondary center of diversity for sweetpotato, with evidence that the species might have reached the highlands there 1,200 years ago. More than 5,000 varieties have been discovered in these isolated ecological conditions, and New Guinea is the only place where sweetpotato can be found growing at altitudes up to 2,800 m. Indeed, recent work using random-amplified polymorphic DNA (RAPD) markers at the International Potato Center (CIP) shows varieties from Papua New Guinea to be substantially different from germplasm in South America (Zhang et al., undated).

Taxonomy and species relationship

Sweetpotato (*Ipomoea batatas* [L.] Lam.) belongs to the Family Convolvulaceae, or the morning glory family. Species from this family have the following distinguishing traits: latex is present in the plant sap; leaves are simple and arranged alternately around the stem; the flowers are complete, have a superior pistil, five stamens and a distinctive trumpet-shaped corolla; and the fruit is a capsule (Lawrence, 1951). Sweetpotato is in the same genus as the leafy vegetable, water convolvulus or water spinach (*I. reptans* = *I. aquatica*), and the various ornamental forms of *Ipomoea* (e.g., *I. purpurea*). Several wild species of *Ipomoea* are important for two reasons: (1) They may be used in grafts with sweetpotato to induce flowering for breeding purposes or to detect virus infection (e.g., *I. nil*, *I. tricolour*, *I. setosa*); and (2) they can harbor certain pests and diseases of sweetpotato, thus acting as alternate hosts.

Austin (1983) listed 11 species that might have contributed to the origin of sweetpotato (Table 11.2). Some of these wild species can also cross with sweetpotato, e.g., *I. trifida*, while *I. tiliacea*, *I. gracilis*, *I. triloba*, *I. lacunosa*, and *I. cordato-triloba* (*I. trichocarpa*) are cross-incompatible with sweetpotato (Teramura, 1979). Most of the hybrids between sweetpotato and diploid accessions of *I. trifida* could produce storage roots (Komaki and Katayama, 1999).

The species name *I. batatas* most probably derives from the Portuguese word for potato, which is *batata*. Interestingly, it might also be possibly derived from the Malay word "*batas*" that means "bed" or "ridge," which is the manner in which the land is prepared for sweetpotato planting. In the U.S., the deep orange, moist-fleshed sweetpotato varieties are often referred to as "yam"—which is a misnomer—not to be confused with *Dioscorea* yams or cocoyam (*Colocasia esculenta*). In New Zealand and in Polynesia, the name used for sweetpotato is kumara.

Table 11.2. Species and hybrids of section Batatas, Genus *Ipomoea.*

Species	Probable chromosome number (2*n*)	Geographical range	Important characteristics
I. trichocarpa Elliott	30	North and South America	Purple; annual or short-lived perennial
I. lacunosa Linn.	30	Eastern U.S.	Small white flowers; annual
I. x leucantha Jacquin	30	Widespread in the tropics and Central America	Flower color variable; principally annual
I. tenuissima Choisy	—	Caribbean	
I. ramosissima (Poir.) Choisy	—	Central and South America	White or lavender flowers; perennial
I. trifida (H.B. & K.) G. Don	30	Central America, Mexico	Lavender flowers; perennial
I. tiliaceae (Willd.) Choisey	60	Caribbean	Lavender flowers; perennial
I. cyanchifolia Meisner	—	Brazil	Lavender flowers; probably perennial
I. x grandiflora (Dammer) O'Donell	—	Southeast South America	Lavender flowers; probably annual
I. peruviana O'Donell	—	Peru	Possibly *I. batatas*
I. gracilis R. Brown	60	Australia	Lavender flowers; perennial
I. littoralis Blume	60	Pacific Basin to Mexico	Lavender flowers; perennial
I. batatas (L.) Lam.	90	American tropics, now widespread	Lavender to white flowers; perennial

Source: Adapted from Austin (1983) by Martin and Jones (1986).

Mating system

The flowers of sweetpotato are borne on erect axillary peduncles. Plants are monoecious and should therefore be largely self-pollinated. However, although the flower bears both male and female reproductive organs, the existence of self-incompatibility precludes self-pollination. Around 90% of sweetpotato varieties are self-incompatible (Fujise, 1964). Among the self-compatible varieties, various degrees of capsule set have been reported, ranging from 1.35–40.80% (Edmond and Ammerman, 1971). So far, 16 cross-incompatible groups classified by observations based on pollen germination on the stigma have been reported among sweetpotato varieties in Japan (Nakanishi and Kobayashi, 1979). Indeed, there are also cross-incompatible systems among groups of genotypes, which somewhat complicates the breeding of sweetpotato.

General biology

Sweetpotato is a twining or trailing plant, with stems that are herbaceous vines. Most domesticated varieties have lost their ability to twine; thus, the vines run over the beds, ridges, or hills on which the plants are established, providing an effective cover against competing weeds. Stems have primary and secondary lateral branches, with either long or short internodes. The different lengths of the main vine, and the primary and secondary laterals give rise to sprawling or erect/bushy plant types. Stem color ranges from green to various degrees of purple to deep purple.

The new shoots may be completely green or purple, or have varying shades of purplish green. Leaves, although simple, have different shapes ranging from cordate (heart-shaped), hastate, to slightly indented, and from shallow to deeply divided blades. Leaf color is typically green, although there are varieties with yellow and tricolor (white, green, purple) leaves which are used as ornamental border plants.

The fruit is typically a small, dark brown capsule bearing a maximum of four seed. The fruit dehisces explosively when ripe. Each seed is small (3 mm) and round with a flattened side. The testa is extremely tough and almost impervious to water and/or oxygen (Edmond and Ammerman, 1971), and seed scarification (e.g., by soaking for 10–20 min in concentrated sulfuric acid) is necessary to promote germination (Martin, 1946).

The storage roots of sweetpotato form from the adventitious roots that arise from the base of the cuttings used in planting. These swollen, succulent, and edible roots range in size and shape (cylindrical, fusiform, spindle, round, and irregular), sometimes even within the same plant and variety. The root skin may be white, yellow, orange, brown, pink, red, or purple in color, while the root flesh may be entirely white, yellow, orange, purple, or have specks of orange and/or purple.

Floral biology and flowering

Flowers are trumpet-shaped and are bisexual. Each flower has five sepals, five petals, five stamens, and a compound pistil. The entire corolla is usually lilac or white in color with a purplish or pink throat. The superior pistil comprises a relatively short style with a relatively broad stigma with elongated papillae. The five stamens are attached to the corolla but are separate from each other. Depending on the variety, they vary in height in relation to the height of the stigma.

The flowers are adapted to self- or cross-pollination by insects, especially bees. They open in the morning and close by afternoon.

Sweetpotato is sensitive to day length. While some varieties are free flowering, some will only flower under conditions of short day length. This may pose problems for a breeding program in the absence of grafting to induce flowering. Other environmental factors that favor flowering seem to be water stress and low nitrogen availability (Jones, 1980).

Genetics and inheritance of important characters

Sweetpotato is hexaploid ($2n = 90$), while the basic chromosome number for *Ipomoea* appears to be $x = 15$. Indeed, *Ipomoea* species vary from being diploid to hexaploid. Kehr et al. (1953) postulated that sweetpotato resulted from a natural cross between a tetraploid and a diploid, followed by chromosome doubling leading to what would usually be a sterile hybrid. Fortunately, in sweetpotato this doubling has given rise to a relatively fertile hybrid (Fig. 11.2).

Nishiyama et al. (1975) have suggested that Species B was a prototype of the *I. trifida* complex (e.g., *I. leucantha*), whereas Species A was a tetraploid from the same source (e.g., *I. littoralis*). They further postulated that the fertile hexaploid was *I. trifida*, and that sweetpotato was derived from this hexaploid through natural selection. This hypothesis is supported by cytological and genetic analysis (Shiotani, 1989; Kumagai et al., 1990). Indeed, crosses of *I. leucantha* with *I. littoralis* have successfully produced hexaploids—the two species being from the same *I. trifida* complex (Kobayashi, 1984). Austin (1983) postulated that sweetpotato might have resulted from autopolyploidy.

Jones (1965), from evidence on meiotic activity in 40 clones, suggested that the

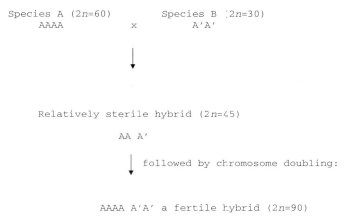

Fig. 11.2. Postulated origin of hexaploid sweetpotato.

ancestors of sweetpotato were not closely related; therefore *I. batatas* may be an allopolyploid.

Nevertheless, none of the primitive *Ipomoea* species forms enlarged storage roots, non-climbing vines, or roots with red skins or orange flesh, unlike in sweetpotato. Thus, the genetic origin of *I. batatas* is still debatable.

Self and cross-compatibility

Self-compatibility, self-incompatibility, cross-compatibility, and cross-incompatibility exist in sweetpotato. Self-incompatibility being the general rule, the selfing programs of sweetpotato have been seriously hindered. The degree of self-incompatibility in a particular variety can be assessed from the percentage of capsule set from self-pollinations, typically numbering 50–250. Self-incompatibility in *Ipomoea* is of the sporophytic multiple-allelic system. Incompatibility is a result of the interaction of two alleles at one locus; thus, all pollen grains show the same incompatibility phenotype (Martin and Jones, 1986). In a diploid *Ipomoea* species, there may be more than two incompatibility groups; there may be unilateral incompatibility between some groups, incompatibility between some parents and their progeny, or

two or four incompatibility groups in a family. Physiologically, incompatibility is the result of inhibition of pollen germination. Such knowledge from the diploid also helps explain incompatibility in sweetpotato, assuming that the incompatibility locus has been duplicated or triplicated in the hexaploid (Martin, 1968).

Nevertheless, genes for self-compatibility do exist in certain varieties (>40.8% capsule set) and can be incorporated into other varieties (Edmond and Ammerman, 1971).

Of course, cross-incompatibility limits possible cross-pollinations among genotypes with desirable traits that could otherwise be used as parents. The degree varies with the variety, and can be checked by hand crosses. Cross combinations can be entirely cross-incompatible (both in the cross and its reciprocal), or successful in one direction, or can be somewhat successful in either direction (cross-compatible). As in the case of self-compatible genes, genes for cross-compatibility also exist (Hernández and Miller, 1962) for incorporation into superior varieties for use in breeding.

An example of how to place genotypes into incompatibility groups is given in Table 11.3 (adapted from Hernández and Miller, 1964). Genotypes in Group VI are regarded

Table 11.3. Example of cross-incompatible and self-incompatible groups in sweetpotato [after Hernandez and Miller (1964)].

Group	♀	1	2	3	4	5	6	7	8	9	10	11	12	13	14	15	16	17	18	19	20	21	22	23	24	25	26	27	28	29	30
I	1	−																													
II	2	+	−	−	−	−	+	−	−	−	−	−	+	−	−	−	−	/	+	+	+	+	+	+	+				+	+	/
	3	+	/	−	+	−	−	−	−	−	−	−	+	−	−	−	−	+	+	+	+	+	+	+	+				+	+	−
	4	−	/	−	+	−	−	−	−	−	−	−							+	+	+	+		+	+				+		
	5	+	−	−	/	−	−	−	−	−	−	−	/	−	−	−	+	+	+	+			+	+	+				+		
	6	+	−	/	/	−	+	−	−	−	−	+	−	−	−	−	+	+	+	+	+	+	+	+	+				+	+	+
	7	+	+	/	/	−	+	−	−	−	−	+	/	−	−	−		+	+	+			+	/					+	+	−
	8	+	−	−	−	−	+	−	−	−	−	−	−	−	−	−	−	+	+	+	+	+	+	+					+	+	−
	9	+	−	−	−	−	−	−	−	−	/	−							+	+	+								+		
	10	+	−			−	−	−	−	−	−	−						−	+	+											
	11	+	−			−	+	−	−	−	+	−				−	+	+	+	+				+					+	+	−
	12																														
	13	+	−	−	−	−	−	−	−	+	−	−	−	−	−	−	−	−	−	+			+	−	+				+	+	−
	14	+	−	−	−	−	+	+	−	+	−	−						+	+	+	+	+	+	+	+				+	+	+
	15	+	−	−	−	−	+	−	−	+	−	−	−			−	+	+	+	+	+	+	+	+	+				+	+	−
	16	+	−	−	−	−	+	/	−	+	−	+	−	−	+	+	+	/	+	−	+	+	+	+	+	+	+		+	+	+
III	17	+	+	+	+	+	+	+	+	+	/	+	/	+	+	+	+	+	+	−	+	+	+	/	+	+	+	+	+	+	/
	18	+	+	+	+	+	+	+	+	+	+	+	−	+	+	+	+	+	+	−	+	+	+	+	+	+	+		+	+	+
	19	+	+	+	+	+	+	+	+	+	+	+						−	−	−	+	+	+	+	+	+	+		+	+	+
IV	20	+	+	+	+	+	+	+	+	+	+	+	+	+	−	−	+	+	+	+	−	−	−	+	+	+	+	+	+		+
	21	+										/								+	+			+			+				
	22	+	+	+	+	+	+	+	+	+		−							+	+		−	−	+			+				
V	23	+	+	+	+	+	+	+	/	+	/	−		+	+	+	+	+	+	+	+	+	+	+	−	−	−	−	+	+	+
	24	+	+	+	+	+	+	+	+	+	+								+	+		+		−	−	−			+	+	+
	25		+					+							+			+	+				+	−							
VI	26		+																				+	+	+	−	+		+	+	+
	27																												+	+	+
	28	+	+	+	+	+	+	+	+	+	+	+		+	−	−	−	/	+	+	+	+	+	+	+	+	+		+	+	+
	29	+	+			−	+	+	+	+	+	+	+	−	−	−	−							+					+	+	+
	30	+	−	/	+	+	+	+	+	+	+	+		−	−	−	−	/	+	+		+	+	+	+	+			+	+	+

+ = cross-compatible; / = low compatibility; − = incompatible.

340

as self-compatible; all of the parents except Genotype 12 were found to be fertile in crosses between compatible parents. Note that Genotype 12 is sterile as the female parent, but is occasionally fertile as the male parent. Genotype 1 was placed in Group 1 because it was cross-compatible as a male parent with genotypes in five other groups, but was self-incompatible. There are 15 genotypes in Group II that are generally self-incompatible and cross-incompatible within the group, but cross-compatible with most of the genotypes in other groups. In Group III, the three genotypes were cross-compatible in crosses and their reciprocals, and there was also cross-compatibility with some genotypes in the other groups. In Group IV, the genotypes were self-incompatible but generally cross-compatible with some genotypes in the other groups. Genotypes in Group V were self- and generally cross-incompatible, but were cross-compatible or of low cross-compatibility with some genotypes in most of the other groups.

Genes in an allelomorphic series S_1, S_2, S_3, S_4, S_5, and S_6 represent Groups I, II, III, IV, V, and VI, respectively, with only one allele per group. Genotypes in Group VI have the fertility gene, *Sf*. However, there are probably more than six incompatibility groups and more complicated cross-incompatible systems because 16 cross-incompatible groups have been observed in Japan (Nakanishi and Kobayashi, 1979).

Another method of checking for self-incompatibility and cross-incompatibility is by means of pollen germination studies under the microscope. This method is less laborious as it involves a smaller number of hand crosses. Two to four hours after pollination, by staining with lactol phenol cotton blue (Wang, 1982) or aniline blue, the stigma may be checked for pollen germination and for pollen tube penetration into the style (Martin, 1959).

The failure of pollen germination post-pollination should not be confused with cross-incompatibility. Martin and Cabanillas (1966) have shown failure in pollen tube growth and embryo development at various times after pollination.

Genetics and heritability

Due to its hexaploid nature, sweetpotato is not a highly suitable candidate for studies on Mendelian inheritance. Each gene may be represented by six alleles, giving rise to very complex segregation ratios. However, where a single dominant allele is present, inheritance may be simple. Thus, simple genetic ratios have indeed been reported.

Poole's (1955) genetic studies showed the inheritance patterns of a number of morphological traits. Simply inherited characters controlled by single genes (following the Mendelian $3:1$ ratio) include the following:

- flowering (F) vs. non-flowering (f)
- red vines (G) vs. green vines (g)

Ridged (N) and smooth (n) root surface, orange (O) vs. cream (o) flesh, and entire (A) vs. serrated (a) leaf margins all follow the $13:3$ ratio. Rooted (R) and non-rooted (r) plants, as well as brown (B) vs. cream (b) skin, give the $9:7$ ratio.

More complex inheritance was shown for characters such as root skin and flesh colors and degree of leaf lobing. Skin and flesh colors appear not to be independent or controlled by single genes. Work by Hernández et al. (1965) showed that white flesh was incompletely dominant over orange. Total carotenoid pigments seem to be controlled by several additive genes (possibly six). Interestingly, reciprocal grafts showed that the ability to synthesize carotenoids and anthocyanins in the roots was governed by genetic factors found in the storage roots themselves (Kehr et al., 1955). A negative association between flesh color and dry matter content has been reported (Jones et al., 1969; Jones, 1977), implying that concurrent selection for high dry matter may be difficult but not impossible.

Kumagai et al. (1990) analyzed ß-amylase activity in the storage root, and found that this activity was controlled by two genes, and was inherited in a hexasomic or tetradisomic manner.

Mainly additive genes control starch content (Sakai, 1964), whereas fiber content (Hammett et al., 1966) seems to involve two sets of genes—one set determining the presence of fiber, while the other set (showing simple dominance) controls the fiber size. Several genes linked to those for fiber size control total fiber content.

The majority of important agronomic characters in sweetpotato are inherited quantitatively. Heritability estimates of some of these traits are given in Table 11.4. These estimates indicate that good genetic advance in the respective traits may be expected following mass selection. Martin and Jones (1986) reported that correlations among the traits indicated that they were independent enough to facilitate the combination of several favorable traits within a single genotype.

Propagation

Botanical or true seed of sweetpotato are, by and large, reserved for breeding work and not used for the planting of the crop. The preferred mode of propagation is by vegetative means, using vine cuttings. While any part of the vine may be used, the best cuttings are from the shoot (apical) portions of the plant as they result in rapid establishment and higher subsequent yields (Tan,

Table 11.4. Broadsense (H) and narrowsense (h^2) heritability estimates for important agronomic traits in sweetpotato.

Trait	Type of heritability	Heritability estimate	Estimation procedure[1]	Reference
Root weight	H	0.71	Variance	Jones et al. (1969)
	h^2	0.57	Variance-Covariance	Li (1975)
	h^2	0.44	(generations 4 and 5)	Li (1975)
	h^2	0.41	Variance-Covariance	Jones et al. (1969)
	h^2	0.44	(generations 6 and 7)	Jones et al. (1978)
	h^2	0.25	Variance-Covariance Regression Regression (after two selection cycles)	Jones (1977)
Number of roots	h^2	0.24	Variance-Covariance (parental partition)	Saladaga and Hernandez (1981)
	H	0.73		
	H	0.40	Variance	Jones et al. (1969)
	H	0.83	Variance	Jong (1974b)
	h^2	0.32	Variance (log transformation of original estimate of 22)	Jong (1974a)
	h^2	0.43	Variance-Covariance Variance-Covariance	Jones et al. (1969) Li (1975)
Dry matter (%)	h^2	0.48	Variance-Covariance	Li (1983)
	h^2	0.65	Regression	Jones (1977)
Crude starch (%)	h^2	0.57	Variance-Covariance	Li (1982)
Fibre	h^2	0.47	Regression	Jones et al. (1978)
Crude protein	h^2	0.57	Variance-Covariance	Li (1982)
Flesh oxidation	H	0.95	Variance	Jones et al. (1969)
	h^2	0.64	Variance-Covariance	Jones et al. (1969)
Root skin color	H	0.97	Variance	Jones et al. (1969)
	h^2	0.81	Variance-Covariance	Jones et al. (1969)

Table 11.4. *Continued*

Trait	Type of heritability	Heritability estimate	Estimation procedure[1]	Reference
Sprouting	h²	0.39	Regression	Jones (1977)
	h²	0.37	Regression	Jones et al. (1978)
	h²	0.06	Variance-Covariance (parental partition)	Saladaga and Hernandez (1981)
Growth cracks	H	0.76	Variance	Jones et al. (1969)
	h²	0.51	Variance-Covariance	Jones et al. (1969)
	h²	0.37	Regression	Jones et al. (1969)
Flesh color	H	0.97	Variance	Jones et al. (1969)
	h²	0.66	Variance-Covariance	Jones et al. (1964)
	h²	0.53	Regression	Jones (1977)
Fusarium wilt	H	0.96	Variance	Collins (1977)
reaction—	h²	0.89	Variance-Covariance (full-sib)	Collins (1977)
(*Fusarium*	h²	0.71		Collins (1977)
oxysporum f.	h²	0.50	Variance-Covariance (diallel)	Collins (1977)
batatas)	h²	0.86	Regression Variance-Covariance	Jones (1969)
Root-knot resistance—egg index	h²	0.75	Regression	Jones and Dukes (1980)
Meloidogyne incognita	h²	0.57	Regression	Jones and Dukes (1980)
M. javanica	h²	0.69	Regression	Jones and Dukes (1980)
Soil insect injury (%)				
WDS complex	h²	0.45	Regression	Jones et al. (1977)
Flea beetle	h²	0.40	Regression	Jones et al. (1977)
African weevil (*Cylas puncticollis*)—				
Root injury	H	0.84	Variance	Hahn and Leuschner (1981)
Shoot injury	H	0.79	Variance	Hahn and Leuschner (1981)
Sweetpotato virus resistance	H	0.82	Variance	Hahn et al. (1981)
	H	0.93	Variance	Hahn et al. (1981)
	H	0.48	Variance (selected population)	Hahn et al. (1981)
	H	0.95	Variance (selected population, improved technique)	Hahn et al. (1981)

[1] Unless noted, estimates of h² were from parents and offspring.
Source: Martin and Jones (1986).

1998). Healthy apical cuttings of 30 cm lengths, which are free from pests and disease, are selected and cut usually just before a crop is harvested. These are raised in a nursery to produce quality planting materials. Cultural practices similar to field production are used except that a fertilizer with a higher proportion of nitrogen will encourage more "above-ground" growth.

Apical cuttings of 30 cm length for planting are ready to be used 2 to $2\frac{1}{2}$ months later.

For a production field with staggered planting, cuttings may be taken every 10–14 days as new sprouts develop. The lower leaves of the cuttings are usually trimmed or pruned to reduce transpiration. The cuttings may then be stored in the shade for three days. Percentage of plant establishment after three days' storage may be improved further if the cuttings are sandwiched between wet gunny sacks and placed on wooden racks for evaporative cooling (Tan, 1998). To eradicate any weevils or stem-borers harboring in the vines, the cuttings are soaked in an insecticide solution for at least 10–15 minutes prior to planting.

In temperate areas, sweetpotato is grown in the summer because winter prevents the propagation of sweetpotato by vine cuttings throughout the year. There, sound storage roots are selected at the end of the growing season and cured at high temperature and high humidity. They are then stored at 12.8°C and high humidity in bunkers during the winter. The storage roots are sprouted in the spring in glasshouses or under plastic. Sprouts reach the desired length of about 20 cm in 5–6 weeks when they are used for planting.

Some attempts have been made to use whole roots or cut root pieces as planting material (Kumagai, 2005). Selection can be made for genotypes that do not have an enlarged misshapen root originating from the seed piece.

Botanical seed has a hard testa which enables the seed to be stored for up to 20 years under proper conditions (Jones and Duke, 1982).

Breeding Goals

Yield maximization

High yield is the ultimate breeding goal where sweetpotato is used as a food staple or is an important component of the daily diet. Breeding for yield improvement is fairly straightforward, but high yield alone may not make much sense if it is not linked to end-use requirements.

For example, where sweetpotato is used as a food staple, there is less emphasis on sweetness—which indeed can be a deterrent—and selection is more toward bland-tasting types with a dry mouth-feel. In the U.S., where sweetpotato is used as an ingredient in pies (especially during Thanksgiving) and in baby food, sweet, orange-fleshed types with moist-mouth-feel are preferred. The varieties 'Centennial,' 'Jewel,' and 'Beauregard' are good examples of such selection. In any case, selection for high yield goes hand in hand, wherever possible, with traits of special interest.

Resistance/tolerance to stresses

Abiotic

Sweetpotato is usually tolerant to abiotic stresses because its relatively long growing period can give it enough time to recover from such stresses. However, water stresses, such as drought and flooding, sometimes cause serious reductions in sweetpotato production and quality. Varietal differences are known for tolerance to water stress that occurs during the growing season, but there have been few systematic breeding programs for enhancing tolerance to abiotic stresses.

Biotic

Biotic stresses include mainly pests (both insects and nematodes) and diseases (of fungal, bacterial, and viral origin). A listing of major pests and diseases of sweetpotato and their distribution is given in Table 11.5. Viruses, particularly sweetpotato virus disease (SPVD), cause the most devastating diseases worldwide. The SPVD is the major cause of cultivar decline and has been found to be the synergistic effect of sweetpotato feathery mottle virus (SPFMV) and sweetpotato chlorotic stunt virus (SPCSV) occurring together (Gutierrez et al., 1999). Unfortunately, breeding for resistance to viruses by conventional means has hitherto not been easy or very successful. However, with biotechnological tools at hand, this problem may prove to be surmountable.

Table 11.5.   Important pests and diseases of sweetpotato and their distribution.

Pest/pathogen	Common name	Distribution
Fungal diseases		
Ceratocystis fimbriata	Black rot	Northern and Central America, parts of South America, West, Central and Southern Africa, India, Southeast Asia, Japan and the Pacific
Cercospora ipomoeae, C. batatas	Cercospora leaf spot	Very common throughout the tropics, USA
Diplodia tubericola	Java black rot	USA
Diaporthe batatatis	*Diaporthe* dry rot and stem rot	USA
Elsinoe batatas	Scab	Brazil, Indonesia, Japan, Malaysia, Pacific Islands
Fusarium oxysporum f. sp. *batatas*	*Fusarium* wilt or stem rot	Mainly in temperate regions—USA, Japan, northern India
F. oxysporum	Surface rot	USA
F. solani	*Fusarium* root rot	USA
Macrophomina phaseoli	Charcoal rot	USA
Monilochaetes infuscans	Scurf	Australia, Brazil, China, Hawaii, Japan, Sierra Leone, USA, Zimbabwe
Phaeoisariopsis bataticola	Cercospora leaf spot	India, USA, Venezuela
Plenodomus destruens	Foot rot	Argentina, Democratic Republic of Congo, Hawaii, Tanzania, USA
Pseudocercospora timorensis	Cercospora leaf spot	Brunei, Fiji, Taiwan, Ghana, Guinea, Hong Kong, India, Malaysia Mauritius, Nepal, New Guinea, New Hebrides, St. Lucia, Sierra Leone, Solomon Islands, Sudan, Tanzania, Uganda
Rhizopus stolonifer, Rhizopus spp.	Soft rots, ring rots	USA
Sclerotium rolfsii	Southern blight, sclerotial blight, circular spot	USA
Bacterial diseases		
Erwinia chrysanthemi	Bacterial stem and root rot	USA
Streptomyces ipomoeae	Pox, soil rot	USA

Pest/pathogen	Genus	Common name	Distribution
Viral diseases			
Sweetpotato Caulimo like Virus	Caulimovirus	SPCaLV	Puerto Rico, New Zealand, Madeira, Solomon Islands, Australia, Papua New Guinea
Sweetpotato Chlorotic Fleck	Potyvirus?	SPCFV	Peru, Japan, Brazil, China, Cuba, Panama, Colombia, Bolivia, Indonesia, Philippines,
Sweetpotato Chlorotic Stunt	Crinivirus	SPCSV, SPSVV	Nigeria, Uganda, Kenya, Democratic Republic of Congo, Taiwan, China, Indonesia, Philippines, USA, Brazil, Argentina, Peru, Israel
Sweetpotato Chlorotic Stunt	Potyvirus	SPCSV???	Caribbean Region, Kenya, Puerto Rico, Zimbabwe
Sweetpotato Feathery Mottle (common strain), Virus A, Chlorotic Leaf Spot	Potyvirus	SPFMV	Worldwide
Sweetpotato Feathery Mottle (Internal cork strain)	Potyvirus	SPFMV	USA
Sweetpotato Feathery Mottle (Russet crack strain)	Potyvirus	SPFMV	USA

continues

Table 11.5. *Continued*

Pest/pathogen		Common name	Distribution
Sweetpotato Feathery Mottle (Vein clearing strain)	Potyvirus	SPFMV	Israel
Sweetpotato Leaf Curl	Badnavirus?	SPLCV	Taiwan, Japan, Egypt, Nigeria
Sweetpotato Leaf Curl	Geminivirus	SPLCV	USA
Sweetpotato Leaf Speckling	Luteovirus	SPLSV	Peru, Cuba
Sweetpotato Mild Mottle, Virus B, Virus T	Ipomovirus	SPMMV, SPV-T	Africa, Indonesia, Papua New Guinea, Philippines, India, Egypt, Peru, USA
Sweetpotato Mild Speckling	Potyvirus	SPMSV	Argentina, Peru, Indonesia, Philippines
Sweetpotato Mosaic		SPMV	Taiwan
Sweetpotato Ring Spot	Nepovirus?	SPRSV	Papua New Guinea
Sweetpotato Vein Mosaic	Potyvirus?	SPVMV	Argentina
Sweetpotato Virus II	Potyvirus	SPV-II	Taiwan
Sweetpotato Virus Disease (SPFMV + SPCSV)		SPVD	Africa, Peru, USA?
Sweetpotato Virus G	Potyvirus	SPVG	Uganda, Egypt, India, China
Sweetpotato Yellow Dwarf	Ipomovirus? Potyvirus?	SPYDV	Taiwan
Sweetpotato Latent	Potyvirus?	SwPLV	Uganda, Kenya, Indonesia, Egypt, China, India, Philippines, Peru, Taiwan
C-6	Potyvirus?		Uganda, Indonesia, Philippines, Peru
Chlorotic Dwarf		SPFMV + SPMSV + SPCSV	
Ilar-like			Guatemala
Ipomoea Crinkle Leaf Curl	Geminivirus?	ICLCV	USA
Reo-like			Asia
Insect pests	Order		
Agrius convolvuli	Lepidoptera	Sweetpotato horn worm, convolvulus hawkmoth	Most of Africa, Australia, Bangladesh, south China, southern Europe, India, Indonesia, Iran, Malaysia, Myanmar, New Zealand, Pacific Islands, Papua New Guinea
Alcidodes dentipes	Coleoptera	Striped sweetpotato weevil	Tropical Africa
A. erroneus	Coleoptera	Weevil	Kenya
A. fabricii	Coleoptera	Weevil	India
Aspidomorpha adhearens, A. australasiae, A. quadriradiata, A. social	Coleoptera	Tortoise beetles	Papua New Guinea
A. miliaris	Coleoptera	Tortoise beetle	India, Papua New Guinea
A. dissentanea	Coleoptera	Tortoise beetle	Angola, Cameroon, Guinea
Bedellia ipomoeae	Lepidoptera	Sweetpotato leaf miner	Fiji
B. somnulentella	Lepidoptera	Sweetpotato leaf miner	Australia, Papua New Guinea
Brachmia convolvuli	Lepidoptera	Leaf folder	Canary Islands, Mauritius, Comoro Islands, Southeast Asia
B. macroscopa	Lepidoptera	Leaf folder	India
Chaetoncnema confinis	Coleoptera	Sweetpotato flea beetle	USA
Conchyloctaenia punctata	Coleoptera	Large spotted tortoise beetle	South Africa
Conoderus falli, C. vespertinus, C. amphicollis	Coleoptera	Wireworms	USA
Cylas formicarius elegantulus	Coleoptera	Sweetpotato weevil	Widely distributed in the tropics
C. puncticollis	Coleoptera	Sweetpotato weevil	Sub-Saharan Africa
C. brunneus	Coleoptera	Sweetpotato weevil	West Africa
Diabrotica balteata	Coleoptera	Banded cucumber beetle	USA
D. undecimpunctata	Coleoptera	Spotted cucumber beetle	USA

Table 11.5. *Continued*

Pest/pathogen		Common name	Distribution
Diacrisia obliqua	Lepidoptera	Bihar hairy caterpillar	South and East Asia
Euscepes postfasciatus	Coleoptera	Scarabee weevil, West Indian sweetpotato weevil	Central America, Bermuda, Brazil, Cook Islands, Fiji, Futuna Island, French Polynesia, Guam, Guyana, New Caledonia, Ryuku Island, Tonga, southern USA, West Indies
Euzophera semifuneralis	Lepidoptera	American plum borer	USA
Megastes grandalis	Lepidoptera	Sweetpotato pyralid moth, stem borer	Brazil, Trinidad
Melanotus communis	Coleoptera	Wireworm	USA
Metriona spp.	Coleoptera	Tortoise beetles	USA
Noxtoxus calcaratus	Coleoptera	Flower beetle	USA
Omphisa anastomosalis	Lepidoptera	Sweetpotato stem or vine borer	Widespread in Southeast Asia, China, India, Japan and Sri Lanka
Peridroma saucia	Lepidoptera	Variegated cutworm	USA
Phyllophaga ephilida	Coleoptera	White grubs	USA
Plectris aliena			
Scolytid	Coleoptera	Ambrosia beetle	USA
Synanthedon dasysceles, S. leptosceles	Lepidoptera	Clearwing moths	Kenya
Systena blanda	Coleoptera	Pale-striped flea beetle	USA
S. elongata	Coleoptera	Elongate flea beetle	USA
S. frontalis	Coleoptera	American flea beetle	USA
Typophorus nitritus viridicyaneus	Coleoptera	Sweetpotato leaf beetle	USA
Nematodes			
Meloidogyne hapla		Northern root-knot nematode	USA
M. incognita, Meloidogyne spp.		Root-knot nematodes	Widespread where sweetpotato grows—Brazil, Côte d'Ivoire, French West Indies, Ghana, Japan, Nigeria, Puerto Rico, Peru, South Africa, Trinidad, USA
M. javanica		Javanese (tropical) root-knot nematode	USA
Rotylenchulus reniformis		Reniform nematode	Reported in Cuba, French West Indies, Ghana, Jamaica, Nigeria, Puerto Rico, Senegal, Trinidad, USA; more widespread

Sources: Anon. (1978).
   Jones et al. (1985).
   Salazar and Fuentes (2000).

Likewise, breeding for resistance to the sweetpotato weevil, *Cylas formicarius elegantulus*, probably the most serious insect pest of sweetpotato in the tropics, has not met with much success. Screening of more than 1,200 sweetpotato accessions failed to identify any usable levels of resistance to the weevil (Talakar, 1989), and an integrated management approach has been advocated.

In the U.S., Fusarium wilt is a major fungal disease. There has been good success in breeding for resistance to this disease after a highly resistant clone, Tinian, was found (Steinbauer, 1948), which provided the source of resistance genes.

There has been more breeding success in the development of resistance in sweetpotato varieties against the root-knot nematode, *Meloidogyne incognita*. Many new varieties developed in the U.S., such as 'Jewel,' 'Pope,' 'Resisto,' 'Regal,' 'Sumor,' and 'HiDry,' have resistance (Martin and Jones, 1986).

Starch production

Sweetpotato is made up of 65–70% water and the rest is carbohydrate, mainly starch. Starch has myriad uses—in both food and non-food applications. Sweetpotato starch does not have specific characteristics distinguishing it from other starches. In Japan, most of the starch is used for the production of liquid sugar, whereas in Korea it is used mainly in making glass noodles.

In Japan, the current sweetpotato varieties for starch production, 'Koganesengan' (Sakai et al., 1967), 'Shirosatsuma' (Sakamoto et al., 1989) and 'Shiroyutaka' (Sakamoto et al., 1987), have starch content around 25% on a fresh weight basis. It should not be a problem to raise starch content to 30% in the future; indeed, recently released varieties, such as 'Satsuma-starch' (Tarumoto et al., 1996), 'Konahomare' (Kumagai et al., 2002), and 'Daichi-no-yume' have starch contents ranging from 28–30%.

To expand sweetpotato starch consumption, there have been attempts to change the amylose content and other properties. Some breeding lines with amylose content which is 50% of the normal level have been identified (Ishiguro and Yamakawa, 2000). Also, large varietal differences in pasting temperature (Katayama et al., 2002; Katayama et al., 2004) and retrogradation of starch (Ishiguro et al., 2001) have been observed. As a result, a new cultivar 'Quick Sweet' with low pasting temperature has been released (Katayama et al., 2003), and collaboration with a starch factory is underway to develop new products using this unique starch property of 'Quick Sweet.' Also, modification of sweet-

potato starch by enzymatic or chemical means will change the physico-chemical properties of the starch, which can pave the way for wider applications.

The production of industrial alcohol (as biofuel) is another possible use of sweetpotato, particularly in light of current rising trends in petroleum prices. Selection for high dry-matter content, together with high root yield, should ensure a high alcohol yield. One such variety, 'HiDry,' with a dry-matter content of 37–40%, has been developed in the U.S. One hundred kilograms of starch typically yields 51 kg ethanol, or 13 liters of ethanol may be expected from 100 kg of sweetpotato.

Table use

Desirable characters for table use are dependent on consumer preference. In the U.S., the sweet, orange-fleshed types with moist mouth-feel are preferred. In Japan, the sweet, yellowish white-fleshed and red-skinned types with dry mouth-feel are favored, in addition to uniform cylindrical or fusiform storage root shape, especially for the fresh market. Where sweetpotato is used as a staple, there is less emphasis on sweetness and selection is aimed at more bland-tasting types with a dry mouth-feel. Popular varieties for table use are 'Centennial,' 'Jewel,' and 'Beauregard' in the U.S., and 'Beniazuma' (Shiga et al., 1985) and 'Kokei 14' in Japan.

Agro-industrial use

Despite its rather small production volume (see Table 11.1), Japan leads the world in developing new industrial uses from sweetpotato. Chief among these is the production of pigments. Sweetpotato produces three main kinds of color pigments in the storage roots: red and purple from anthocyanin, and orange from carotenoids. The natural pigment market is set to become larger, as consumers shift their preference from artificial pigments to natural ones as food ingredients. Hitherto, most natural red pigments had been extracted from insects (lac, cochineal), microorganisms (red rice yeast), plants such as red cabbage and red beet, and agro-wastes such as grape skins (from the red wine industry).

Anthocyanin extracted from a native Japanese sweetpotato cultivar, 'Yamagawamurasaki,' was found to be as good as that of red cabbage for color quality and stability; however, the root yield of 'Yamagawamurasaki' is too low to justify its use as a raw material in pigment production. 'Ayamurasaki,' with extremely high anthocyanin content and a higher root yield, was released in 1995 (Yamakawa et al., 1997). Cell lines from the purple callus taken from 'Ayamurasaki' storage root explants have been selected to produce plenty of anthocyanin without light or hormone (Konczak-Islam et al., 2000). This can pave the way for production of purple/red pigments from sweetpotato in bioreactors.

Flour of different colors can also be processed from highly colored sweetpotato. For example, 'Ayamurasaki' has not only high anthocyanin content but also about 35% dry matter, which is an acceptable level for processing into purple flour. Varietal improvement for orange flour production is less straightforward because the dry matter content of sweetpotato varieties containing high amounts of carotenoids tends to be very low, usually less than 30%. Low dry matter content results in a poor conversion rate to flour. As a compromise, selection for high carotenoid content (above 10 mg/100 g FW) must be done in tandem with a reasonable level of dry matter content, e.g., above 30% (Komaki and Yamakawa, 2006). Mixes of colored sweetpotato flour with wheat flour are good materials for processing such products as bread, noodles, snacks, and cakes that will be similarly colored.

Juice has been developed from highly colored sweetpotato, too. Some orange-fleshed varieties contain more beta-carotene than carrot (Takahata et al., 1993). High-quality juice from colored sweetpotato must have a pleasant aroma and suffer no color degradation. Also, varieties with low starch content are preferred for juice. 'J-Red,' which fits the bill for low starch, high β-carotene content, low color degradation, and high yield, was released in 1997 in Japan for making juice (Yamakawa et al., 1998). This cultivar can also be used as a raw material in making snack food, or as a vegetable for cooking—in a manner similar to pumpkin and carrot because of its low starch content and relatively low sugar content.

Alcoholic beverages, such as beer and spirits (*shochu*), have also been developed in Japan. Yellow, red, and dark colored beverages like beer and sparkling liquor from sweetpotato are being sold in Kagoshima prefecture. Selection for suitably colored varieties, identification of the enzymes to decompose starch to sugars, as well as the yeast strain for fermentation are key technologies needed to ensure the success of developing these new beverages.

Nutritional quality

Roots

Sweetpotato is known to be a low protein crop, in which crude protein content ranges usually from 4–7% on a dry weight basis. The major storage protein is sporamin (ipomoein). Genetic variation is known to be large for crude protein content and individual amino acids content (Toyama et al., 2005). Thus, it is believed that sweetpotato protein quantity and quality may be genetically improved. Sweetpotato contains about 10 mg/100 g FW of dietary fiber, 30 mg/100 g FW of Vitamin C, and large amount of minerals (e.g., 400–500 mg/100 g FW of potassium) in its storage roots.

Tops

The tops of sweetpotato are usually fed to animals or are returned to the soil as green manure at harvest. Sweetpotato tops are rich in nutrients, such as protein, vitamins, and minerals. It would be advantageous to develop new technologies to use the tops more efficiently and economically as raw materials for food processing.

Sweetpotato tops, especially the leaf blades, have a large amount of polyphenols, mainly composed of chlorogenic acid and its relatives, which have strong physiological functions, e.g., suppressive effects to some food poisoning microorganisms (Yoshimoto et al., 2002).

The new cultivar 'Suioh' was released in 2001 in Japan specifically for exploiting its tops. 'Suioh' produces a total of 200 t/ha of tops on a fresh weight basis and 20 t/ha on a dry weight basis. The production system involves placing the seed storage roots in nursery beds and harvesting the tops once a month. After harvest by machine, the tops are washed, chopped up and dried, then milled. The dark green leaf blades are separated from the light green leaf petioles and stems by sieves. The dark green parts, which are rich in protein and polyphenols, can be used as a green ingredient for food, green drinks like tea, or medicine, if the medicinal effects are proven and approved (Ishiguro et al., 2004).

Breeding strategies

Maintenance and utilization of genetic resources

Genetic resources of sweetpotato are mostly maintained in the field and/or through in vitro conservation. In most tropical countries, field conservation is very popular, while in the U.S. and International Potato Center (CIP) most accessions are conserved under in vitro conditions. China has adopted both conservation methods. Field conserva-tion requires a lot of labor, and frequently encounters contamination and loss of genetic resources. It is necessary to characterize the accessions and to select specific germplasm for core collections to reduce the cost and labor for maintaining sweetpotato genetic resources. Then, when breeders attempt to improve a specific character in their breeding programs, they can choose the parents based on the characterized data.

The CIP-AVRDC-IBPGR descriptor list for sweetpotato (Huaman, 1991) has been used to characterize germplasm accessions at CIP and several other national research programs. It is a useful guide, relying largely on color and shape to describe the leaves, vines, roots, and flowers of an accession. These qualitative traits, including size (e.g., diameter and length), are all assigned numerical codes to characterize them into distinct categories. An example of a data sheet for morphological characterization is given in Table 11.6. The traits from the third column onward are: twining, plant type, vine diameter, vine length, predominant vine color, secondary vine color, vine tip pubescence, general leaf outline, type of leaf lobe, number of leaf lobes, shape of central leaf lobe, mature leaf size, abaxial leaf vein pigmentation, mature leaf color, immature leaf color, petiole pigmentation, petiole length, storage root shape, storage root defects, storage root cortex thickness, predominant skin color of storage root, intensity of predominant skin color, secondary skin color, predominant flesh color of storage root, secondary flesh color, distribution of secondary flesh color, and storage root arrangement.

The numerical coding for size, shape, and color for vine diameter, leaf outline, and root flesh color is as follows:

Vine internode diameter (taken as the mean of at least three internodes located in the middle section of the vine).
1   Very thin (<4 mm)
3   Thin (4–6 mm)
5   Intermediate (7–9 mm)

7   Thick (10–12 mm)

9   Very thick (>12 mm)

Leaf outline

1   Rounded

2   Reniform (kidney-shaped)

3   Cordate (heart-shaped)

4   Triangular

5   Hastate (trilobular, spear-shaped, with the basal lobes more or less divergent)

6   Lobed

7   Almost divided

Storage root flesh color (describes the predominant color, secondary color, and the distribution of the secondary color, from cross and longitudinal sections made at about the middle of freshly harvested storage roots).

Predominant flesh color

1   White

2   Cream

3   Dark cream

4   Pale yellow

5   Dark yellow

6   Pale orange

7   Intermediate orange

8   Dark orange

9   Strongly pigmented with anthocyanins (purple)

Secondary flesh color

0   Absent

1   White

2   Cream

3   Yellow

4   Orange

5   Pink

6   Red

7   Purple-red

8   Purple

9   Dark purple

Distribution of secondary flesh color

0   Absent

1   Narrow ring in cortex

2   Broad ring in cortex

3   Scattered spots

4   Narrow ring in flesh

5   Broad ring in flesh

6   Ring and other areas in flesh

7   In longitudinal sections

8   Covering most of the flesh

9   Covering all flesh

The CIP-AVRDC-IBPGR descriptor list for sweetpotato (Huaman, 1991) may be referred to for further details on morphological characterization.

Induction of flowering

As mentioned before, the breeding of sweetpotato can be hampered by self-incompatibility and cross-incompatibility systems. There is also the lack of, or poor, flowering in some genotypes. Various methods of flower induction have been used. One is to train the normally sprawling sweetpotato vine to climb up a trellis. The improvement in light penetration together with a regime of controlled watering and the addition of P fertilizer seemed to induce flowering and seed set (Miller, 1937). Another more commonly used method is to graft sweetpotato onto a free-flowering *Ipomoea* species, such as *I. nil*, or to a free-flowering sweetpotato genotype.

Controlled hand-crosses are carried out in the following manner (adapted from Edmond and Martin, 1946): Flower buds of the male parent are selected in the late afternoon and kept closed with a paper clip placed over the tip. Likewise, flowers meant for selfing may be protected from foreign pollen in this manner. At the same time, emasculation of the female parent is accomplished by selecting buds on the female parent that are ready to open the next day (easily recognizable by their size and color). These buds are slit open vertically to remove the corolla. This action removes the attached stamens as well. The exposed pistil is then enclosed with a short length of drinking straw, which is then folded over at the top to prevent entry of insects.

The clip is removed from the flower the next morning to collect (with forceps) the

Table 11.6.   Example of some important CIP-AVRDC-IBPGR descriptors used in the characterization of sweetpotato germplasm.

DATA SHEET FOR CHARACTERIZATION OF SWEETPOTATO GERMPLASM

				Vine characters						Mature leaf characters						
Plot	ID No.	Twining	Plant type	Diameter	Length	Pre-dom color	Secon color	Tip pubes	Out line	Lobe type	No. lobes	Centr lobe shape	Leaf size	Abaxl vein pigmt	Leaf color

stamens from the male parent. The straw covering the flower of the female parent is removed, and the anther is gently tapped to dust the dehisced pollen onto the stigma. The straw length is replaced and the flower tagged with details of the cross.

Two days later, the straw is removed so as not to impede the development of the fertilized ovary. The dark brown fruit are collected when they reach maturity, and they are dried/stored in paper envelopes so that on dehiscence the seed is not lost.

To overcome known incompatibility groups among the accessions in the working germplasm, an alternative to hand-crosses has been used quite successfully by sweetpotato breeders. This is the polycross system (Hernández et al., 1969). Breeding parents from as many incompatibility groups as possible are selected on the basis of field performance and genotype (as expressed by the phenotype). Desirable characters include yield, good root shape, flesh and skin colors, palatability, and disease and/or pest resistance. These parents are planted in an isolated area to allow random pollinations (usually by bees). To improve flowering, they may be trained on a wire trellis or some

Table 11.6. *Continued*

Date of data collection: _____

Yg Leaf color	Petiole		Root characters										
	Pigmt	Length	Root shape	Root defect	Cortx thickn	Pred skin color	Color intens	Secon skin color	Pred flesh color	Secon flesh color	Distr color	Root arrang

other support system. Resulting seed are identified only on the basis of the maternal parent.

Breeding systems

Simple selection

Where there is sufficient genetic variation, either through introductions and collection of germplasm, or generated by natural crossing, simple selection can be employed profitably and rapidly for improving sweetpotato. This is, of course, facilitated by the fact that the crop can be propagated vegetatively, and thus is able to "lock in" the favorable combination of genes.

Pedigree selection

Rather than leaving selection to chance, parental genotypes—each with contrasting but desired traits—may be selected for use in controlled hybridizations. This allows better opportunities for selecting desired recombinations among the resulting progeny. However, because only a maximum of four seed (usually fewer) are produced per cross, hand-crosses are very labor-intensive. By planting two parental accessions in isolation, insects (bees) can be relied on to carry out pollinations. The constraint is the limited number of parents used each time.

Mass selection and the polycross system

Several sweetpotato accessions with one or more desired characteristics are selected and planted together in a polycross block where natural pollinations (by bees) are allowed to occur. The only obstacles are: (1) the selected parents must flower more or less simultaneously (or at least induced to do synchronized flowering); and (2) the parental cultivars must be cross-compatible.

Seedlings from the polycross block are evaluated for the desired characteristics, and the best are used (with or without the best parents) to initiate a new polycross block. This results in rapid accumulation of favorable genes, and it may help in breaking unfavorable gene linkages. The polycross system has worked well for weakly expressed traits controlled by many genes, e.g., resistance to insect pests and diseases.

Mutation breeding

Generating variability through the use of chemical mutagens or irradiation has not been widely used in sweetpotato breeding. Nevertheless, it must be mentioned that sweetpotato has a peculiarity in that it produces "bud sports" (Harter, 1926), which are spontaneous mutations. Some of these mutations have been found to be useful and have given rise to new cultivars. The frequency and type of mutations depend on the cultivar, e.g., 'Centennial' was reported to produce 23 mutants (Love et al., 1977). Most conspicuous are mutations that change the skin or flesh color of the storage roots; some have changed growth habit.

Evaluation and selection system

Figure 11.3 represents an example of an evaluation and selection system practiced by the Malaysian Agricultural Research & Development Institute. Seeds resulting from controlled hand-crosses, from polycrosses, or from introductions from other sources are scarified (as described earlier) before planting in individual small polybags filled with seedling mix. When the vines of the germinated seedlings reach 30 cm in length, they are transplanted into the field, and multiplied. Five cuttings are taken from each seedling for single-row evaluation with a plant-to-plant spacing of 25 cm on ridges that are built 1 m apart. Single-rows of check varieties are included at random spots for comparison of performance parameters, such as marketable root number and yield, total root yield, and harvest index. Characteristics such as resistance to disease (e.g., scab), root shape, and absence of growth cracks are also important.

Selected lines are multiplied sufficiently to carry out two seasons of yield trials, i.e., plot size of 3 m × 4 m (4 rows of 12 plants per row) with four replications. Check varieties are included, and a randomized complete block design is adopted. Dry matter content (also, computed dry root yield), root flesh color and palatability of boiled roots may be included as selection criteria in the yield trials. Checks are used for eating quality (potential as table variety), yield, high dry matter content, high carotenoid content or high anthocyanin content, as the case may be.

After two seasons' testing at the main station, the shortlisted clones are multiplied for testing in multi-location trials (MLTs) sited on contrasting agro-ecologies or production areas, usually on-station. The same plot size and design as in the yield trials are used.

The final two or three selected clones are then tested in local verification trials (LVTs)

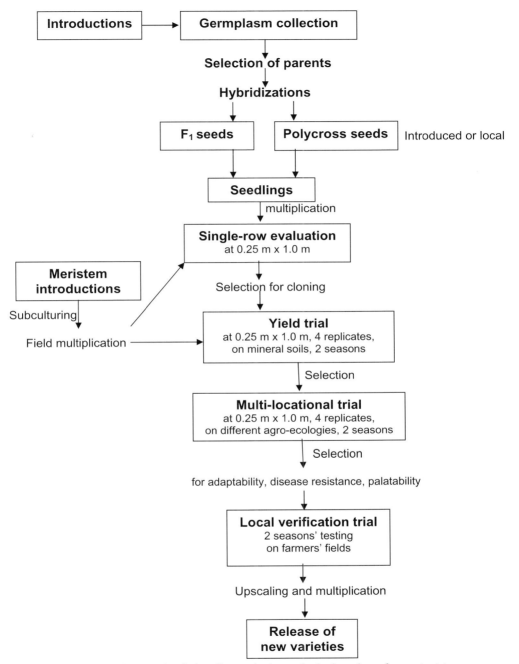

Fig. 11.3.   An example of a breeding, evaluation, and selection scheme for sweetpotato.

in the same agro-ecologies/production areas, but this time in farmers' fields. The best current variety in a particular area is included together with the official checks. Each set of materials is tested on at least three farms (i.e., three replications) per agro-ecology, across two seasons. Clones found to be superior to the checks in yield, quality

Table 11.7. Some outstanding commercial varieties.

Variety	Country	Significant traits
Ayamuraski	Japan	Anthocyanin content
Beauregard	USA, New Zealand	Table quality
Beniazuma	Japan	Table quality
Benihayato	Japan	Carotene content
Canó	Dominican Republic	Cosmetics, yield, table quality
Centennial	USA	Yield, table quality, *Fusarium* resistance
Jewel	USA	Yield, wide adaptation, *Fusarium* resistance
Koganesengan	Japan	Starch
Miguela	Puerto Rico	Table quality
Owairaka Red	New Zealand	Table quality
Picadito	Cuba	Table quality, wide adaptation
Shiroyutaka	Japan	Starch
Sumor	USA	Multiple resistances to diseases and insect pests
Suwon 147	Korea	Yield
Tainan 57	Taiwan	Yield, table quality
Tucumana Lisa	Argentina	Earliness, yield, wide adaptation

(carotenoid, anthocyanin, dry matter content or palatability), disease resistance, or other desirable trait/s are then multiplied for release as new varieties.

In the case of materials introduced in vitro from other sources, the subcultured plantlets are field-multiplied, and may then enter the evaluation scheme (see Fig. 11.3) either at the single-row stage or the yield trial stage, depending on the availability of planting materials.

Some outstanding commercial varieties

Table 11.7 contains some outstanding varieties that have performed well commercially. This list does not include the many landraces produced by farmers in developing countries, which are not marketed by name.

True seed propagated varieties—a step into the future

Attempts to cultivate sweetpotato via true seed have been made in Japan since the 1980s. These attempts indicate that true seed

cultivation could produce 20–30 t/ha of storage roots using methods to prolong the growing period (such as sowing true seed in small plastic pots in the green house, transplanting the seedlings into the field, and then covering the hills with plastic film) (Komaki et al., 1994). However, as plantlets from true seed are very susceptible to both abiotic and biotic stresses, true seed cultivation cannot as yet be regarded as an alternative cultivation technology that would reduce production costs.

Biotechnology tools for genetic improvement

Tissue culture

Sweetpotato virus disease (SPVD) is believed to be a synergistic relationship between sweetpotato chlorotic stunt virus (SPCSV) and sweetpotato feathery mottle virus (SPFMV). Recently, a synergistic effect between sweetpotato leaf curl virus (SPLCV) and SPFMV was also found to cause "little leaf" or "witches' broom" symptoms in

Malaysia (Noor Suhana et al., 2005). This disease complex has of late become a fairly serious problem in both research and production fields in the country. In the course of "cleaning up" planting material of important varieties and clones by means of meristem tissue cultures, both viruses have been detected by PCR and the genes for coat protein and replicase have been cloned (Noor Shuhana et al., 2005). This may provide the springboard from which resistance to SPVD can be incorporated into sweetpotato by genetic engineering.

Somaclonal variation

Various somaclonal variations are known to occur as a result of tissue culture. The most likely cause of these is transposable elements. Transposable elements have been found in sweetpotato (Tahara et al., 2004). They are "silent" (latent) under normal conditions; however, they become active and can move through the genome during the dedifferentiation that occurs in tissue culture, thus producing somaclonal variations. As mentioned earlier, somaclonal variation in the field has also been reported as "bud sports."

Molecular biology

Molecular genetics

Studies on molecular genetics in sweetpotato have made limited progress because of the hexaploidy of the species and its complicated genome system. So far, molecular markers, such as restriction fragment length polymorphism (RFLP), random amplified polymorphic DNA (RAPD), simple sequence repeats (SSR), amplified fragment length polymorphism (AFLP), and inter-simple sequence repeats (ISSR) markers, have been developed and are used for gene pool analysis and genome analysis. Taxonomic and phylogenetic studies have been carried out using RFLP (Jarret et al., 1992), RAPD

(Jarret and Austin, 1994; He et al., 1995; Komaki et al., 1998) and ISSR (Huang and Sun, 2000; Hu et al., 2003). Thompson et al. (1997), using RAPD markers, and Kriegner et al. (2001), using AFLP markers, have demonstrated the feasibility of constructing genetic linkage maps for sweetpotato. Recently, transposable elements have been shown to play a great role in enhancing sweetpotato variation (Tahara et al., 2004). These molecular techniques will no doubt also provide the definitive genetic fingerprinting of germplasm accessions hitherto not always possible with morphological characterization. Genetic modification by insertion of transgenes has been demonstrated (Otani et al., 1998), and analysis of transgene expression is currently being undertaken.

A collaborative project initiated in 1991 between the Kenyan Agricultural Research Institute (KARI) and Monsanto has succeeded in several areas: in obtaining a CP-gene construct for SPFMV resistance; in generating transgenic sweetpotato; and in conducting on-station greenhouse and field transgenic trials. Commercialization and regional testing will follow (Wambugu, 2001; Odame et al., 2004).

Marker-assisted selection

Marker-assisted selection is a strong tool to select for "invisible" characters as well as quantitative characters, such as storage root yield. RAPD markers related to soil rot disease have been developed in Japan, and their efficiency has been evaluated. However, more detailed analyses/investigations with marker-assisted selection are needed.

Future breeding and prospects for sweetpotato

Although ranking seventh in world production, sweetpotato has a limited market when it is sold fresh—largely due to its high perishability, in contrast to grains. Prospects for transforming it into a raw material for agro-based industries are bright. In this role, sweetpotato may be used as a feasible tropical substitute for temperate raw materials (such as the potato and wheat currently used in myriad food processing industries).

Being highly perishable, sweetpotato storage roots must first be processed into an intermediate storable form, such as flour or puree. It is possible to substitute from 50–100% sweetpotato flour for wheat flour in bakery products such as bread, buns, muffins, biscuits, and cakes. Similarly, when sweetpotato is used in the production of extruded products, it can substitute for wheat and maize in snack and breakfast foods.

Sweetpotato has also been found to be an excellent raw material for making french fries, a favorite among the youth. They are already on the market in New Zealand, Taiwan, and the U.S.

With the breeding emphasis shifting to sweetpotato varieties with highly nutritious characteristics, such as high contents of ß-carotene, anthocyanin, or other phytochemicals with anti-oxidative properties, wholesome food and beverage products may be developed from these new varieties. Some are being commercialized or are already on the market in countries such as Japan and New Zealand; products include pure and mixed juices, ice cream, jams, candies, noodles, nuggets, croquettes, soup mixes, as well as serving as ingredients in pies, salads, baby food, and ready-to-eat foods, such as TV dinners (Duell, 2006; Lewthwaite, 2006).

International and national research programs

Most sweetpotato research activity is dependent on national programs. In China, the Xuzhou Sweetpotato Research Center is focused on sweetpotato breeding, and provincial agricultural experiment stations also

have breeding programs—mostly dedicated to developing cultivars for table use or starch production. In Japan, the National Institute of Crop Science (NICS) and the National Agricultural Research Center for Kyushu and Okinawa Region (KONARC) are engaged in sweetpotato breeding. Cultivars are developed for table use, starch production, and suitability for use in food processing. Physiological functions of color pigments in the storage roots (anthocyanin, carotenoids and unidentified flavonoids), e.g., anti-oxidation, anti-cancer and protection against liver injury, are also investigated. Flour, juice, and alcoholic beverages have been developed using colored sweetpotato cultivars.

In Korea, the breeding programs are carried out at the Mokpo Experiment Station of the National Institute of Crop Science, Rural Development Administration. Most sweetpotatoes are consumed fresh, i.e., for table use, via the fresh market. Breeding objectives include the development of cultivars having small storage roots with red skin, yellowish and dry type flesh, and for roots rich in carotene or anthocyanin for use as health foods.

In India, mainly the Central Tuber Crops Research Institute (CTCRI) in Kerala carries out sweetpotato breeding. The Vietnam Agricultural Science Institute, the Philippines Root Crops Research and Training Center (PhilRootcrops), the Indonesian Legumes and Tuber Crops Research Institute (ILETRI), and the Malaysian Agricultural Research and Development Institute (MARDI) take responsibility for sweetpotato breeding in their respective countries. These breeding programs are supported by their governments, sometimes in collaboration with the International Potato Center (CIP) (especially for germplasm acquisition). Cultivars for both table and processing use have been successfully developed.

In the U.S., there are breeding programs in the major sweetpotato production states, e.g., North Carolina, Louisiana, and California. North Carolina State University, the Vegetable Laboratory of the Agricultural Research Service, USDA, and Louisiana State University lead the breeding network. Sweetpotato collaborators' meetings and sweetpotato growers' meetings are each held once a year, and newly recommended cultivars are registered following discussions at these meetings.

CIP in Peru is given the mandate of sweetpotato research under the CGIAR (Consultative Group on International Agricultural Research) system. In 1985, CIP took over this role previously held by the Asian Vegetable Research and Development Center (AVRDC) in Taiwan. The CIP holds a vast sweetpotato germplasm collection of about 6,500 accessions, including wild relatives. The focus of current sweetpotato research projects includes:

- Sweetpotato improvement and virus control
- Integrated pest management
- Post-harvest quality, nutrition, and market impact
- Biodiversity and genetic resources
- Gene discovery, evaluation and mobilization for crop improvement
- Global commodity analysis and impact assessment

Active sweetpotato collaboration between CIP and national programs exists in Asia and Africa.

The Asian Network for Sweetpotato Genetic Resources (ANSWER) was established in 1996, supported by the International Plant Genetic Resources Institute (IPGRI) and CIP. China, India, Indonesia, Japan, Korea, Malaysia, Papua New Guinea, the Philippines, and Vietnam are members of ANSWER. ANSWER meetings are held from time to time to discuss common problems and strategies for the conservation of genetic resources.

References

Anon. 1978. Pest control in tropical root crops. PANS Manual No. 4, 235 pp. Centre for Overseas Pest Research, London.

Austin, D.F. 1978. The *Ipomoea batatas* complex-I. Taxonomy. Bull. Torrey Bot. Club 105:114–129.

Austin, D.F. 1983. Variability in sweet potatoes in America. In: Breeding new sweet potatoes for the tropics (Martin, F.W., ed.), pp. 15–25. Proc. Am. Soc. Hort. Sci., Tropical Region 27(B).

Burkill, I.H. 1935. A Dictionary of the Economic Products of the Malay Peninsula. Reprinted in 1966. Ministry of Agriculture and Co-operatives, Kuala Lumpur.

Collins, W.W. 1977. Diallel analysis of sweet potatoes for resistance to fusarium wilt. J. Amer. Soc. Hort. Sci. 102:109–111.

De Candolle, A. 1886. Origin of cultivated plants. 2nd ed., reprinted in 1959. Hafner, New York.

Duell, B. 2006. The role of the Kawagoe Friends of Sweetpotato in popularizing the crop in Japan. Proc. 2nd International Symposium on Sweetpotato and Cassava, 14–17 June 2005, Kuala Lumpur. Acta Horticulturae 703 (in press).

Edmond, J.B. and Ammerman, G.R. 1971. Sweet potatoes: Production, processing, marketing. AVI Publ. Co., Westport, Connecticut.

Edmond, J.B. and Martin, J.A. 1946. The flowering and fruiting of the sweet potato under greenhouse conditions. Proc. Am. Soc. Hort. Sci. 47:391–399.

Engel, E. 1970. Exploration of the Chilca Canyon. Current Anthropology 17:55–58.

FAO 2005. Available at: http://faostat.fao.org/site/395/default.aspx

Fujise, K. 1964. Studies on flowering, fruit setting, and self and cross incompatibility in sweet potato varieties. Bull. Kyushu Agric. Exp. Sta. 9:123–146.

Gutierrez, D., Fuentes, S., Molina, J. and Salazar, L.F. 1999 Sweetpotato virus disease (SPVD) in Peru. Fitopatología 34: 143.

Hahn, S.K. and Leuschner, K. 1981. Resistance of sweet potato cultivars to African sweet potato weevil. Crop Sci. 21:499–503.

Hahn, S.K., Terry, E.R. and Leuschner, K. 1981. Resistance of sweet potato to virus complex. HortScience 16:535–537.

Hammett, H.L., Hernandez, T.P. and Miller, J.C. 1966. Inheritance of fiber content in the sweet potato, *Ipomoea batatas*. Proc. Am. Soc. Hort. Sci. 88:428–433.

Harter, L.L. 1926. Bud sports in sweet potato. J. Agric. Res. 33:523–525.

He, G., Prakash, C.S. and Jarret, R.L. 1995. Analysis of genetic diversity in a sweetpotato (*Ipomoea batatas*) germplasm collection using DNA fingerprinting. Genome 38:938–945.

Hernández, T.P. and Miller, J.C. 1962. Self and cross incompatibilities in the sweet potato varieties and seedlings. Proc. Am. Soc. Hort. Sci. 81:428–433.

Hernández, T.P. and Miller, J.C. 1964. Further studies on incompatibility in the sweet potato. Proc. Am. Soc. Hort. Sci. 85:426–429.

Hernández, T.P., Hernández, T.P., Constantin, R. and Miller, J.C. 1965. Inheritance of and method of rating flesh color in Ipomoea batatas. Proc. Am. Soc. Hort. Sci. 87:387–390.

Hernández, T.P., Hernández, T.P., Constantin, R.J. and Kakar, R.S. 1969. Improved techniques in breeding and inheritance of some of the characters in the sweet potato, Ipomoea batatas (L.). Proc. International Symposium on Tropical Root Crops, St. Augustine, Trinidad, 2–8 April 1967 (Tai, E.A. et al., eds.), Vol. I, Section I (Breeding and improvement), pp. 31–40. University of West Indies, St. Augustine, Trinidad.

Heyerdahl, T. 1952. The Kon-Tiki Expedition: By Raft Across the South Seas. The Reprint Society, London.

Hu, J., Nakatani, M., Lalusin, A.G., Kuranouchi, T. and Fujimura, T. 2003. Genetic analysis of sweetpotato and wild relatives using Inter-simple Sequence Repeats (ISSRs). Breed. Sci. 53:297–304.

Huaman, Z. (ed.). 1991. Descriptors for sweet potato. CIP, AVRDC and IBPGR, Rome.

Huang, J.C. and Sun, M. 2000. Genetic diversity and relationships of sweetpotato and its wild relatives in Ipomoea series Batatas (Convolvulaceae) as revealed by inter-simple sequence repeat (ISSR) and restriction analysis of chloroplast DNA. Theor. Appl. Genet. 100:1050–1060.

Ishiguro, K. and Yamakawa, O. 2000. Selection of low and high amylose sweetpotato lines. Proc. 12th Symp. Intl. Soc. Trop. Root Crops, Tsukuba, Japan 10–16 Sept. 2000. pp. 220–224.

Ishiguro, K. and Yoshimoto, M. 2006. Content of the eye-protective nutrient lutein in sweetpotato leaves. Proc. 2nd International Symposium on Sweetpotato and Cassava, 14–17 June 2005, Kuala Lumpur. Acta Horticulturae 703 (in press).

Ishiguro, K., Toyama, J. and Yoshimoto, M. 2004. Portional differences of nutrition and physiological function in "Suioh", a sweetpotato cultivar for tops utilization. Rept. Kyushu Branch of the Crop Sci. Soc. Japan 70:36–39.

Jarret, R.L. and Austin, D.F. 1994. Genetic diversity and systematic relationships in sweetpotato [Ipomoea batatas (L.) Lam.] and related species as revealed by RAPD analysis. Genet. Res. Crop Evol. 41:165–173.

Jarret, R.L., Gawel, N. and Whittemore, A. 1992. Phylogenetic relationships of the sweetpotato [Ipomoea batatas (L.) Lam.]. J. Am. Soc. Hortic. Sci. 117:633–637.

Jones, A. 1965. Cytological observations and fertility measurements of sweet potato (Ipomoea batatas (L.) Lam.). Proc. Am. Soc. Hort. Sci. 86:527–537.

Jones, A. 1977. Heritabilities of seven sweet potato root traits. J. Am. Soc. Hort. Sci. 102(4):440–442.

Jones, A. 1980. Sweet potato. In: Hybridization of crop plants (W.R. Fehr and H.H. Radley, eds.), pp. 645–655. Amer. Soc. Agron. and Crop Sci. Soc. Amer., Madison, Wisconsin, USA.

Jones, A. and Dukes, P.D. 1982. Longevity of stored seed of sweet potato. HortScience 17:756–757.

Jones, A. and Dukes, P.D. 1980. Heritabilities of sweet potato resistances to root knot caused by Meloidogyne incognita and M. javanica. J. Amer. Soc. Hort. Sci. 105:154–156.

Jones, A., Hamilton, M.G. and Dukes, P.D. 1978. Heritability estimates for fiber content, root weight, shape, cracking and sprouting in sweet potato. J. Amer. Hort. Sci. 103:374–376.

Jones, A., Steinbauer, C.E. and Pope, D.J. 1969. Quantitative inheritance of ten root traits in sweet potatoes. J. Amer. Soc. Hort. Sci. 94:271–275.

Jones, A., Dukes, P.D. and Schalk, J.M. 1985. In: Sweet Potato Breeding (Bassett, M.J., ed.), Breeding vegetable crops, pp. 1–35. AVI Publ. Co., Westport, Connecticut.

Jong, S.K. 1974a. Study of the sequential characteristics of sweet potato populations. Intern. Inst. Trop. Agr., Ibadan, Nigeria (mimeo).

Jong, S.K. 1974b. Genotype × season interaction and heritability in sweet potato, Ipomoea batatas (L.) Lam., selection experiments. Intern. Inst. Trop. Agr., Ibadan, Nigeria (mimeo).

Katayama, K., Komae, K., Kohyama, K., Kato, T., Tamiya, S. and Komaki, K. 2002. New sweet potato line having low gelatinization temperature and altered starch structure. Starch 54:51–57.

Katayama, K., Tamiya, S., Kuranouchi, T., Komaki, K. and Nakatani, M. 2003. New sweet potato cultivar "Quick Sweet". Bull. Natl. Inst. Crop Sci. 3:35–52.

Katayama, K., Tamiya, S. and Ishiguro, K. 2004. Starch properties of new sweet potato lines having low pasting temperature. Starch 56:563–569.

Kehr, A.E., Ting, Y.C. and Miller, J.C. 1953. Induction of flowering in the Jersey type sweet potatoes. Proc. Am. Soc. Hort. Sci. 62:437–442.

Kehr, A.E., Ting, Y.C. and Miller, J.C. 1955. The site of carotenoid and anthocyanin synthesis in sweet potatoes. Proc. Am. Soc. Hort. Sci. 65:396–398.

Kobayashi, M. 1984. The Ipomoea trifida complex closely related to sweet potato. In: Proc. Symposium of the International Society of Tropical Root Crops. Lima, Peru, Feb. 21–26, 1983 (Shideler, S.F., and Rincon, H., ed.), pp. 561–568. CIP, Lima, Peru.

Komaki, K. 2001. Phylogeny of Ipomoea species closely related to sweetpotato and their breeding use. Bull. Natl. Inst. Crop Sci. 1:1–56.

Komaki, K. and Katayama, K. 1999. Root thickness of diploid *Ipomoea trifida* (H.B.K.) G. Don and performance of progeny derived from the cross with sweetpotato. Breed. Sci. 49:123—129.

Komaki, K. and Yamakawa, O. 2005. R&D collaboration with industry—The Japanese sweetpotato story. Proc. 2nd International Symposium on Sweetpotato and Cassava, 14–17 June 2005, Kuala Lumpur. Acta Horticulturae 703 (in press).

Komaki, K., Yamakawa, O., Kukimura, H., Yoshinaga, M. and Hidaka, M. 1994. Yield improvement of a true seed sweetpotato population through recurrent selection and seed planting techniques. Bull. Kyushu Natl. Agric. Exp. Sta. 28:107–118.

Komaki, K., Regmi, H.N., Katayama, K. and Tamiya, S. 1998. Morphological and RAPD pattern variations in sweetpotato and its closely related species. Breed. Sci. 48:281–286.

Konczak-Islam, I., Yoshinaga, M., Nakatani, M., Terahara, N. and Yamakawa, O. 2000. Establishment and characteristics of an anthocyanin-producing cell line from sweet potato storage root. Plant Cell Reports 19:472–477.

Kriegner, A., Cervantes, J.C., Burg, K., Mwanga, R.O. and Zhang, D.P. 2001. A genetic linkage map of sweetpotato (*Ipomoea batatas* (L.) Lam.) based on AFLP markers. In: Scientist and Farmer, Partners in Research for the 21st Century, Program Report 1999–2000, pp. 303–313. International Potato Center, Lima, Peru.

Kumagai, T. 2005. Selection of sweetpotato lines for direct planting cultivation in Japan. In: Concise papers of the 2nd International Symposium on Sweetpotato and Cassava, 14–17 June 2005, Kuala Lumpur (Tan, S.L. et al., ed.), pp. 149–150. MARDI, Kuala Lumpur.

Kumagai, T., Umemura, Y., Baba, T. and Iwanaga, M. 1990. The inheritance of beta-amylase null in storage roots of sweet potato, *Ipomoea batatas* (L.) Lam. Theor. Appl. Genet. 79:369—376.

Kumagai, T., Yamakawa, O., Yoshinaga, M., Ishiguro, K., Hidaka, M. and Kai, Y. 2002. 'Konahomare': New sweetpotato cultivar for starch production. Bull. Natl. Agri. Res. Ctr. for Kyushu Okinawa Region 40:1–16.

Lawrence, G.H.M. 1951. Taxonomy of vascular plants. Macmillan Co., New York.

Lewthwaite, S.L. 2006. Sweetpotato products in a modern world: The New Zealand experience. Proc. 2nd International Symposium on Sweetpotato and Cassava, 14–17 June 2005, Kuala Lumpur. Acta Horticulturae 703 (in press).

Li, L. 1983. The inheritance of crude protein content and its correlation with root yield in sweet potatoes (in Chinese with English summary). J. Agr. Assoc. China (Taiwan) 100:78–87.

Li, L. 1982. The inheritance of crude starch percentage and its correlation with root yield and other traits in sweet potatoes. Chiayi Agr. Exp. Sta., Taiwan Agr. Res. Inst. 10:765–772.

Li, L. 1975. The inheritance of qualitative characters in a randomly intermating population of sweet potatoes, Ipomoea batatas (L.) Lam. (in Chinese with English summary). J. Taiwan Agr. Res. 24:32–42.

Love, J.E., Hernández, T.P. and Marziah, M. 1977. Mutation studies with the Centennial sweet potato. HortScience 12:405 (Abstr.).

Mai, T.H. 2006. Ensiling sweetpotato roots and vines for pigs in Vietnam. Proc. 2nd International Symposium on Sweetpotato and Cassava, 14–17 June 2005, Kuala Lumpur. Acta Horticulturae 703 (in press).

Martin, J.A., Jr. 1946. Germination of sweet potato seed as affected by different methods of scarification. Proc. Am. Soc. Hort. Sci. 47:387–390.

Martin, F.W. 1959. Staining and observing pollen tubes by means of fluorescence. Stain Technol. 34:125–128.

Martin, F.W. 1968. The system of self-incompatibility in *Ipomoea*. J. Hered. 59:263–267.

Martin, F.W. and Cabanillas, E. 1966. Post-pollen germination barriers to seed set in sweet potato. Euphytica 15:404–411.

Martin, F.W. and Jones, A. 1986. Breeding sweet potatoes. In: Plant breeding reviews, Vol. 4(10): 313–345. Avi Publ. Co., Westport, Connecticut.

Menzes, G. 2002. 1421: The year china discovered the world. Bantam Press, London.

Miller, J.C. 1937. Inducing the sweet potato to bloom and set seed. J. Hered. 28:347–349.

Nakanishi, T. and Kobayashi, M. 1979. Geographic distribution of cross incompatibility group in sweet potato. Incompatibility Newsletter 11:72–75.

Nishiyama, I., Fujise, K., Teramura, T. and Miyazaki, T. 1961. Studies of sweet-potato and its related species. I: Comparative investigation on the chromosome numbers and the main plant characters of *Ipomoea* species section *Batatas*. Japan. J. Breed. 11:37–43 (in Japanese).

Nishiyama, I., Miyazaki, T. and Sakamoto, S. 1975. Evolutionary autoploidy in the sweet potato (*Ipomoea batatas* (L.) Lam.) and its progenitors. Euphytica 24:197–208.

Noor Suhana, A., Habibuddin, H. and Hamidah, G. 2005. PCR detection of sweetpotato leaf curl and sweetpotato feathery mottle viruses. In: Concise papers of the 2nd International Symposium on Sweetpotato and Cassava, 14–17 June 2005, Kuala Lumpur (Tan, S.L. et al., ed.), pp. 103–104. MARDI, Kuala Lumpur.

Odame, H., Kameri-Mbote, P. and Wafula, D. 2004. The role of innovation in policy and institutional change: The case of transgenic sweetpotato in Kenya. International Environmental Law Research Centre. Available at: http://www.ielrc.org/content/n0206.htm.

Onjeda, G., Freyre, R. and Iwanaga, M. 1990. Use of *Ipomoea trifida* germplasm for sweet potato improvement. 3. Development of 4x interspecific hybrids between *Ipomoea batatas* (L.) Lam (2n = 6x = 90) and *I. trifida* (H.B.K.) G. Don. (2n = 2x = 30) as storage-root initiators for wild species. Theor. Appl. Genet. 83:159–163.

Otani, M., Shimada, T., Kimura, T. and Saito, A. 1998. Transgenic plant production from embryogenic callus of sweet potato (*Ipomoea batatas* (L.) Lam.) using *Agrobacterium tumefaciens*. Plant Biotechnol. 15:11–16.

Poole, C.F. 1955. Sweet potato genetic studies. Hawaii Agricultural Experiment Station Technical Bulletin No. 27, 19 pp. University of Hawaii, College of Agriculture, Honolulu.

Sakai, K. 1964. Studies on the enlargement of variations and the improvement of selection methods in sweet potato breeding. Bull. Kyushu Agric. Exp. Sta. 9:207–297.

Sakai, K., Marumine, S., Hirosaki, S., Kikukawa, S., Ide, Y. and Shirasaka, S. 1967. On the new variety of sweet potato, Koganesengan. Bull. Kyushu Agri. Exp. Sta. 13:55–68.

Sakamoto, S., Marumine, S., Ide, Y., Yamakawa, O., Kukimura, H., Yoshida, T. and Tabuchi, S. 1987. 'Shiroyutaka': A new sweet potato cultivar registered. Bull. Kyushu Natl. Agri. Exp. Sta. 24:279–305.

Sakamoto, S., Shiga, T., Ishikawa, H., Kato, S., Takemata, T., Umehara, M. and Ando, T. 1989. New sweet potato cultivar 'Shirosatsuma'. Bull. Natl. Agri. Res. Ctr. 15:1–13.

Salazar, L.F. and Fuentes, S. 2000. Current knowledge of major virus diseases of sweet potatoes. In: Proc. International workshop on Sweetpotato Cultivar Decline Study, 8–9 Sept. 2000, Miyakonojo, Japan (Nakazawa, Y. and Ishiguro, K., ed.), pp. 14–19. Kyushu Natl. Agri. Expt. Stn., Miyakonojo, Japan.

Saladaga, F.A. and Hernandez, T.P. 1981. Heritability and expected gain from selection for yield, weight loss in storage and sprouting in field bed of sweet potato. Ann. Trop. Res. 3:1–7.

Shiga, T., Sakamoto, S., Ando, T., Ishikawa, H., Kato, S., Takemata, T. and Umehara, M. 1985. On a new sweet potato cultivar 'Beniazuma'. Bull. Natl. Agri. Exp. Sta. 3:73–85.

Shiotani, I. 1989. Genomic structure of the sweetpotato and hexaploids in *Ipomoea trifida* (H.B.K) G. Don. Japan. J. Breed. 39:57–66.

Steinbauer, C.E. 1948. A sweet potato from Tinian Island highly resistant to fusarium wilts. Proc. Am. Soc. Hort. Sci. 52:304–306.

Tahara, M., Aoki, T., Suzuka, S., Yamashita, H., Tanaka, M., Matsunaga, S. and Kokumai, S. 2004. Isolation of an active element from a high-copy-number family of retrotransposons in the sweetpotato genome. Mol. Gen. Genomics 272:116–127.

Takahata, Y., Noda, T. and Nagata, T. 1993. HPLC determination of β-carotene content of sweet potato cultivars and its relationship with color values. Japan. J. Breed. 43:421–427.

Talakar, N.S. 1989. Development and testing of an integrated pest management technique to control sweet potato weevil. In: Improvement of sweet potato (*Ipomoea batatas*) in Asia, report of a "Workshop on sweet potato improvement in Asia," 24–28 October 1988, ICAR, India, pp. 117–126. CIP, Lima, Peru.

Tan, S.L. 1998. Agronomic modifications for mechanized planting of sweet potato. J. Trop. Agric. and Fd. Sc. 26(1): 25–33.

Tarumoto, I., Katayama, K., Tamiya, S., Ishikawa, H., Komaki, K. and Kato, S. 1996. New sweetpotato cultivar 'Satsuma-starch'. Bull Natl. Agri. Exp. Ctr. 25:1–20.

Teramura, T. 1979. Phylogenetic study of *Ipomoea* species in the section *Batatas*. Mem. Col. Agr. Kyoto Univ. 114:29–48.

Thompson, D.G., Hong, L.L., Ukoskit, K. and Zhu, Z. 1997. Genetic linkage of randomly amplified polymorphic DNA (RAPD) markers in sweetpotato. J. Am. Soc. Hort. Sci. 122:79–82.

Toyama, J., Yoshimoto, M. and Yamakawa, O. 2005. Varietal differences in trypsin inhibitor activity of sweetpotato roots. Breed. Res. 7:17–23.

Wambugu, F.M. 2001. Virus resistant sweetpotato project in Kenya. Paper presented at a conference "Agricultural Biotechnology—The Road to Improved Nutrition and Increased Production," Boston, Massachusetts, USA, 1–2 Nov. 2001.

Wang, H. 1982. The breeding of sweet potatoes for human consumption. In: Sweet potato: Proc. 1st international symposium, Shanhua, Tainan, Taiwan (Villareal, R.L. and Griggs, T.D., ed.), AVRDC Publ. No. 82–172, pp. 297–311. AVRDC, Tainan.

Yamakawa, O., Yoshinaga, M., Hidaka, M., Kumagai, T. and Komaki, K. 1997. 'Ayamurasaki': A new sweetpotato cultivar. Bull. Kyushu Natl. Agri. Exp. Sta. 31:1–22.

Yamakawa, O., Yoshinaga, M., Kumagai, T., Hidaka, M., Komaki, K., Kukimura, H. and Ishiguro, K. 1998. 'J-Red': A new sweetpotato cultivar. Bull. Kyushu Natl. Agri. Exp. Sta. 33:49–72.

Yen, D.E. 1971. Construction of the hypothesis for distribution of the sweet potato. In: Man across the sea. (Riley, C.L. et al., ed.), pp. 328–342. Univ. Texas Press, Austin, USA.

Yoshimoto, M., Yahara, S., Okuno, S., Islam, M.D., Ishiguro, K. and Yamakawa, O. 2002. Antimutagenicity of mono-, di-, and tricaffeoylquinic acid derivatives isolated from sweetpotato leaf. Biosci. Biotechnol. Biochem. 66:2336–2341.

Yoshimoto, M., Kurata, R., Okuno, S., Ishiguro, K., Yamakawa, O., Tsubata, M., Mori, S. and Takagaki, K. 2006. Nutritional value of and product development from sweetpotato leaves. Proc. 2nd International Symposium on Sweetpotato and Cassava, 14–17 June 2005, Kuala Lumpur. Acta Horticulturae 703 (in press).

Zhang, D.P., Ghislain, M., Huamán, Z., Golmirzaie, A. and Hijmans, R. (accessed Aug. 2005). RAPD variation in sweetpotato varieties from South America and Papua New Guinea. Program 2: Germplasm management and enhancement, CIP, Lima, Peru. http://www.cipotato.org/market/PgmRprts/pr95-96/program2/prog26.htm.

Chapter 12

Cassava Genetic Improvement

Hernán Ceballos, Martin Fregene, Juan Carlos Pérez, Nelson Morante, and Fernando Calle

Introduction of the crop

Cassava (*Manihot esculenta* Crantz), along with maize (*Zea mays* L.), sugarcane (*Saccharum* spp.), and rice (*Oryza sativa* L.), constitute the most important sources of energy in the human diet in most tropical countries of the world. The species originated in South America (Allem, 2002) and was domesticated less than 10,000 years ago. Early European sailors soon recognized the advantages of the crop and carried it to Africa. From there, traders later introduced it to Asia. Until recently, cassava and its products were little known outside the tropical and subtropical regions where it grew. No major scientific efforts had been made to improve the crop. However, with the creation of the International Institute of Tropical Agriculture (IITA) in Nigeria and the International Center of Tropical Agriculture (CIAT) in Colombia in the early 1970s, a new era began for cassava that included the implementation of successful breeding projects, modernization of cultural practices, and the development of new processing methods (Cock, 1985; Jennings and Iglesias, 2002). National research centers in Brazil, Colombia, Cuba, India, Thailand, Vietnam, and other countries have conducted successful research on cassava as well.

Plant breeding has one of the highest rates of return among the investments in agricultural research. The remarkable increase in the productivity of many crops during the twentieth century can be attributed to genetic gains achieved through crop breeding. Cassava has also benefited from technological inputs in the area of breeding (Kawano, 2003). New varieties in Africa, Asia, and Latin America and the Caribbean satisfy the needs of farmers, processors, and consumers, bringing millions of dollars in additional income to small farmers. New technologies in the area of tissue culture, genetic transformation, and molecular biology have also made positive contributions (Calderón-Urrea, 1988; DeVries and Toenniessen, 2001; Fregene et al., 1997, 2000; Puonti-Kaerlas et al., 1997).

Currently, cassava is an important crop in regions at latitudes between 30° N and 30° S, and from sea level up to 1800 meters above sea level (Fig. 12.1). Although its most common product is the starchy root, cassava foliage has excellent nutritional quality for animal and human consumption and offers great potential. Cassava is the fourth most important basic food after rice, wheat (*Triticum aestivum* L.), and maize; it is a fundamental component in the diet of millions of people. Scott et al. (2000) estimated that for the year 1993, annual production of cassava was about 172.4 million tons, with a value of approximately \$US9.31 billion. Between 1961–63 and 1995–97, cassava production increased at a rate of 2.35%/year (Scott et al., 2000), a trend comparable to that found in other crops, such as wheat (4.32%), potato (4.00%), maize

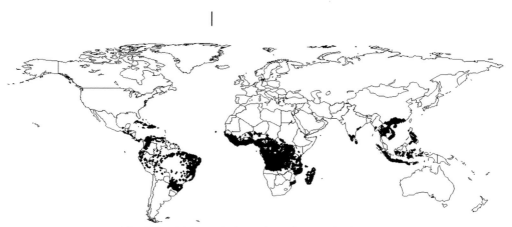

Fig. 12.1.   Major producing regions of cassava in the world.

(3.94%), yams (3.90%), rice (2.85%), and sweetpotato (1.07%). Between 1994 and 2005, cassava productivity was expected to increase at 1.1%/year.

Cassava is a very rustic crop that grows well under marginal conditions where few other crops could survive. A large proportion of cassava varieties are drought-tolerant, can produce in degraded soils, and are resistant to the most important diseases and pests. The crop is naturally tolerant to acidic soils, and offers a convenient flexibility in that it can be harvested when the farmers need it.

The plant and its uses

Cassava is a perennial shrub. Basically, every part of the plant can be used, but the starchy roots are by far the most commonly used product. There are many different uses of cassava for human and animal consumption as well as for products for industrial processes (Balagopalan, 2002). Cyanogenic glucosides (CG) are found in every tissue except in the cassava seed. The most abundant CG is linamarin (about 85%), with lesser amounts of lotaustralin. The CG, synthesized in the leaves and transported to the roots, are broken down by the enzyme linamarase to produce hydrogen cyanide (HCN), a volatile poison (Andersen et al., 2000; Du et al., 1995; McMahon et al., 1995;

Wheatley and Chuzel, 1995). Linamarin and linamarase accumulate in different parts of the cell, thus preventing the formation of free cyanide. However, most processing methods disrupt the tissues, allowing the enzyme to act on the substrate for a rapid release of cyanide. CG accumulation varies with genotypes, environments, agronomic practices, age of the plant, and plant tissue; the highest accumulation levels are found in the leaves and peel of roots (Cock, 1985).

The starchy roots are a valuable source of energy and can be boiled or processed in different ways for human consumption. Roots can also be used for obtaining native or fermented starches and as dried chips, meal, or pellets for animal feed. Cultivars with less than $100\,\mathrm{mg\,CG\,kg^{-1}}$ fresh weight in the roots are considered "sweet." Above this level cassava roots are considered "bitter." Depending on the processing methodologies, bitter or sweet clones may be preferred. In addition to the cyanogenic potential, other relevant traits for the roots are dry matter content, percentage of amylose in the starch composition, and protein and carotenoid contents. There is variation in starch quality in relation to its amylose percentage with a mean around 15% (Wheatley et al., 1993). Cassava roots are low in protein content, with an average around 2% (dry weight basis). There have been, however, some preliminary results that suggest that protein content in the roots can be considerably higher (6–8%) in some landraces, particularly from Central America (CIAT, 2002). Yellow cassava roots have considerable amounts of carotenes (Chávez et al., 2000, 2005; Iglesias et al., 1997).

Cassava roots are not tubers and, therefore, cannot be used for reproductive purpose. A major consequence of this situation is short shelf life (Beeching et al., 1998); within one or two days after harvest, there is a rapid initiation of post-harvest physiological deterioration (PPD). With the exception of a delay observed in yellow roots (Sánchez et al., 2006), little useful genetic variation to delay or reduce PPD has been found, and the solution to this problem remains one of the most important goals for cassava breeding.

Cassava stems are the most important source of planting material to propagate the crop. Cassava foliage is not widely exploited in spite of its high nutritive value, although consumption of leaves by human populations is relatively common in certain countries of Africa and Asia. Foliage is also used for feeding animals. Crude protein content in leaves typically ranges from 20–25% of dry weight (Babu and Chatterjee, 1999; Buitrago, 1990; Gomez et al., 1983), but levels as high as 30% (dry basis) have been identified (Buitrago, 1990). Content of cyanogenic glucosides in the foliage is markedly higher (3,800–5,900 mg HCN/kg^{-1} fresh weight) than in roots (4–113 mg HCN/kg fresh weight). Exploitation of cassava foliage is expected to increase because of the recent developments and testing of mechanical harvesters and the development of alternative cultural practices. (Cadavid López and Gil Llanos, 2003).

Reproduction in cassava

Cassava can be propagated either by stem cuttings or from botanical seed (true seed) (Fig. 12.2). However, the former is the most common practice used by farmers for multiplication and planting purposes. Propagation from botanical seed occurs occasionally in farmers' fields and, as such, is a starting point for the generation of useful genetic diversity (Alves, 2002). Most breeding programs generate seed through crossing, a means of creating new genetic variation. Occasionally, botanical seed has been used in commercial propagation schemes (Iglesias et al., 1994; Rajendran et al., 2000).

Cassava is monoecious, with female flowers opening 10–14 days before the male flowers on the same branch (Fig. 12.3). Self-pollination can occur because male and female flowers on different branches or on different plants of the same genotypes can open simultaneously (Jennings and Iglesias, 2002). Flowering depends on the genotype and the environmental conditions. Branching occurs when an inflorescence is formed. Because erect, non-branching types are frequently preferred by farmers, crossing elite clones in certain regions may become more difficult; this type of plant tends to produce fewer flowers relatively late in the season (and sometimes it produces no flowers at all).

Synchronization of flowering remains a difficult issue in cassava breeding. Some clones flower relatively early at 4 or 5 months after planting, whereas others flower only at 8 to 10 months after planting. Because of this, and the time required for the seed to mature, it generally takes at least a year to obtain seeds of a planned cross. On average, between one and two seeds (out of the three possible formed in the trilocular fruit) per pollination are obtained. Several publications illustrate the procedures for controlled pollinations in cassava (Kawano, 1980; Jennings and Iglesias, 2002). Seeds often have a dormancy period of a few months after maturity, and they require relatively high temperatures (30–35°C) for optimum germination (Ellis et al., 1982).

Fig. 12.2. Vegetative and botanical seed of cassava. A: cutting the stems; B: storage of stems under a tree until planting time; C: preparing the stems for transport; D: botanical seed of cassava.

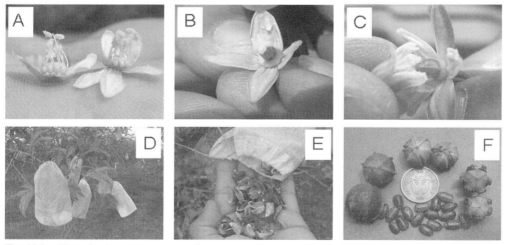

Fig. 12.3. Flowering in cassava. A: Male flower; B: Female flower; C: Hand pollination for directed crosses; D: Cloth bags placed around the maturing fruit to capture the seeds; E: Harvest of seed; F: Fruits at different stages of maturity and botanical seed.

Breeding objectives

Breeding objectives depend on the ultimate use of the crop. Productivity is the focus for industrial uses of cassava (i.e., starch production and dried roots for animal feed), whereas stability of production is fundamental in the many regions where cassava is the main subsistence crop. Industrial uses of cassava require high dry matter content as the main quality trait for the roots, whereas human consumption needs frequently emphasize cooking quality or starch characteristics as a determining trait. Good cooking quality is usually associated with other morphological traits, such as the color of the peel of the roots, the leaf petiole, or the shoot. Farmers frequently reject newly introduced varieties that show changes in such morphological traits, although they may have little or no correlation with actual cooking quality. Participatory research and breeding approaches have been developed and used to incorporate the diversity of criteria used by farmers and consumers in selecting the best cassava varieties (DeVries and Toenniessen, 2001; Gonçalvez Fukuda et al., 2000; Gonçalvez Fukuda and Saad, 2001).

Other root quality traits relevant to cassava breeding programs are the cyanogenic potential in the roots (Dixon et al., 1994), early bulking capacity, higher protein content in the roots, and reduced post-harvest physiological deterioration. Unfortunately, the genetic variability for the latter two traits is limited in *M. esculenta* and, therefore, interspecific crosses with other *Manihot* species have been attempted to introgress useful alleles (Ceballos et al., 2006).

Stability of production is associated with resistance or tolerance to major biotic and abiotic stresses; the emphases vary with the target environments. In Africa, cassava mosaic disease (CMD) and cassava brown streak disease (CBSD) are important constraints. A disease similar to CMD is also present in southern India. In certain regions of Latin America and the Caribbean (LAC), frogskin disease causes roots to become "corky" and commercially unusable. The causal agent has not yet been identified, although it has been suspected for many years that it may be a virus. Bacterial blight, induced by *Xanthomonas axonopodis* pv. *Manihotis* (also known as *X. campestris* pv. *manihotis*), is found in Asia, Africa, and LAC, and can have devastating effects on yield and the availability of planting material, particularly in Africa and LAC (Hillocks and Wydra, 2002). Several fungal diseases may also affect cassava productivity. Super-elongation disease, induced by *Sphaceloma manihoticola* (teleomorph: *Elsinoe brasiliensis*), is widespread in the Americas, from Mexico to Southern Brazil. In tropical lowlands with high rainfall, *Cercospora, Cercosporidium, Phaeoramularia* or *Colletotrichum* species can affect cassava productivity (Jennings and Iglesias, 2002). *Phoma* species cause leaf and stem lesions in the tropical highlands. Several species of *Phytophthora* induce root rot. Root rots are also induced by different species of the genera *Sclerotium, Armillaria,* and *Fusarium.* Fortunately, there are sources of genetic resistance to most of these diseases (CIAT, 2001; Hillocks and Wydra, 2002).

Several arthropod pests feed on cassava and can reduce productivity. The green mite (*Mononychellus tanajoa*) devastated cassava fields upon its introduction in Africa in the 1970s (Nyiira, 1975). Other mites important for cassava are *Tetranychus urticae, T. cinnabarinus, Mononychellus caribbeanae* and *Oligonychus peruvianus* (Bellotti et al., 2002). The mealybugs *Phenacoccus manihotis* and *P. herreri* feed on cassava fields of Africa and LAC, respectively. Thrips (particularly *Frankliniella williamsi* and *Scyrtotrips manihoti*) considerably reduce yields of susceptible genotypes. Clones with pubescent leaves in their early stages of development offer excellent levels of resistance to these insects (Bellotti, 2002), and this trait has been broadly incorporated into improved varieties.

Whiteflies are among the most widespread pests in cassava. *Aleurotrachelus socialis* is the predominant species in northern South America, where it causes considerable crop damage through direct feeding. *Bemisia tabaci* is widely distributed in tropical Africa and several Asian countries. Until 1990, *B. tabaci* biotypes found in the Americas did not feed on cassava. The major effect of *B. tabaci* is through its role as a vector of the devastating CMD disease in Africa. Several other species of whiteflies affect cassava in different regions. Genetic resistance to whiteflies in cassava has been found, particularly for *A. socialis,* in several germplasm accessions from the CIAT collection (Bellotti, 2002). Based on breeding work at CIAT, Colombia released the first whitefly-resistant variety of any crop in 2002, targeted toward the Tolima Valley, where whiteflies typically devastate plantations.

There are several other arthropod pests affecting cassava roots, foliage and/or stems, particularly Lepidoptera, Diptera, and Hemiptera. There is little or no genetic resistance to these pests and their management is commonly achieved through biological control measures. Attempts to produce transgenic cassava have succeeded with the introduction of *cry* genes encoding insect-specific endotoxins (Bt toxins) from *Bacillus thuringiensis* (Fregene and Puonti-Kaerlas, 2002; Ladino et al., 2001; Talylor et al., 2004).

There are a variety of abiotic factors limiting cassava productivity. The crop is frequently grown in drought-prone regions and/or on low fertility soils. It can be found in alkaline or acidic soils, most frequently the latter. Some traits associated with adaptation to these conditions have been suggested (Jennings and Iglesias, 2002), such as leaf longevity (CIAT, 2001; Fregene and Puonti-Kaerlas, 2002; Lenis et al., 2006), optimum leaf area index, and ideal plant architecture

(Hanh et al., 1979; Kawano et al., 1998; Kawano, 2003). The capacity of the stems to withstand long storage periods (sometimes up to two months) from harvest to planting is an important trait. This characteristic affects final density of established plants and is fundamental for areas with relatively long dry spells or erratic rainfall, because the storage period may extend to the point that it compromises their viability. While there is known genetic variation for stem storability, it has not been a major breeding objective of any program so far.

During the 1990s, there was a drastic change in the economies of tropical and subtropical countries where cassava is grown. As a result of the globalization of the economies, it became obvious that tropical production of maize was not competitive compared with that from temperate regions. Several factors explain this situation (Pandey and Gardner, 1992). There has been an increased volume of temperate maize imported by tropical countries. This, in turn, has opened an opportunity never before available to cassava because both governments and private sectors realized that the crop was a key, but underutilized, commodity (Ceballos et al., 2004). These changes made clear that, in addition to high and stable productivity, the cassava-breeding project had the opportunity to expand and exploit genetic variability to generate clones with increased value for industrial processes in which cassava can be a strategic raw material. Examples of key traits for the different industries are given below:

Animal Feed Cassava is an important commodity as source of energy in diets for people and different animals. However, it has low levels of protein and, therefore, when using it as a staple, it must be supplemented with another source of protein (typically soybean derivatives). For this reason, the rule of thumb is that cassava cannot be more than 70% of the price of maize (Tewe, 2004). Key qualitative traits for this industry would be finding cassava clones with higher levels of protein in the roots. In addition, including other nutritional traits, such as pro-vitamin A carotenoids, would be beneficial.

Starch Industry Cassava starch has properties of its own, which make it particularly adapted (or not adapted) to certain uses. This sector has always requested novel cassava starch types to diversify its uses.

Ethanol and bioplastics This is a relatively new demand for cassava products, which was accentuated with the recent increases in the price of oil. A "sugary" cassava, such as the one recently reported (Carvalho et al., 2004), would make the process of fermentation to produce ethanol or lactic acid (an alternative product in the pathway for the production of bioplastics) economically and environmentally less expensive.

Processed food Acyanogenesis (roots without even traces of cyanogenic glucosides) has been a trait requested by this sector. Another characteristic that affects the efficiency of this industry is the need for a longer shelf life of the roots (delayed PPD) and the usual need for high dry-matter content.

Breeding scheme

As in most crop breeding activities, cassava genetic improvement starts with the assembly and evaluation of a broad germplasm base, followed by production of new recombinant genotypes derived from selected elite clones. Scientific cassava breeding began only a few decades ago and, therefore, the divergence between landraces and improved germplasm is not as wide as in crops with a more extensive breeding history. As a result, landrace accessions play a more relevant role in cassava than in other crops. Parental lines are selected based mainly on their *per se* performance and little progress has been made in using general combining ability

(Hallauer and Miranda, 1988) as a criterion of parental selection. Crossing can be by controlled pollinations, done manually, to produce full-sib families or in polycross nurseries where open pollination results in half-sib families.

For open pollinations, a field planting design developed by Wright (1965) is followed to maximize the frequency of crosses of all the parental lines incorporated in the nursery. Knowledge about flowering capacity is important for the selection of a group of materials with synchronized flowering. When there are considerable differences in flowering habit, a delayed planting and/or pruning of the earliest flowering genotypes may be required. At harvest, the seed harvested from each clone are bulked to form a half-sib family. Seeds from full-sib families can be obtained in isolated open pollination plots where two clones are planted together and one of them, chosen to act as female progenitor, is emasculated. Alternatively, several male-sterile clones are identified, which can act as female parents.

The botanical seed obtained by the different crossing schemes (Kawano, 1980) may then be planted directly in the field (as done at IITA) to take advantage of the availability of irrigation and high temperatures. At CIAT, seeds are germinated in greenhouse conditions and the resulting seedlings transplanted to the field when they are about 20–25 cm tall (Jennings and Iglesias, 2002). Root systems in plants derived from botanical seed or vegetative cuttings may differ considerably. The taproots from seedlings tend to store fewer starches than roots from cuttings (Rajendran et al., 2000). Because of this, it is difficult, if not impossible, to correlate the root yield of clones at later stages in the evaluation/selection process (for example in Advanced Yield Trials) with results generated early in the selection process when evaluating the plants produced from botanical seeds (F_1 stage). However, when seeds are germinated in containers and later transplanted, the taproot often does not develop, and the seedling-derived plant may be more similar to subsequent stake-derived plants in starchy root conformation.

The vegetative multiplication rate of cassava is low. From one plant, 5–10 cuttings typically can be obtained, although this figure varies widely by genotype. This situation implies a lengthy process to reach the point where replicated evaluations across several locations can be conducted. It takes about 5–6 years from the time the botanical seed is germinated until the evaluation/selection cycle reaches the regional trial stage in which several locations can be included. One further complication in a cassava program is the number of factors that can affect quality of planting material. For example, the original positioning of the vegetative cutting along the stem considerably affects the performance of the plant it produces. Cuttings from the mid-section of the stems usually produce better performing plants than those at the top or the bottom. This variation in the performance of the plant, depending on the physiological status of the vegetative cutting, results in larger experimental errors and undesirable variation in the evaluation process.

Table 12.1 illustrates a typical selection cycle in cassava. It begins with the crossing of elite clones and ends when a few clones surviving the selection process reach the

stage of regional trials across several locations. There is some variation among different cassava-breeding programs regarding the numbers of genotypes and plants representing them through the different stages; however, the numbers presented in Table 12.1 are fairly common and illustrate the different stages required to complete a selection cycle and the kind of selection pressures that are generally applied (Ceballos et al., 2004).

The first selection is conducted in the second year in the nurseries, with plants derived from botanical seed (F_1 in Table 12.1). Because of the low correlations between the performance at this early stage of selection and the time that the genotypes reach replicated trials, the early selections are based on high-heritability traits, such as plant type, branching habits and, particularly, reaction to diseases (Hahn et al., 1980a, 1980b; Hershey, 1984; Iglesias and Hershey, 1994). At IITA, combined selection for resistance to CMD and bacterial blight (*Xanthomonas axonopodis* pv. *manihotis*) begins with about 100,000 seedlings, and only about 3,000 genotypes survive this first stage of selection, which is based on single plant performance. At CIAT, Colombia, smaller F_1 nurseries (up to 50,000 seedlings) have typically been planted since there is no single trait, such as CMD resistance, which drastically reduces the number of selections.

Table 12.1. A typical selection cycle in cassava, beginning with the crossing of elite clones, through the different stages of the selection process (from Jennings and Iglesias, 2002).

Year	Activity	Number of genotypes	Plants per genotype
1	Crosses among elite clones	Up to 100,000	
2	F_1: Evaluation of seedlings from botanical seeds. Strong selection for ACMV in Africa.	100000[a]; 50000[b]; 17500[c]	1
3	Clonal Evaluation Trial (CET)	2000–3000[a, b] 1800[c]	6–12
4	Preliminary Yield Trial (PYT)	100[a]; 300[b]; 130[c]	20–80
5	Advanced Yield Trial (AYT)	25[a]; 100[b]; 18–20[c]	100–500
6–8	Regional Trials (RT)	5–30[a, b, c]	500–5000

Figures for the cassava breeding at [a] IITA (Ibadan, Nigeria); [b] CIAT (Cali, Colombia) and [c] CIAT-Rayong Field Crops Research Center (Thailand). Averages from data in Kawano, 2003.

The second stage of selection is the clonal evaluation trial (CET). The few surviving genotypes from the single-plant selection conducted during the F_1 stage produce the 6–10 vegetative cuttings required for this second step. The capacity to produce this number of cuttings is in fact another selection criterion used at the F_1 stage. The number of clones used in CETs usually range from 2,000 to 3,000. Within a given trial, however, the same number of plants is used to avoid the confounding effects between number of plants and genotypic differences. Because the competition between neighboring genotypes in the CET may favor more vigorous plant architectures, selection at this stage relies heavily on high heritability traits, such as harvest index (Kawano et al., 1998; Kawano, 2003). Plant type is an important selection criterion at early stages of selection; plants whose main stem does not branch until it reaches about 1 m are preferred (Kawano et al., 1978; Hahn et al., 1979). Other selection criteria at this stage include high dry matter and cyanogenic potential (Iglesias and Hershey, 1994). Between 100 and 300 clones survive the CET. For most programs, a common feature is selection based just on visual evidence in the first two stages of selection, with no data recording. This allows the screening of a large number of segregating progenies at low costs. Figure 12.4

Fig. 12.4. Harvest of a CET. A: The 8-plant plot harvested; B: Cutting the roots from the crown; C: Weighing above ground biomass; D: Weighing roots; E: A harvested trial ready for weighing. Paper bags are used to collect about 5 kg of roots for dry matter content measurements.

illustrates the activities during the harvest of a CET, as currently practiced, including data recording.

One important trait that makes the harvest of large trials, such as a CET, expensive and time-demanding is the measurement of dry matter content (DMC) in the roots. The productivity of cassava depends ultimately on the number of fresh roots produced and the DMC of those roots. It is feasible to have excellent dry matter yields based on high production of fresh roots, even if they have below-average DMC. This situation is gen-

erally not acceptable because the transport and processing costs are too high. Figure 12.5 illustrates the process of measuring DMC. A sample of about 5 kg of roots is weighed with a hanging scale (Fig. 12.5C) and then, the same sample of roots is weighed with the roots immersed in water (Figs. 12.5A and B). The relationship between the two weights provides an accurate estimate of DMC. A simple modification of the system reduced the time required to quantify DMC: A few years ago, a three-beam scale was used to weigh roots in water

Fig. 12.5. Steps required for measuring dry matter content in the roots by the gravimetric method. Roots are weighed in hanging scale (C), and then immersed in water (A and B). The use of an electronic scale allows one person to attend to two scales simultaneously (D).

Fig. 12.6. Preliminary yield trials (PYT): The photograph on the left shows two 2-row plots with a total of 10 plants per plot. The photograph on the right shows the kind of contrast in root yields that can be observed in these PYTs.

(Fig. 12.5A), which required about one minute per sample. In 2000, an electronic scale was introduced, which required only a few seconds to stabilize (Fig. 12.5B). Furthermore, since the electronic scale does not require any human intervention, two scales can be set up, so while one is stabilizing, the operator records the reading from the other scale (Fig. 12.5D). This has improved considerably the volume of clones that can be evaluated and reduced the costs of harvest.

The following stage of selection, the preliminary yield trials (PYT), at CIAT are currently based on the evaluation of 10 plants in three replications. The 10 plants in each replication are planted in two 5-plant rows. Rows are spaced only 0.8 m apart instead of the standard 1.0 m, and one empty row is left between plots to increase within-clone competition and reduce between-clone competition. Figure 12.6 illustrates a typical PYT as currently conducted. Large genetic variability occurs among clones, even within the same family. Although poorly performing clones are mostly eliminated at the CET stage, there is still a considerable variation in the PYT trials. This highlights the need for a gradual process of selection and the need to avoid strong selection pressures.

With the initiation of replicated trials, the emphasis of selection shifts from high-heritability traits to those of low heritability, such as yield. Starting with PYT and increasingly during the advanced yield trials (AYT) and the regional trials (RT), a greater weight is given to yield and its stability across locations. Cooking quality, "poundability" (IITA), and "farinha" quality (Brazil) trials also begin at these stages, when the number of genotypes evaluated is more manageable. The AYTs are typically grown in one or two locations for two consecutive years. They have three replications per location, and plots contain four rows with five plants per row. Yield data are taken from the six central plants of the plot and the remaining 14 plants are used as source of planting material for the next season. The RTs are conducted for at least two years in 4–10 locations each year. Plots have five rows with five plants per row. Yield data are taken from nine plants from the center.

The clones that show outstanding performance in the RTs are released as new varieties and, often, incorporated as parents in the crossing nurseries. This completes a selection cycle and a new one begins. It should be pointed out that the selection scheme described above has the following characteristics:

- The process is, indeed, a mass phenotypic recurrent selection, because no family data are involved in the selection process.
- Few data are taken in early stages of selection, especially on genotypes that can be readily discarded by visual evaluation. Therefore, no data regarding general combining ability effects (≈breeding value) are available for more effective selection of parental materials.
- There is no proper separation between general (GCA ≈ additive) and specific (SCA ≈ heterotic) combining ability effects. The outstanding performance of selected materials is likely to depend substantially on positive heterotic effects, which cannot be transferred to the progenies sexually derived from them.
- Inbreeding has been intentionally omitted in the breeding scheme. Therefore large genetic loads are likely to remain hidden in cassava populations and useful recessive traits are difficult to detect.
- Two or more stages of selection may be based on non-replicated trials. A large proportion of genotypes are eliminated without a proper evaluation set up.

Because of the foregoing, there are some clear opportunities to improve the efficiency and effectiveness of cassava breeding. Kawano et al. (1998) mentioned that during a 14-year period, about 372,000 genotypes, derived from 4,130 crosses, were evaluated at CIAT-Rayong Field Crop Research Center. Only three genotypes emerged from the selection process to be released as official varieties. However, these varieties have achieved remarkable success in Asia, occupying more that one million hectares. Similar outcomes have been observed at IITA, CIAT-Colombia and Brazil. The resulting increases in productivity account for higher incomes (about one billion $US annually) for the poor farmers who grow the improved germplasm (Kawano, 2003).

Inheritance of relevant traits

Little progress has been achieved in understanding the inheritance of traits with agronomic relevance. Only a few articles on the inheritance of quantitative traits have been published (Easwari et al., 1995; Easwari and Sheela, 1993, 1995, 1998). In this regard, cassava is an interesting case. Although a molecular map has already been developed (Fregene et al., 1997; Mba et al., 2001) and many studies of genetic variability among and within landraces have been conducted (Asante and Offei, 2003; Elias et al., 2001a, 2001b; Elias et al., 2004; Peroni and Hanazaki, 2002; Sambatti et al., 2001; Zaldivar et al., 2004), little knowledge on quantitative genetics has so far been generated. Losada presented a Ph.D. thesis on quantitative genetics of cassava in 1990, using a diallel scheme. This work, however, was not published in a scientific journal despite its value. The heterozygous nature of cassava complicates work on improving the existing molecular map and implementing marker-assisted selection. Different authors have suggested that cassava is a segmental allotetraploid (Umanah and Hartman, 1973; Magoon et al., 1969), which would further increase the complexities of gene interactions within and between loci and between homolog and homeolog genomic components.

Cassava is an interesting crop because its vegetative propagation allows the estimation of within-family genetic variation and, indirectly, the relative importance of epistatic effects. Genetic studies analyzing the importance of epistatic effects are not very common, particularly in annual crops. Accurate measurement of epistatic effects for complex traits, such as yield, is difficult and expensive. Reports in the literature on the relevance of epistasis are not as frequent as those estimating additive and dominance variances or effects, and generally they take advantage of the vegetative multiplication that some species offer (Comstock et al., 1958; Stonecypher and McCullough, 1986; Foster and Shaw, 1988; Rönnberg-Wästljung et al., 1994; Rönnberg-Wästljung and Gullberg, 1999; Isik et al., 2003). In many cases, these reports are on forest trees. Also, because of the complexities of these analyses and the costs involved, reports in the literature related to epistatic effects are frequently based on a limited number of genotypes.

Holland (2001) published a comprehensive review on epistasis and plant breeding. Several cases of significant epistasis have been reported in self-pollinated (Brim and Cockerham, 1961; Busch et al., 1974; Gravois, 1994; Hanson and Weber, 1961; Pixley and Frey, 1991; Orf et al., 1999) and cross-pollinated (Ceballos et al., 1998; Eta-Ndu and Openshaw, 1999; Lamkey et al., 1995; Melchinger et al., 1986; Wolf and Hallauer, 1997) crops. According to Holland (2001), finding significant epistasis seems to be easier in self- than in cross-pollinated species and in designs based in the contrasts of means, rather than analyses of variance.

Three different diallel studies were conducted using three different sets of cassava parents targeting specific environments: sub-humid, acid soil, and mid-altitude valley environments (Cach et al., 2005a, 2005b; Calle et al., 2005; Jaramillo et al., 2005; Pérez et al., 2005a; 2005b). These studies took advantage of the vegetative propagation of cassava that allows separation of the genetic from the environmental variation within a family. The within-family analysis allows estimation of the relative importance of epistatic effects (Hallauer and Miranda, 1988).

Sub-humid environment

The analysis of variance and other details of this study have been published (Cach et al., 2005a, 2005b). Based on the magnitude of the estimates for between- and within-family genetic variances, a large proportion of the genetic variability (79–93%) was within-family variation (Table 12.2). As expected, the lowest within-family variation (79% of total genetic variance) was measured for a relatively simply inherited trait, e.g., reaction to thrips (Bellotti, 2002), which showed only additive variance to be statistically significant. The tolerance/resistance in outstanding parents that was transmitted to the progeny tended to accentuate differences between families and reduce the variability among sister clones. However, it is clear that

Table 12.2.   Variances and test for epistasis from the evaluation of a diallel set combining data from two locations (Pitalito and Sto. Tomás) in Atlántico Department, Colombia. The standard error for each estimate is shown in parentheses.

Genetic parameter	Thrips (1–5)	Fresh root yield	Fresh foliage yield	Harvest index	Dry matter content	Dry matter yield
σ^2_G between F_1	0.225	13.09	11.53	0.0010	0.772	0.694
σ^2_G within F_1	0.641	127.21	131.86	0.0037	5.556	9.977
σ^2_A	0.419	17.82	11.93	0.0009	1.452	0.741
	(0.211)	(13.75)	(12.59)	(0.0010)	(0.985)	(0.933)
σ^2_D	0.231	23.87	27.02	0.0027	0.765	1.589
	(0.068)	(11.15)	(10.00)	(0.0011)	(0.497)	(0.919)
Epistasis Test*	0.259	100.40	105.64	0.0013	4.257	8.414
	(0.119)	(12.74)	(11.84)	(0.0009)	(0.673)	(0.990)

* Test for epistasis = $\sigma^2_{c/F1}$ − 3 Cov. FS + 4 Cov. HS.

Table 12.3.   Variances and test for epistasis from the evaluation of a diallel set from ten parents combining data from two different edaphic environments at CORPOICA—La Libertad (Villavicencio) in Meta Department, Colombia. The standard error for each estimate is shown in parentheses.

Genetic parameter	Fresh root yield	Fresh foliage yield	Harvest index	Dry matter content	Plant type score	SED[†] score
σ^2_G between F_1	1.649	1.325	0.0010	1.600	0.089	0.237
	(2.954)	(3.094)	(0.0006)	(0.664)	(0.039)	(0.055)
σ^2_G within F_1	21.082	38.557	0.0030	3.216	0.121	0.088
	(2.297)	(3.242)	(0.0003)	(0.169)	(0.012)	(0.066)
σ^2_A	−1.485	1.172	0.0015	3.379	0.160	0.523
	(6.321)	(8.035)	(0.0016)	(2.399)	(0.144)	(0.234)
σ^2_D	9.028	3.384	0.0011	0.873	0.096	0.092
	(7.930)	(6.594)	(0.0013)	(0.666)	(0.033)	(0.050)
Epistasis Test*	15.054	35.433	0.0014	0.872	−0.031	−0.242
	(6.740)	(6.858)	(0.0012)	(1.294)	(0.077)	(0.139)

* Test for epistasis = σ^2_{G/F_1} − 3 Cov. FS + 4 Cov. HS.
[†] SED = super elongation disease induced by the fungus *Sphaceloma manihoticola.*

a considerable within-family variation still remained even for the reaction to thrips. On the other hand, complex traits, such as root and foliage yields, showed a larger partitioning of the total genetic variance (>90%) into within-family variation, suggesting that there were, comparatively, smaller differences in the breeding values of the progenitors.

Dominance effects were very important for thrips, harvest index, and root and foliage yields, with variance estimates significantly different from zero (estimates two times or more the size of the respective standard error). Only the score for thrips and dry matter content showed larger estimates for the additive, compared with the dominance, variance (Table 12.2). This highlights the importance of heterosis in cassava breeding for many relevant traits, which in turn justifies the implementation of a reciprocal recurrent selection scheme for cassava genetic improvement. Epistatic effects were significant for all variables, except harvest index, based on the test for epistasis.

Acid-soil savannas

Results of the diallel for this ecosystem have been published (Calle et al., 2005; Pérez et al., 2005b). As in the previous diallel, a

large proportion of the genetic variability was detected as within-family variation for fresh root and foliage yields (Table 12.3). The within-family genetic variances for harvest index, dry matter content, and plant type score were larger than for between-family variation, but the difference was not as large as for the root and foliage yields. On the other hand, the score for super-elongation disease (SED), induced by the fungus *Sphaceloma manihoticola,* showed larger variation in the between- compared with the within-family component. Larger between-family variation was observed for reactions to thrips, white flies, and mites in the diallels for the sub-humid and mid-altitude valley environments and with a different set of parental lines (Cach et al., 2005b; Pérez et al., 2005a). The inheritance of resistance to diseases or pests (with a strong dominance component) is relatively simple. It generates large variation between the averages of progenies involving one or two resistant parents compared with those from susceptible ones. In addition, they have relatively little or no variation among the individual genotypes or clones within each family.

The cassava-breeding project at CIAT has recently started to generate data from the earlier phases of the selection process that allow an estimation of the breeding value of parents used in generating these trials. In general, these estimations of breeding values based on the CET will be effective for traits in which the genetic variation is concentrated in the between-family component or shows strong additive effects. Selection of outstanding parents for a given trait, such as SED score, will tend to generate uniform progenies also outstanding for that trait. This, in turn, could allow the implementation of the backward GCA selection described by Mullin and Park in 1992. For characteristics such as fresh root yield, with strong non-additive effects and large within-family variation, the selection of outstanding parents would not be sufficient and individual clone analysis, within a given family, would be required. In this environment, epistatic effects were important only for fresh root and foliage yields. These results agree with those observed in similar studies conducted for the other environments.

Mid-altitude valleys

Results from this diallel have been published (Jaramillo et al., 2005; Pérez et al., 2005a). In this diallel, as in the previous ones, a large proportion of the genetic variability was found to be within-family variation

(Table 12.4). It was surprising to find such a large variation for the within-family component for reaction to the two pests, which was mostly attributable to additive variation.

The magnitude and generalized significance of dominance variance (σ^2_D) highlights the importance of non-additive genetic effects (heterosis) in this allogamous species. Only the reactions to pests showed significant estimates for σ^2_A, not only in this diallel but in the other two, as well. In this third diallel study, fresh-root yield was the only trait showing significant conditioning by epistatic effects, following the same trend observed in the previous studies.

Attempts to quantify epistatic effects frequently fail to reach statistical significance, in part because of the size of the standard errors typical for complex linear functions (Hallauer and Miranda, 1988; Holland, 2001). In these three diallel studies, however, this was not the case, and epistasis was consistently important for complex traits such as fresh-root yield.

The phenotypic clonal selection used for cassava breeding takes advantage of the vegetative reproduction of the crop. In selecting outstanding clones, all genetic effects (additive, dominance, and epistatic) are exploited (Hershey, 1984; Mullin and Park, 1992; Jennings and Iglesias, 2002; Ceballos et al.,

Table 12.4. Variances and test for epistasis from the evaluation of a diallel set from nine parents combining data from two different mid-altitude valleys environments at CIAT, Palmira and Jamundí in Valle del Cauca Department, Colombia. The standard error for each estimate is shown in parentheses.

Genetic parameter	Fresh root yield	Harvest index	Dry matter content	Reaction to mites	Reaction to whiteflies
σ^2_G between F_1	42.8	0.0016	1.19	0.271	0.345
	(13.3)	(0.0004)	(0.43)	(0.067)	(0.115)
σ^2_G within F_1	288.9	0.0029	2.25	0.188	0.119
	(19.2)	(0.0002)	(0.21)	(0.107)	(0.120)
σ^2_A	11.9	0.0029	1.43	0.571	0.994
	(24.7)	(0.0015)	(1.33)	(0.271)	(0.467)
σ^2_D	152.1	0.0018	2.47	0.170	−0.210
	(49.1)	(0.0008)	(0.89)	(0.065)	(0.132)
Epistasis test*	168.9	0.0001	−0.32	−0.225	−0.221
	(40.2)	(0.0010)	(0.92)	(0.179)	(0.279)

* Test for epistasis = $\sigma^2_{c/F1}$ − 3 Cov. FS + 4 Cov. HS.

2004). However, the current recurrent selection system lacks the capacity to direct genetic improvement in such a way that the frequency of favorable genetic combinations (within or between loci) is maximized. To achieve this, special efforts to design parental clones that produce better crosses are required. The development of clones, specifically designed for their use as parents in breeding nurseries, would be one alternative that offers interesting advantages. Introduction of inbreeding in these parental clones would facilitate the gradual and consistent assembly of favorable gene combinations, which in the current system occur just by chance. Inbreeding would also facilitate the reduction of the genetic load of this crop, which is expected to be relatively large at this stage.

One major constraint for the introduction of inbreeding in cassava is the time required for it. The production of doubled haploids through anther or microspore culture is an interesting approach that would reduce the time required to obtain homozygous genotypes. The use of inbred parental lines will facilitate the gradual exploitation of dominance and epistatic genetic variation, which were found to be significant in the diallel studies mentioned above. The strategy would be to implement some scheme (not yet fully defined) of reciprocal recurrent selection. CIAT is currently executing a project financed by the Rockefeller Foundation to develop a protocol for the production of doubled-haploids in cassava.

Approaches to improve the efficiency of cassava genetic improvement

Adequate screening of available genetic variability

Genetic variability available within *Manihot* has not been fully explored and screened. This genetic wealth has not been fully

exploited and, therefore, should offer interesting possibilities for the future. In part, the limited evaluation of cassava genetic variability is due to the fact that collection and maintenance of cassava germplasm is difficult, cumbersome, and expensive. Compared with other crops, a relatively smaller number of accessions is maintained in the germplasm collections (5,728 for cassava vs. 32,445 for beans, in the collections at CIAT). The remarkable genotype-by-environment interaction detected in cassava (Bueno, 1986) complicates the interpretation of evaluations across different environments. Furthermore, detection of some of the economically important traits in the roots is difficult. For instance, the many different starch mutants in maize (popcorn, sweet, floury, waxy corn, etc.) are easily recognizable. No equivalent mutant has so far been reported for cassava.

Nutritional quality factors studied to-date also show relatively low genetic variation, with the exception of the high carotene levels found in yellow cassava roots (Iglesias et al., 1997). However, as a result of new initiatives, an aggressive screening of cassava germplasm allowed Chávez et al. (2005) to report not only interesting variation in carotenoids, but also for crude protein content in the roots. Further analyses (Ceballos et al., 2006) have confirmed the occurrence of cassava clones with 2–3 times higher crude protein contents ($\approx$6–8%) compared with the typical levels found in cassava roots ($\approx$2%).

Another activity that is relevant to the proper screening of genetic variability is the introduction of inbreeding, which allows for the identification of useful recessive traits. CIAT started to systematically self-pollinate cassava germplasm (elite improved clones and materials from the germplasm collection) in 2004. As a result, in early 2006, an interesting mutation was found (Fig. 12.7) in a self-pollinated plant that possesses a starch with smaller granules in its roots. This discovery is

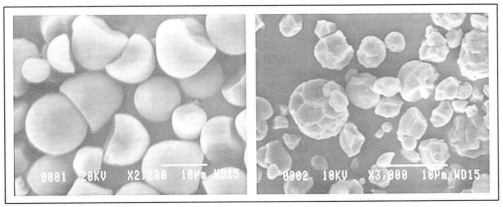

Fig. 12.7. Electron microscope photographs of cassava starch, showing normal granule size (left) and a mutation with smaller granule size (right).

important not only because of the economic value of such trait, but also because it proves the usefulness of introducing inbreeding in cassava genetic improvement.

Stratification of selection in the large Clonal Evaluation Trials (CET)

A major problem with the CET is its large size (easily 2 ha in size) and the unavoidable environmental effect in the selection. This problem is particularly relevant in the case of cassava, because the target environments for cassava are typically in "marginal" agriculture conditions and prone to large variation. Since CETs are frequently the first stage of evaluation, only a few stakes (<10) are available for trials. So the introduction of replications that could help overcome this problem is not practical.

The simple principles suggested by Gardner in 1961 were recently introduced into the evaluations of CETs. The plot where the CET is to be planted is divided into three "blocks" of about equal size, with an attempt to maximize differences among blocks and minimize variation within each block. The replication of each clone is difficult to implement because of the lack of sufficient planting material available for CETs. On the other hand, clones are grouped in either full-

or half-sib families. Generally, since many clones are available from each family, they are randomly allocated to one of these three "blocks." In other words, instead of planting all the clones from a given family together, one after the other, they are split in three groups (Fig. 12.8). This approach provides two interesting advantages:

- There is a replication effect for the families because all the clones from a given family are scattered in three "repetitions" in the field. The averages from all the clones are less affected by environmental variation in such a large experiment.
- Selection is made within each block. This is similar to the stratified mass selection suggested by Gardner. This approach effectively overcomes the environmental variation that can be measured by comparing the means of each block.

Because all the clones from the CET are sub-divided, the average performance of each family can be estimated more precisely, because each family is scattered in three different parts of the field, whereas before it was concentrated in just one sector (Fig. 12.8). As a consequence, the estimates of GCA for each family (described in detail below) are much more precise.

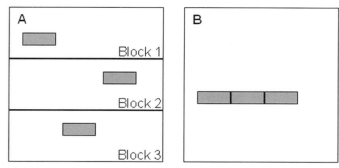

Fig. 12.8. Advantage of splitting each family of clones into three groups that were randomly assigned to each of three blocks in the CET. (A—current procedure; B—previous situation).

Table 12.5. Results of the Clonal Evaluation Trials for the three main target environments harvested in May 2003. Data present the variation between the three blocks into which each CET was divided.

	Yield (t/ha)		Harvest index	Plant type	Dry matter	Selection
Block	Fresh roots	Dry matter	(0 to 1)*	(1 to 5)†	content (%)	Index
	(averages of the 412, 412 and 411 clones in Blocks 1, 2 and 3, respectively from the CET targeting the acid-soil savannas)					
Block 1	20.88	6.66	0.50	3.33	31.59	0.00‡
Block 2	21.73	6.88	0.49	3.35	31.24	0.00‡
Block 3	22.30	7.28	0.50	3.48	32.44	0.00‡
	(averages of the 749, 746 and 705 clones in Blocks 1, 2 and 3, respectively from the CET targeting the sub-humid conditions)					
Block 1	14.19	3.70	0.50	2.87	26.09	0.00‡
Block 2	14.37	3.91	0.46	2.88	27.21	0.00‡
Block 3	12.89	3.38	0.44	2.87	26.26	0.00‡
	(averages of the 605, 588 and 568 clones in Blocks 1, 2 and 3, respectively from the CET targeting the mid-altitude valleys)					
Block 1	24.05	8.86	0.63	2.68	36.61	0.00‡
Block 2	28.08	10.21	0.57	2.63	36.02	0.00‡
Block 3	27.51	9.76	0.54	2.97	35.09	0.00‡

*The harvest index is obtained by dividing the production of commercial roots by total biomass (roots + aerial parts). Preferred harvest indexes are >0.5.
†Plant type integrates under one value, plant architecture, leaves health, and capacity to produce stakes on a scale where 1 = excellent and 5 = very poor is used.
‡Average selection index within blocks must be zero, because it is based on a combination of standardized variables.

A summary of the results from the CET harvested in 2003 for the three main target environments (sub-humid, acid soils, and mid-altitude valleys) is presented in Table 12.5. The benefit of the introduction of stratified selection is directly proportional to the differences between the mean performances in each of the strata. In general, variations in the order of 10–20% have been observed among average performances of the three blocks. These are, in other words, gains in precision attained by introducing the stratification of the CETs. Currently, the possibility of increasing the number of blocks to four or five is under consideration.

Estimation of breeding values from CET data

One of the major decisions made by any breeder is the selection of parents used to produce a new generation of segregating progenies. In cassava, this decision has been mainly based on the *per se* performance of each clone. Nonetheless, some empirical knowledge of the quality of progenies produced by different parents could be developed. The lack of organized information on the breeding values of parental lines used in the breeding projects was partially due to the fact that no data were taken and recorded during the first stages of selection (CET and PYT in Table 12.1), or the data were incomplete. Therefore, it was not possible to generate a balanced set of data that would allow the breeder to have an idea of the relative performance of the progeny of each elite parental line. In other words, no formal process to assess the breeding values of the progenitors used in the cassava-breeding projects was available.

To overcome this problem, the decision was made to record data and to introduce the use of selection indexes. Selection is made within each stratum, as explained in the previous section. Data from each family are then pooled across the three blocks in which it was planted. The stratification means that, in a way, there is a replication effect at the family level. Since a given progenitor may be used more than once, data from all the families in which each progenitor participated are pooled to obtain an idea of the general performance of all the progenies from a given parental clone.

Results from the CET harvested in 2003 for the sub-humid environment have been chosen as an example of the kind of information that the current evaluation system allows. These results are summarized in Table 12.6. A total of 39 parents participated in generating all the progenies evaluated in

Table 12.6. Number of progenies evaluated and selected from each progenitor. Data from the Clonal Evaluation Trial for the sub-humid environment (Santo Tomás, Atlántico Department, Colombia) harvested in 2003.

	Progenitor	Family size	Progenies selected (number)		Progenitor	Family size	Progenies selected (number)
1	R 90	73	45	21	CM 4365-3	41	4
2	KU 50	64	30	22	SM 1657-14	21	2
3	MTAI 8	73	34	23	SM 1210-10	83	7
4	R 5	32	13	24	SM 1201-5	37	3
5	SM 1068-10	68	20	25	SM 1422-4	51	4
6	SM 2192-6	50	12	26	CM 7389-9	103	8
7	SM 1411-5	97	23	27	SM 1521-10	42	3
8	CM 7514-8	118	24	28	SM 1754-21	28	2
9	SM 1657-12	52	10	29	SM 1210-10	101	7
10	SM 643-17	32	6	30	SM 1619-3	29	2
11	MVEN 25	53	9	31	CM 8027-3	46	3
12	SM 1665-2	57	9	32	MNGA 19	215	12
13	CG 1141-1	33	5	33	CM 2772-3	28	1
14	SM 1511-6	87	13	34	SM 1600-4	61	2
15	SM 890-9	69	10	35	CM 7395-5	42	1
16	SM 1433-4	213	26	36	SM 805-15	73	1
17	SM 1565-17	108	13	37	CM 6438-14	53	0
18	CM 3372-4	52	6	38	CM 7514-7	56	0
19	CM 6754-8	49	5	39	SM 1431-2	33	0
20	SM 1438-2	109	11				

that CET. Some parents were used considerably more than others, to a large extent because of their flowering habit in Palmira, Colombia, where the crosses were made. MNGA 19 and SM 1433-4 were used as parents in 215 and 213 clones, respectively. On the other hand, SM 1657-14 and SM 1754-21 were the parents of only 21 and 28 clones, respectively.

The interesting information from Table 12.6 comes from the proportion of clones selected from each half-sib family. For instance, the four best parents, according to the proportion of their progenies being selected, were Rayong 90, KU50 (Kasetsart University 50), Rayong 60, and Rayong 5. All of these clones were developed in Thailand and showed excellent adaptation to the subhumid environment of Colombia. More than 40% of the progenies from each of these parents were selected (see Table 12.6). On the other hand, none of the progenies from CM 6438-14, CM 7514-7 and SM 1431-2 was selected, even though they were not particularly small families (53, 56, and 33 clones, respectively).

This system allows not only the ability to know the proportion of clones derived from a given parent that has been selected, but also the reasons behind these selections. Since there are phenotypic data recorded for each genotype in the CET, an average across all the progeny from a given progenitor for all the variables is available. It was possible, for example, to determine which progeny tended to have above-average fresh root yield or to conclude that the progeny from CM 6438-14 had an unacceptably low DMC (21.9%). This information is very valuable for defining which parents should stay in the crossing blocks, which should be removed, and to suggest crosses that may result in better progenies because the progenitors better complement one another.

Use of molecular markers

A molecular genetic map has been developed for cassava (Fregene et al., 1997).

Cassava genetic improvement can be made more efficient through the use of easily assayable molecular genetic or DNA markers (MAS) that enable the precise identification of genotype without the confounding effect of the environment, by increasing heritability. MAS can also contribute toward the efficient reduction of large breeding populations at the seedling stage based upon a "minimum selection criterion." This is particularly important given the length of the growing cycle of cassava and the expenses involved in the evaluation process. Therefore, a pre-selection at the F_1 phase could greatly enhance the efficiency of the CET trials. The selection of progenies based on genetic values derived from molecular marker data substantially increases the rate of genetic gain, especially if the number of cycles of evaluation or generations can be reduced (Meuwissen et al., 2001).

Another application of MAS in cassava breeding is reducing the length of time required for the introgression of traits from wild relatives. Wild relatives are an important source of genes for pest and disease resistance in cassava (Hahn et al., 1980a, b; Chavarriaga et al., 2004), but the need to reduce or eliminate undesirable donor genome content or linkage drag can lengthen the process, making it an unrealistic approach for most breeders. Simulations by Stam and Zeven (1981) indicate that markers could reduce linkage drag and would reduce the number of generations required in the backcross scheme. Hospital et al. (1992) corroborated this by achieving a reduction of two backcross generations with the use of molecular marker selection. Frisch et al. (1999), through a simulation study, found that use of molecular markers for the introgression of a single target allele saved two to four backcross generations. They inferred that MAS had the potential to reach the same level of recurrent parent genome in generation BC_3 as reached in BC_7 without the use of molecular markers. Below, a few examples of the use of molecular markers implemented for cassava genetic improvement are given.

Selection for casssava mosaic disease An ideal target for MAS is breeding for disease resistance in the absence of the pathogen. This is the case for the cassava mosaic disease (CMD) in the Americas, where the disease does not occur. Molecular markers that allow selection of segregating progenies that carry the resistance to CMD have been successfully utilized (Fregene et al., 2006).

Introgression of useful traits from wild relatives Wild *Manihot* germplasm offers a wealth of useful genes for the cultivated *M. esculenta* species, but its use in regular breeding programs is restricted due to linkage drag and the long reproductive breeding cycle. However, the use of molecular markers can facilitate the introgression of a single target region (Frisch et al., 1999). It has been shown in several crops that the "tremendous genetic potential" locked up in wild relatives can be released more efficiently through the aid of new tools of molecular genetic maps and the advanced backcross QTL mapping scheme (ABC-QTL). CIAT is currently implementing a modification of ABC-QTL to introgress genes for high protein content, waxy starch, delayed PPD, and resistance to whiteflies and the hornworm.

Mutation breeding and DNA TILLING The Targeted Induced Local Lesions in Natural Genomes (TILLING) is a technique based on single nucleotide polymorphism (SNP) detection via hetero-duplex analysis using the nuclease *Cel I* (McCallum et al., 2000). In 2004 the National University of Colombia and CIAT initiated a collaborative project on mutation breeding and DNA TILLING to identify novel starches, including waxy starch. More than 2,000 sexual seeds were irradiated using gamma rays (a Cobalt-60 source) and fast neutrons at the International Atomic Energy Agency (IAEA), Vienna, Austria and shipped back to CIAT. The seeds were germinated, transferred to the field and more than 1,000 genotypes were self-pollinated (to overcome the problem of chimeras present in the M_1 generation) and produced more than 3,000 S_1 plants. Work will focus on starch biosynthetic genes: the granule-bound starch synthase (GBSSI) for the production of amylose, and the soluble starch synthase genes (SSSI, SSSII, SSSIII) for the production of amylopectin. DNA TILLING will be applied to these plants to detect sense point mutations and deletions in these genes.

Estimation of average heterozygosity during inbreeding of cassava Cassava genotypes are heterozygous, and very little inbreeding has been practiced until now in cassava breeding. But inbred lines are better as parents because they do not have the confounding effect of dominance, and they carry lower levels of genetic load (undesirable alleles). Considerations in inbreeding heterozygous crops are arerage heterozygosity of the original parental lines, homozygosity level of selected genotypes at the end of self-pollinating phase, and the speed of self-pollination process (Scotti et al., 2000). It is expected that the first few cycles of self-pollination will result in marked reduction of vigor (inbreeding depression associated with the genetic load of the parental lines). Therefore, selection for tolerance to inbreeding depression must be exerted. However, this selection is biased by the differences in homozygosity levels of segregating partially inbred genotypes. This highlights the need for a method to measure the heterozygosity level in these partially inbred individuals to be used in a co-variance correction in the selection of phenotypically vigorous genotypes. Eventually, molecular markers can also be used for determining regions in the genome that are particularly related to the expression of heterosis and for measuring genetic distances among inbred lines to direct crosses with higher probabilities of expression of a high level of heterosis.

Other potential MAS targets There are several additional work activities based on molecular markers for the genetic improvement of cassava. The most important of these activities are the delineation of heterotic groups, β-carotene content, cyanogenic potential, and dry matter content in the roots.

Introduction of inbreeding

In different sections of this chapter, the usefulness of the introduction of inbreeding in cassava genetic improvement has been mentioned. Below, a summary of all these advantages is provided:

- Reduction of genetic load
- Discovery of useful recessive traits
- Possibility of better exploitation of dominance and epistatic effects
- Possibility of implementing the traditional backcross scheme to facilitate the transfer of simply inherited traits
- Germplasm exchange and storage
- Possibility of making planting material pathogen-free

Genetic transformation

Genetic transformation of cassava was first achieved by Calderón-Urrea in 1988. Since then, several laboratories have improved the protocol for the transformation and/or regeneration of calli (Taylor et al., 2004). Transformation has been achieved in a variety of cassava germplasm and the introduction of several different genes has been accomplished. Traits to be improved include modification of starches, introduction of enhanced protein content, and silencing the genes related to the molecular pathway for the production of cyanogenic glucosides.

Prospects and challenges

During the past 30–40 years, significant progress has been achieved in the initial phase of the scientific genetic improvement

of cassava. One could say that the adaptation of the crop to more intensive cultivation systems has been completed. This process involved assembling major traits such as improved yield (mainly through a higher harvest index), low cyanogenic content (when desirable), improved plant architecture, and resistance/tolerance to the major diseases and pests.

Future activities involve an increasing emphasis on complex traits, such as higher yield and dry matter content in the roots, early bulking, etc., which are more difficult to improve. It is critical for cassava that efficient methods for the improvement of these complex traits are found to maintain the competitive edge that this crop currently has in tropical regions as an alternative to imported carbohydrate sources from temperate regions. Several approaches have been used to address this situation in recent years. Modifications of the breeding scheme have been implemented for a more dynamic recurrent selection system and for obtaining valuable information on the breeding value of parental clones. Biotechnology tools have been adapted to cassava and are currently incorporated in different projects for its genetic improvement. A molecular map has been developed (Fregene et al., 1997, 2000; Mba et al., 2001) and marker-assisted selection is currently used for key traits. Genetic transformation protocols are available and have been used successfully for the incorporation of different genes (Taylor et al., 2004). Tissue culture techniques can also benefit cassava through the production of doubled-haploid lines.

Challenges for the crop include increasing the number of accessions in germplasm collections and developing approaches that will allow for an efficient evaluation of such germplasm. In this regard, tools for rapid identification of novel starch types are needed. The lack of genetic variability for overcoming the problem of post-harvest physiological deterioration remains a major bottleneck for cassava utilization and commercialization. For logistical reasons, the unresolved problem of frog skin disease remains a frustrating bottleneck for researchers at CIAT Experimental Station in Palmira, Colombia. The inherent potential of cassava, its capacity to grow in marginal environments, and the incorporation of new, powerful biotechnology tools as described in several of the references provided in this review, offer bright prospects for the crop and the people that depend on it.

References

Allem, A.C. 2002. The origins and taxonomy of cassava. In: Hillocks, R.J., Tresh, J.M. and Bellotti, A.C. (eds.). Cassava: Biology, production and utilization. CABI Publishing, pp. 1–16.

Alves, A.A.C. 2002. Cassava botany and physiology. In: Hillocks, R.J., Tresh, J.M. and Bellotti, A.C. (eds.). Cassava: Biology, production and utilization. CABI Publishing, pp. 67–89.

Andersen, M.D., Busk, P.K., Svendsen, I. and Møller, B.L. 2000. Cytochromes P-450 from cassava (*Manihot esculenta* Crantz) catalyzing the first steps in the biosynthesis of the cyanogenic glucosides linamarin and lotaustralin. The Journal of Biological Chemistry 275(3): 1966–1975.

Asante, I.K. and Offei, S.K. 2003. RAPD-based genetic diversity study of fifty cassava (*Manihot esculenta* Crantz) genotypes. Euphytica 131:113–119.

Balagopalan, C. 2002. Cassava utilization in food, feed and industry. In: Hillocks, R.J., Tresh, J.M. and Bellotti, A.C. (eds.). Cassava: Biology, production and utilization. CABI, Publishing, pp. 301–318.

Babu, L. and Chatterjee, S.R. 1999. Protein content and amino acid composition of cassava tubers and leaves. J. Root Crops 25(20): 163–168.

Beeching, J.R., Yuanhuai, H., Gómez-Vázquez, R., Day, R.C. and Cooper, R.M. 1998. Wound and defense responses in cassava as related to post-harvest physiological deterioration. In: Romeo, J. T., Downum, K.R. and Verpporte, R. (eds.) Recent advances in phytochemistry: Phytochemical signals in plant-microbe interactions. Plenum Press, New York and London. 32:231–248.

Bellotti, A.C. 2002. Arthropod pests. In: Hillocks, R.J., Thresh, J.M. and Bellotti, A.C. (eds.) Cassava: Biology, production and utilization. CABI Publishing, pp. 209–235.

Bellotti, A.C., Arias, V.B., Vargas, H.O., Reyes, Q.J.A. and Guerrero, J.M. 2002. Insectos y ácaros dañinos a la yuca y su control. In: Ospina, B. and Ceballos,

H. (eds.) La yuca en el tercer milenio: Sistemas modernos de producción, procesamiento, utilización y comercialización. CIAT Publication No. 327. Apartado Aéreo 6713, Cali, Colombia, pp. 160–203.

Brim, C.A. and Cockerham, C.C. 1961. Inheritance of quantitative characters in soybeans. Crop Sci. 1:187–190.

Busch, R.H., Janke, J.C., Frohberg, R.C. 1974. Evaluation of crosses among high and low yielding parents of spring wheat (*Triticum aestivum* L.) and bulk prediction of line performance. Crop Sci. 14:47–50.

Bueno, A. 1986. Avaliacão de cultivares de mandioca visando a seleção de progenitores para cruzamentos. R. Bras. Mand. 5(1):23–54.

Buitrago, A.J.A. 1990. La yuca en la alimentación animal. Centro Internacional de Agricultura Tropical (CIAT), Cali, Colombia, 446 p.

Cadavid López, L.F. and Gil Llanos, L. 2003. Investigación en producción de yuca forrajera en Colombia. Informe annual de Actividades CLAYUCA. Apdo Aéreo 6713, Cali Colombia, pp. 266–275.

Cach, N.T., Pérez, J.C., Lenis, J.I., Calle, F., Morante, N. and Ceballos, H. 2005a. Epistasis in the expression of relevant traits in cassava (*Manihot esculenta* Crantz) for subhumid conditions. Journal of Heredity 96(5): 586–592.

Cach, N.T., Lenis, J.I., Pérez, J.C., Morante, N., Calle, F. and Ceballos, H. 2005b. Inheritance of relevant traits in cassava (*Manihot esculenta* Crantz) for sub-humid conditions. Plant Breeding 124:1–6.

Calderón-Urrea, A. 1988. Transformation of *Manihot esculenta* (cassava) using *Agrobacterium tumefaciens* and expression of the introduced foreign genes in transformed cell lines. MSc thesis. Vrije University. Brussels. Belgium.

Calle, F., Pérez, J.C., Gaitán, W., Morante, N., Ceballos, H., Llano, G. and Alvarez, E. 2005. Diallel inheritance of relevant traits in cassava (*Manihot esculenta* Crantz) adapted to acid-soil savannas. Euphytica 144(1–2): 177–186.

Carvalho, L.J.C.B., de Souza, C.R.B., Cascardo, J.C.M., Junior, C.B. and Campos, L. 2004. Identification and characterization of a novel cassava (*Manihot esculenta* Crantz) clone with high free sugar content and novel starch. Plant Molecular Biology 56:643–659.

Ceballos, H., Fregene, M., Lentini, Z., Sánchez, T., Puentes, Y.I., Pérez, J.C., Rosero, A. and Tofiño, A.P. 2006. Development and Identification of High-Value Cassava Clones. Acta Horticulturae (in press).

Ceballos, H., Iglesias, C.A., Pérez, J.C. and Dixon, A. G.O. 2004. Cassava breeding: Opportunities and challenges. Plant Molecular Biology 56:503–515.

Ceballos, H., Pandey, S., Narro, L. and Pérez, J.C. 1998. Additive, dominance, and epistatic effects for maize grain yield in acid and non-acid soils. Theor. Appl. Genet. 96:662–668.

Ceballos, H., Sánchez, T., Chávez, A.L., Iglesias, C., Debouck, D., Mafla, G. and Tohme, J. 2006. Variation in crude protein content in cassava *(Manihot esculenta* Crantz) roots. J. Sci. Food Agric. (in press).

Chavarriaga, P., Prieto, S., Herrera, C.J., López, D., Bellotti, A. and Tohme, J. 2004. Screening transgenics unveils apparent resistance to hornworm (*E. ello*) in the non-transgenic, African cassava clone 60444. In: Alves and Tohme (eds.) Adding value to a small-farmer crop: Proceedings of the Sixth International Scientific Meeting of the Cassava Biotechnology Network. 8–14 March 2004, CIAT, Cali Colombia. Book of Abstracts pp. 4.

Chávez, A.L., Sánchez, T., Jaramillo, G., Bedoya, J.M., Echeverry, J., Bolaños, E.A., Ceballos, H. and Iglesias, C.A. 2005. Variation of quality traits in cassava roots evaluated in landraces and improved clones. Euphytica 143:125–133.

Chávez, A.L., Bedoya, J.M., Sánchez, T., Iglesias, C.A., Ceballos, H. and Roca, W. 2000. Iron, carotene, and ascorbic acid in cassava roots and leaves. Food and Nutrition Bulletin. 21:410–413.

CIAT (Centro Internacional de Agricultura Tropical), 2002. Project IP3, Improved Cassava for the Developing World, Annual Report 2002. Apdo Aéreo 6713, Cali, Colombia.

CIAT (Centro Internacional de Agricultura Tropical), 2001. Project IP3, Improved Cassava for the Developing World, Annual Report 2001. Apdo Aéreo 6713, Cali, Colombia.

Cock, J. 1985. Cassava: New potential for a neglected Crop. Westview Press. Boulder, CO., USA, 240 pp.

Comstock, R.E., Kelleher, T. and Morrow, E.B. 1958. Genetic variation in an asexual species: The garden strawberry. Genetics 43:634–646.

DeVries, J. and Toenniessen, G. 2001. Securing the harvest: Biotechnology, breeding and seed systems for African crops. CABI Publishing Oxon, UK and New York, USA, pp. 147–156.

Dixon, A.G.O., Asiedu, R. and Bokanga, M. 1994. Breeding of cassava for low cyanogenic potential: Problems, progress and perspectives. Acta Horticulturae 375:153–161.

Du, L., Bokanga, M., Møller, B.L. and Halkier, B.A. 1995. The biosynthesis of cyanogenic glucosides in roots of cassava. Phytochemistry 39(2): 323–326.

Easwari Amma, C.S. and Sheela, M.N. 1998. Genetic analysis in a diallel cross of inbred lines of cassava. Madras Agric. J. 85:264–268.

Easwari Amma, C.S. and Sheela, M.N. 1995. Combining ability, heterosis and gene action for three major quality traits in cassava. J. Root Crops 21:24–29.

Easwari Amma, C.S. and Sheela, M.N. 1993. Heterosis in cassava: Nature and magnitude. Proc. Tropical Tuber Crops Problems, Prospects and Future Strategies, ISRC, Trivandrum, India. 6–9 November, pp. 88–94.

Easwari Amma, C.S., Sheela, M.N. and Thankamma Pillai, P.K. 1995. Combining ability analysis in cassava. J. Root Crops 21:65–71.

Elias, M., Mühlen, G.S., McKey, D., Roa, A.C. and Tohme, J. 2004. Genetic diversity of traditional South American landraces of cassava (*Manihot esculenta* Crantz): An analysis using microsatellites. Economic Botany 58:242–256.

Elias, M., McKey, D., Panaud, O., Anstett, M.C. and Robert, T. 2001a. Traditional management of cassava morphological and genetic diversity by the Makushi Amerindians (Guyana, South America): Perspectives for on-farm conservation of crop genetic resources. Euphytica 120:143–157.

Elias, M., Penet, L., Vindry, P., McKey, D., Panaud, O. and Robert, T. 2001b. Unmanaged sexual reproduction and the dynamics of genetic diversity of a vegetatively propagated crop plant: Cassava (*Manihot esculenta* Crantz) in a traditional farming system. Molecular Ecology 10:1895–1907.

Ellis, R.H., Hong, T.D. and Roberts, E.H. 1982. An investigation of the influence of constant and alternating temperature on the germination of cassava seed using a two-dimentional temperature gradient plate. Annals of Botany 49:241–246.

Eta-Ndu, J.T., Openshaw, S.J. 1999. Epistasis for grain yield in two F2 populations of maize. Crop Sci. 39:346–352.

Foster, G.S. and Shaw, D.V. 1988. Using clonal replicates to explore genetic variation in a perennial plant species. Theor. Appl. Genet. 76:788–794.

Fregene, M., Okogbenin, E., Marin, J., Moreno, I., Ariyo, O., Akinwale, O., Barrera, E., Ceballos, H. and Dixon, A. 2006. Molecular Marker-assisted Selection (MAS) of resistance to the cassava mosaic disease (CMD). Theor. Appl. Genet. (in press).

Fregene, M. and Puonti-Kaerlas, J. 2002. Cassava biotechnology. In: Hillocks, R.J., Thresh, J.M. and Bellotti, A.C. (eds.), Cassava: Biology, production and utilization. CABI Publishing, pp. 179–207.

Fregene, M., Bernal, A., Duque, M., Dixon, A. and Tohme, J. 2000. AFLP analysis of African cassava (*Manihot esculenta* Crantz) germplasm resistant to the cassava mosaic disease (CMD). Theor. Appl. Genet. 100:678–685.

Fregene, M., Angel, F., Gomez, R., Rodríguez, F., Chavarriaga, P., Roca, W. and Tohme, J. 1997. A molecular genetic map of cassava (*Manihot esculenta* Crantz). Theor. Appl. Genet. 95:431–441.

Frisch, M., Bohn, M. and Melchinger, A.E. 1999. Comparison of Seletion Strategies for Marker-Assisted Backcrossing of a Gene. Crop Sci. 39:1295–1301.

Gardner, C.O. 1961. An evaluation of effects of mass selection and seed irradiation with thermal neutrons on yields of corn. Crop Sci. 1:241–245.

Gomez, G., Santos, J. and Valdivieso, M. 1983. Utilización de raíces y productos de yuca en alimentación animal. In: Domínguez, C.E. (ed.) Yuca: Investigación, producción y utilización. Working Document No. 50. Centro Internacional de Agricultura Tropical (CIAT), Cali, Colombia.

Gonçalvez Fukuda, W.M. and Saad, N. 2001. Participatory research in cassava breeding with farmers in Northeastern Brazil. Document CNPMF No. 99. EMBRAPA, Cruz das Almas. Bahia, Brazil.

Gonçalvez Fukuda, W.M., Fukuda, C., Leite Cardoso, C.E., Lima Vanconcelos, O. and Nunes, L.C. 2000. Implantação e evolução dos trabalhos de pesquisa participativa em melhoramento de mandioca no nordeste Brasileiro. Documento CNPMF No. 92. EMBRAPA, Cruz das Almas. Bahia, Brazil.

Hahn, S.K., Terry, E.R., Leuschner, K., Akobundu, I.O., Okali, C. and Lal, R. 1979. Cassava improvement in Africa. Field Crops Research 2:193–226.

Hahn, S.K., Terry, E.R. and Leuschner, K. 1980a. Breeding cassava for resistance to cassava mosaic disease. Euphytica 29:673–683.

Hahn, S.K., Howland, A.K. and Terry, E.R. 1980b. Correlated resistance of cassava to mosaic and bacterial blight diseases. Euphytica 29:305–311.

Hallauer, A.R. and Miranda Fo, J.B. 1988. Quantitative Genetics in Maize Breeding. Second Edition. Iowa State University Press. USA, pp. 45–114.

Hanson, W.D. and Weber, C.R. 1961. Resolution of genetic variability in self-pollinated species with an application to the soybean. Genetics 46:1425–1434.

Hershey, C.H. 1984. Breeding cassava for adaptation to stress conditions: Development of a methodology. In: Proceedings of the 6th Symposium of the International Society for Tropical Root Crops. Lima, Peru. 20–25 February, 1983.

Hillocks, R.J. and Wydra, K. 2002. Bacterial, fungal and nematode diseases. In: Hillocks, R.J., Thresh, J.M. and Bellotti, A.C. (eds.), Cassava: Biology, production and utilization. CABI Publishing, pp. 261–280.

Hospital, F., Chevalet, C. and Mulsant, P. 1992. Using markers in gene introgression breeding programs. Genetics 132:1119–1210.

Iglesias, C.A., Mayer, J., Chávez, A.L. and Calle, F. 1997. Genetic potential and stability of carotene content in cassava roots. Euphytica 94:367–373.

Iglesias, C.A. and Hershey, C. 1994. Cassava breeding at CIAT: Heritability estimates and genetic progress in the 1980s In: Ofori, F. and Hahn, S.K. (eds.) Tropical root crops in a developing economy. ISTRC/ISHS, Wageningen, Netherlands, pp. 149–163.

Iglesias, C.A., Hershey, C., Calle, F. and Bolaños, A. 1994. Propagating cassava (Manihot esculenta Crantz) by sexual seed. Exp. Agric. 30:283–290.

Jaramillo, G., Morante, N., Pérez, J.C., Calle, F., Ceballos, H., Arias, B. and Bellotti, A.C. 2005. Diallel analysis in cassava adapted to the midaltitude valleys environment. Crop Sci. 45:1058–1063.

Jennings, D.L. and Iglesias, C.A. 2002. Breeding for crop improvement. In: Hillocks, R.J., Thresh, J.M. and Bellotti, A.C. (eds.) Cassava: Biology, production and utilization. CABI Publishing, pp. 149–166.

Kawano, K. 1980. Cassava. In: Fehr, W.R. and Hadley, H.H. (eds.) Hybridization of crop plants. ASA, CSSA. Madison, Wisconsin, pp. 225–233.

Kawano, K. 2003. Thirty years of cassava breeding for productivity—biological and social factors for success. Crop Sci. 43:1325–1335.

Kawano, K., Daza, P., Amaya, A., Ríos, M. and Gonçalvez, M.F. 1978. Evaluation of cassava germplasm for productivity. Crop Sci. 18:377–380.

Kawano, K., Narintaraporn, K., Narintaraporn, P., Sarakarn, S., Limsila, A., Limsila, J., Suparhan, D., Sarawat, V. and Watananonta, W. 1998. Yield improvement in a multistage breeding program for cassava. Crop Sci 38(2): 325–332.

Ladino, J., Mancilla, L.I., Chavarriaga, P., Tohme, J. and Roca, W.M. 2001. Transformation of cassava cv. TMS60444 with A. tumefaciens carrying a cry 1Ab gene for insect resistance. Proceeding of the Fifth International Scientific Meeting of the Cassava Biotechnology Network, Donald Danforth Plant Science Center. St. Louis, Missouri, USA. November 4–9, 2001.

Lamkey, K.R., Schnicker, B.J. and Melchinger, A.E. 1995. Epistasis in an elite maize hybrid and choice of generation for inbred line development. Crop Sci. 35:1272–1281.

Lenis, J.I., Calle, F., Jaramillo, G., Pérez, J.C., Ceballos, H. and Cock, J. 2006. Leaf retention and cassava productivity. Field Crops Res. 95(2–3): 126–134.

Magoon, M.L., Krishnan, R. and Vijaya Bai, K. 1969. Morphology of the pachytene chromosomes and meiosis in Manihot esculenta Crantz. Cytologia 34:612–625.

Mba, R.E.C., Stephenson, P., Edwards, K., Melzer, S., Mkumbira, J., Gullberg, U., Apel, K., Gale, M., Tohme, J. and Fregene, M. 2001. Simple sequence repeat (SSR) markers survey of the cassava (Manihot esculenta Crantz) genome: Towards an SSR-based molecular genetic map of cassava. Theor. Appl. Genet. 102:21–31.

McCallum, C.M., Comai, L., Greene, E.A. and Henikoff, S. 2000. Targeting induced local lesions in genomes (TILLING) for plant functional genomics. Plant Physiol. 123:439–442.

McMahon, J.M., White, W.L.B. and Sayre, R.T. 1995. Cyanogenesis in cassava (Manihot esculenta Crantz). Journal of Experimental Botany 46:731–741.

Melchinger, A.E., Geiger, H.H. and Schnell, F.W. 1986. Epistasis in maize (*Zea mays* L.). 2. Genetic effects in crosses among early flint and dent inbred lines determined by three methods. Theor. Appl. Genet. 72:231–239.

Meuwissen, T.H.E., Hayes, B.J. and Goddard, M.E. 2001. Prediction of Total Genetic Value Using Genome-Wide Dense Marker Maps. Genetics 157:1819–1829.

Mullin, T.J. and Park, Y.S. 1992. Estimating genetic gains from alternative breeding stratregies for clonal forestry. Can. J. For. Res. 22:14–23.

Nyiira, Z.M. 1975. Advances in research on the economic significance of the green cassava mite *Mononychellus tanajoa* Bondar in Uganda. International exchange and testing of cassava germplasm in Africa. In: Terry, E.R. and MacIntyre, R. (eds.) Proceedings of an interdisciplinary workshop. Ibadan, Nigeria, 17–21 November, 1975. IDRC-063e, Ottawa, Canada, pp. 22–29.

Orf, J.H., Chase, K., Adler, F.R., Mansur, L.M. and Lark, K.G. 1999. Genetics of soybean agronomic traits: II. Interactions between yield quantitative trait loci in soybean. Crop Sci. 39:1652–1657.

Pandey, S. and Gardner, C.O. 1992. Recurrent selection for population, variety and hybrid improvement in tropical maize. Advances in Agronomy 48:1–87.

Pérez, J.C., Ceballos, H., Jaramillo, G., Morante, N., Calle, F., Arias, B. and Bellotti, A.C. 2005a. Epistasis in cassava adapted to mid-altitude valley environments. Crop Sci. 45:1491–1496.

Pérez, J.C., Ceballos, H., Calle, F., Morante, N., Gaitán, W., Llano, G. and Alvarez, E. 2005b. Within-family genetic variation and epistasis in cassava (*Manihot esculenta* Crantz) adapted to the acid-soils environment. Euphytica 145 (1–2):77–85.

Peroni, N. and Hanazaki, N. 2002. Current and lost diversity of cultivated varieties, especially cassava, under swidden cultivation systems in the Brazilian Atlantic Forest. Agriculture, Ecosystems and Environment 92:171–183.

Pixley, K.V. and Frey, K.J. 1991. Combining ability for test weight and agronomic traits of oat. Crop Sci. 31:1448–1451.

Puonti-Kaerlas, J., Li, H.Q., Sautter, C. and Potrykus, I. 1997. Production of transgenic cassava (*Manihot esculenta* Crantz) via organogenesis and *Agrobacteruim*-mediated transformation. African Journal of Root and Tuber Crops 2:181–186.

Rajendran, P.G., Ravindran, C.S., Nair, S.G. and Nayar, T.V.R. 2000. True cassava seeds (TCS) for rapid spread of the crop in non-traditional areas. Central Tuber Crops Research Institute (Indian Council of Agricultural Research). Thiruvananthapuram, 695 017, Kerala, India.

Rönnberg-Wästljung, A.C. and Gullberg, U. 1999. Genetics of breeding characters with possible effects on biomass production in *Salix viminalis* (L.). Theor. Appl. Genet. 98:531–540.

Rönnberg-Wästljung, A.C., Gullberg, U. and Nilsson, C. 1994. Genetic parameters of growth characters in *Salix viminalis* grown in Sweden. Can. J. For. Res. 24:1960–1969.

Sambatti, J.B.M., Martins, P.S. and Ando, A. 2001. Folk taxonomy and evolutionary dynamics of cassava: A case study in Ubatuba, Brazil. Economic Botany 55:93–105.

Sánchez, T., Chávez, A.L., Ceballos, H., Rodriguez-Amaya, D., Nestel, P. and Ishitani, M. 2006. Reduction or delay of post-harvest physiological deterioration in cassava roots with higher carotenoid content. Journal of the Science of Food and Agriculture 86(4): 634–639.

Scott, G.J., Rosegrant, M.W. and Ringler, C. 2000. Roots and tubers for the 21st centruy. Trends, projections, and policy options. International Food Policy Research Institute (IFPRI)/Centro Internacional de la papa (CIP). Wasington, USA, 64 p.

Scotti, C., Pupilli, F., Salvi, S. and Arcioni, S. 2000. Variation in Vigot and in RFLP-estimated heterozygosity by selfing tetraploid alfalfa: New perspectives for the use of selfing in alfalfa breeding. Theor. Appl. Genet. 101:120–125.

Stam, P. and Zeven, A.C. 1981. The theoretical proportion of donor genome in near-isogenic lines of self-fertilizers bred by backcrossing. Euphytica 30:227–238.

Stonecypher, R.W. and McCullough, R.B. 1986. Estimates of additive and non-additive genetic variance from a cloneal diallel of Douglas-fir *Pseudotsuga menziesii* (Mirb.) Franco. In: Proc. Int. Union for Res. Org., Joint Mtg. Working Parties Breed Theor., Prog. Test, Seed Orch. Williamsburg, VA. Publ by NCSU-Industry Coop Tree Imp. Pgm., pp. 211–227.

Taylor, N., Cavarriaga, P., Raemakers, K., Siritunga, D. and Zhang, P. 2004. Development and application of transgenic technologies in cassava. Plant Molecular Biology 56:671–688.

Tewe, O. 2004. Cassava for livestock feed in Sub-Saharan Africa. The Global Cassava Development Strategy. NeBambi, L. (Coordinator). Food and Agriculture Organization of the United Nations (FAO), Rome, Italy.

Umanah, E.E. and Hartmann, R.W. 1973. Chromosome numbers and karyotypes of some *Manihot* species. J. Amer. Soc. Hort. Sci. 8(3): 272–274.

Wheatley, C.C. and Chuzel, G. 1995. Cassava: The nature of the tuber and use as a raw material. In: Macrae, R., Robinson, R.K. and Sadler, M.J. (eds.) Encyclopaedia of Food Science, Food Technology and Nutrition. Academic Press, San Diego, California, pp. 734–743.

Wheatley, C.C., Sanchez, T. and Orrego, J.J. 1993. Quality evaluation of the core cassava collection at

CIAT. In: Roca, W.M. and Thro, A.M. (eds.) Proc. of the First Intl. Scientific Meeting of the Cassava Biotechnology Network, Cartagena, Colombia, August 1992. CIAT, Cali, Colombia, pp. 255–264.

Wolf, D.P. and Hallauer, A.R. 1997. Triple test cross analysis to detect epistasis in maize. Crop Sci. 37:763–770.

Wright, C.E. 1965. Field plans for a systematically designed polycross. Record of Agricultural Research 14:31–41.

Zaldivar, M.E., Rocha, O.J., Aguilar, G., Castro, L., Castro, E. and Barrantes, R. 2004. Genetic variation of cassava (*Manihot esculenta* Crantz) cultivated by Chibchan Amerindians of Costa Rica. Economic Botany 58:204–213.

Chapter 13

Banana Breeding

Michael Pillay and Leena Tripathi

Origin and classification of *Musa*

Bananas and plantains (*Musa* spp.), hereafter referred to as banana, originated in the Southeast Asian and Western Pacific regions where wild species are still in existence in natural forests (Robinson, 1996). Consequently, this region is considered to be the place where bananas were first domesticated and is known as the primary center of *Musa* diversity. Centers of diversity are areas where the ancestors of contemporary crop species developed and adapted themselves over a long period of time. Secondary centers of diversity are also recognized for some of the subgroups of *Musa*. The humid lowlands of West and Central Africa are considered to be the secondary center of diversification for plantains (Swennen and Rosales, 1994), whereas the Highlands of East Africa are recognized as the secondary center of diversity for the East African Highland cooking and beer bananas (De Langhe 1961, 1964; Stover and Simmonds, 1987). In both cases, the diversity observed in these subgroups resulted from an accumulation of somatic mutations, which was enhanced by the long history of cultivation of these banana groups. These centers contain the total gene pool available for research and breeding purposes in *Musa*.

Banana belongs to the genus *Musa*, which is one of the two genera in the family Musaceae and order Zingiberales. *Musa* is a rela-

tively small genus with about 30–40 species. The number of species in the genus *Musa* is under constant revision; a number of new species have been recently identified (Hakkinen, 2004; Hakkinen and Meekiong, 2005; Valmayor et al., 2004). On the basis of morphology and basic chromosome numbers, the species in the genus *Musa* were divided into five sections, viz., *Eumusa*, *Rhodochlamys*, *Australimusa*, *Callimusa*, and *Ingentimusa*. Species in sections *Eumusa* and *Rhodochlamys* have 2n = 22 chromosomes, whereas those in *Australimusa* and *Callimusa* have 2n = 20 chromosomes. A single species, *M. ingens* with 2n = 14 chromosomes, is recognized in the section *Ingentimusa*. Most edible bananas are in section *Eumusa*, which is the largest and most widely distributed section. Simmonds (1962) reported that all edible bananas except for a small group called "Fei" bananas were derived from inter- and/or intra-specific hybridization from two wild diploid species, *M. acuminata* and *M. balbisiana* that contain the AA and BB genomes, respectively. *Musa acuminata* is a complex of subspecies that have been grouped in several ways by taxonomists. The general consensus is that there are nine subspecies in the *M. acuminata* complex. These are subspecies: *banksii*, *burmannica*, *burmannicoides*, *errans*, *malaccensis*, *microcarpa*, *siamea*, *truncata*, and *zebrina* (De Langhe and Devreux, 1960; Shepherd, 1988; Tezenas du Montcel, 1988). On the other hand, molecular data suggest that there are fewer subspecies, perhaps three, viz., *burmannica*, *malaccensis*, and *microcarpa* within the *M. acuminata* complex (Ude et al., 2002a). If information on all the accessions from the different subspecies were available, further research could accurately determine subspecies classification in this complex.

Musa balbisiana, the other progenitor of modern bananas, is widely distributed in tropical and subtropical Asia, including China, India, Indonesia, Malaysia, Myanmar, Nepal, Papua New Guinea, the Philippines, Sri Lanka, Thailand, and Vietnam (Ge et al., 2005). No subspecies classification is present in *M. balbisiana* although new research findings show that there is substantial molecular and morphological differentiation within the species (Ude et al., 2002b; Ge et al., 2005). *Musa balbisiana* has numerous valuable agronomic traits, e.g., disease and pest resistance and the ability to thrive in dry and cold environments.

Four genomes, A, B, S, and T, are known to be present in cultivated bananas (Pillay et al., 2004). The A, B, and S genomes are representative of the section *Eumusa*. The S genome is present in a single species, *M. schizocarpa*, while the T genome is characteristic of species in the *Australimusa*. The S and T genomes are known to exist in a few cultivars, whereas the majority of cultivated bananas contain the A and B genomes. Genome composition has played a major role in the classification of bananas. The major genomic groups include diploids (AA, AB, BB), triploids (AAA, AAB, ABB), and tetraploids (AAAA, AAAB, AABB, ABBB). Recently, via in situ hybridization, the S and T genomes have been identified in some cultivars in Papua New Guinea (D'Hont et al., 2000). Genome groups that include the S genome are AS, AAS, and ABBS whereas those with the T genomes are represented by AAT, AAAT, and ABBT. Of the four genomes present in *Musa*, molecular markers have been identified for the A and B genomes

(Howell et al., 1994; Pillay et al., 2000; Nwakanma et al., 2003).

One of the major disadvantages of genomic classification is that it groups together bananas with different characteristics. For example, the plantains and the "Pome" group are placed in the same group because they both have the AAB genome. But the groups are very different from one another in that plantains are usually cooked before consumption whereas the Pome subgroup is eaten as a dessert banana. Similarly, the dessert and East African Highland bananas, both with AAA genome, would be classified together. But dessert bananas are generally eaten raw when they are ripe, whereas the East African Highland bananas are steamed and mashed before eating. To accommodate such anomalies, the genus *Musa* was divided into subgroups and clones (De Langhe and Valmayor, 1980; Swennen and Vuylsteke, 1987). Examples of subgroups are plantains, Mysore, Silk, and Pome. Subgroups are named after the best-known clone in a subgroup, e.g., Cavendish subgroup. For more details of genome groups and subgroups in *Musa*, the interested reader is referred to Jones (2000). Currently, there is a general acceptance of genome groups and subgroups in *Musa*, but there is no clarity on categories at the subgroup level and below. Karamura (1998) introduced the term "clone set" to a group of similar clones of East Africa Highland bananas. A clone set is a higher order category, above the clone, but below the subgroup.

The importance of bananas

Bananas are one of the most important subsistence crops grown in more than 120 tropical and subtropical countries of the world (Jones, 2000). It is grown widely in all types of agricultural systems, ranging from small, mixed subsistence gardens to large, multinational commercial plantations. Bananas constitute one of the major staple foods as well as providing a valuable source of income through local and international trade for millions of people. In gross production value, banana is the fourth most important food crop after rice (*Oryza sativa* L.), wheat (*Triticum aestivum* L.), and maize (*Zea mays* L.) and is second to cassava (*Manihot esculenta* Crantz) in sub-Saharan Africa. The total world production of banana is currently about 97 million tons and only 10% of this makes up the export trade (FAO STAT, 2003). Over 90% of the bananas produced in many countries are consumed locally in many different forms, with each country having its own traditional dish and method of processing (Frison and Sharrock, 1998). Banana consumption rates are highest in the Great Lakes region of East Africa, and in Uganda they average 200–250 kg per capita/ year. Plantains (AAB) are one of the main types of cooking bananas in West and Central Africa and Latin American-Caribbean regions. Plantains account for 23% of the world's banana production. The East African Highland bananas that make up 18% of all banana fruit produced are either cooked to prepare "matooke" or used to make banana beer. The ABB genomic group produces starchy fruit that is cooked in many countries and also used as a beer-producing banana in East Africa. The cooking banana types generally contain more starch when ripe and are boiled, fried, or roasted to make them palatable. Bananas that are introduced into a country for the first time appear to find uses that are different from the traditional uses in the places from where they originated. For example, 'Yangambi km5' has been used as a dessert, for cooking, and for the preparation of flour and "nsima" in Malawi where it was introduced for the first time in 2004. There is evidence that the male bud is also being used as a vegetable in Malawi. Male buds are commonly used as a vegetable in Asian countries. The dissemination of post-harvest training techniques in different countries can encourage multiple uses of banana, as in the case of Malawi. There is an increasing demand for finding new value-added products from banana. Novel post-harvest technologies are being developed for banana since it is a highly perishable crop. In this regard, many alternative edible forms of banana, such as sun-dried banana figs, fried chips, flour, and banana sauce, are being processed in many countries. In addition to the fruit, other parts of the banana plant, such as the leaves, serve many purposes, including use as animal feed, as plates, for food preparation, and even for construction of temporary homes.

Banana and plantain are important staple food crops for millions of people in the tropical and subtropical regions of the world, providing 25% of their carbohydrate requirement and 10% of their daily calorie intake. Consequently, they form an important component in the farming systems of the tropical and subtropical regions. Although bananas are rich in carbohydrate, fiber, protein, fat, vitamins A, C, and B_6 (Marriott and Lancaster, 1983; Robinson, 1996), they are largely deficient in iron, iodine, and zinc. Indeed, reports from communities where banana is a major source of food indicate high levels of malnutrition and vitamin A deficiency, especially among pregnant women and preschool children, when banana is missing in their diets. Kikafunda et al. (1996) observed that cooking bananas are served as weaning foods to children in banana-growing regions of Uganda, and this has exposed a majority of the children to diseases associated with iron, zinc, vitamin A, and iodine deficiencies. These deficiencies not only compromise the immune system but can irreversibly retard brain development in utero and in infancy. Thus, given the role banana plays as staple food crop for millions of people, it is necessary to increase its nutritional value. By nutritionally enhancing bananas and plantains, severe deficiencies can be eliminated in developing countries where diets are largely banana-based. Banana biofortification can be achieved through breeding and utilization of the genetic variation for micronutrients in the germplasm. Graham et al. (1999) and Graham and Rosser (2000) noted that exploiting the genetic variation in crop plants for micronutrient density is one of the most powerful tools that can change the nutrient balance of a given diet on a large scale. A number of banana cultivars from Papua New Guinea and the Federated States of Micronesia (FSM) have been observed with a deep yellow or orange color associated with carotenoids (Pillay, unpublished). Engelberger (2003) demonstrated that cultivars from FSM contain enough provitamin A carotenoids to meet half of the estimated vitamin A requirements for pregnant women. Therefore, a food-based solution to vitamin A deficiency can be obtained from carotenes rather than vitamin A itself. Bananas may be able to provide such a solution.

The need to address micronutrient deficiencies has been brought into sharp focus by statistics from the World Health Organization and the World Bank (FAO, 1992; World Bank, 1994). The impact of micronutrient malnutrition is primarily seen among women, infants, and children from impoverished families in developing countries (WHO, 1992; Mason and Garcia, 1993). Indeed, women and children in sub-Saharan Africa, South and Southeast Asia, Latin America, and the Caribbean are especially at risk of disease, premature death, and impaired cognitive abilities because of diets poor in crucial nutrients, particularly iron, vitamin A, iodine, and zinc. However, micronutrient deficiencies are not restricted to developing countries alone. Australia, for example, has up to 5% anemia overall and almost a staggering 10% in teenage girls (Cobiac and Bathurst, 1993). Additionally, 41% of all poor, pregnant African-American women in the United States are anemic. A food system that delivers all required nutri-ents in adequate amounts is being sought through biofortification. Biofortification is a process of improving the health of the poor by providing them with staple foods that are rich in micronutrients.

Major production constraints

A complex of foliar diseases, nematodes, viruses, and pests threatens the production of bananas worldwide. The use of resistant varieties is considered to be the most effective, economical, and environment-friendly approach to controlling diseases and pests. Breeding programs have been established in a few countries to develop resistant varieties. Two of the most important diseases of banana are the fungal diseases black Sigatoka (*Mycosphaerella fijiensis* Morelet) and Fusarium wilt (*Fusarium oxysporum* Schlect. f.sp. *cubense* (E.F. Smith). The main pests of bananas include a complex of nematodes (*Radopholus similis*, *Pratylenchus* spp. *Helicotylenchus*) and the banana weevil (*Cosmopolites sordidus* Germar). New diseases, such as banana Xanthomonas wilt (BXW), have been recently identified in East Africa. Although currently identified in four countries, BXW could soon spread to neighboring countries since most of these countries are major banana producers and informal trade links exist among the East and Central African countries. The disease was initially identified in Ethiopia in the 1960s (Yirgou and Bradbury, 1968). This bacterial disease has quickly become the most serious threat to the production of bananas in East and Central Africa.

Breeding objectives

The main objective in banana breeding is the development of cultivars with high consumer acceptability and resistance to the major pests and diseases. Consumer acceptability has been a major stumbling block in the adoption of new hybrids produced in breeding programs. However, this scenario is changing, and farmers in some countries are now adopting new hybrids. One of the major problems in the adoption of the new varieties has been lack of farmers' accessibility to new hybrids. The widespread distribution of hybrids has been rather slow. In cases where the hybrids have been able to reach farmers, the adoption rate appears to be high, and there is an increased demand for cultivars. The question often raised is, why has banana breeding yielded only limited results despite more than 80 years of work in this area. The answer to this question lies in the several difficulties and challenges involved in banana breeding.

One of the major difficulties in banana breeding is male and female infertility due to triploidy. Indeed, some groups and subgroups of *Musa* are totally sterile and seedless. Consequently, it is not possible to improve such varieties by conventional breeding techniques. In the future, non-conventional breeding techniques, such as genetic transformation, may play a role in improving such sterile varieties. However, some banana groups do produce a few seeds when pollinated, which opens the window for further improvement of bananas. Breeders can produce hybrid seeds through a process of embryo rescue to produce plantlets (Ssebuliba et al., 2006). Robinson (1996) noted banana breeding is a very slow process, taking about three years from seed to seed. Besides male and female infertility, there is a lack of genetic variation in the triploid, edible varieties that greatly impedes the breeding process. Yet success in plant breeding is directly dependent on the genetic variability in the source populations (Borojevic, 1990). Breeders are therefore obliged to create new genetic variability by recombining genes from different parents.

Variation for pest and disease resistance in *Musa* is found primarily in wild diploid species. Although many diploids with useful sources of resistance are available for breeding, many other diploid species existing in the wild have not been collected and evaluated for their breeding value. There is a danger that some of these accessions with useful breeding traits may be lost due to deforestation and other disasters. Exploration, collection, and characterization of wild diploids are consequently important breeding objectives. Introgression of genes from wild bananas to cultivated species, however, requires a series of crosses and backcrosses.

Bananas are highly heterozygous and progenies segregating from breeding populations lead to segregation for too many characters. Consequently, it is not possible to easily detect superior genotypes in breeding populations. To increase the frequency of desired genotypes in breeding populations, superior parents are required. Therefore, improving the diploids has been a key objective in banana breeding (Tezenas du Montcel et al., 1996; Rowe, 1984).

A Short history of banana breeding

Banana breeding began in Trinidad in 1922 and in Jamaica in 1924 with the goal of producing 'Gros Michel'-type cultivars with fusarium resistance. The first hybrid from this program, 'IC 2', an AAAA tetraploid obtained from Gros Michel × *M. acuminata*, was released in 1928. IC 2 was widely distributed but soon became infected with Fusarium wilt. In 1933, the program began concentrating its efforts on producing black Sigatoka resistant hybrids. This breeding program eventually slowed down due to financial difficulties and finally ceased operating in the early 1980s.

The United Fruit Company began breeding bananas in the 1920s, but accelerated its program in the late 1950s at La Lima in Honduras. Although the initial goal of this program was to develop fusarium-resistant dessert bananas, the objectives later changed to incorporate all possible desirable qualities of banana into one commercial variety. The program was the first to develop agronomically improved diploids with resistance to Fusarium wilt, the burrowing nematode, and black Sigatoka. In 1984, the United Fruit Company donated the program to the Fundacion Hondurena de Investigacion Agricola (FHIA) in Honduras. The FHIA program has developed hybrids of a range of banana types and was the first breeding program to test promising black Sigatoka-resistant hybrids in several countries.

In 1982, a banana breeding program, Empresa Braslieira de Pesquisa Ágropecuaria (EMBRAPA), was initiated in Brazil. The aim of the program was to improve local landraces, such as 'Prata' (AAB, Pome subgroup) and 'Maca' (AAB, known as 'Silk'), especially for Fusarium wilt and yellow and black Sigatoka. When black Sigatoka reached West Africa and threatened the livelihoods of millions of subsistence plantain producers in the 1980s, two conventional breeding programs were established for the development of plantains resistant to Sigatoka. One began at the Centre Africain de Recherches sur Bananiers et Plantains (CARBAP, [formerly CRBP]) at Nyombe in Cameroon and the other at the International Institute of Tropical Agriculture (IITA) at Onne in Nigeria. The two programs have developed a number of plantain and cooking banana hybrids with resistance/tolerance to black Sigatoka.

In 1994, IITA established a new breeding program at Namulonge in Uganda specifically for improving East African Highland bananas in collaboration with the country's National Banana Breeding Program. The program has developed a number of black Sigatoka-resistant/tolerant hybrids, with some also showing resistance to the burrowing nematode, *Radopholus similis*. Farmers are now evaluating these hybrids.

The Centre de Coopération Internationale en Recherche Agronomique pour le Dével-oppement (CIRAD) established a banana breeding program in Guadeloupe in the 1980s. CIRAD's efforts were aimed at developing new triploid dessert bananas for export by exploiting the diversity found in existing diploid varieties (Bakry et al., 1997).

Musa breeding strategies

Bananas and plantains are difficult to breed because most of the important and popular genotypes are highly sterile and do not produce seeds (Adeleke et al., 2004). Compared with other food crops, there is also a relative lack of knowledge about *Musa* genetics and cytology (Pillay et al., 2002). Nevertheless, *Musa* breeding has made significant progress in recent years and new improved varieties are now emerging. Two major breeding schemes, namely conventional and biotechnological, are used in banana improvement.

Until the beginning of the 1970s, all the breeding programs based their strategy on the residual female fertility found in some triploid cultivars. During the first steps in the development of new varieties, breeding programs aimed to develop hybrids using wild species in conventional crosses. Wild species known for their resistance to pests and diseases were crossed with triploid landraces. This is referred to as the 3x-2x breeding strategy (Tomekpe et al., 2004). The 3x-2x crosses produce a substantial number of tetraploids in the progeny due to unreduced triploid gametes that preserve all the characteristics of the female parent in the offspring. This strategy has been used to create tetraploid hybrids with disease and pest resistance and good agronomic value by crossing susceptible triploid landraces with fertile diploid males that harbor resistance genes. Many breeding programs have successfully bred a number of tetraploids with good bunch characteristics and resistance to black Sigatoka by using the 3x-2x strategy. Tetraploid hybrids from the FHIA program include FHIA 17, 23, and 25, those from IITA include BITA 3, PITA 16, and others, all of which are being adopted in certain countries. One of the weaknesses of the tetraploid hybrids is that they inherit male and female fertility from the male parents and are capable of producing seeds, which is an undesirable characteristic in edible banana. This reduces the consumer quality of the tetraploids. In view of these problems, selected tetraploids obtained from the initial 3x-2x crosses are hybridized with improved diploids to produce secondary triploids. However, the genetic gain obtained from the nuclear restitution of the preceding stage is reduced by recombination during meiosis in the tetraploid (Tomekpe et al., 2004). The secondary triploid is a combination of a new set of genes and could lose desirable characteristics. The choice of the male parent in the development of secondary triploids is therefore very important since it contributes to the quality of the secondary triploid.

This 4x-2x strategy has enabled breeding programs to develop larger breeding populations from which good quality secondary triploids are selected. Consequently, improvement of the diploids to be used as male parents for creating secondary triploids is an integral part of any banana breeding program. Diploid bananas are highly fertile, with considerable genetic diversity and often possess one or more sources of resistance/tolerance to diseases and pests. Most commonly used fertile diploids with resistance genes, e.g., 'Calcutta 4,' often have very poor bunch characteristics that are passed on to the progeny. Therefore, breeders sought to improve the diploids to take into account their agronomic characteristics and fruit quality traits. Good-quality diploids have been developed by many breeding programs and include plantain and banana-derived diploids (Vuylsteke and Ortiz, 1995; Tenkouano et al., 2003) from IITA and SH2989, SH3142, SH3217, and SH2095 from FHIA.

A variation of the 4x-2x strategy is being used at the breeding programs in CIRAD and CARBAP. This strategy imitates the natural process of triploid evolution from ancient diploids by the production of unreduced gametes in one of the diploid parents during hybridization (Simmonds, 1962). To imitate the meiotic error that produced the unreduced gamete, colchicine is used to double the chromosome number of one of the parents. The resulting tetraploids are then used as male parents and crossed with diploids (wild or bred) to produce triploids. The main difference between this strategy and those described above is that it does not try to improve existing varieties but tries instead to create new improved varieties by using ancestral varieties. Using this strategy, CARBAP has been able to obtain large populations of several hundred individuals from which several triploid AAA *acuminata* hybrids resistant to black Sigatoka are being evaluated (Tomekpe et al., 2004).

A new strategy involving 4x-4x crosses has been initiated at IITA (Pillay, unpublished). One of the striking features of the progenies obtained from the 4x-4x crosses as compared with progenies obtained from other breeding strategies (3x-2x, 4x-2x) is their marked resemblance in morphology, bunch characteristics, and palatability to their female grandparents. A successful breeding program should aim at gradually incorporating resistance to pests and diseases in the existing cultivars, rather than aiming for genetic materials that are drastically different from the existing cultivars. The development of disease-resistant hybrids with consumer characteristics that match the traditionally grown cultivars is still the major challenge to banana breeders (Jones, 2000). Preliminary evidence shows that the 4x-4x breeding strategy may provide a solution to this challenge. The 4x-4x breeding strategy is advantageous because of the male and female fertility of the starting material in the hybridization process.

Genetic resources and gene pools for banana improvement

Genetic resources encompass all the various forms of germplasm that are available for collection, storage, and use. They comprise the diversity of genetic material contained in traditional varieties, modern cultivars, wild relatives, and other wild species. Genetic diversity provides farmers and plant breeders with options to develop, through selection and breeding, new and more productive crops that are resistant to virulent pests and diseases and adapted to changing environments. For thousands of years, farmers have used the genetic variation in wild and cultivated plants to develop new cultivars that are adaptable to changing environments, including new pests and diseases and climatic conditions. The centers of diversity are the rich sources of genes that can be exploited to the benefit of the researchers and farmers. The wild species, primitive edible cultivars, landraces, materials developed and used in ongoing breeding programs, and advanced cultivars in particular, constitute the gene pool for breeding. The wild species and primitive cultivars are useful for conferring resistance characteristics, landraces for their quality characteristics, and hybrids for superior agronomic characteristics.

A gene pool is simply the totality of genes within a population. Gene pools can be categorized as primary, secondary, or tertiary. The primary gene pool corresponds to the traditional concept of a biological species and the secondary gene pool includes all species that will cross with that species. The tertiary gene pool represents the maximum reach of high-tech breeding (via embryo rescue, protoplast fusion, etc.), whereas the quaternary gene pool contains heritable material from all other species. Genetic material is the source of new information for the development of new products and the improvement of existing ones. Plant and animal breeders, to develop new genotypes, regularly use genetic information stored in different species. Indeed, the development of gene banks and their use indicate the importance of genetic resources. In fact, as early as the 1930s, well before the Green Revolution, the potentially damaging consequences of planting large areas to single uniform crop cultivars were recognized by agricultural scientists.

Modern crop cultivars are more genetically uniform than the landrace cultivars. Breeders maintain a continual search for novel genetic combinations from which to select plants with superior traits, such as crop quality, yield, regional performance, and tolerance to pests and diseases. A primary source of genetic variation is the wide array of germplasm within each crop species and closely related wild species that are capable of interbreeding. For many crops, such as banana, breeders have relied heavily on the introduction of genes from closely related wild plants to increase genetic variation in the crops. In banana and plantains, landraces, wild relatives and hybrids constitute the genetic pool used in banana breeding. However, the ease of crossing depends on how closely they are related. Breeding programs select offspring displaying different degrees and combinations of the characteristics of the parents. Offspring are crossed through a number of generations to recover the required characteristics of the landrace while incorporating the desired trait of the wild relative. Advances in genetic engineering, however, are blurring the distinction between individual gene pools, providing ways to identify, isolate, and move genes using biotechnological methods. In fact, biotechnology is dramatically changing the accessibility of individual genes to breeders. Breeders can reduce the many years needed for conventional breeding by technically manipulating and transferring genes between plants. The current use of biotechnology in *Musa* and its implications for germplasm utilization are discussed later.

Breeding for disease resistance in *Musa*

Black sigatoka

Of all the fungal leaf diseases affecting bananas and plantain, black Sigatoka caused by *M. fijiensis* is the most economically important disease (Carlier et al., 2000). Until the discovery of black Sigatoka, yellow Sigatoka caused by *M. musicola* was the most important foliar disease of bananas. Sigatoka disease was recorded in the Sigatoka valley on the island of Viti Levu in Fiji (Philpott and Knowles, 1913), hence the name Sigatoka. *Mycosphaerella fijiensis* and *M. musicola* can be differentiated microscopically by characteristics of their conidia and conidiophores. Conidia of *M. fijiensis* are longer than those of *M. musicola* and have a thickened basal hilum. Additionally, conidiophores of *M. fijiensis* possess conidial scars that are absent in *M. musicola* (Meredith and Lawrence, 1969). Yellow Sigatoka has been largely supplanted by black Sigatoka in many banana producing areas, but remains a significant problem at higher altitudes and cooler temperatures (Mouliom-Perfoura et al., 1996). Although Sigatoka disease does not kill the plant, it causes heavy loss of leaves leading to reduced photosynthesis, suppression of fruit filling, and reduced bunch sizes. Yield losses due to black Sigatoka range from 33% to 76% during the first and second crop cycle in plantains (Mobambo et al., 1996). Indeed, black Sigatoka is more damaging and more difficult to control than the related yellow Sigatoka disease, and it has a wider host range that includes the plantains, dessert and ABB cooking bananas that are usually not affected by yellow Sigatoka.

Although chemical control is effective against black Sigatoka, it is expensive and so not a feasible approach for the resource-poor, small-scale farmers who are the major producers of the crop. Large plantations are sprayed aerially at a cost of US\$400–1400 ha^{-1} year^{-1} (Jones, 2000). The production and cultivation of resistant cultivars is generally regarded as the most appropriate and feasible approach for controlling black Sigatoka. Consequently, many banana breeding programs were initiated with the objective of developing hybrids with resistance to black Sigatoka. However, prior to the development of resistant hybrids, resistant or less susceptible cultivars were deployed in areas where black Sigatoka was responsible for serious yield losses and where banana was the staple food. The ABB cultivars, such as 'Bluggoe,' 'Pelipita,' and 'Saba,' were tested in Central America and West Africa as substitutes for the plantain. 'Mysore' (AAB), 'Saba,' 'Bluggoe,' and 'Rokua Mairana' replaced many other cultivars in the Pacific islands. While some of these cultivars did not match the cooking properties and taste of the plantains, others were adopted by some farmers and are now regularly seen in the local markets. In Cuba, large areas of the country are now planted with improved black Sigatoka-resistant material from FHIA. Some of the IITA cooking banana hybrids, such as BITA 3, are now found in Cuba, India, and Malawi.

Sources of resistance to black Sigatoka

The major source of resistance for black Sigatoka is the diploid male fertile clone 'Calcutta 4,' a wild seeded banana belonging to the *Musa acuminata* ssp. *burmannicoides* complex (De Langhe and Devreux, 1960). This accession has been widely used in the FHIA program and has been referred to as FHIA accession IV-9. The IITA breeding program has also used 'Calcutta 4' as a source of resistance to black Sigatoka. Other sources of resistance include 'Pisang lilin,' a parthenocarpic diploid (AA) accession related to *M. acuminata* ssp. *malaccensis* and 'Tuu Gia.' One of the most serious problems associated with 'Calcutta 4' as a male parent is that it is non-parthenocarpic and has a bunch size of less than 2 kg. In an effort to produce improved diploid parents, breeding programs engage in recurrent diploid breeding. By crossing 'Calcutta 4' with other bred diploids, the FHIA program was able to develop several diploids with exceptional bunch sizes and resistance to diseases and pests. Notable diploids with resistance to black Sigatoka include SH2829, SH3437, SH3142, and SH3217.

Fusarium wilt

Fusarium wilt, caused by a soil borne fungus (*Fusarium oxysporum* Schlect. f.sp. *cubense* (E.F. Smith) Snyder and Hansen), is one of the most destructive and catastrophic diseases of banana. The disease was responsible for the destruction of the 'Gros Michel' export industry in many countries and for its replacement by Cavendish varieties. Four races (race 1, 2, 3m and 4) of *F. oxysporum* were recognized by host differential testing (Ploetz, 1994). 'Gros Michel' (AAA) and 'Pisang awak'(ABB) are susceptible to race 1 and 'Bluggoe' (ABB) to race 2, whereas race 4 affects Cavendish cultivars as well as all genotypes susceptible to race 1 and 2. Race 3 reportedly affects *Heliconia* spp. and is weakly pathogenic to 'Gros Michel' and seedlings of *M. balbisiana* (Waite, 1963).

Somatic compatibility has also been used to identify genetically isolated populations of *Fusarium*. Vegetative compatibility groups (VCGs) are identified by heterokaryon formation between non-nitrate-utilizing auxotrophic mutants. There are about 16 VCGs or VCG complexes in *F. oxysporum* f.sp. *cubense* (Ploetz, 1990; Ploetz et al., 1997). The typical symptoms of Fusarium wilt include yellowing of older leaves followed by wilting and collapse of the leaves at the petiole base. Longitudinal splitting of the outer sheaths of the lower pseudostem is also typical of the disease. However, similar symptoms are often observed during weevil infestation and infection by banana streak virus. Therefore, all the symptoms should be considered carefully for correct identification of Fusarium wilt. Internally, a reddish to dark brown discoloration of the vascular system is often observed in *Fusarium*-affected plants. The pathogen is commonly spread through infected planting material, but running water, farm implements, and machinery (Jones, 2000) can also spread the disease.

Micropropagated plantlets produced through tissue culture are considered the most reliable source of clean planting material and consequently one of the alternatives for controlling the disease. However, the cost of tissue culture material is too high for subsistence farmers. Besides, planting clean material in diseased soils is not an answer to Fusarium wilt, as the fusarial spores can remain viable in the soil for more than 30 years. Other control measures, such as injection of chemicals into soils, soil treatments such as fumigation, and the incorporation of soil ameliorants/amendments, were shown to reduce the severity of the disease, but these methods are not commercially applicable since they rarely provide long-term control (Hwang and Ko, 1987; Pegg, 1996). The only solution, therefore, appears to be breeding for resistance. Fortunately, all races of *Fusarium* are rarely present in all countries and there is a preponderance of one or two races in most countries. While this is an advantage on one hand, it also presents handicaps for a breeding program that aims to develop resistant hybrids.

Establishing genotypes in an infested field and assaying disease symptoms among susceptible clones over a period of time can best be achieved by screening clones for resistance. Field screening is relatively expensive when one considers the large size and long life cycle of banana. One has to ensure that the *Fusarium* inoculum is also well dispersed in the testing field to prevent "escapes." Factors affecting disease expression, such as inoculum concentration, edaphic conditions, temperature, and other variables, are difficult to control under field conditions. For these reasons, researchers have devised screening techniques whereby seedlings at the nursery stage can be tested for *Fusarium* resistance. Mak et al. (2004) described a double-tray technique. One of the criticisms of laboratory or screenhouse testing is the durability of resistance when such plants are transferred to a disease-infected field (Stover and Buddenhagen, 1986). Mak et al. (2004) pointed out that plantlets that survived *Fusarium* in the double-tray method also maintained their tolerance in a *Fusarium* "hot spot" after a year.

Sources of resistance to Fusarium wilt

Shepherd et al. (1994) and Rowe and Rosales (1996) reported that the wild diploids *M. acuminata* ssp. *malaccensis* and *M. acuminata* ssp. *burmannicoides* 'Calcutta 4' and the edible diploid 'Pisang lilin' are good sources of resistance to races 1 and 2 of *Fusarium.* Consequently, the first banana-breeding program in Trinidad used 'Pisang lilin' as the source of resistance to develop a Gros Michel-like cultivar with resistance to Fusarium wilt. The program released two cultivars, 'Bodles Altafort' or '1847' from a 'Gros Michel' × 'Pisang lilin' cross and '2390–2' from the 'Highgate' × 'Pisang lilin' cross. Later, male parents were selected from crosses between *M. acuminata* ssp. *banksii* and 'Paka', an AA diploid from Zanzibar. When these male parents were crossed with 'Highgate,' two tetraploid (AAAA) selections T6 and T8 with good resistance to

black Sigatoka were produced. In the 1960s, a male parent from a cross between *M. acuminata* ssp. *malaccensis* and *M. acuminata* ssp. *banksii* was used for breeding. Later, improved diploids were obtained by crossing diploids obtained from *M. acuminata* ssp. *malaccensis* × ssp. *banksii* with others, including those derived from *M. acuminata* ssp. *banksii* × 'Paka.'

The FHIA diploids SH3142, SH3362, and SH3437 are resistant to races 1 and 2, and SH3362 is also resistant to race 4 (Rowe and Rosales, 1996). Armed with these diploids, FHIA has developed two tetraploid hybrids, FHIA 17 and FHIA 23, that are resistant to race 1. In addition, FHIA 01 has been shown to be resistant to race 1 and race 4. The Brazilian breeding program at EMBRAPA also has bred clones with resistance to *Fusarium* (Shepherd et al., 1994). These include PA 03–22 (AAAB), PV 03–44 (AAAB), and PC12-05 (AAAB) that are derived from 'Prata Ana,' 'Pacovan,' and 'Prata,' respectively. These clones are resistant to race 1 and are being tested in many countries. Fusarium wilt is one of the major constraints affecting 'Pisang awak' (ABB) and 'Sukali Ndizi' (AAB) in Eastern and Southern Africa. Nine genotypes, comprising 'Pisang lilin,' 'Fougamou,' 'Pisang awak,' 'SH3217,' 'SH3142,' 'SH3362,' 'Yangambi Km5,' 'Calcutta 4,' and 'Kikundi,' were tested for resistance in a *Fusarium*-infested field in Uganda. Of the nine genotypes, 'Pisang awak' and 'Fougamou' showed, within six months, typical symptoms of Fusarium wilt, including leaf discoloration and wilting, pseudostem splitting, and browning of the vascular tissue. These clones are known to be susceptible to race 1 of *Fusarium.* The genotypes 'Pisang lilin,' 'SH3217,' 'SH3142,' 'SH3362,' 'Yangambi Km5,' and 'Kikundi' have not displayed any symptoms of Fusarium wilt after being in the infested field for 36 months. The latter genotypes should be regarded as resistant.

Banana Xanthomonas wilt

Xanthomonas wilt is a devastating bacterial disease that can spread rapidly and result in total yield loss once the disease establishes itself in an area. Bacterial wilt was initially reported as a disease of the closest relative of banana, *Ensete*, in Ethiopia (Yirgou and Bradbury, 1968; 1974). The disease has now been identified in Uganda, Rwanda, Eastern Democratic Republic of Congo, and Tanzania (Tushemereirwe et al., 2003; Ndungo et al., 2005).

Typical symptoms of the disease include wilting, premature yellowing, and rotting of fruits (Tushemereirwe et al., 2003). Even in the absence of apparent external symptoms, wilting and death of banana plants are frequently observed. Ratoon crops arising from infected mats are severely diseased and often wilt before producing bunches or produce bunches with rotted fruits.

The disease causes an estimated annual loss of US$360 million to the Ugandan economy. If unchecked, the disease could cause massive losses in the Uganda's western districts and in other neighboring countries where intensive banana cultivation is practiced.

Field observations in Uganda suggest that *Xanthomonas* infects banana plants either through the lower parts of the plant (roots and mats) possibly from soil-borne inoculum and through the inflorescence from inoculum dispersed primarily by insects. Though no data are available, it appears that inflorescence infection is much more predominant and important than soil-borne infection in spreading the disease.

Xanthomomas wilt is similar to Moko disease of banana with respect to sites of infection, disease development, dissemination patterns, and other epidemiological features (Ivan Buddenhagen, personal communication). *P. solanacearum* is the causal agent of Moko disease, which is transmitted from infected inflorescence to healthy inflorescence by the bees, flies, and wasps that normally visit banana flowers (Buddenhagen and Elsasser, 1962). The insects pick up the bacteria from bacterial ooze that exudes from flower cushions of infected inflorescences. Infection by Moko disease takes place when contaminated insects visit flower cushions of the male bud and deposit the bacterium on moist cushions on the peduncle from which male flowers have recently abscised. The bacterium penetrates into the peduncle through the moist cushions, invades the vascular bundles, colonizes and rots developing fruits, and subsequently moves downwards to invade the corm and new suckers arising from it. In the susceptible clones, such as 'Pisang awak', fresh cushions are exposed everyday for nearly 60 days until fruit maturity, providing a long susceptibility period.

Banana clones with hermaphrodite flowers and non-dehiscent male flower bracts do not expose moist cushions to contaminated insects and therefore escape Moko disease. Avoidance of inflorescence infection would largely protect the mother plant and the subsequent suckers from being systemically infected unless the inoculum is introduced into the plants through contaminated cutting tools. Cultural control measures are useful in curbing the spread of the disease but need to be adopted by all farmers for effective containment of the disease (Eden-Green, 2004; Blomme et al., 2005).

Sources of resistance to Xanthomonas

Germplasm screening for the identification of resistance to Xanthomonas wilt in banana is still in early stages. An interesting observation from such screening trials is that a few genotypes, such as 'Yangambi km5,' 'Kikundi,' 'Kisubi,' 660-K-1,1201K, 'Yalim,' 'Nakasabira,' 1438k-1, 7197-2, SH3217, SH3362, SH3640-9 and 'Nakitengu,' do not appear to get infected naturally although they are surrounded by infected plants. This may suggest the presence of minor genes for resistance to Xanthomonas wilt in these varieties and is being investigated further.

Viral diseases

Four well-characterized viruses affect bananas and other *Musa* species. These are banana bunchy top virus (BBTV), banana streak virus (BSV), cucumber mosaic virus (CMV), and banana bract mosaic virus (Pietersen and Thomas, 2001). Recently, new virus-like particles have been observed in bananas from Africa, the Americas, Southeast Asia, and Australia. The virus has been isolated and named "banana mild mosaic virus." Another new virus, banana die-back virus, has also been described from Nigeria. Of all the viral diseases, banana bunchy top is the most serious and devastating disease. Once established, it is extremely difficult to eradicate or manage. It is widely distributed and thus a major production constraint in many banana-producing areas of the world, especially Southeast Asia and the Pacific (Thomas et al., 1994). BBTV is also prevalent in the Philippines, Taiwan, most of the South Pacific islands, and parts of India and Africa. BBTV does not occur in Central or South America. Infected suckers, corms, and tissue culture plantlets transmit BBTV from place to place; plant-to-plant transmission being via aphid vectors, *Pentalonia nigronervosa*. Early identification and eradication of infected plants are the best means of controlling the disease. This appears to be a problem in developing countries where the skills for identification of the disease are lacking and farmers do not have the funds to destroy infected plants. Occasionally, a diseased mat might produce healthy plants with a bunch. Farmers are, therefore, reluctant to destroy infected mats. The characteristic symptom of the disease is continuous or broken yellow streaks running across the leaf blade.

BSV has a worldwide distribution and was not considered to be a problem until recently. The disease was first identified in Cote d'Ivoire in 1966 (Lassoudiere, 1974) and the causal agent was isolated in 1986 (Lockhart, 1986). New outbreaks of BSV are continuously being reported from different countries. The disease is spread mainly through planting material and may also be transmitted via mealy bug (Lockhart and Jones, 2000). A high degree of heterogeneity has been identified among BSV isolates differing serologically, genomically, and biologically (Lockhart and Olszewski, 1993; Geering et al., 2000). BSV genomic sequences are integrated into the genome of *Musa* (La Fleur et al., 1996). Two integrated BSV sequences have been well characterized so far. One of these sequences appears to be incapable of giving rise to episomal BSV infection. There is evidence that the second integrated sequence is the source of *de novo* episomal BSV infection. It appears that stress factors, such as plants in tissue culture and cold temperatures, can trigger episomal BSV symptoms. The most effective means of controlling BSV is to ensure that planting materials are virus-free.

The CMV is a disease of worldwide importance because it is widely distributed and infects more than 800 plant species; it is an emerging threat to banana production in parts of India (Estelitta et al., 1996). In bananas, CMV and BSV symptoms are similar and difficult to differentiate by a non-specialist. CMV is characterized by inter-veinal chlorosis of the leaves and, in severe cases, is accompanied by rotting of the cigar leaf and central cylinder. Other characteristic symptoms of CMV include chlorosis, mosaic, and heart rot. Generally, CMV does not have a serious impact on banana production. Consequently, the virus may be present in the plant but may not cause visible symptoms and damage. In the case of serious outbreaks, CMV is normally controlled by elimination of alternative hosts, especially cucurbits, tomatoes, and tobacco.

Other viruses affecting bananas include banana-die back virus (BDBV), which was recorded in Nigeria (Hughes et al., 1998), banana mild mosaic virus (BanMMV), and banana bract mosaic virus (BBrMV). The first virus has not been associated with a specific disease in banana but may increase the severity of BSV when it co-infects plants (Tushemereirwe et al., 1996). BBrMV was first observed in the Philippines in 1979 (Thomas and Magnaye, 1996). It is now widespread in India and Sri Lanka (Rodoni et al., 1997). Generally, viruses are becoming more important in banana cultivation due to the free movement of germplasm and the discovery of new viruses. The most effective control for all virus diseases is the eradication of infected plants. Screening procedures for viruses are not well developed, although virus diagnosis techniques have been improved recently. Germplasm screening for viruses has relied on field scoring of virus incidence based on visual symptoms. Large scale screening of germplasm for virus resistance is limited. Sources of resistance for the different viruses affecting banana are not known. This lack of suitable parental material limits progresss in breeding for virus resistance.

Breeding for pest resistance in *Musa*

Weevils

The banana weevil, *Cosmopolites sordidus,* is the most important pest of banana and plantain. Banana weevils are host-specific, attacking plants only in the genera *Musa* and *Ensete* (family Musaceae). Adult weevils lay eggs at the base of the plant and upon hatching, the larvae bore into the corm to feed and develop. Extensive tunneling of the corm by the larvae causes reduced water and mineral uptake by the plant, weakening of

the plant and yield (bunch weight) reduction. In severe cases of weevil infestation, toppling and premature death of the plant are common. Pseudostem splitting at the base and distortion of young leaves are the other observed symptoms of extreme weevil damage. Weevil damage is often severe in newly planted fields, where heavy infestation can damage the young corms and result in total failure of the crop (Mitchell, 1980; Ambrose, 1984).

Weevils enter new fields either by movement of infested planting material or migration from established neighboring fields (Gold et al., 1999). Weevil populations build up in plantations and damage becomes more pronounced in ratoon crops (Mitchell, 1980; Rukazambuga, 1996). Several strategies, including cultural control, host plant resistance, and biological control have been used in management of the banana weevil (Gold, 1998). Cultural control practices include selection of clean planting material, paring of suckers, hot-water treatment, deep planting, mulching, trapping, intercropping, application of organic-based manures, weeding, field sanitation, desuckering, propping, varietal mixtures, and crop rotation (Reddy et al., 1999). The fungal species *Beauveria bassiana* and *Metarhizium anisopliae* have also been used as biological control agents against the banana weevil (Nankinga et al., 1999). While these fungi have the potential of significantly reducing weevil populations and damage to planting material, further work is necessary to develop a delivery system and integrate it with other control measures.

Breeding for weevil resistance

Few studies have evaluated banana germplasm for resistance to weevils. Fogain and Price (1994) screened 52 genotypes for weevil resistance and found that plantains were the most susceptible, whereas the AAA genotypes were generally less susceptible.

Similarly, Ittyeipe (1986) observed a high level of weevil damage among plantains in Jamaica compared with diploid (AA) cultivars. Pavis (1991) demonstrated that cultivars from the 'Pisang awak' group exhibit high levels of tolerance despite heavy tunneling, whereas 'Yangambi Km5' was almost free from attack. 'Pisang awak' and 'Bluggoe' (AABs) are generally considered more resistant than plantains (AAB) and the East African Highland bananas (Abera et al., 1997).

Breeding for resistance to weevils is hampered by the absence of good sources of resistance and the lack of a simple screening technique to identify resistance. Crosses between plantains and 'Calcutta 4' demonstrated that most of the diploid hybrids were resistant to weevils, whereas the majority of the polyploids were susceptible (Ortiz et al., 1995). On this basis, 'Calcutta 4' was considered to be a source of resistance for weevils. Other genotypes that can impart weevil resistance include 'Yangambi Km 5,' 'Pisang awak,' 'Sannachenkadali' (AA), 'Sakkaki' (ABB), 'Senkadali' (AAA), 'Elacazha' (BB), 'Njalipovan' (AB), FHIA03, TMBx612-74, TMB2x6142-1, TMB2x8075-7, TMB2x7197-1, 'Long Tavoy,' 'Njeru,' 'Muraru,' 'Bluggoe,' and *M. balbisiana* (Kiggundu et al., 2003). The weevil resistance shown by hybrid TMB2x 6142-1, derived from a cross between 'Nyamwhiagora,' an East African Highland banana, and 'Long Tavoy,' suggests that there may be sources of resistance among the East African Highland banana germplasm despite their high levels of susceptibility to weevils.

Nematodes

In most regions of the world, nematodes are recognized as important pests of bananas. Crop losses due to nematodes are very high, with average annual yield losses estimated at about 20% worldwide (Sasser and Freckman, 1987). Since nematodes attack the root and/or corm tissues, the reduction in the mechanical (anchorage) and physiological (uptake and transportation of water and nutrients) functions of the root affect plant growth and yield.

The most important and widespread banana nematodes are *Radopholus similis* (burrowing nematode), *Helicotylenchus multicinctus* (spiral nematode) *Pratylenchus coffeae* and *P. goodeyi* (root lesion nematodes) and *Meliodogyne incognita* (root knot nematode). In addition to the major nematodes, minor nematodes, such as *Heterodera oryzicola* (India), *Radopholus reniformis, Helicotylenchus dihystera* (Malaysia), *Rotylenchus* (Indonesia), have been observed on bananas in some countries. The most damaging nematodes are the migratory endoparasitic lesion-nematodes *R. similis* and *O. coffeae* (Sarah et al., 1996).

Disruption of the anchorage system by nematodes causes seriously infected plants to topple or uproot even from mild winds. Plant toppling is more pronounced in older plantations where the banana mat has moved to the surface, providing less anchorage. Economically, nematodes affect bunch size, leading to reduced yields. When use of clean planting material in nematode-free soil and growing the plants under strict quarantine conditions cannot prevent nematode attack, nematode management in bananas mainly relies on crop rotation and chemical control. However, in areas where bananas are cropped continuously, crop rotation is obviously precluded; and the price of chemical nematicides is often prohibitive for small-scale farmers (Luc et al., 1990). It is important to note that most nematicides are extremely toxic to the environment. Although naturally occurring nematode resistance and tolerance have long been exploited for many agricultural crops (De Waele, 1996), these means of nematode management have so far been neglected in bananas. This is despite the evidence, albeit limited, that nematode resistance and tolerance sources are present in the *Musa* genepool (Pinochet, 1996).

Sources of nematode resistance

Resistance to nematodes has been identified in wild species, landraces, and new synthetic plants developed by breeding programs. The wild species *M. acuminata* spp. *burmannicoides* 'Calcutta 4' possesses field resistance to nematodes, especially *R. similis*. Its resistance to *R. similis* was confirmed in pot trials (Viaene et al., 2003). Some of the other subspecies of *M. acuminata* (*malaccensis, microcarpa,* and *zebrina*) have been found to have moderate to good resistance to *R. similis* (Wehunt et al., 1978; Fogain, 1996). *Musa acuminata* ssp. *malaccensis* is also considered to be resistant to nematodes and has been used by FHIA as a new source of resistance. Some diploid cultivars, such as 'Pisang Mas' and 'Pisang Lidi,' have moderate resistance to *R. similis* (Wehunt et al., 1978; Fogain, 1996). 'Pisang jari buaya' (AA) and 'Pisang Batuau' are reported to be resistant to *R. similis* (Wehunt et al., 1978; Fogain, 1996; Elsen and De Waele, 2004). The FHIA breeding program used 'Pisang jari buaya' as a female parent to produce SH 3142, a synthetic diploid with resistance to burrowing-nematode. The diploid AB clone 'Kunnan' has consistently shown high levels of resistance to *R. similis* and *P. coffeae* (Collingborn and Gowen, 1997). 'Yangambi Km 5' (AAA) is reportedly resistant to *R. similis* (Pinochet et al., 1998).

IITA's banana breeding program has developed a number of diploid and tetraploid hybrids that are resistant to *R. similis*. These include TMB2x 9128-3, TMB2x 5265S-1, TMB2x 3107S-4, TMB2x 2582S-1, TMB2x 4443S-1, TMB2x 2569S-1, and TMB2x 5105-1. Among the Fe'i varieties, Stoffelen et al. (1999) identified two cultivars 'Rimina' and 'Menei' as being resistant to *R. similis*. Most accessions of *M. balbisiana* have been found to be resistant to *R. similis* (Fogain, 1996), whereas the ABB cultivar 'Pelipita' is reported to be moderately resistant (Price and McLaren, 1996).

Low cost multiplication methods for banana

Cultivated banana and plantain are highly female-sterile, i.e., they produce no seeds and cannot reproduce sexually. Thus, only clonal (vegetative) propagation is possible, implying that survival in nature and geographical dispersal are not possible without human intervention. The lack of clean and healthy planting material is one of the most important constraints for large-scale banana production in many countries. Several types of planting materials, which vary in their degree of suitability, include the maiden sucker, water sucker, sword suckers, butt, peeper, and bits. These can be used for the establishment of new banana plantations (Ndubizu and Obiefuna, 1982; Baiyeri and Ndubizu, 1994). However, planting material of this nature tends to perpetuate pests (weevils, nematodes), wilts (Fusarium, Xanthomonas), and diseases (banana bunchy top, BSV). This planting matierial has been responsible for the spread of many pests and diseases from existing to newly established fields.

Various techniques have been used to produce large amounts of planting material in a relatively short time. The underground banana stem (rhizome) has a large number of axillary buds at the base of the leaf sheaths and provides a source of a large number of plantlets if they are allowed to develop (Barker, 1959). One of the first experiments in the vegetative multiplication of bananas involved splitting the rhizome and planting the pieces under nursery conditions (Anon, 1955). Navarre (1957) reported that more than 180 plants were obtained from one rhizome by initiating callus formation. The technique involved digging, cleaning, and injuring of the rhizome to initiate callus formation. The callus tissue gave rise to new plants that were removed and rooted. Hamilton (1965) used the same method and obtained over 150 plants from a single rhizome in five to seven months (Hamilton, 1965). Barker (1959) used the stripping technique in which the outer leaf sheaths of the pseudostem were removed to expose the axillary buds. Soil was then mounded around the buds to initiate development of suckers. When a sucker reached 1 m in height, it was removed and the process was repeated. About 20 plants were obtained from a single parent plant with this method.

In 1974, Loor carried out experiments to try and improve vegetative multiplication of bananas. One experiment established the most suitable kind of corm to use and another experiment the kind of wound that produced the best results. A third experiment examined the action of hormones and cytokinins in stimulating the growth of adventitious buds.

Cook first studied in vitro banana cell culture work in Jamaica (Menendez and Loor, 1979). Later, in vitro work was carried out in Honduras (Berg and Bustamante, 1974), Taiwan (Su-Shien and Shii, 1974; Ma, 1974), and in the Philippines (Guzman et al., 1976). Micro-propagation using meristem/tissue culture can rapidly multiply and produce disease-free planting materials. Modern methods, especially micro-propagation using meristem/tissue culture, are very efficient for rapid multiplication and production of healthy, vigorous, and disease-free planting materials (Swennen, 1990). The in vitro method however, requires a more sophisticated technique, requiring higher levels of skill and precision (Vuylsteke and Talengera, 1998). Tissue culture as a method of generating planting materials is still poorly developed in many developing countries, thus it is unavailable to the subsistence farmers who are the major stakeholders in the production of bananas and plantain. Though effective and fast, tissue culture (in vitro multiplication) is not an option for traditional producers.

Currently, tissue culture plants are very expensive for the small-scale farmer, with prices averaging about $1.00 per plant. However, a number of robust low-cost means of multiplying bananas are now available (Figs. 13.1–5). These include false decapitation (Fig. 13.1a, 13.1b), complete decapitation (Figs. 13.2a, 13.2b), excised corm (Figs. 13.3a, 13.3b), split corm (Figs. 13.4a, 13.4b), and a bud manipulation technique (Figs. 13.5a–i). These techniques, known as macro-propagation, are easy, affordable alternatives to tissue culture and

Figure 13.1a and b. False decapitation method.

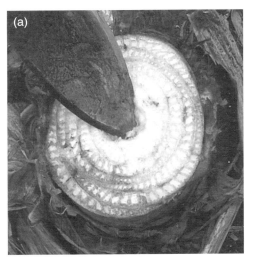

Figure 13.2a and b. Complete decapitation method.

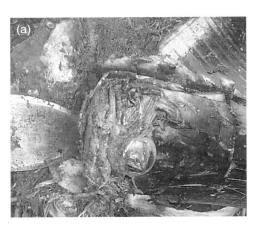
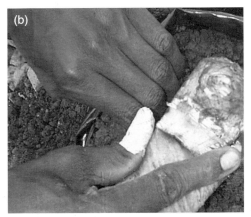

Figure 13.3a and b. Excised corm method.

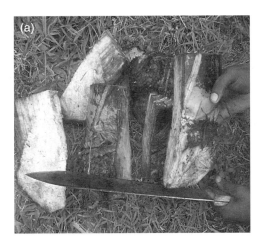

Figure 13.4a and b. Split corm method.

410

Figure 13.5a–i. Bud manipulation method.

Figure 13.5a–i. Bud manipulation method. (*Continued*)

provide a means for large-scale sucker production at the farm level. These methods can generate from 16 to 50 or more plantlets from one sword-sucker corm utilizing sawdust as the plantlet-initiation medium.

False decapitation destroys the actively growing region (meristem) of the mother plant by carefully cutting a hole through the pseudostem base with a knife or other sharp instrument (Fig. 13.1a). New plantlets appear from the axillary buds at the base of the plant (Fig. 13.1b). When the suckers reach a height of 30–40 cm, they are detached carefully with a sharp instrument and planted directly in the field. Complete decapitation differs from false decapitation in that the pseudostem is completely cut down and the meristematic tissue is removed or destroyed. (Fig. 13.2a.). The destruction of the apical meristem stimulates the growth of the axillary buds (Fig. 13.2b) that are detached and planted in a field. One has to be aware that both false and complete decapitations produce best results when there is no competing ratoon sucker growing in the same mat. A sucker in the same mat will exert apical dominance on the axillary buds and retard their growth. In the excised corm technique, the large, visible buds are carefully removed together with some of the surrounding corm tissue and then sown in suitable sized polythene bags (Figs. 13.3a, b). Just enough water is applied to keep the soil moist.

The split corm technique, as the name suggests, involves splitting the corm into two or more fragments. The pieces are planted into soil with the axillary buds facing down (Figs. 13.4 a, b). Sprouted plantlets are detached and potted before transplantation to the field.

The technique that has the highest potential for producing the largest number of suckers is the bud manipulation technique. This is a modification of the Barker's technique (1959): Medium-sized corms are obtained from healthy mats (Fig. 13.5a); the

sucker is pared with a sharp knife to remove the roots and old dead tissue (Fig. 13.5b) (paring also gets rid of superficial weevil and nematode eggs); the pseudostem is then cut about 20 cm above the collar (Fig. 13.5c); the leaf sheaths are removed individually to expose the axillary buds (Fig. 13.5d); and the apical meristem is removed. A well prepared corm in the bud manipulation technique is shown in Fig. 13.5e. After preparation, the bud is carefully placed in sawdust or other appropriate medium in a plastic growth chamber as shown in Fig. 13.5f. The axillary buds begin to develop after approximately three weeks (Fig. 13.5g). This technique is capable of producing 16–30 plantlets in 10–18 weeks (Fig. 13.5h). The plantlets are detached, rooted in soil in a low-light chamber and then hardened under direct sunlight (Fig. 13.5i).

Delivery of hybrid bananas

Pest and disease pressures, coupled with other factors, such as soil fertility and poor agronomic practices, are considered to be responsible for serious yield losses in banana and plantain. Consequently, the major banana breeding programs at IITA, FHIA, CIRAD-FLHOR, EMBRAPA and CARBAP have developed a range of improved hybrids in response to the major production challenges. However, with the exception of Cuba and Tanzania, these hybrids have not been widely adopted in banana-producing countries. A common limiting factor to large-scale adoption and expansion of existing farms is difficulty in obtaining planting materials. Thus, the IITA and the International Network for the Improvement of Banana and Plantain (INIBAP), in collaboration with NARs, NGOs and community-based organizations, set up a program of delivering and distributing banana and plantain hybrids to small-scale farmers in sub-Saharan Africa. The ultimate goal of hybrid delivery schemes is to improve the food and income security of small-scale farmers, which will also lead to economic growth of rural communities in Africa. In addition to providing improved hybrids, the program trains farmers on improved practices of banana production, utilization, and post-harvest technologies and value addition whereby novel, high-value products are produced from banana.

Example of hybrid delivery implementation

The IITA/INIBAP program has a centralized facility, which is equipped to receive the in vitro plants and carry out the weaning and hardening. The centralized facility serves as the meeting point for bringing together the NARS, NGOs, and CBOs involved in the project. The central point also provides a location for training extension workers and representatives in a range of skills and technologies related to improved banana and plantain production and marketing. Participants learn about the production of clean planting material, integrated pest management, post-harvest processing, and the use of banana fiber and other by-products. The main idea is to transfer technologies that are already available and tested, but which are not yet being widely used by farmers. These representatives, together with the community extension workers, are expected to transfer the new technologies and skills to participating farmers at the village level.

A baseline survey is used to identify cultivar needs for each country. Each farmer is provided with plants that are ready for field establishment. Farmers are trained in handling tissue culture material and taught how to plant and manage their fields. Further training is provided in the area of disease and pest management, particularly black Sigatoka, nematodes, and weevils. Finally, farmers are trained in post-harvest issues and marketing.

Biological control for diseases and pests in *Musa*

To reduce the harmful effects of pesticides on humans and the environment, the natural and biological control of pests and diseases has gained much attention in the last 20 years. One recently developed alternative to pesticides is biological enhancement of banana planting material with beneficial microorganisms to increase plant resistance to infection. Since most of the major pests and diseases of banana attack the plant via the roots or corm, technologies using mutualistic fungal endophytes to manage nematodes, fungal wilts, and weevils have been advanced (Sikora et al., 2004). Endophytes are organisms that inhabit the interior of plants and cause no apparent harm to the host. Endophytic fungi have several advantages over other biocontrol agents that are applied directly to the soil. Since biological control agents are generally used in large amounts to treat the soil, they are costly. They have to be applied frequently and their efficacy is usually strongly influenced by the environment (Niere et al., 2004). Endophytes, however, live in plant tissue and reduce the risk of side effects on non-target organisms.

The most frequently isolated and, perhaps, most common fungal endophyte in banana appears to be in the genus *Fusarium* followed by those in the genus *Acremonium*. The most common species is *Fusarium oxysporum*. This species is also reported as an endophyte in many crop species as an effective colonizer of the roots. Other typical soil fungi belonging to the genera *Penicillium, Aspergillus,* and *Gongronella* have also been found to inhabit the banana roots and corm. Therefore, it appears that a few fungal genera and species are dominant as endophytic fungal communities in banana (Niere et al., 2004).

We know that *Fusarium oxysporum* f.sp. *cubense* is the causal agent of Fusarium wilt in banana. *Fusarium oxysporum* f.sp. *cubense* is divided into vegetative compatibility groups. Reportedly, none of the fungal endophytes in the genus *Fusarium* isolated from banana is vegetatively compatible with the pathogenic form of *Fusarium*. Different strains of fungal endophytes can also be found in the same plant. In testing banana, fungal endophytes are isolated at random from healthy roots or corm tissue. They are cultured, identified, mass cultured, inoculated into tissue culture plantlets, allowed to colonize, and then the plant is challenged with target organisms, such as pathogenic *Fusarium,* nematodes and weevils. The antagonistic activity is measured. A large percentage of the isolates obtained from the endorrhiza have demonstrated significant antagonistic activity toward the burrowing nematode, *R. similis* (Niere et al., 1998; Pocasangre, 2000). Niere et al. (2004) found that nematode numbers in endophyte-treated plants were reduced by more than 30% compared with the controls in some banana clones. Nematode damage, expressed as a percentage of necrotic root tissue, also showed a reduction in the experimental plants compared with the controls. This result was also clone-dependent. Since nematode numbers were reduced in plants inoculated with endophytes, it may be speculated that the endophytes affect the development of juvenile nematodes into adults. Some isolates were shown to be pathogenic on eggs of the banana weevil, *Cosmopolites sordidus,* and to have negative effects on the growth of weevil larvae (Griesbach, 1999; Gold et al., 2003). Endophyte-treated plants also showed less weevil damage. The activity of fungal endophytes against Fusarium wilt has so far been shown to be inadequate (Pocasangre, 2000). However, investigations at IITA seem to have provided promising results.

Genetic transformation of *Musa*

Genetic transformation has become an important tool for crop improvement. Genetic transformation can be defined as the transfer of any foreign gene(s) isolated from plants, viruses, bacteria, or animals into a new genetic background. Successful genetic transformation in plants requires the production of normal, fertile plants expressing the newly inserted gene(s). The process of genetic transformation involves several distinct steps, namely: identification of a useful gene, the cloning of the gene into a suitable plasmid vector, and delivery of the vector into the plant cell (insertion and integration). This is followed by expression and inheritance of the foreign DNA encoding a protein.

Gene insertion in plants can be achieved by direct gene transfer through particle bombardment or electroporation or through biological vectors such as a disarmed Ti (tumor-inducing) plasmid of *Agrobacterium tumefaciens.*

Recently, some success in genetic engineering of bananas and plantains has been achieved to enable the transfer of foreign genes into plant cells. Genetic transformation using microprojectile bombardment of embryogenic cell suspension is now a routine process in bananas (Becker et al., 2000; Sági et al., 1995). An efficient method for direct gene transfer via particle bombardment of embryogenic cell suspension has been reported in the cooking banana cultivar 'Bluggoe' and plantain (Sági et al., 1995; Becker et al., 2000). Although *Musa* was generally regarded as recalcitrant for *Agrobacterium*-mediated transformation, Hernandez (1999) demonstrated its compatibility with *A. tumefaciens* and highlighted its potential for genetic transformation. The recovery of transgenic plants of banana obtained by means of *A. tumefaciens*-mediated transformation has been reported. A protocol for *Agrobacterium*-mediated transformation of embryogenic cell suspensions of the banana cultivars 'Rasthali,' 'Cavendish,' and 'Lady finger' has been developed (Ganapathi et al., 2001; Khanna et al., 2004). At present, most of the transformation protocols use cell suspensions. However, establishing a cell suspension is a lengthy process and is cultivar-dependent. Therefore, alternative methods, such as transformation with shoot tips, has been attempted and reported from various cultivars of *Musa* (May et al., 1995; Tripathi et al., 2002; Tripathi et al., 2005a). This technique is applicable to a wide range of *Musa* cultivars regardless of ploidy or genotype (Tripathi et al., 2003; Tripathi et al., 2005a). This process does not incorporate steps using disorganized cell cultures but uses micropropagation, which has the advantage of allowing regeneration of homogeneous populations of plants within a short time. This procedure offers several potential advantages over the use of embryogenic cell suspensions (ECS) as it allows for rapid transformation of *Musa* species. The status of research on genetic engineering of banana for disease resistance and future possibilities for the field have been reviewed (Sági et al., 1998; Sági, 2000; Tripathi, 2003; Tripathi et al., 2004, Tripathi et al., 2005b). Currently, the transgenes used for banana improvement have been exclusively isolated from heterologous sources such as other plant species, insects, microbes, and animals (Tripathi, 2003; Tripathi et al., 2004).

Potential strategies for developing resistance to diseases

One of the major handicaps in banana breeding is the apparent lack of genes for resistance in the cultivated varieties. As an alternative, genetic-transformation technology with fungicidal or bactericidal transgenes and other relevant genes may offer a solution to these problems. The most attractive strategy for serious fungal disease like black Sigatoka control in *Musa* is probably the production of disease-resistant plants through the transgenic approach, including the expression of genes encoding plant, fungal, or bacterial hydrolytic enzymes

(Lorito et al., 1998), genes encoding elicitors of defense response (Keller et al., 1999), and antimicrobial peptides (Cary et al., 2000; Li et al., 2001).

Antimicrobial peptides (AMPs) have a broad-spectrum antimicrobial activity against fungi as well as bacteria, and most are non-toxic to plant and mammalian cells. Examples of AMPs are magainin from the African clawed frog (Bevins and Zasloff, 1990; Zasloff, 1987), cecropins from the giant silk moth (Boman and Hultmark, 1987), mammalian defensins (Ganz and Lehrer, 1994) and plant defensins (Broekaert et al., 1995). The cecropin (Alan and Earle, 2002; De Lucca et al., 1997) and its derivatives (D4E1: Cary et al., 2000; Rajasekaran et al., 2001) have been found to inhibit the in vitro growth of several important bacterial and fungal pathogens. Transgenic tobacco plants expressing cecropins have increased resistance to *Pseudomonas syringae* pv. *tabaci,* the cause of tobacco wildfire disease (Huang et al., 1997). Similarly, magainin is effective against plant pathogenic fungi (Kristyanne et al., 1997; Zasloff, 1987). Li et al. (2001) reported enhanced disease resistance in transgenic tobacco expressing Myp30, a magainin analogue. Chakrabarti et al. (2003) reported successful expression of this synthetic peptide and enhanced disease resistance in transgenic tobacco and banana. Using the broad-spectrum activity against fungal pathogens, individual or combined expression of cecropin, magainin and their derivatives, banana breeders may be able to develop increased resistance to several pathogens.

Another source of antimicrobial proteins has been lysozyme, either from bacteriophage, chicken eggs, or bovine. The lysozyme attacks the murein layer of bacterial peptidoglycan, resulting in cell wall weakening and eventually leading to lysis of both Gram-negative and Gram-positive bacteria. The lysozyme genes have been used to confer disease resistance against plant pathogenic bacteria in transgenic tobacco (Trudel et al., 1995), potato (Düring et al., 1993), and apple (Ko, 1999). Human lysozyme transgenes have conferred disease resistance in tobacco through inhibition of fungal and bacterial growth, suggesting the possible use of the human lysozyme gene for controlling plant disease (Nakajima et al., 1997).

There are many reports on the application of plant proteins with distinct antimicrobial activities (Broekaert et al., 1997). Thionins are highly abundant polypeptides with anti-fungal activities. There are a number of known plant defensins that are known to protect against plant pathogens. The radish defensin Rs-AFP2 conferred partial resistance to the tobacco pathogen *Alternaria longipes* (Terras et al., 1995), whereas a defensin from alfalfa (alfAFP) has been expressed in potato in the greenhouse as well as in the field (Gao et al., 2000). Kanzanki et al. (2002) reported the over-expression of the WTI defensin from wasabi bean conferring enhanced resistance to blast fungus in transgenic rice.

The AMPs of plant origin may be potential candidates for fungal resistance in *Musa* as they possess high in vitro activity to *M. fijiensis* and *F. oxysporum f.sp. cubense,* two major fungal pathogens of *Musa;* they are also non-toxic to humans and banana cells. Several hundreds of transgenic lines of *Musa,* especially plantains expressing defensin type AMPs, have been developed at the Katholic University at Leuven, Belgium (Remy, 2000).

Plants have their own networks of defense against plant pathogens that include a vast array of proteins and other organic molecules produced prior to infection or during pathogen attack. Pathosystem-specific plant resistance (R) genes have been cloned from several plant species against many different pathogens (Bent, 1996). These include *R* genes that mediate resistance to pathogens, such as bacteria, fungi, viruses, and nematodes. Many of these *R* gene products share

structural motifs, which indicates that disease resistance to diverse pathogens may operate through similar pathways. In tomato (*Lycopersicon esculentum*), the R gene *Pto* confers resistance against strains of *Pseudomonas syringae* pv. *tomato* (Martin et al., 1993; Kim et al., 2002). *Pto*-overexpressing plants show resistance not only to *P. syringae* pv. *tomato* but also to *Xanthomonas campestris* pv. *vesicatoria* and to the fungal pathogen *Cladosporium fulvum* (Mysore et al., 2003). Similarly, the *Arabidopsis* gene *RPS4* specifies disease resistance to *Pseudomonas syringae* pv. *tomato* strain DC3000 expressing *avrRps4* (Gassmann et al., 1999). The *Bs2* resistance gene of pepper specifically recognizes and confers resistance to strains of *X. campestris* pv. *vesicatoria* (Tai et al., 1999). Transgenic tomato plants expressing the pepper *Bs2* gene suppress the growth of *Xcv.* The *Xa21* gene isolated from rice has been shown to confer resistance against many isolates of *X. oryzae* pv. *oryzae* (Song et al., 1995; Wang et al., 1996). Transgenic plants expressing *Xa* 21 under the control of the native promoter of the genomic fragment of the *Xa* 21 gene showed enhanced resistance to bacterial leaf blight caused by most *Xoo* races. The *Xa1* gene, also isolated from rice, confers resistance to Japanese race 1 of *X. oryzae* pv. *oryzae,* the causal pathogen of bacterial blight (Yoshimura et al., 1998).

Successful transfer of resistance genes to heterologous plant species provides another option for developing disease-resistant plants. R gene-mediated resistance has several attractive features for disease control. When induced in a timely manner, the concerted responses can efficiently halt pathogen growth with minimal collateral damage to the plant. No input is required from the farmer and there are no adverse environmental effects. Unfortunately, *R* genes are often quickly defeated by co-evolving pathogens (Pink, 2002). Many *R* genes recognize only a limited number of pathogen strains and therefore do not provide broad-spectrum resistance. Also, efforts to transfer *R* genes from model species to crops, or between distantly related crops, could be hampered due to restricted taxonomic functionality.

Plants also use a wide array of defense mechanisms against pathogen attack. Among these, hypersensitive response (HR) is an induced-resistance mechanism, characterized by rapid, localized cell death upon a plant's encounter with a microbial pathogen. Several defense genes have been shown to delay the hypersensitive response induced by bacterial pathogens in non-host plants through the release of proteinaceous elicitors. Elicitor-induced resistance is not specific against particular pathogens. Hence, manipulation of such defense genes may be more useful.

The ferredoxin-like amphipathic protein (*pflp*, formerly called AP1) and the hypersensitive response-assisting protein (*hrap*) isolated from the sweet pepper, *Capsicum annuum*, are novel plant proteins that can intensify the harpinPSS-mediated hypersensitive response (Lin et al., 1997; Chen et al., 2000). These proteins have a dual-function: iron-depletion antibiotic action, and harpin-triggered HR enhancing. The transgenes have been shown to delay the hypersensitive response induced by various pathogens (including *Erwinia, Pseudomonas, Ralstonia, and Xanthomonas* spp.) in non-host plants through the release of the proteinaceous elicitors and harpinPSS in crops, including dicotyledons such as tobacco, potato, tomato, broccoli, and orchids, and monocotyledons such as rice (Tang et al., 2001; Lu et al., 2003; Huang et al., 2004). Because elicitor-induced resistance is not specific against particular pathogens, it could be a very useful strategy (Wei and Beer, 1996). Research using *pflp* and *hrap* genes is in progress at IITA with the objective of producing varieties resistant to bacterial wilt disease caused by *Xanthomonas* spp.

The most promising transgenic strategy for ssDNA viruses such as BBTV is expressing a defective gene that encodes an essential virus life cycle activity. For instance, the replication of the virus is encoded in the replication gene or genes (*Rep*). The resultant Rep protein may retain the ability to bind to its target viral DNA but lack the functions of the *Rep* gene (Brunetti et al., 2001). The defective Rep protein binds to the invading viral DNA and is thought to out-compete the native viral Rep protein, thus reducing or eliminating virus DNA replication. Lucioli et al. (2003) expressed the first 630 nucleotides of the *Rep* gene of tomato yellow leaf curl Sardinia virus to generate resistance (Brunetti et al., 2001). The duration of the resistance was related to the ability of the invading virus to switch off transgene expression through post-transcriptional gene silencing (PTGS). Many researchers are trying to develop transgenic plants of *Musa* resistant to BBTV by targeting the PTGS mechanism using mutated or antisense *Rep* genes. Unfortunately, there appear to be no strategies that have been developed that generate high-level resistance to the plant dsDNA or pararetroviruses, including the badnaviruses. Researchers at the IITA, Nigeria, in collaboration with the John Innes Centre (JIC), UK, are attempting to generate transgenic resistance to BSV based on PTGS. This approach involves the specific silencing of a viral gene or genes known to be involved in replication or pathogenesis.

Potential strategies for developing resistance to pests

There are several possible approaches for developing transgenic bananas with improved weevil and nematode resistance. A variety of genes are available for genetic engineering for pest resistance (Sharma et al., 2000). Among these are proteinase inhibitors, *Bacillus thuringiensis* (Bt) toxins, plant lectins, vegetative insecticidal proteins (VIPs), and alpha-amylase inhibitors (AI). Proteinase inhibitors contribute to host-plant resistance against pests and pathogens (Green and Ryan, 1972). They operate by disrupting protein digestion in the insect mid-gut via inhibition of proteinases. The two major proteinase classes in the digestive systems of phytophagus insects are the serine and cysteine proteinases. Coleopteran insects, including the banana weevil, mainly use cysteine proteinases (Murdock et al., 1987) and recent studies indicate a combination of both serine and cysteine proteinases are useful (Gerald et al., 1997). These inhibitors have already been used for insect control in genetically modified (GM) plants (Leple et al., 1995). Presently, cysteine proteinase activity has been identified in the mid-gut of the banana weevil, and in vitro studies have shown that cysteine proteinases are strongly inhibited by both a purified recombinant rice (oryzacystatin-I [OC-I]) and papaya cystatin (Abe et al., 1987; Kiggundu et al., 2003).

The use of proteinase inhibitors (PIs) as nematode antifeedants is an important element of natural plant defense strategies (Ryan, 1990). This approach offers prospects for novel plant resistance against nematodes and the reduced use of nematicides. The potential of PIs for transgenic crop protection in banana is supported by the lack of harmful effects when humans consume naturally occurring PIs in grains such as rice and cowpea. Cysteine PIs (cystatins) are inhibitors of cysteine proteinases and have been isolated from seeds of a wide range of crop plants consumed by people, including sunflower, cowpea, soybean, maize, and rice (Atkinson et al., 1995). Transgenic expression of PIs provides effective control of both cyst and root–knot nematodes. The cystatins Oc-I and an engineered variant Oc-IδD86 were shown to mediate nematode resistance when expressed in tomato hairy root (Urwin et al., 1995), *Arabidopsis* plants (Urwin et al., 1997), rice (Vain et al., 1998) and pineapple (Urwin et al., 2000). The partial resistance was conferred in a small-scale potato field trial on a susceptible cultivar by expressing cystatins under control of the cauliflower mosaic virus 35S promoter (Urwin et al., 2001). The enhanced transgenic plant resistance to nematodes has been demonstrated by using dual proteinase inhibitor transgenes (Urwin et al., 1998). Since cystatins have been shown to function in rice, which, like *Musa*, is a monocotyledon and also to have clear efficacy against a wide range of nematode species, their usefulness as transgenes for development of transgenic *Musa* for resistance to nematodes is promising.

The expression and biological activity of the Bt toxins has been investigated in GM plants for insect control. The *Bt* gene technology is currently the most widely used system for lepidopteran control in commercial GM crops (Krettiger, 1997). The expression of a selected *Bt* gene for weevil resistance may be a rather long-term strategy since no potential *Bt* gene with significant toxic effects against the banana weevil has yet been identified (Kiggundu et al., 2003). Some Bt proteins are also effective against saprophagous nematodes (Borgonie et al., 1996). The Cry5B protein is toxic to wild type *Caenorhabditis elegans;* other *C. elegans* mutants are resistant to Cry5B but susceptible to the Cry6A toxin (Marroquin et al., 2000). The approach using *cry* genes has potential for plant nematode control (Wei et al., 2003).

Other strategies for creating nematode resistance include the use of natural plant resistance genes (R-genes) and lectins. Several *R* genes are targeted against nematodes. The Hs1pro-1 from a wild species of beet confers resistance to the cyst nematode *Heterodera schachlii* (Cai et al., 1997). Plant lectins confer a protective role against a range of organisms (Sharma et al., 2000). Lectins have been isolated from a wide range of plants, including snowdrop, pea, wheat, rice, and soybean. Their carbohydrate-binding capability renders them toxic to insects. The snowdrop lectin (*Galanthus nivalis* agglutinin: GNA) is toxic to several insect pests in the orders Homotera, Coleoptera, and Lepidoptera (Tinjuangjun, 2002). Some lectins, such as GNA, have biological activity against nematodes (Burrows et al., 1998). However, many lectins have toxic effects on insects and mammals, which raise concerns regarding toxicological safety. This may prove a substantial limitation to the future commercial development of lectins.

Alpha-amylase inhibitors (AI) and chitinase enzymes might also have future potential for weevil control in *Musa*. Alpha-amylase inhibitors operate by inhibiting the enzyme alpha-amylase that breaks down starch to glucose in the insect gut (Morton et al., 2000). Transgenic adzuki beans are produced with enhanced resistance to bean bruchids that are Coleopterans like the banana weevil (Ishimoto et al., 1996). Therefore, alpha-amylase inhibitors may be of interest for banana weevil control. Chitinase enzymes are produced as a result of invasion either by fungal pathogens or insects. Transgenic expression of chitinase has shown improved resistance to Lepidopteran insect pests in tobacco (Ding et al., 1998).

The future of biotechnology in the genetic enhancement of *Musa*

Plant biotechnology has the potential to play a key role in the sustainable production of *Musa*. Currently, no genetically transformed bananas and plantains are commercially available. However, there is enormous potential for genetic manipulation of *Musa* species for disease and pest resistance using the existing transformation systems. Using molecular techniques, novel genes encoding agronomically important traits can be identified, isolated, characterized, and introduced into cultivars via a combination of genetic transformation and in vitro regeneration. The use of appropriate gene constructs may allow the production of nematode-, fungus-, bacteria-, and virus-resistant plants in a sig-

nificantly shorter period of time than they are by use of conventional breeding, especially if several traits can be introduced at the same time. It may also be possible to incorporate other characteristics, such as drought tolerance, thus extending the geographic spread of banana and plantain production and contributing significantly to food security and poverty alleviation in developing countries. Long-term and multiple disease resistance can be achieved by integrating several genes with different targets or modes of action into the plant genome. Technically, this can be done either in several consecutive steps or simultaneously.

Conclusion

Thus far the major emphasis in plant breeding has been the continuous improvement in the productivity of crop plants. Breeding for yield was a major target, followed by breeding for resistance to diseases and pests that impact yield. The pressure of an increasing population and consequent increase in demand for food on the one hand and the depletion of arable land, on the other hand, have placed new emphases in plant breeding (Swaminathan, 1969). The world population is expected to reach 7 billion within 25 years. At the same time, agricultural production is growing at a slower rate of about 1.8% annually (Altman, 1999). In addition to yield, disease and pest resistance, nutritive and organoleptic qualities, and other considerations will have to be added to breeding parameters. These parameters, according to Sinha and Swaminathan (1984), include: increasing the nutritive value of crops by breeding for enhanced proteins, vitamins, and other essential minerals; improving the efficiency of food production for each unit of cultural and solar energy invested; producing stable crops across a wide range of environments with resistance to damage by weeds, pests and pathogens; improving and identifying plants as sources of biomass and renewable energy; breeding genotypes for optimum response to high, low, and zero inputs; breeding efficient plant types for crop-livestock production systems; and breeding for suitability of crops in drought- and flood-prone areas. Plant breeding aims to improve crop performance and/or quality and create new cultivars. However, new cultivars must preserve the quality characteristics and meet consumer demands. The process of developing a new cultivar can take up to 14 years in the case of banana from the first cross to the cultivar entering the marketplace. Traditional plant breeding cannot meet all the food requirements of human beings. Therefore breeding through plant biotechnology is a necessity. The effective merging of classical breeding with modern plant biotechnology and the novel tools it provides are the gateway for the continuous increase in agricultural productivity and human survival (Altman, 1999). In this chapter we have highlighted the gains made in banana breeding via classical methods, discussed the challenges that remain in banana breeding, and showed how biotechnology can be used to complement classical breeding.

References

Abe K, Kondo H, Arai S. 1987. Purification and characterization of a rice cysteine proteinase inhibitor. Agr. Biol. Chem. 51:2763–2768.

Abera A, Gold CS, Kyamanywa S. 1997. Banana weevil oviposition and damage in Uganda. African Crop Science Conference Proceedings 3:1199–1205.

Adeleke MTV, Pillay M, Okoli BE. 2004. Meiotic irregularities and fertility relationships in diploid and triploid *Musa* L. Cytologia 69:387–393.

Alan AR, Earle ED. 2002. Sensitivity of bacterial and fungal pathogens to the lytic peptides, MSI-99, magainin II, and cecropin B. Mol Plant-Microbe Interactions 15:701–708.

Altman A. 1999. Plant biotechnology in the 21st century: The challenges ahead. Electronic Journal of Biotechnology. 2(2).

Ambrose E. 1984. Research and development in banana crop protection (excluding Sigatoka) in the English speaking Carribean. Fruits 39:234–247.

Anon. 1955. La multiplication vágátative du bananier. Bull. Inf. I.N.E.A.C. 4:49–52.

Atkinson HJ, Urwin PE, Hansen E, McPherson MJ. 1995. Designs for engineered resistance to root-parasitic nematodes. Trends Biotechnol 13:369–374.

Baiyeri KP, Ndubizu, TOC. 1994. Variability in growth and field establishment of Falsehorn plantain suckers raised by six cultural methods. MusAfrica 4:1–3.

Bakry F, Carreel F, Caruana M-L, Cote FX, Jenny C, Tezenas du Montcel H. 1997. Les bananiers. In: Charrier A, Jacquot M, Hamon S, Nicolas D (eds). L'ameloiration des plantes tropicales. CIRAD and ORSTOM.

Barker WG. 1959. A system of maximum multiplication of the banana plant. Tropical Agriculture, Trin. 36:275–284.

Becker DK, Dugdale B, Smith MK, Harding RM, Dale JL. 2000. Genetic transformation of Cavendish banana (*Musa* sp. AAA group) cv. Grand Nain via microprojectile bombardment. Plant Cell Rep. 19:229–234.

Bent AF. 1996. Plant disease resistance genes: Function meets structure. Plant Cell 8:1757–1771.

Berg LA, Bustamante M. 1974. Heat treatment and meristem culture for production of virus-free bananas. Phytopathology 64:320–322.

Bevins CL, Zasloff M. 1990. Peptides from frog skin. Ann. Rev. Biochem. 59:395–414.

Blomme G, Mukasa H, Ssekiwoko F, Eden-Green S. 2005. On-farm assessment of banana bacterial wilt control options. African Crop Science Society Conference Proceedings, Entebbe, Uganda, 5–9 December 2005. pp. 317–320.

Boman HG, Hultmark D. 1987. Cell free immunity in insects. Ann. Rev. Microbiol. 41:103–126.

Borgonie G, Claeys M, Leyns F, Arnaut G, De Waele D, Coomans A. 1996. Effect of nematicidal *Bacillus thuringiensis* strains on free-living nematodes. Fund. App. Nematol. 19:391–398.

Borojevic S. 1990. Principle and methods of plant breeding. Elsevier, Amsterdam. pp. 368.

Broekaert WF, Terras FRG, Cammue BPA, Osborn RW. 1995. Plant defensins: Novel antimicrobial peptides as components of the host defense system. Plant Physiol 108:1353–1358.

Broekaert WF, Cammue BPA, De Bolle M, Thevissen K, De Samblanx G, Osborn RW. 1997. Antimicrobial peptides from plants. Critical Review of Plant Science 16:297–323.

Brunetti A, Tavazza R, Noris E, Luciolo A, Accotto G, Tavazza M. 2001. Transgenically expressed T-Rep of tomato yellow leaf curl Sardinia virus acts as a trans-dominant-negative mutant, inhibiting viral transcription and replication. J. Virol. 75:10573–10581.

Buddenhagen IW, Elsasser TA 1962. An insect-spread bacterial wilt epiphytotic of Bluggoe banana. Nature 194:164–165.

Burrows RP, Barker ADP, Newell CA, Hamilton WDO. 1998. Plant-derived enzyme inhibitors and lectins for resistance against plant-parasitic nematodes in transgenic crops. Pesticide Sci. 52:176–183.

Cai DG, Kleine M, Kifle S, Harloff HJ, Sandal NN. 1997. Positional cloning of a gene for nematode resistance in sugar beet. Sci. 275:832–834.

Carlier J, Foure E, Gauhl F, Jones DR, Lepoivre P, Mourichon X, Pasberg-Gauhl C. 2000. Fungal diseases of the foliage. In: Jones DR (ed). Diseases of banana, Abaca and Enset. CABI Publishing Wallingford, UK. pp. 37–141.

Cary JW, Rajasekaran K, Jaynes JM, Cleveland TE. 2000. Transgenic expression of a gene encoding a synthetic antimicrobial peptide results in inhibition of fungal growth in vitro and in planta. Plant Sci. 154:171–181.

Chakrabarti A, Ganapathi TR, Mukherjee PK, Bapat VA. 2003. MSI-99, a magainin analogue, imparts enhanced disease resistance in transgenic tobacco and banana. Planta 216:587–596.

Chen CH, Lin HJ, Ger MJ, Chow D, Feng TY. 2000. The cloning and characterization of a hypersensitive response-assisting protein that may be associated with the harpin-mediated hypersensitive response. Plant Mol. Biol. 43:429–438.

Cobiac L, Bathurst K. 1993. Supplement to Food Australia, April, 1993. CSIRO Div. Human Nutr., Adelaide. pp. S1–24.

Collingborn FMB, Gowen SR. 1997. Screening of banana cultivars for resistance to Radopholus similis and Pratylenchus coffeae. International Network for the Improvement of Banana and Plantain (INIBAP), Montpellier, France. InfoMusa 6:3.

D'Hont A, Paget-Goy A, Escoute J, Carreel F. 2000. The interspecific genome structure of cultivated bananas, Musa spp. revealed by genomic DNA in situ hybridization. Theor. Appl. Genet. 100:177–183.

De Langhe E, Devreux M. 1960. Une sous-espece nouvelle de Musa acuminata colla. Bull. Jard. Bot. Brux. 30:375–388.

De Langhe E, Valmayor RV. 1980. French plantains in Southeast Asia. IBPGR/SEAN 4:3–4.

De Langhe E. 1961. La taxonomie du bananier plantain en Afrique Equatoriale. Journal d'Agriculture Tropicale et de Botanique Appliquee (Brussels) 8:419–449.

De Langhe E. 1964. The origin of variation in the plantain banana. Ghent, Belgium, State Agricultural University of Ghent 39:45–80.

De Lucca AJ, Bland JM, Jacks TJ, Grimm C, Cleveland TE, Walsh TJ. 1997. Fungicidal activity of cecropin A. Antimicrobial Agents in Chemotherapy 41:481–483.

Ding X, Gopalakrishnan B, Johnson LB, White FF, Wang X, Morgan TD, Kramer KJ, Muthukrishnan S. 1998. Insect resistance of transgenic tobacco expressing an insect chitinase gene. Transgenic Res 7:77–84.

Düring K, Porsch P, Fladung M, Lörz H. 1993. Transgenic potato plants resistant to the phytopathogenic bacterium Erwinia carotovora. Plant J. 3:587–598.

Eden-Green S. 2004. How can the advance of banana Xanthomonas wilt be halted? InfoMusa 13(2):38–41.

Elsen A, De Waele D. 2004. Recent developments in early in vitro screening for resistance against migratory endoparasitic nematodes in Musa. In: Mohan Jain S, Swennen R (eds). Banana Improvement: Cellular, molecular biology, and induced mutations. Science Publishers, Inc., NH, USA. pp. 193–208.

Engelberger L. 2003, Carotenoid-rich bananas in Micronesia. InfoMusa 12:2–5.

Estelitta S, Radhakrishnan TC, Paul TS. 1996. Infectious chlorosis disease of banana in Kerala. InfoMusa 5:25–26.

FAO. 1992. International Conference on Nutrition: World Declaration and Plan of Action for Nutrition Food and Agricultural Organization of the United Nations and World Health Organization, Rome. pp. 1–42.

FAO STAT Agriculture Data. 2003. Available at: http://apps.fao.org.

Fogain R, Price NS. 1994. Varietal screening of some Musa cultivars for susceptibility to the banana weevil, Cosmopolites sordidus (Coleoptera: Curculionidae). Fruits 49:247–251.

Fogain R. 1996. Screenhouse evaluation of Musa for susceptibility to Radopholus similis: Evaluation of plantains AAB and diploids AA, AB and BB. In: Frison EA, Horry JP, De Waele D (eds). New frontiers in resistance breeding for nematodes, fusarium and Sigatoka. Proceedings of the workshop held in Kuala Lumpur, Malaysia, 25 October 1995. INIBAP, Montpellier, France. pp. 79–88.

Frison E, Sharrock S. 1998. Banana weevil: Ecology pest status and prospects for integrated control with emphasis on East Africa. In: Saini SK (ed). Proceedings of a Symposium on Biological Control in Tropical Habitats: Third International Conference on Tropical Entomology. ICIPE, Nairobi, Kenya. pp. 49–74.

Ganapathi TR, Higgs NS, Balint-Kurti PJ, Arntzen CJ, May GD, Van Eck JM. 2001. Agrobacterium-mediated transformation of the embryogenic cell suspensions of the banana cultivar Rasthali (AAB). Plant Cell Rep. 20:157–162.

Ganz T, Lehrer RI. 1994. Defensins. Cur. Opin. Immunol. 6:584–589.

Gao A, Hakimi SM, Mittanck CA, Wu Y, Woerner BM, Stark DM, Shah DM, Liang J, Rommens CMT. 2000. Fungal pathogen protection in potato by expression of a plant defensin peptide. Nature Biotechnol. 18:1307–1310.

Ge XJ, Liu MH, Wang WK, Schaal BA, Chiang TY. 2005. Population structure of wild bananas, *Musa balbisiana*, in China determined by SSR fingerprinting and cpDNA PCR-RFLP. Molecular Ecology 14:933–944.

Geering ADW, McMichael LA, Dietzgen RG, Thomas JE. 2000. Genetic diversity among banana streak virus isolates from Australia. Phytopathology 90:921–927.

Gerald RR, Kramer KJ, Baker JE, Kanost JF, Behke CA. 1997. Proteinase inhibitors and resistance of transgenic plants to insects. In: Carozi N, Koziel M (eds). Advances in Insect Control: The Role of Transgenic Plants, Taylor and Francis, London. pp. 157–183.

Gold CS. 1998. Banana weevil: Ecology, Pest status and Prospectus of Integrated Control with emphasis on East Africa. In: Siani RK (ed). Proceedings of the Third International Conference on Tropical Entomology, 30 October–4 November 1994, Nairobi. International Centre for Insect Physiology and Ecology, Nairobi. pp. 49–74.

Gold CS, Bagabe MI, Sendege R. 1999. Banana weevil, *Cosmopolites sordidus* (Germar) (Coleoptera: Curculionidae): Tests for suspected resistance to carbofuran and dieldrin in the Masaka District, Uganda. African Entomology 7:189–196.

Gold CS, Nankinga C, Niere B, Maestrie Godonou. 2003. IPM of banana weevil in Africa with emphasis on microbial control. In: Neuenschwander P, Borgemeister B, Langewald J (eds). Biological control in IPM systems in Africa, CAB International, Wallingford, UK.

Gowen SR, Queneherve P. 1990. Nematode parasites of banana, plantain and abaca. In: Luc M, Sikora RA, Bridge J. (eds). Plant parastic nematodes in subtropical and tropical agriculture. CAB International, Wallingford. pp. 431–461.

Graham RD, Rosser JM. 2000. Carotenoids in staple foods: Their potential to improve human nutrition. Food Nutr. Bull. 21:404–409.

Graham RD, Senadhira D, Beebe SE, Iglesias C, Ortiz-Monasterio I. 1999. Breeding for micronutrient density in edible portions of staple food crops: Conventional approaches. Field Crop Res. 60:57–80.

Green TR, Ryan CA. 1972. Wound-induced proteinase inhibitors in plant leaves: A possible defence mechanism against insects. Sci. 175:776–777.

Griesbach M. 1999. Occurrence of mutualistic fungal endophytes in banana (*Musa* spp.) and their potential as biocontrol agents of the banana weevil *Cosmopolites sordidus* (Germar) (Coleoptera: Curculionidae) in Uganda. Ph.D thesis, University of Bonn, Germany.

Guzman E de, Ubalde E, Rosario AG. 1976. Banana and coconut in vitro cultures for induced mutation studies. Improvement of vegetatively propagated plants and tree crops through induced mutations. Vienna, Austria, International Atomic Energy Agency. pp. 33–54.

Hakkinen M, Meekiong K. 2005. A new species of the wild banana genus, *Musa* (*Musaceae*), from Borneo. Systematics and Biodiversity 2:169–173.

Hakkinen M. 2004. *Musa voonii*. A new *Musa* species from Northern Borneo and discussion of the section *Callimusa* in Borneo. Acta Phytotax. Geobot. 55:79–88.

Hamilton KS. 1965. Reproduction of banana from adventitious buds. Tropical Agriculture. Trin. 42: 69–73.

Hernandez JBP, Remy S, Sauco VG, Swennen R, Sági L. 1999. Chemotactic movement and attachment of *Agrobacterium tumefaciens* to banana cells and tissues. J. Plant Physiol. 155:245–250.

Howell E, Newbury HJ, Swennen RL, Withers LA, Ford-Lloyd BV. 1994. The use of RAPD for identifying and classifying *Musa* germplasm. Genome 37:328–332.

Huang SN, Chen CH, Lin HJ, Ger MJ, Chen ZI, Feng TY. 2004. Plant ferredoxin-like protein AP1 enhances *Erwinia*-induced hypersensitive response of tobacco. Physiological and Molecular Plant Pathology 64:103–110.

Huang Y, Nordeen RO, Di M, Owens LD, McBeth JH. 1997. Expression of an engineered cecropin gene cassette in transgenic tobacco plants confers disease resistance to *Pseudomonas syringae* pv. *tabaci*. Phytopathol. 87:494–499.

Hughes Jd'A, Speijer PR, Olatunde O. 1998. Banana die back virus: A new virus infecting banana in Nigeria. Plant Disease 82:129.

Hwang S, Ko WH. 1987. Somaclonal variation of bananas and screening for resistance to Fusarium wilt. In: Persley GJ, De Langhe EA (eds). Banana and plantain breeding strategies. Proceedings of a workshop held at Cairns, Australia, 13–17 Oct. 1986. ACIAR Proceedings No. 21. pp. 157–160.

Ishimoto M, Sato T, Chrispeels MJ, Kitamura K. 1996. Bruchid resistance of transgenic azuki bean expressing seed α-amylase inhibitor of common bean. Entomol. Exp. Appl. 79:309–315.

Ittyeipe K. 1986. Studies on host preference of banana weevil borer, *Cosmopolites sordidus* Germ. (*Curculionidae: Coleoptera*). Fruits 46:375–379.

Jones DR. 2000. Introduction to banana, abaca and enset, In: Jones DR (ed). Diseases of banana, abaca and enset, CABI Publishing, Wallingford, UK. pp.1–36.

Kanzanki H, Nirasawa S, Saitoh H, Ito M, Nishihara M, Terauchi R, Nakamura I. 2002. Over-expression of the wasabi defensin gene confers enhanced resistance to blast fungus (*Magnaporthe griea*) in transgenic rice. Theor. App. Gen. 105:809–814.

Karamura D. 1998. Numerical taxonomic studies of the East African highland bananas (*Musa* AAA-EA) in Uganda. INIBAP, France.

Keller H, Pamboukdjian N, Ponchet M, Poupet A, Delon R, Verrier JL, Roby D, Ricei P. 1999. Pathogen-induced elicitin production in transgenic tobacco generates a hypersensitive response and non-specific disease resistance. Plant Cell 11:223–235.

Khanna H, Becker D, Kleidon J, Dale J. 2004. Centrifugation Assisted *Agrobacterium tumefaciens*-mediated Transformation (CAAT) of embryogenic cell suspensions of banana (*Musa* spp. Cavendish AAA and Lady finger AAB). Mol. Breed. 14:239–252.

Kiggundu A, Pillay M, Viljoen A, Gold C, Tushemereirwe W, Kunert K. 2003. Enhancing banana weevil (*Cosmopolites sordidus*) resistance by plant genetic modification: A perspective. African J. Biotechnol. 2:563–569.

Kikafunda JK, Walker AF, Kajura BR, Basalirwa R. 1996. The nutritional status and weaning foods of infants and young children in Central Uganda. The Proceedings of the Nutrition Society 56:1A–16A.

Kim YJ, Lin N-C, Martin GB. 2002. Two distinct *Pseudomonas* effector proteins interact with the Pto kinase and activate plant immunity. Cell 109:589–598.

Ko K. 1999. Attacin and T4 lysozyme transgenic 'Galaxy' apple: Regulation of transgene expression and plant resistance to fire blight (*Erwinia amylovora*). Ph.D dissertation, Cornell University, Ithaca, NY.

Krettiger AF. 1997. Insect resistance in crops: A case study of *Bacillus thuringiensis* (Bt.) and its transfer to developing countries. International Service for the Acquisition of Agri-biotech Applications, Ithaca, NY.

Kristyanne ES, Kim KS, Stewart JM. 1997. Magainin 2 effects on the ultrastructure of five plant pathogens. Mycologia 89:353–360.

La Fleur DA, Lockhart BEL, Olszewski NE. 1996. Portions of the banana streak badnavirus genome are integrated into the genome of its host *Musa* spp. Phytopathology 86:S100.

Lassoudiere A. 1974. La mosaïque dite "à tirets" du bananier 'Poyo' en Côte d'Ivoire. Fruits 29:349–357.

Leplá JC, Bonadá-Bottino M, Augustin S, Pilate G, Lê Tân VD, Delplanque A, Cornu D, Jouanin L. 1995. Toxicity to *Chrysomela tremulae* (Coleoptera, Chrysomelidae) of transgenic poplars expressing a cysteine proteinase inhibitor. Mol. Breed. 1:319–328.

Li Q, Lawrence CB, Xing HY, Babbitt RA, Bass WT, Maiti IB, Everett NP. 2001. Enhanced disease resistance conferred by expression of an antimicrobial magainin analogue in transgenic tobacco. Planta 212:635–639.

Lin H J, Cheng HY, Chen CH, Huang HC, Feng TY. 1997. Plant amphipathic proteins delay the hypersensitive response caused by harpinPss and *Pseudomonas syringae* pv. *syringae,* Physiol. Mol. Plant Pathol. 51:367–376.

Lockhart BEL, Jones DR. 2000. Banana streak. In: Jones DR (ed). Diseases of banana, Abaca and Ensete. CAB International, Wallingford, UK.

Lockhart BEL, Olszewski NE. 1993. Serological and genomic heterogeneity of banana streak badnavirus: Implications for virus detection in *Musa* germplasm. In: Ganry J (ed). Breeding banana and plantain for resistance to diseases and pests. Proceedings of the International Symposium on Genetic Improvement of bananas for resistance to diseases and pest organized by CIRAD-FLHOR, Montpellier, France, 7–9 September 1992. CIRAD, Montpellier, France. pp. 105–113.

Lockhart BEL. 1986. Purification and serology of a bacilliform virus associated with a streak disease of banana. Phytopathology 76:995–999.

Loor F. 1974. Tecinas para la induccion de plantas adventicias de banano y su posible aplicacion en mejoramiento genetico, Tesis de grado Facultad de Ingeniria agronomica, universidad Tecnica de Manabi, Portoviejo, Ecuador.

Lorito M, Woo SL, Garcia I, Collude G, Harman GE, Pintor-Toro JA, Filippone E, Muccifora S, Lawrence CB, Zoina A, Tuzun S, Scala F, Fernandez IG. 1998. Genes from mycoparasitic fungi as a source for improving plant resistance to fungal pathogens. Proc. Nat. Acad. Sci. 95:7860–7865.

Luc M, Sikora RA, Bridge J. 1990. Plant Parasitic nematodes in subtropical and tropical agriculture. CABI, Wallingford, UK. pp. 431–460.

Lucioli A, Noris E, Brunetti A, Tavazza R, Ruzza V, Castilla A, Bejarano E, Accotto G, Tavazza M. 2003. Tomato yellow leaf curl Sardinia virus Rep-derived resistance to homologous and heterologous geminiviruses occurs by different mechanisms and is overcome if virus-mediated transgene silencing is activated. J. Virol. 77:6785–6798.

Ma SS. 1974. Micropropagation of bananas. In: Theoretical and practical uses of plant cell and tissue culture. USA and Rep. of China Co-operative Program Seminar, Taipei.

Mak, C., Mohamed AA, Liew KW, Ho YW. 2004. Early screening technique for Fusarium wilt resistance in banana micro propagated plants. In: S Mohan J, Swennen R (ed). Banana Improvement Cellular, Molecular Biology, and Induced Mutations, Science Publishers, Inc. Enfield, New Hampshire, USA. pp. 219–227.

Marriott J, Lancaster PA. 1983. Bananas and Plantains. In: Harvey TC Jr. (ed). Handbook of Tropical Foods. Marcel Dekker, Inc. pp. 85–142.

Marroquin LD, Elyassnia D, Griffits JS, Aroian RV. 2000. *Bacillus thuringiensis* (Bt) toxin susceptibility and isolation of resistance mutants in the nematode *Caenorhabditis elegans*. Genetics 155:1693–99.

Mason JB, Garcia M. 1993. Micronutrient deficiency—the global situation. SCN News 9:11–16.

May GD, Rownak A, Mason H, Wiecko A, Novak FJ, Arntzen CJ. 1995. Generation of transgenic banana (*Musa acuminata*) plants via *Agrobacterium*-mediated transformation. Bio/Technol. 13:486–492.

Menendez T, Loor FH. 1979. Recent advances in vegetative propagation and their application to banana breeding. In: Reunion da Acorbat: 4, 1979, UPEB. Anais, Panamá. pp. 211–222.

Meredith DS, Lawrence JS. 1969. Black leaf streak disease of bananas (*Mycosphaerella fijiensis*): Symptoms of disease in Hawaii, and notes on the conidial state of the causal fungus. Transactions of the British Mycological Society 52:459–476.

Mitchell G. 1980. Banana entomology in the Windward Island. Final report, Center for Overseas Pest Research, London UK. pp. 1974–1978.

Mobambo KN, Gauhl F, Swennen R, Pasberg-Gauhl C. 1996. Assessment of the cropping cycle effects on black leaf streak severity and yield decline of plantain and plantain hybrids. International Journal of Pest Management 42:1–8.

Morton RL, Schroeder HE, Bateman KS, Chrispeels MJ, Armstrong E, Higgins TJV. 2000. Bean α-amylase inhibitor-I in transgenic peas (*Pisum sativum*) provided complete protection from pea weevil (*Bruchus piorum*) under field conditions. Proc. Nat. Acad. Sci. 97:3820–3825.

Mouliom-Perfoura A, Lassoudiere A, Foko J, Fontem DA. 1996. Comparison of development of *Mycosphaerella fijiensis* and *Mycosphaerella musicola* on banana and plantain in the various ecological zones in Cameroon. Plant Disease 80:950–954.

Murdock LL, Brookhart G, Dunn PE, Foard DE, Kelley S, Kitch L, Shade RE, Shuckle RH, Wolfson JL. 1987. Cysteine digestive proteinases in Coleoptera. Comp. Biochem. Physiol. 87:783–787.

Mysore KS, D'Ascenzo MD, He X, Martin GB. 2003. Overexpression of the disease resistance gene *pto* in tomato induces gene expression changes similar to immune responses in human and fruitfly. Plant Physiol. 132:1901–1912.

Nakajima H, Muranaka T, Ishige F, Akutsu K, Oeda K. 1997. Fungal and bacterial disease resistance in transgenic plants expressing human lysozyme. Plant Cell. Rep. 16:674–679.

Nankinga CM, Moore D, Bridge P, Gowen S. 1999. Recent advances in microbial control of banana weevil. In: Frison EA, Gold CS, Karamura EB, Sikora RA (eds). Mobilizing IPM for sustainable banana production in Africa. Proceedings of a workshop on banana IPM held in Nelspruit, South Africa, 23–28 November 1998.

NARO. 2003. National Agricultural Research Organization (NARO) Bulletin, May–June 2003.

Navarre E. 1957. Multiplication des musa a feuilles rouges. Revue Horticole (Francia) 29:712.

Ndubizu, TOC, Obiefuna JC. 1982. Upgrading inferior plantain propagation material through dry-season nursery. Scientia Horticulturae 18:31–37.

Niere B, Gold CS, Coyne D. 2004. Can fungal endophytes control soil borne pests in banana? In: Sikora RA, Gowen S, Hauschild R, Kiewnick S (eds). Multitrophic Interactions in soil and integrated control. IOBC/WPRS Bulletin 27:203–210.

Niere BI, Speijer PR, Gold CS, Sikora RA. 1998. Fungal endophytes from banana for the biocontrol of *Radopholus similis* In: Frison E, Gold CS, Karamura EB, Sikora RA (eds). Mobilizing IPM for sustainable banana production in Africa. Proceedings of a workshop on Banana IPM held in Nelspruit, South Africa, 23–28 November 1998. pp. 313–318.

Nwakanma DC, Pillay M, Okoli BE, Tenkouano A. 2003. PCR-RFLP of the ribosomal DNA internal transcribed spacers (ITS) provides markers for the A and B genomes in *Musa* L. Theor. Appl. Genet. 108:154–159.

Pavis C. 1991. Etude des relations plante-insecte chez le charançon du bananier *Cosmopolites sordidus* Germar (Coleoptera:Curculionidae). In: Gold CS, Gemmill, B (eds). Biological and Integrated control of highland Banana and Plantain pests and diseases. Proceedings of a research coordination meeting 12–14 November 1991, Cotonou, Benin IITA, Ibadan, Nigeria. pp. 171–181.

Pegg KG. 1996. Fusarium wilt of banana in Australia: A review. Australian Journal of Agricultural Research 47:637–650.

Philpott J, Knowles CH. 1913. Report on a visit to Sigatoka. Pamphlet of the Department of Agriculture, Fiji, 3, Suva.

Pietersen G, Thomas JE. 2001. Overview of *Musa* virus diseases. In: Plant virology in sub-Saharan Africa. Available at: www.iita.org/cms/details/virology/pdf_files/50-60.pdf.

Pillay M, Nwakanma DC, Tenkouano A. 2000. Identification of RAPD markers linked to A and B genome sequences in *Musa*. Genome 43:763–767.

Pillay M, Tenkouano A, Hartman J. 2002. Future challenges in *Musa* breeding. In: Kang MS (ed). Crop improvement: Challenges in the twenty-first century, Food Products Press, Inc. New York. pp. 223–252.

Pillay M, Tenkouano A, Ude G, Ortiz R. 2004. Molecular characterization of genomes in *Musa* and its applications. In: Mohan Jain S, Swennen R (eds). Banana improvement: Cellular, molecular biology, and induced mutations. Science Publishers, Inc., Einfield, New Hampshire, USA. pp. 271–286.

Pink DAC. 2002. Strategies using genes for non-durable resistance. Euphytica 1:227–236.

Pinochet J, Jaizme MC, Fernandez C, Jaumot M, De Waele D. 1998. Screening banana for root-knot (*Meliodogyne* spp.) and lesion nematode (*Pratylenchus goodeyi*) resistance for the Canary Islands. Fund. Appl. Nematol. 21:17–23.

Pinochet J. 1996. Review of past research on *Musa* germplasm and nematode interactions. In: Frison EA, Horry JP, De Waele D (eds). New frontiers in resistance breeding for nematodes, fusarium and Sigatoka. Proceedings of the workshop held in Kuala Lumpur, Malaysia, 25 October 1995. INIBAP, Montpellier, France. pp. 32–44.

Ploetz RC, Vazquez A, Nagel J, Benscher D, Sianglew P, Srikul S, Kooariyakul S, Wattanachaiyingcharoen W, Lertrat P, Wattanachaiyingcharoen D. 1997. Current status of Panama disease in Thailand. Fruits 51:387–395.

Ploetz RC. 1990. Population biology of *Fusarium oxysporum* f.sp. *cubense*. In: Ploetz RC (ed). Fusarium wilt of banana. APS Press, St Paul, Minnesota, USA. pp. 63–76.

Ploetz RC. 1994. Panama disease: Return of the first banana menace. International Journal of Pest Management 40:326–336.

Pocasangre L. 2000. Biological enhancement of banana tissue culture plantlets with endophytic fungi for the control of the burrowing nematode *Radopholus similis* and Panama disease (*Fusarium oxysporum* f.sp. *cubense*). Ph.D thesis, University of Bonn, Gemany. pp. 263–274.

Price NS, McLaren CG. 1996. Techniques for field screening of *Musa* germplasm. In: Frison EA, Horry JP, De Waele D (eds). New frontiers in resistance breeding for nematodes, fusarium and Sigatoka. Proceedings of the workshop held in Kuala Lumpur, Malaysia, 25 October 1995. INIBAP, Montpellier, France. pp. 87–107.

Rajasekaran K, Stromberg KD, Cary JW, Cleveland TE. 2001. Broad-spectrum antimicrobial activity in vitro of the synthetic peptide D4E1. J. Agr. Food. Chem. 49:2799–2803.

Reddy KVS, Gold CS, Ngode L. 1999. Cultural control strategies for banana weevil, *Cosmopolites sordidus* Germar In: Frison EA, Gold CS, Karamura EB, Sikora RA (eds). Mobilizing IPM for sustainable banana production in Africa. Proceedings of a workshop on banana IPM held in Nelspruit, South Africa, 23–28 November 1998. pp. 51–57.

Remy S. 2000. Genetic Transformation of Banana (*Musa* sp.) for disease Resistance by Expression of Antimicrobial Proteins. Ph.D thesis KUL, Belgium.

Robinson JC. 1996. Bananas and plantains. Crop Production Science in Horticulture 5. CAB International, Wallingford, UK.

Rodoni BC, Ahlawat YS, Varma A, Dale J, Harding RM. 1997. Identification and characterization of banana bract mosaic virus in India. Plant Disease 81:669–672.

Rowe P. 1984. Breeding bananas and plantains. In: Janick J (ed). Plant Breeding reviews Vol 2: AVI Publishing Company, Westport, Connecticut. pp. 135–155.

Rowe P, Rosales FE (1996) Bananas and plantains. In: Fruit Breeding Vol I: Tree and tropical fruits, Janick J and Moore JN (eds) pp. 167–211. John Wiley and Sons. Inc. New York.

Rukazambuga NDTM. 1996. The effects of banana weevil (*Cosmopolites sordidus* Germar) on the growth and productivity of banana (*Musa* AAA EA) and the influence of host vigor on attack. Ph.D thesis. University of Reading. UK. 249 pages.

Ryan CA. 1990. Proteinase inhibitors in plants: Genes for improving defences against insects and pathogens. Ann. Rev. Phytopath 28:425–449.

Sági L, Remy S, Cammue BPA, Maes K, Raemaekers T, Panis B, Schoofs H, Swennen R. 2000. Production of transgenic banana and plantain. Acta Hort. 540:203–206.

Sági L, May GD, Remy S, Swennen R. 1998. Recent developments in biotechnological research on bananas (*Musa* spp.). Biotechnol. Genet. Eng. Rev. 15:313–327.

Sági L, Panis B, Remy S, Schoofs H, De Smet K, Swennen R, Cammue B. 1995. Genetic transformation of banana (*Musa* sp.) via particle bombardment. Bio/Technol. 13:481–485.

Sarah JL, Pinochet J, Stanton J. 1996. The burrowing nematode of bananas, *Radopholus similis* Cobb, 1913. *Musa* pest fact sheet, No. 1, INIBAP, Montpellier, France.

Sasser, JN, Freckman DW. 1987. A world perspective on nematology: The role of the society. In: Veech JA, Dickson DW (eds). Vistas on nematology, Society of Nematologists, Hyattsville, Maryland, USA. pp. 7–14.

Sharma HC, Sharma KK, Seetharama N, Ortiz R. 2000. Prospects for using transgenic resistance to insects in crop improvement. Biotech 3, Available at: www.ejbiotechnology.info/content/vol3/issue2/full/3.

Shepherd K. 1988. Observation on *Musa* taxonomy. In: Jarrett R (ed). Identification of genetic diversity in the genus *Musa*. Proceedings of an international workshop held at Los Banos, Philippines 5–10 September 1988. INIBAP, Montpellier, France. pp. 158–165.

Shepherd K, Dantas JLL, de Oliveira e Silva S. 1994. Breeding Prata and Maca cultivars in Brazil. In: Jones DR (ed). The improvement and testing of *Musa*: A global partnership, Proceedings of the first Global Conference of the International Testing Program held at FHIA, Honduras, 27–30 April

1994. INIBAP, Montpellier, France. pp. 157–168.

Sikora RA, Pocasangre LE. 2004. New technologies to increase root health and crop production. InfoMusa 13:25–29.

Simmonds NW. 1962. The evolution of the bananas. Longmans Green & Co., London.

Sinha SK, Swaminathan MS. 1984. New parameters and selection criteria in plant breeding In: Vose PB, Blixt SG (eds). Crop breeding, Pergamon Press, New York. pp. 1–31.

Song WY, Wang GL, Chen L, Kim HS, Pi LY, Gardner J, Wang B, Holsten T, Zhai WX, Zhu LH, Fauquet C, Ronald PC. 1995. A receptor kinase-like protein encoded by the rice disease resistance gene Xa21. Science 270. 1804–1806.

Stoffelen R, Verlinden R, Xuyen NT, Swennen R, DeWaele D. 1999. Screening of Papua New Guinea bananas to root-lesion and root- knot nematodes. InfoMusa 8:12–15.

Stover RH, Buddenhagen IW. 1986. Banana breeding: Polyploidy, disease resistance and productivity. Fruits 41:175–1919.

Stover RH, Simmonds NW. 1987. Bananas. 3rd Edition. Longman, London.

Ssebuliba R, Talengera D, Makumbi D, Tenkouano A, Pillay M. 2006. Reproductive efficiency and breeding potential of East African highland banana. Field Crops Research 95:250–255.

Su-Shien M, Shii C. 1974. Growing banana plantlets from adventitious buds. J. Chinese Hort. Sci. 20:1–7.

Swaminathan MS. 1969. Mutation breeding. Proc XII Congr Genetics. C. Oshima (ed). 3:327–347.

Swennen R, Rosales F. 1994. Bananas. Encyclopedia of Agricultural Science 1:215–232.

Swennen R. 1990. Plantain Cultivation under West African Conditions: A Reference Manual. IITA, Ibadan, Nigeria. 24 pages.

Swennen R, Vuylsteke D. 1987. Morphological taxonomy of plantain Musa cultivars (AAB) in West Africa. ACIAR Proc. 21:165–71.

Tai TH, Dahlbeck D, Clark ET, Gajiwala P, Pasion R, Whalen MC, Stall RE, Staskawicz, BJ. 1999. Expression of the Bs2 pepper gene confers resistance to bacterial spot disease in tomato. Proceedings of the National Academy of Sciences USA 23:14153–14158.

Tenkouano A, Vuylsteke D, Okoro J, Makumbi D, Swennen R, Ortiz R. 2003. Diploid Banana hybrids TMB2x5105-1 and TMB2x9128-3 with good combining ability, resistance to black Sigatoka and nematodes. Hortscience 38:468–472.

Terras FRG, Eggermont K, Kovaleva V, Raikhel NV, Osborn RW, Kester A, Rees SB, Torrekens S, Van Leuven F, Vanderleyden J, Cammue BPA, Broekaert WF. 1995. Small cysteine-rich antifungal proteins from radish: Their role in host defense. Plant Cell 7:573–588.

Tezenas du Montcel H, Carreel F, Bakry F. 1996. Improve the diploids: The key for banana breeding. In: Frison EA, Horry JP, De Waele D (eds). New frontiers in resistance breeding for nematodes, fusarium and Sigatoka. Proceedings of the workshop held in Kuala Lumpur, Malaysia, 25 October 1995. INIBAP, Montpellier, France. pp. 119–128.

Tezenas du Montcel H. 1988. Musa acuminata subspecies banksii: Status and diversity. In: Jarrett R (ed). Identification of genetic diversity in the genus Musa. Proceedings of an International workshop held at Los Banos, Philippines 5–10 September 1988. INIBAP, Montpellier, France. pp. 211–218.

Thomas JE, Iskra-Caruana ML, Jones DR. 1994. Banana bunchy top disease. Musa Disease Fact Sheet no. 4. International Network for the Improvement of banana and plantain (INIBAP), Montpellier, France.

Thomas JE, Magnaye LV 1996. Banana bract mosaic virus disease. Musa Disease Fact Sheet no.7. International Network for the Improvement of banana and plantain (INIBAP), Montpellier, France.

Tinjuangjun P. 2002. Snowdrop lectin gene in transgenic plants: Its potential for Asian agriculture. Available at: Agbiotechnet.com.

Tomekpe K, Jenny C, Escalant JV. 2004. A review of conventional improvement strategies for Musa. InfoMusa 13:2–6.

Tripathi, L. 2003. Genetic Engineering for improvement of Musa production in Africa. African J. Biotechnol. 2:503–508.

Tripathi L, Hughes Jd'A, Tenkouano A. 2002. Production of transgenic Musa varieties for sub-Saharan Africa. 3rd International Symposium on the Molecular and Cellular Biology of Banana. Available at: http://www.promusa.org/publications/leuven-abstracts.pdf.

Tripathi L, Tripathi JN, Hughes Jd'A. 2005a. Agrobacterium-mediated transformation of plantain cultivar Agbagba (Musa spp.). African J. Biotechnol. (in press).

Tripathi L, Tripathi JN, Tushemereirwe WK, Tenkouano A, Pillay M, Bramel P. 2005b. Biotechnology for improvement of banana production in Africa. Proceedings of 9th ICABR International Conference on Agricultural Biotechnology: Ten years later, Ravello (Italy), July 6–10. Available at: economia.uniroma2.it/conferenze/icabr2005/papers/Tripathi_paper.pdf.

Tripathi L, Tripathi JN, Oso RT, Hughes Jd'A, Keese P. 2003. Regeneration and transient gene expression of African Musa species with diverse genomic constitution and ploidy levels. Tropical Agr. 80:182–187.

Tripathi L, Tripathi JN, Tushemereirwe WK. 2004. Strategies for resistance to bacterial wilt disease of banana through genetic engineering. African J. Biotechnol. 3:688–692.

Trudel J, Potvin C, Asselin A. 1995. Secreted hen lyso-zyme in transgenic tobacco: Recovery of bound enzyme and in vitro growth inhibition of plant pathogens. Plant Sci. 106:55–62.

Tushemereirwe WK, Karamura EB, Karyeija R. 1996. Banana streak virus (BSV) and an associated fila-mentous virus (unidentified) disease complex of highland bananas in Uganda. InfoMusa 5:9–14.

Tushemereirwe W, Kangire A, Smith J, Ssekiwoko F, Nakyanzi M, Kataama D, Musiitwa C, Karyaija R. 2003. An outbreak of bacterial wilt on banana in Uganda. InfoMusa 12:6–8.

Ude G, Pillay M, Nwakanma D, Tenkouano A. 2002a. Analysis of genetic diversity and sectional relation-ships in Musa using AFLP markers. Theor. Appl. Genet. 104:1239–1245.

Ude G, Pillay M., Nwakanma D, Tenkouano A. 2002b. Genetic diversity in Musa acuminata Colla and Musa balbisiana Colla and some of their natural hybrids using AFLP markers. Theor. Appl. Genet. 104:1246–1252.

Urwin PE, Atkinson HJ, Waller DA, McPherson MJ. 1995. Engineered oryzacystatin-I expressed in transgenic hairy roots confers resistance to Glo-bodera pallida. Plant J. 8:121–131.

Urwin PE, Levesley A, McPherson MJ, Atkinson HJ. 2000. Transgenic resistance to the nematode Roty-lenchulus reniformis conferred by A. thaliana plants expressing protease inhibitors. Mol. Breed 6:257–64.

Urwin PE, Lilley CJ, McPherson MJ, Atkinson HJ. 1997. Resistance to both cyst and root knot nema-todes conferred by transgenic Arabidopsis express-ing a modified plant cystatin. Plant J. 12:455–461.

Urwin PE, McPherson MJ, Atkinson HJ. 1998. Enhanced transgenic plant resistance to nematodes by dual protease inhibitor constructs. Planta 204:472–79.

Urwin PE, Troth KM, Zubko EI, Atkinson HJ. 2001. Effective transgenic resistance to Globodera pallida in potato field trials. Mol. Breed. 8:95–110.

Vain P, Worland B, Clarke MC, Richard G, Beavis M. 1998. Expression of an engineered cysteine prote-ase inhibitor (Oryzacystatin-I1D86) for nematode resistance in transgenic rice plants. Theor. Appl. Genet. 96:266–271.

Valmayor R, Danh LD, Hakkinen M. 2004. Rediscov-ery of Musa splendida A. Chevalier and description of two new species (Musa viridis and Musa lutea). The Philippine Agricultural Scientist 87:110–118.

Viaene N, Duran LF, Rivera JM, Duenas J, Rowe P, De Waele D. 2003. Responses of banana and plantain cultivars, lines and hybrids to the burrowing nema-tode Radopholus similis. Nematology 5:85–98.

Vuylsteke D, Ortiz R. 1995. Plantain-derived diploid hybrids (TMP2x) with black Sigatoka resistance. Hortscience 30:147–149.

Vuylsteke D, Talengera D. 1998. Postflask manage-ment of micropropagated bananas and plantains. IITA, Ibadan, Nigeria.

Waite BH. 1963. Wilt of Heliconia spp. caused by Fusarium oxysporum f.sp. cubense Race 3. Tropical Agriculture 40:299–305.

Wang GL, Song WY, Ruan DL, Sideris S, Ronald PC. 1996. The cloned gene, Xa 21, confers resistance to multiple Xanthomonas oryzae pv. Oryzae isolates in transgenic plants, Mol. Plant-Microbe Interact. 9:850–855.

Wehunt EJ, Hutchinson DJ, Edwards DI. 1978. Reac-tion of banana cultivars to the burrowing nematode (Radopholus similis). Journal of Nematology 10:368–370.

Wei ZM, Beer SV. 1996. Hairpin from Erwinia amy-lovora induces plant resistance. Acta Hort. 411:427–431.

Wei JZ, Hale K, Cara L, Platzer E, Wong C, Fang SC, Raffi V. 2003. Bacillus thuringiensis crystal pro-teins that target nematodes. Proc. Nat. Acad. Sci. 10:2760–2765.

WHO. 1992. Second Report on the World Nutrition Situation. Vol. I. Global and Regional Results. United Nations Administrative Committee on Coor-dination, Subcommittee on Nutrition, Washington, D.C.

World Bank. 1994. The challenge of dietary deficien-cies of vitamins and minerals. In: Enriching lives: Overcoming vitamin and mineral malnutrition in developing countries. World Bank, Washington D. C. pp. 6–13.

Yirgou D, Bradbury JF. 1968. Bacterial wilt of Ensete (Ensete ventricosum) incited by Xanthomonas campestris pv. musacearum. Phytopathology 58:111–112.

Yirgou D, Bradbury JF. 1974. A note on wilt of banana caused by the Enset wilt organism, Xanthomonas musacearum. East African Agric. Forest. J. 40:111–114.

Yoshimura S, Yamanouchi U, Katayose Y, Toki S, Wang ZW, Kono I, Kurata N, Yano M, Iwata N, Sasaki T. 1998. Expression of Xa1, a bacterial blight-resistance gene in rice, is induced by bacte-rial inoculation. Proc. Nat. Acad. Sci. 95:1663–1668.

Zasloff M. 1987. Magainins, a class of antimicrobial peptides from Xenopus skin: Isolation, characteriza-tion of two active forms and partial cDNA sequence of a precursor. Proc. Nat. Acad. Sci. 84:5449–5453.

Index